BRITISH COLUMBIA »

A NEW HISTORICAL ATLAS

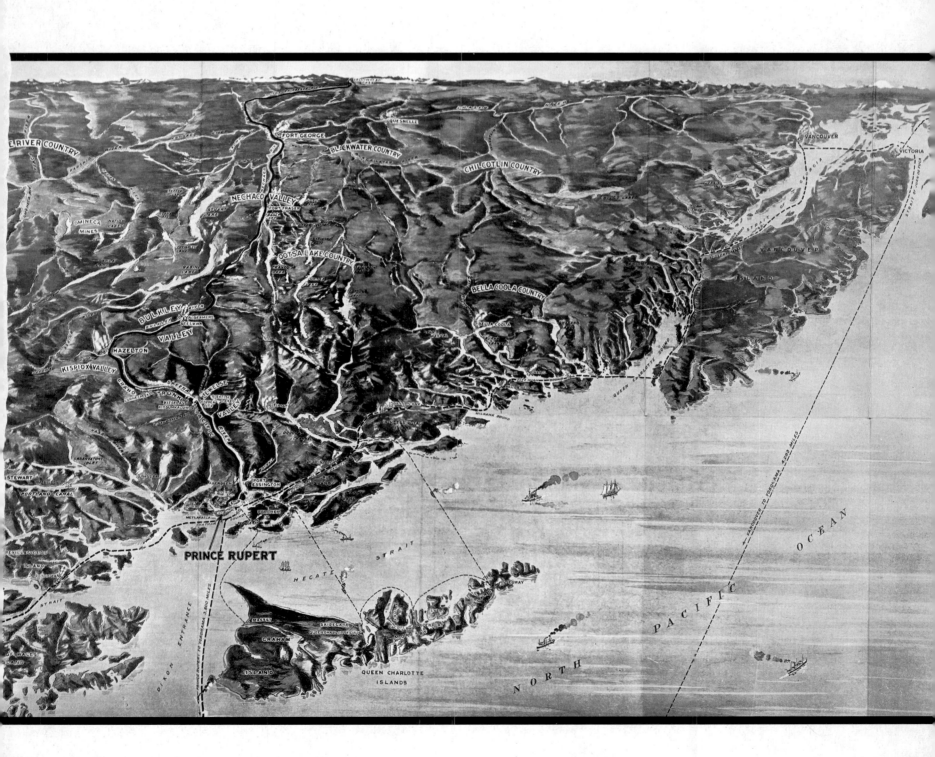

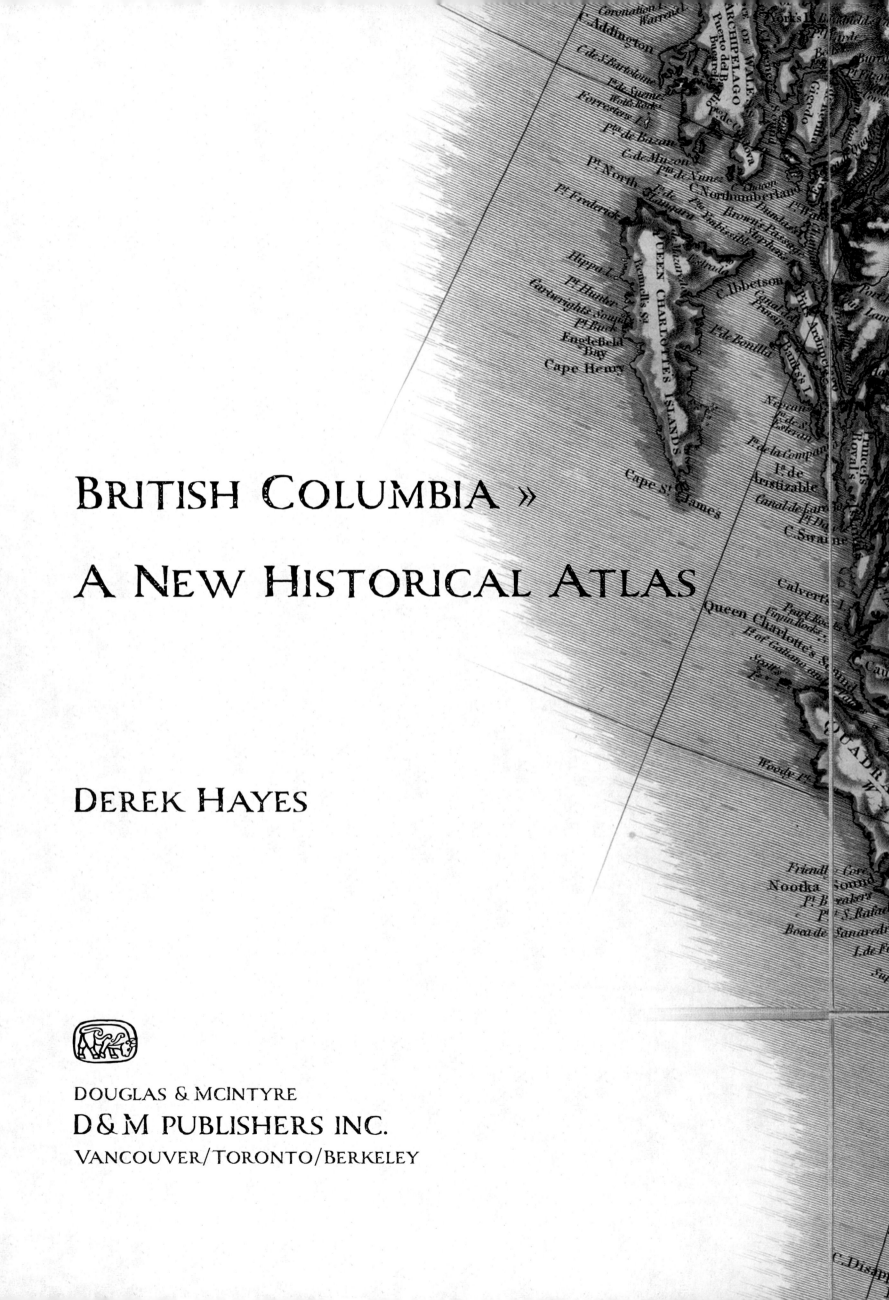

BRITISH COLUMBIA »

A NEW HISTORICAL ATLAS

DEREK HAYES

DOUGLAS & MCINTYRE
D&M PUBLISHERS INC.
VANCOUVER/TORONTO/BERKELEY

Douglas & McIntyre
An imprint of D&M Publishers Inc.
2323 Quebec Street, Suite 201
Vancouver, British Columbia V5T 4S7
www.douglas-mcintyre.com

Cataloguing data available from Library and Archives Canada

ISBN 978-1-926812-57-1

Editing and copyediting by Iva Cheung
Interior design and layout by Derek Hayes
Image research, acquisition, and modern photography
by Derek Hayes
Index by Karen Griffiths
Jacket design by Setareh Ashrafologhalai
Printed and bound in China by C & C Offset Printing Co., Ltd.
Printed on acid-free paper
Distributed in the U.S. by Publishers Group West

We gratefully acknowledge the financial support of the Canada Council for the Arts, the
British Columbia Arts Council, the Province of British Columbia through the Book Publishing
Tax Credit, and the Government of Canada through the Canada Book Fund for our
publishing activities.

Books by Derek Hayes can be seen at: www.derekhayes.ca

Derek Hayes's photos can be seen at: www.derekhayesphotography.com

Comments, further information, and corrections are welcome.
Please contact the author by emailing: derek@derekhayes.ca

MAP 1 (half-title page).
Part of a fine bird's-eye map of British Columbia from an unusual angle
published in a brochure titled *Prince Rupert and a Dawning Empire* by
the Prince Rupert Publicity Club and Board of Trade in 1910. See page 252.

MAP 2 (title page).
Part of a map of North America published by British mapmaker James
Wyld in 1823 displays the beginning knowledge of the northern interior
of British Columbia from the North West Company's New Caledonia. See
page 44.

MAP 3 (right).
This fine map with neoclassical pretensions formed the front cover of
the 1912 *Annual British Columbia Development Edition* of *B.C. Saturday Sunset*, a weekly magazine devoted to promoting and reporting the
growth of the province during the real estate promoter's halcyon years
of 1907 to 1913. The magazine is an excellent source for advertisements
and reports of the dreams and schemes of a period when the building
of the Panama Canal seemed to promise endless growth for West Coast
cities. The mirage dissipated in a hurry with the coming of World War I,
when patriotic British Columbians marched off to fight for the Empire.
A woman with flowing robes points to the jewel of the province's ambitions—*Vancouver*. Railways abound, despite the fact that most did not
exist at that date, and some never would.

MAP 4 (contents page).
Part of a fine map of the province compiled by surveyor Edward Mohun
for the British Columbia Department of Lands and Works. It was published in 1884.

Acknowledgements

A book of this size and scope would have been impossible to create without the willing assistance of many individuals and institutions in British Columbia and elsewhere, and I thank
all of them and especially their typically very helpful and co-operative staff members.

I am particularly indebted to Mike Thomson, surveyor general of British Columbia, for
arranging photographic access to the vault of historical maps stored at the Land Title and
Survey Authority office in Victoria. Well over a hundred of the maps in the book come from
this source. Calvin Woelke and Brenda Hansen assisted me in the navigation of this vast and
valuable treasure trove of maps.

Another major source of maps was the Rare Books and Special Collections Department
at the University of British Columbia, where Katherine Kalsbeek, Sarah Romkey, and Ralph
Stanton have been most helpful. Other major sources of maps have been Library and Archives Canada (Jean Matheson, Ellyse Rupert, and many others); the United Kingdom National Archives; Vancouver Public Library Special Collections (Andrew Martin and all the
staff); the British Columbia Archives (Margaret Hutchison, Marion Tustanoff, Diane Wardle,
Kelly-Ann Turkington, and many others); the Hudson's Bay Company Archives (Michelle
Rydz, Marcia Stentz, and Debra Moore); and the City of Vancouver Archives (Carol Haber
and Chak Yung).

Members of the Historical Map Society of British Columbia, including Robin Inglis, Frances
Woodward, Jim Foulkes, David Malaher, Bruce Watson, Ken Martin, Gary Little, and the
late Bruce Ward and John Crosse have been helpful in tracking down maps and information
about them from a myriad of sources.

As always, David Rumsey permitted me to use a number of maps from his wonderful
map website. I also thank (in no particular order) the following individuals and institutions
from which the maps in the book are drawn: John Cloud, National Oceanic and Atmospheric Administration (NOAA) Central Library, Washington, D.C.; Susan Cross, City of
Kamloops Archives; Elisabeth Duckworth, Kamloops Museum; Ron Hatch, Thompson Rivers History and Heritage Society; Ken Favrholdt, Osoyoos & District Museum and Archives;
Nancy Anderson; John Keenlyside; Henry Ewert; Frank Leonard; James Walker; Richard Goodacre; Ross Hayes; Vern Byberg; Greg Walter; Sharon Keen; Ian Smith, Canadian Railroad
Historical Association; Jeanne Boyle and Peter Ord, Penticton Museum and Archives, and
Randy Manuel, former curator; Tara Hurley, Kelowna Public Archives; Barbara Bell, Greater
Vernon Museum and Archives; Lynn Couch-Alaric, Oliver Archives; Emily Yearwood-Lee,
British Columbia Legislative Library; Susan Kooyman, Glenbow Archives; Mike Pennock, Fernie Museum; Joyce Austin, Rossland Historical Museum; Sarah Benson, Trail Archives, Trail
Historical Society; Laura Fortier and Shawn Lamb, Shawn Lamb Archives, Touchstones Nelson; Judith Maltz and Andrew Rhodes, Sandon Museum, Sandon Historical Society; Bridget
Watson and Kit Willmot, Ladysmith and District Historical Society; Kristin Schachtel, Fort
Steele Heritage Town Museum and Archives; Marie Stang, Kimberley District Heritage Society; Lena Goon, Whyte Museum of the Canadian Rockies; Terri Lang, Senior Meteorologist, Environment Canada, Kelowna; Hugh Ellenwood, White Rock Museum and Archives;
Cathy English, Revelstoke Museum; Alex, Windermere Valley Museum, and Chris Prosser,
Invermere chief executive; Milton Parent, Arrow Lakes Historical Society; Jo Atkinson, Nicola Valley Museum and Archives; Kathy Bossort and Catharine McPherson, Delta Museum
and Archives; Andy Korsos; the late Lieutenant-Colonel Vic Stevenson, Royal Canadian
Artillery Museum; Priscilla Lowe, Cowichan Valley Museum & Archives; Valda Stefani and
Trevor Livelton, City of Victoria Archives; Dave Brown and Iris Morgan, University of Calgary Library; Janis Ringuette; Donald Luxton; David Hill-Turner, Nanaimo Museum; Valerie
Billesberger, Mission Community Archives; Chelsea Pielou and Catherine Gilbert, Museum
at Campbell River; Adrian Mitescu, *Vancouver Sun/Province* Library; Randy Bouchard; Patrick McGeer; Charles Campbell; Heather Stephens, Vanderhoof Community Museum; Bob
Campbell and Alisha Rubadeau, Fraser–Fort George Regional Museum (Exploration Place);
Jake Jacobs, B.C. Ministry of the Environment; Archie Miller, New Westminster Historical Society; Ethel Field and Maureen Arvanitidis, New Westminster Heritage Preservation Society;
Trevor Mills, Archivist, West Coast Railway Association; Lara Gerrits, Vancouver International Airport; Chris McLeod, Whistler/Blackcomb; John Broadhead, Gowgaia Institute; and
Brian Wilson, Okanagan Archive Trust Society.

I am grateful to Jean Barman, who kindly read the first draft and made many useful
corrections and suggestions that have been incorporated into the book. I also thank Ken
Favrholdt, who read the second draft and did likewise. Very special thanks are due to my
ever eagle-eyed and dedicated editor, Iva Cheung, who is responsible for a great deal of the
book's being better than it was at the beginning. I emphasize, however, that any remaining
errors are my responsibility alone. Finally, thanks are due once more to D&M's CEO, Scott
McIntyre, for his continuing enthusiasm for my books.

C O N

TENTS

A Geographical View of British Columbia's History

Contemporary maps can yield a tremendous amount of information to the historian, yet for some reason old maps are often not consulted at all. This book, a history principally illustrated with contemporary maps, hopefully goes some way to redress this imbalance.

Although there are many maps in the book, they are but a tiny fraction of what was produced, though a slightly larger proportion of those that actually survived. As is to be expected in a book that tries to cover the whole extent of history in a single volume, the choice of maps is necessarily a personal one. The selection is of course constrained by what is available, and one becomes aware of the vast numbers of maps—and other types of historical documents, for that matter—that simply did not survive. Historical documents and artifacts have to make it through the period when they are simply old or old-fashioned before they enter a phase where they might—if they are lucky—be felt worth preserving. Many maps that were known to have been made are no longer to be found; the map drawn by Hudson's Bay Company captain Aemilius Simpson as he approached the site that was to become Fort Langley in 1827—a map that in many ways can be considered to be the founding document of modern British Columbia—simply does not exist; only a copy remains (MAP 132, *page 49*). There are similarly almost no maps created by British Columbia's Aboriginal peoples because the media on which most would have been drawn were ephemeral (see page 15).

Nevertheless, enough maps did survive to permit many events in British Columbia history to be documented, and that is what this book uses to illustrate those events. Maps have been selected for their historical significance but also for their interest. Many maps are works of art in their own right, pictorial, whimsical, and colourful; there are bird's-eye maps from before the advent of aviation, maps based on satellite photos, and everything in between. A map is always only a representation of what is on the ground, and the interpretations are many.

British Columbia was among the last temperate places on earth to appear on the map of the world—a function of its position on the "backside of America," as the Elizabethans would have termed

Above. In 1915 Alberta boundary commissioner Richard W. Cautley places a theodolite over a British Columbia boundary marker. The instrument is used to measure angles for a triangulation survey (see page 285).

MAP 5 (*below*).

This is the first map to depict the new colony of *British Columbia*. It was published in 1858 just weeks after the proclamation of the colony, complete with the name just chosen by Queen Victoria. It seems that mapmaker Edward Weller had a map of New Caledonia ready for the press and was able to alter it quickly to incorporate the new name. In the south the colony had boundaries close to those of today—the exception being that the San Juan Islands were included; they would not be awarded to the United States until 1872

(see page 101). In the north the boundary confined the colony to roughly half of its extent today, being drawn at the *Finlay R.*, a tributary of the Peace, and *Simpsons R.*, a non-existent conflation of the Skeena and the Nass Rivers (see page 47). This mistake shows the limitations of geographical knowledge of the north that applied in 1858 despite many years of occupation and explorations by the North West Company and the Hudson's Bay Company. The map also shows *Vancouver Island*, a separate colony created in 1849 (see page 53).

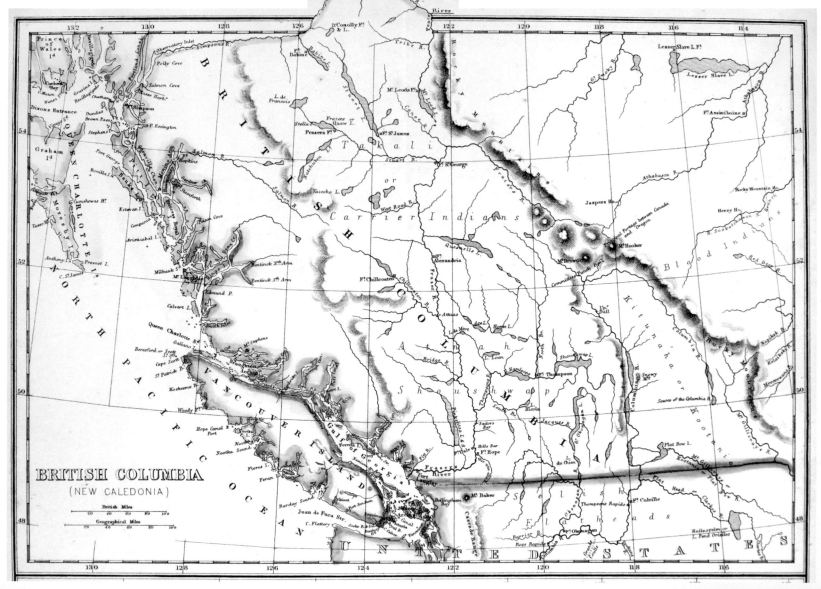

BRITISH COLUMBIA
(NEW CALEDONIA)

it, far from Europe whether travelling east or west, and lacking until the mid-nineteenth century the suggestion of gold that drove men much earlier to colonize warmer lands to the south.

The text, which includes many sometimes extensive captions that are integral to the book, is generally concise so as to allow more room for the book's central feature—the maps. More details can be found in the list of books included in the bibliography (page 356).

Here, then, is an attempt to summarize the span of British Columbia's history, all illustrated with contemporary maps. Here are maps Aboriginal and maps European—British, Spanish, French, Russian—and American, the story of exploration and exploitation, fur trade empires and the seekers of gold. Here are the maps of the engineers, the advertisers, the promoters, the diplomats, the explorers, the soldiers, the surveyors, and the road builders. The maps include those of the great railways—British Columbia went railway mad during one phase of its history—the Canadian Pacific and the Grand Trunk Pacific, the Great Northern and the Canadian Northern, the Pacific Great Eastern Railway to nowhere and the Coast-to-Kootenay railway to the mineral wealth of the interior.

And the mines, exploited by those eager for gold and copper and a host of other minerals. Above-ground maps and below-ground plans, working maps and ornamental maps, the plans of the dreamers and schemers of a thousand projects, some built, many shelved. There are maps of the coastal defences during World War II and maps of the great power projects that came after. And even maps of murder.

Many of the maps have never been published before. They have been collected from museums, archives, and private individuals all over the province and many from other parts of the world.

I trust and hope that you will have as much fun reading my book as I have had collecting the maps, doing the research, and writing and preparing it. Happy reading!

In this book the term *EuroCanadian* has been used to describe those of European origin that appear as explorers, exploiters, or settlers, and the term *Aboriginal people* is used to describe those who arrived by migration long ago and were living in the region we now call British Columbia when the EuroCanadians arrived. *Aboriginal people* appears to be the emerging Canadian term of choice for those in the past referred to as Native people or Indians.

> Please note that, in addition to conventional usage, *italics* are used in the map captions to refer to names or features that can be found *on the map* and in the main text refer to the location of maps or other illustrations *on the pages* of the book.

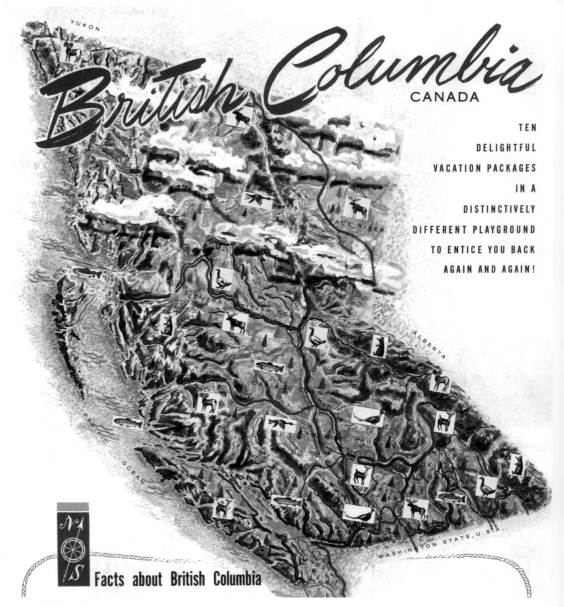

MAP 6 (*above*).
This rather attractive map of the province complete with clouds and little pictures of wildlife was published in a tourist brochure in the 1950s by the British Columbia Travel Bureau. Maps can be artistic as well as functional.

MAP 7 (*above*).
Maps can also be entirely functional, showing us locational detail that would be unavailable from any other source. Drawn up in 1925 to plan the *removal* of some tracks from the Great Northern Railway's depot at Carrall and Pender Streets in Vancouver, this map shows us where the tracks and other facilities such as the *Freight House, Passenger Platform,* and *Ash Dump* (for steam locomotives) were located. Before the infilling of the eastern part of False Creek, the railway had crossed False Creek on a trestle located just west of today's Main Street and terminated at this depot deep in Vancouver's Chinatown district, a location distinguished today by the newer buildings of the Chinese Cultural Centre and Dr. Sun Yat-Sen Garden. This roundabout entry into Vancouver had been made necessary by the monopolistic position of the Canadian Pacific Railway and was finally overcome by creating land in False Creek where none had existed before (see page 209).

First Inhabitants—First Nations

Until recently archaeologists believed that humans had first entered North America about 13,000 years ago, a date coinciding with a lowering of sea levels and the creation of a land bridge across Bering Strait about that time. This theory was corroborated by human artifacts of that date first found at Clovis, New Mexico. However, in 2011 researchers announced the discovery of a site in Texas showing human artifacts dating to about 15,500 years ago; nearer to British Columbia archaeologists have recently redated a spear point embedded in mastodon bones found at Sequim, Washington, to 13,800 years ago, and others have found human evidence dating to 14,100 years ago in an Oregon cave, and so the date of human arrival now seems certain to have been at least 15,000 years ago. The theory is now that humans arrived from Asia by gradually working their way along the coast in small boats despite inland ice cover, and some researchers have suggested that this could have occurred as long as 30,000 years ago or even earlier.

Adding to the confusion is the discovery of a possibly 40,000-year-old footprint near Mexico City and the discovery of the bones of the so-called Kennewick Man on the banks of the Columbia River near Kennewick, Washington. These bones have been dated at about 9,300 years old but, interestingly, have proven to be most closely related to the Ainu of northeast Asia. He could have been the victim of a shipwreck, for it is known that currents tend to push boats towards the Northwest Coast, and previous shipwrecks of Japanese fishermen have been documented, such as three who were rescued by the Hudson's Bay Company near the entrance to the Strait of Juan de Fuca in 1833 and returned to Japan via London.

However, the exact date of the occupation of British Columbia by Aboriginal peoples is not significant, for there is no question that they arrived long before the first Europeans in 1774. Indeed, by that time the region that is now the coast of British Columbia was quite densely populated.

British Columbia has an extraordinary diversity of Aboriginal peoples, principally the effect of an especially fecund coastal environment created by the historically easy availability of food supplies, especially salmon. Compared with anywhere else in North America, British Columbia has a multitude of different groups within a relatively small area.

The maps on these pages that show the distribution of various Aboriginal groups indicate the situation as it was recorded by EuroCanadians generally in the second half of the 1800s and as such do not display any distribution as it may have been before that. If definitive lines between Aboriginal groups ever existed at all, they have certainly been blurred by forays beyond them, including by migration and intermarriage, such that the boundaries depicted on the maps shown here are at best broad zones of transition, lines intended more for the information of EuroCanadians and latterly for the processing of land claims, which in any case often overlap each other.

The homelands of the original inhabitants of British Columbia, although mostly similar in broad terms to those we see today, are not exactly the same in all cases. For example, the Haida had displaced southern Tlingit just before the arrival of Europeans.

Map 8 (left).

The British Arrowsmith firm of mapmakers published an *Aboriginal Map of North America* in 1857, which attempted to delineate the locations and boundaries of "various Indian Tribes." This is the British Columbia part of that map, but, a year before the creation of the mainland colony, there was not much detail known, and all of it was derived from the Hudson's Bay Company. Far more detail was given on the map in the eastern part, not shown here. The blue area was "Kolooch," and the red area Athabascan or Chipewyan. Within these broad bands there are some recognizable groups (though not all in their correct locations): *Carrier* (Dakelh and Wet'suwet'en), *Kootanie* (Ktunaxa), *Sikani* (Sekani), and *Haidah* (Haida), for example. Many of the spellings derive from European attempts to write names. Later maps would be much more accurate than this very early attempt. Indeed, many of the Hudson's Bay Company's own maps, a number of which are reproduced in this book, more accurately located different Aboriginal groups, right down to showing villages and the population (see, for example, Map 124, *page 46*). Fur traders, of course, had a vested interest in the location of Aboriginal people: they needed to trade with them in order to accumulate furs. The changing names by which Aboriginal groups are known is a direct result of their lack of a written language and their earlier EuroCanadian interpretations.

Above, left. A wonderful shorefront display of totem poles and longhouses in the Haida village of Skidegate, on Haida Gwaii, a photo taken by geologist George Mercer Dawson in 1878. He was so captivated by Haida art and social organization that he developed a lifelong interest in ethnology as a result of his 1878 visit.

Above, right. A Nicola Valley (Nlaka'pamux) basket from about 1907, displayed at the Nicola Valley Museum in Merritt.

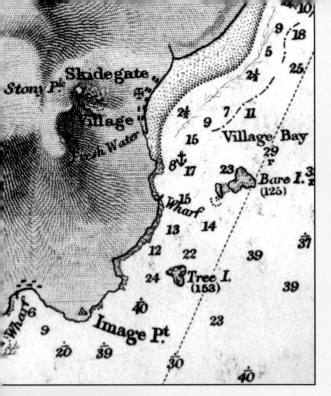

MAP 10 (*right*) and
MAP 11 (*right, centre*).
In the Lower Mainland there are many sites that were occupied by Aboriginal peoples. One that shows up many times is a Cowichan summer village on the Fraser, on the south side of Lulu Island in Richmond. The village was known as Klik-a-teh-nus. The Cowichan part of the Hul'qumi'num group, in turn generally classified as part of the Coast Salish, had permanent villages in the Cowichan Valley of Vancouver Island but set up other encampments over the summer to take advantage of the vast runs of salmon entering the Fraser. *Kawitchen Vill[s].* is noted on MAP 10, from 1858, and *Villages* on MAP 11, from 1849. Howe Sound and Burrard Inlet are noted as *Inhabited*. The village is more precisely located on MAP 132, *page 49*. Other settlements in the Lower Mainland include one in what is now Stanley Park—Xwáýxway, or Whoiwhoi—which was a Squamish (Skwxwú7mesh) Coast Salish village; and of course Musqueam, or XwMuthkwium, another Hul'qumi'num and Coast Salish settlement, perhaps better known because of the visit by Simon Fraser in 1808 (see page 40).

MAP 9 (*above*).
Corresponding to the photo, *left,* is this map of the Haida *Skidegate Village,* as surveyed by naval hydrographic surveyor Daniel Pender in 1866 and published in a British Admiralty chart, *Skidegate Inlet,* in 1872. The longhouses seen in the photo can be seen on the map, arrayed along the high-tide mark.

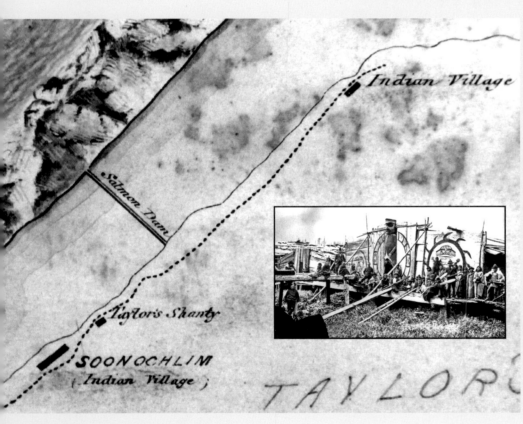

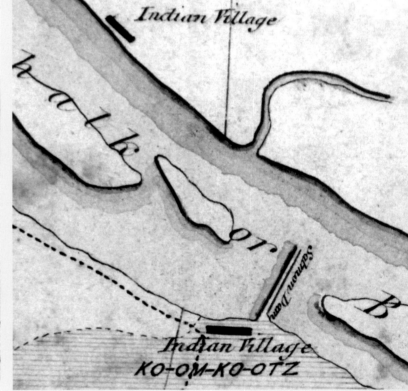

MAP 12 (*above, in two parts*).
The Aboriginal settlements of the Bella Coola Valley were visited by Alexander Mackenzie in 1793 (see page 38). They were documented on maps after the valley's exploration as a possible route inland in 1862 (see page 78), the date of these map details. The valley is the home of the Nuxalk people, also Coast Salish, whose economy was based on the salmon. Some salmon was caught using a trap, shown here as *Salmon Dam.* The rich ornamentation of the buildings in the villages is shown well in the photo, *inset, above.* At *right* is an 1873 photo of Bella Coola, taken during the visit of Indian Commissioner Israel Powell. On the map the settlement is shown as *Indian Village Ko-om-ko-otz* (Qomq'-ts). The whole map is shown as MAP 212, *page 78.*

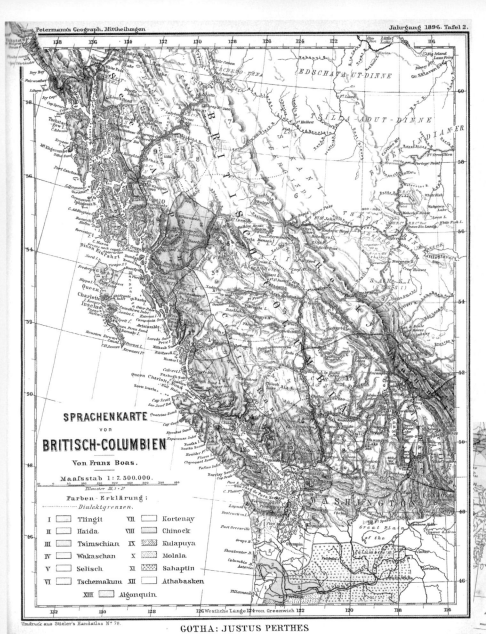

GOTHA: JUSTUS PERTHES
1896.

Map 13 (*left*).
Germans in particular seemed fascinated by the Aboriginal peoples of the American and Canadian West. So fascinated, indeed, that their studies were not always objective. German-American anthropologist Franz Boas, for example, was known to have added a feather to the head of a Nuxalk man before photographing him, just because he thought it looked better. In 1896 German geographer Augustus Petermann published this map of the linguistic families of British Columbia as researched by Boas. They are broadly correct. The area of southern British Columbia and a pocket on the mid-coast coloured pink is Salishan ("Selisch" on the map key), the yellow in the north with a pocket centred on Princeton is Athabascan ("Athabasken"), while the green-coloured areas of western Vancouver Island and the mid coast are Wakashan ("Wakaschan"). To its north is Tsimshian ("Tsimschian"). The yellow-coloured area of the southeast is Kootenay or Kutenai ("Kortenay").

Map 14.
An ethnological map of British Columbia published in 1900 by the provincial museum. Note that the boundaries of the various groups correspond to the language groups shown in Map 13.

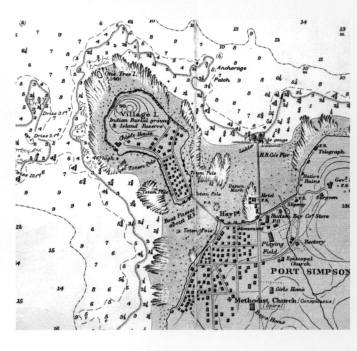

Map 15 (*below, centre*).
A survey by the British surveying ship *Egeria* in 1901. The chart shows the Kwakwaka'wakw (Kwakiutl) *Indian Village* at Alert Bay, on Cormorant Island near Port McNeill on Vancouver Island. A number of EuroCanadian buildings are also shown, including a *Cannery*; the Kwakwaka'wakw economy, like that of all coastal peoples, revolved around fishing. Wealth and status, however, were measured by how much you gave away (rather than by how much you had, as was the case in most societies), and Alert Bay was well known for its large potlatches. The potlatch, basically a big gift-giving party to display wealth, was forbidden by federal law in 1884 and allowed again only in 1951. It was seen as a barrier to the process of "civilizing the Indians." In 1921 hundreds of artifacts such as masks, costumes, and copper shields were confiscated by the Canadian government and were only regained after the 1970s. Alert Bay is home to a 56.4-m-high totem pole, claimed to be the highest in the world. The much-reproduced staged photo, *above, right,* is of a Kwakwaka'wakw canoe. The photo was taken about 1890 by pioneer American photographer Edward Curtis, who photographed many Aboriginal peoples in the West.

Map 16 (*below*).
Another 1901 *Egeria* survey shows the extensive Tsimshian settlement on *Village I.* at *Port Simpson* (Lax Kw'alaams) and indicates the locations of the *Chief's House, Indian Burial ground,* and two *Totem Pole*(s).

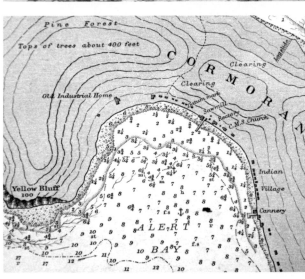

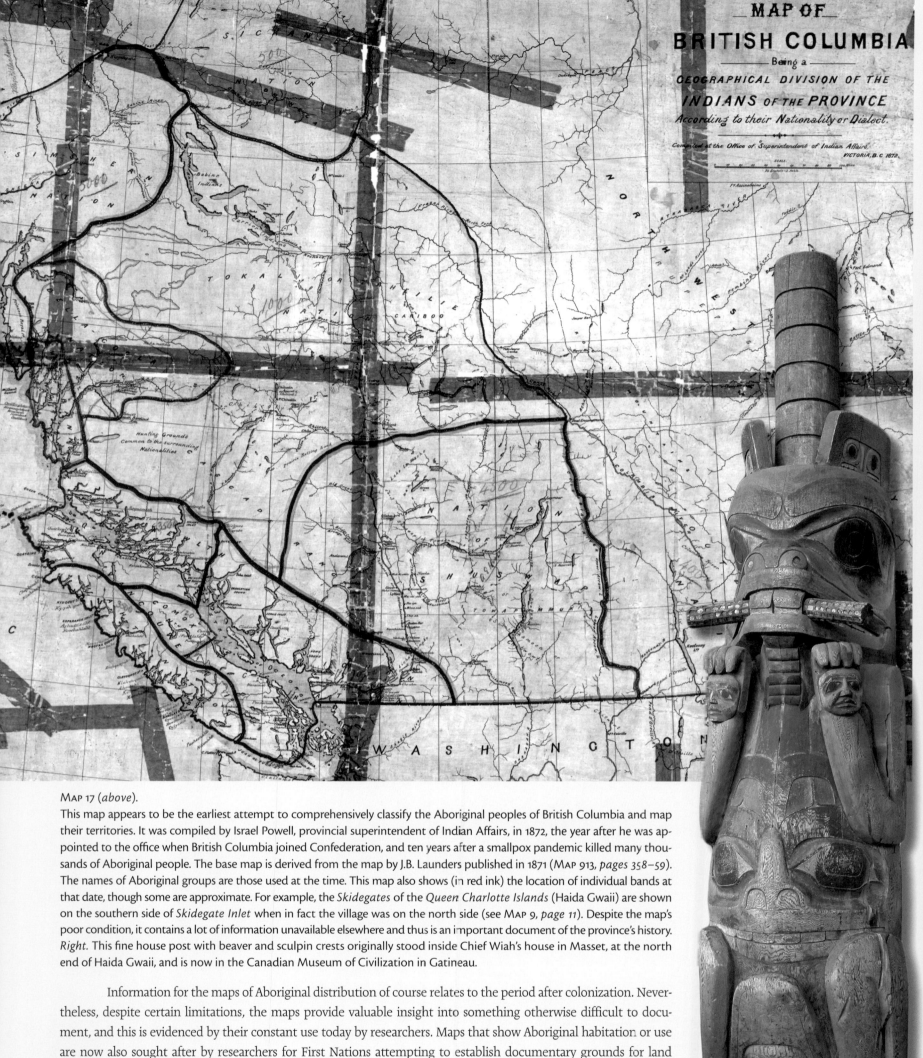

Map 17 (*above*).

This map appears to be the earliest attempt to comprehensively classify the Aboriginal peoples of British Columbia and map their territories. It was compiled by Israel Powell, provincial superintendent of Indian Affairs, in 1872, the year after he was appointed to the office when British Columbia joined Confederation, and ten years after a smallpox pandemic killed many thousands of Aboriginal people. The base map is derived from the map by J.B. Launders published in 1871 (Map 913, *pages 358–59*). The names of Aboriginal groups are those used at the time. This map also shows (in red ink) the location of individual bands at that date, though some are approximate. For example, the *Skidegates* of the *Queen Charlotte Islands* (Haida Gwaii) are shown on the southern side of *Skidegate Inlet* when in fact the village was on the north side (see Map 9, *page 11*). Despite the map's poor condition, it contains a lot of information unavailable elsewhere and thus is an important document of the province's history. *Right.* This fine house post with beaver and sculpin crests originally stood inside Chief Wiah's house in Masset, at the north end of Haida Gwaii, and is now in the Canadian Museum of Civilization in Gatineau.

Information for the maps of Aboriginal distribution of course relates to the period after colonization. Nevertheless, despite certain limitations, the maps provide valuable insight into something otherwise difficult to document, and this is evidenced by their constant use today by researchers. Maps that show Aboriginal habitation or use are now also sought after by researchers for First Nations attempting to establish documentary grounds for land claims. All sorts of historic maps, not just those that were created to show Aboriginal occupation, are used today for this purpose, an attempt to satisfy the EuroCanadian requirement for documentary evidence by a society that was itself at the time purely oral.

Population levels were likely also reduced by the time these maps were made, owing to the introduction by Europeans of diseases such as smallpox, against which the Aboriginal people had no immunity. A particularly rampant smallpox pandemic raged through the Aboriginal population in 1862–63, killing an estimated 30,000 or more

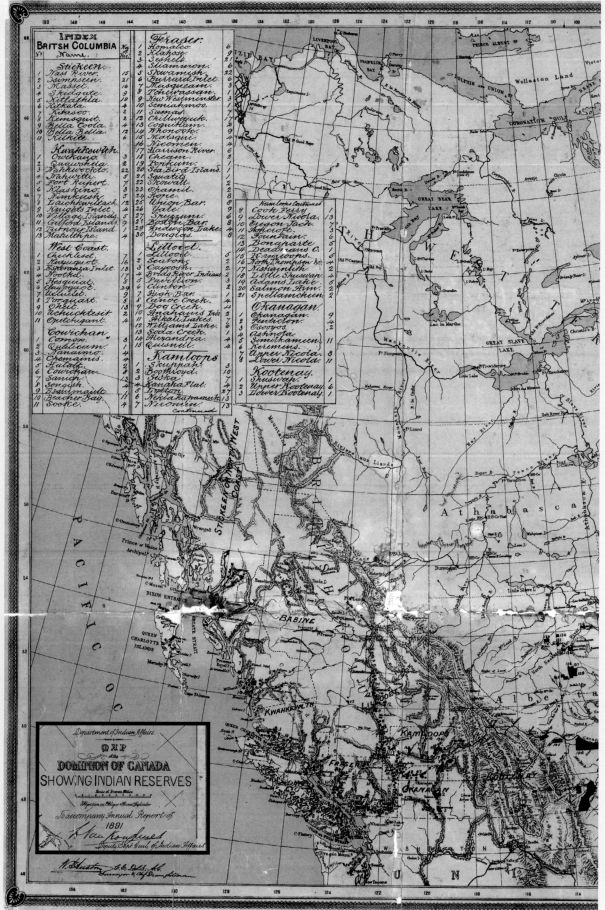

MAP 18 (left).
Since the eighteenth century Canada has negotiated—though sometimes coerced more than negotiated—with its Aboriginal peoples with the intent of clearing them off land required by EuroCanadian settlers and confining them to much smaller tracts of land set aside for them. In British Columbia between 1850 and 1854 Governor James Douglas purchased fourteen tracts of land around Vancouver Island that were required for settlers. The sale documents are usually known as the Douglas Treaties. No more treaties were made owing to lack of funds, and for a while Aboriginal peoples were able to maintain their existing settlements and Euro-Canadians were forbidden to pre-empt them. Then, in 1864, Joseph Trutch became Commissioner of Lands and Works and Surveyor General (and in 1871 the first lieutenant governor of the new province of British Columbia) and initiated a policy of reducing the size of Aboriginal lands. Aboriginal peoples were allocated roughly 10 acres (4 ha) as compared with the typical 160 acres (65 ha) for EuroCanadian settlers. In 1876 a Joint Commission on Indian Reserves was created, and over the next thirty years or so most of the Indian reserves in the province were created. (There are now approximately 1,600 reserves owned by about 189 bands.) Some reserves in the northeastern part of the province were created or confirmed by Treaty 8 in 1899. The latter was one of a series of numbered treaties that were negotiated right across the Prairies (see page 233). Otherwise, while Aboriginal peoples have been confined to defined reserves, there has been little negotiation or consent (with a few notable exceptions, such as the treaty signed with the Nisga'a of the Nass Valley in 1999 after interminable years of negotiation; see page 335), and the question of Indian title was circumvented by what was considered a generous allocation of reserves. On this map, dated 1891, the black dots represent Indian reserves, and the main groups of Aboriginal peoples and bands are listed in the table at top left.

MAP 20 (below).
This 1873 map shows the mouth of the Campbell River in South Surrey, with a settlement of *Indians*—the Semiahmoo, a Coast Salish people. Dislocated by a sawmill, the Semiahmoo First Nation Reserve is now on the other side of the river. The basalt arrowhead, *right*, typical of those fashioned by Aboriginal peoples, was found in 1970 on the beach near the old encampment, which was also the location of the boundary surveyors' camp in 1858 (see MAP 172, *page 60*).

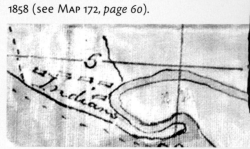

MAP 19. An 1878 map of an area near Alkali Lake, south of Williams Lake, shows an *I. Village* and *Indⁿ field* near a *Catholic Church.* This was the land of the Secwepemc (Shuswap) people. The river is Canoe Creek, a tributary of the Fraser.

people. The population of the Hudson's Bay Company territory west of the Rocky Mountains is stated in the margins of MAP 8, *page 10*, to be about 80,000 (in 1857), but even that is almost certainly a reduction from the levels that would have existed at contact.

There is an extreme paucity of maps of anywhere in British Columbia that were drawn before they were likely influenced by seeing EuroCanadian maps. This is because, before explorers appeared holding paper and pen, the only means Aboriginal peoples had of making maps were relatively ephemeral—such as charcoal on a skin or tree bark or a stick in dust on the ground—all far less likely to survive than a carefully filed map perhaps taken back to an office in Britain. The Aboriginal maps that exist today tend to be those that Europeans asked Aboriginal people to draw to aid them in their task or to find their way. The maps drawn for the American explorers Meriwether Lewis and William Clark during their 1804–06 venture to the mouth of the Columbia River are the classic Western example of this. From Alexander Mackenzie on, European explorers' wayfinding was made much easier with Aboriginal help.

Aboriginal peoples tended to live relatively local lives and knew how to get around in their environment, and thus they had no regular need for maps, in the same way that today's GPS user has no need of the device near home.

Aboriginal maps, of course, did not follow European cartographic conventions, such as putting north at the top. The size at which Aboriginal mapmakers drew a feature was typically related to the importance of that feature to their people. MAP 21, *above*, is principally a map of the highways that mattered—the river, streams, and lakes that enabled easy movement and communication.

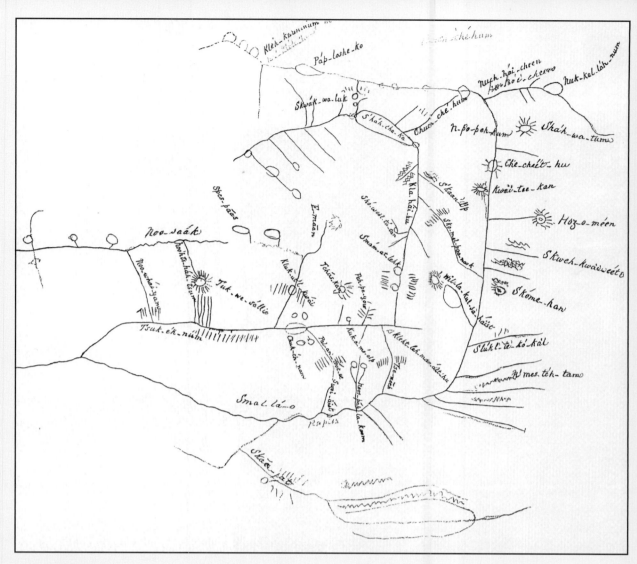

MAP 21 (*above*).
This rare example of a British Columbian Aboriginal map was drawn about 1859 by Somena chief Thiusoloc at the request of Henry Custer, a surveyor with the U.S. Boundary Commission (see page 60). It shows the river and lake system of part of the eastern end of the Lower Fraser Valley and the drainage basins of the *Noo-saák* (Nooksack) River and other rivers in the northwest part of what is now Washington State. The lake labelled *S'háh-cha-ka* seems to be Cultus Lake, and the river immediately above it the Chilliwack-Vedder River. The map suggests that Thiusoloc personally knew the rivers and lakes he drew, although, as a chief, he might simply have been kept informed by others. Whether he would have drawn such a map for his own use is doubtful, since it was clearly in his head in any case.

Left and *right*.
The Nuu-chah-nulth (Nootka), like the Haida, had a tradition of erecting totem poles and, as with their northern kin, left them on the ground to return to the earth once they fell. But they were replaced with new poles. This pole, in two views, was lying on the ground at Yuquot in the summer of 1998—but no new pole had been carved to replace it.

A Coast Unknown

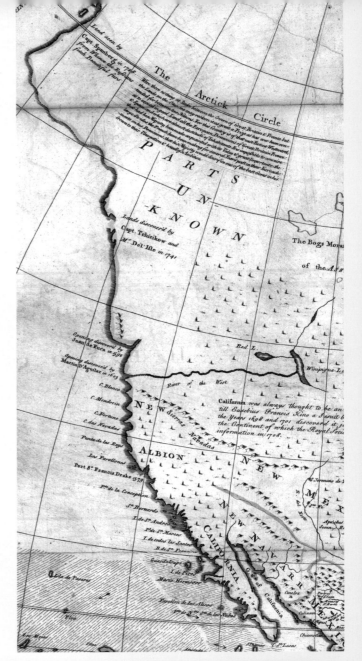

Until the early sixteenth century, the entire west coast of the Americas was completely unknown to Europeans. Early landfalls in the Americas were invariably shown on maps as landings on the eastern extremity of Asia. But the West Coast appeared first on a map some time before any recorded navigators reached those shores, accessible from Europe only by dangerous and immensely long voyages.

In 1513 Vasco Núñez de Balboa crossed the Isthmus of Panama and saw the Pacific Ocean. Seven years later Fernão de Magalhães, better known as Ferdinand Magellan, sailed through the southern strait that bears his name and emerged into an unusually calm sea he named Mar Pacifica. The Spanish invaded the Aztec Empire of Mexico in 1519 and two years later established their colony of New Spain, the region from which further exploration would take place, both north and south. But southward there was the lure of Inca gold; to the north—nothing. Although some almost half-hearted coastal voyages of exploration were conducted northward, it would take another two and a half centuries for the Spanish to reach British Columbia, and only then because they feared Russian incursion into territory they considered their own. It was this primary lack of motivation that left the Northwest Coast as a blank on the map of the world until 1774—a coast unknown.

Map 22 (*right*).
When this map was published in London in 1750, the southern half of the West Coast was shown with tolerable accuracy, but north of Francis Drake's *New Albion* a hypothetical coast was depicted, broken only by the notation of Aleksei Chirikov and Vitus Bering's landfalls in 1741 (see page 21; the M^r. *De L'Isle* referred to on the map was Joseph-Nicolas De L'Isle, geographer to the expedition but not actually with Bering). Here all British Columbia is *Parts Unknown*. In Bering Strait, to the north, the notation *Land seen by Capt. Spanberg* refers to Bering's first voyage in 1728 (Martin Spanberg was one of Bering's officers), but the first sighting of the Northwest Coast is usually attributed to another Russian, Mikhail Gvozdev, four years later. The *Opening discovered by Juan de Fuca in 1592* is shown as a major indentation of the coast (see page 18).

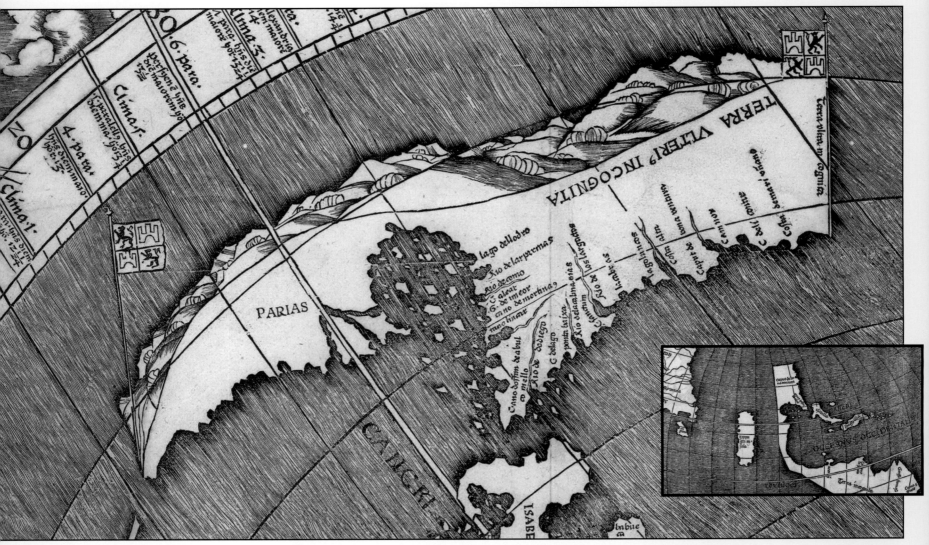

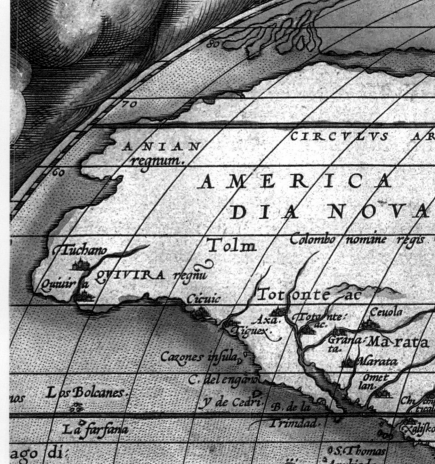

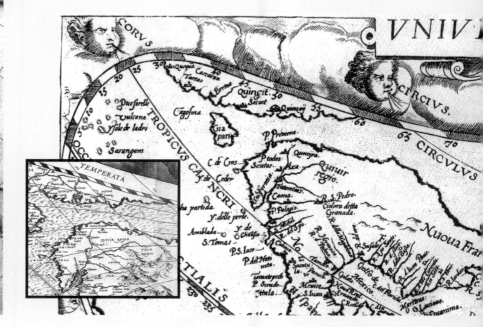

Map 23 (*left*).

The first map to show a west coast—indeed the first to show the Americas as a separate continent—was this map, part of a very large world map published in 1507. It was the work of Martin Waldseemüller and Matthias Ringmann. Before this map Asia was thought to be the first land west of Europe. The inset map was part of a smaller world map, on a different projection, shown at the top of the main map, and the fact that the west coast is here but a straight line strongly suggests it was little more than a guess. On the main map *Terra Vlteri' Incognita* is written—unknown land. The map, purchased a few years ago by the Library of Congress for $10 million, is kept in a special argon-filled case. The presence of the word "America" near the tip of South America has led to the dubbing of this map as "America's birth certificate."

Map 24 (*above, top left*).

North America is not separated from Asia on this nicely illustrated map by Venetian mapmaker Paolo Forlani, drawn in 1565. The land that would become British Columbia is included but only by default; much of the map north of Baja California, found in 1533 by Hernán Cortés, is a figment of the mapmaker's imagination, but it certainly looks like quite a reasonable generalized representation of the actual coastline.

Map 25 (*above, top right*) and **Map 26** (*above, left*).

Two maps by famous Dutch mapmaker Abraham Ortelius, both from his 1570 atlas *Theatrum Orbis Terrarum*, show that different versions of unknown and little-known geographies can come from the same source at the same time. Ortelius has drawn quite a different west coast of North America on these maps. **Map 25** refers to North America as *America [In]dia Nova*—New India. The Marco Polo–inspired *Anian* is also shown on this map, as are the mythical cities of gold in the Southwest. This map is an interesting extrapolation of known geography to the unknown and results in a surprisingly good approximation of reality. **Map 26** shows a large *Iapan* (Japan) sitting in the middle of the Pacific Ocean.

Map 27 (*above, right*) and **Map 28** (*above, right, inset*).

These two maps on similar projections show the west coast of North America on the edge of the world. **Map 27** is from the 1578 atlas of Dutch mapmaker Gerard de Jode, while **Map 28** is from the 1564 map by Abraham Ortelius. The latter shows small islands off the coast in the position of British Columbia—islands that have been erroneously interpreted as Vancouver Island, the Olympic Peninsula, Haida Gwaii, and Prince of Wales Island, drawn there as a result of Francis Drake's voyage. This could not possibly be the case, since Drake arrived on the West Coast fifteen years later than the date of this map. Mapmakers, even famous ones like Ortelius, often added details of their own contrivance when no other information was available.

Were the Chinese First?

A great Chinese encyclopedia compiled between 502 and 556 documents a voyage by a Buddhist monk, Hui Shen, in 499, to a place "where the sun rises." This has been interpreted as the west coast of North America but could just as easily have been a more local coast, such as that of Japan or Sakhalin. However, Fusang, or Fousang, as the place the Chinese monk reached was called, was placed on a map of North America in 1752 by French geographer Philippe Buache. His map found its way into some volumes of the authoritative *Encyclopédie* of Denis Diderot, published in several editions in the second half of the eighteenth century. This gave the idea wider circulation, as it did the notion of vast inland seas (see page 22). But no actual evidence of any early Chinese presence has ever been found, and the idea remains a myth.

MAP 29 (*above*).
Chinese world maps such as this one from about 1650 are used as evidence that Chinese navigators found the west coast of North America at some undefined earlier date. China is at centre, where it was usually placed by Chinese mapmakers, attached to the rest of Asia, India, and Africa, with a ring of land perhaps representing North and South America, east coast on the left and west coast on the right, the latter said to show a sequoia tree as evidence of its American location.

MAP 31 (*right, top*).
Fou-Sang des Chinois is located exactly where British Columbia is, straddling the 50°N line of latitude, on this 1774 map of the Northwest Coast that follows Buache's fantastic geography (see page 22).

MAP 32 (*right*).
On this 1778 map, just north of the *Entrance found by Juan de Fuca*, is the notation *this Land is supposed to be the Fou-Sang of the Chinese Geographers*.

MAP 30 (*left*).
Alaska and part of the British Columbia coast are shown on this map of the Pacific Rim countries held at the Library of Congress, where it is dated circa 1300. It has been attributed to Marco Polo, though this seems unlikely. If the date is anywhere near correct, it does indicate early Chinese knowledge of the Northwest Coast. The map, a hard-to-reproduce palimpsest, has a ship on the left, giving it its popular name of "Map with ship." The map has a long history, at one time even being given to J. Edgar Hoover of the FBI, who tried, not very successfully, to determine the age of the inks; the map remains an enigma.

Juan de Fuca's Strait

MAP 33 (*above*).
An *Entrance discover'd by Juan de Fuca in 1592* is noted on this 1768 map.

Another unproven voyage, that of the elusive Juan de Fuca, nevertheless had a more lasting presence on the maps of the Northwest Coast, because when the Strait of Juan de Fuca was finally found by Europeans—Charles Barkley in 1787 (see page 29)—it was given that name because of Barkley's understanding that it was Juan de Fuca who had found it first, almost two centuries before.

The story originated with Michael Lok, an English promoter of voyages to find a Northwest Passage, who said he met Apostolos Valerianos, or Juan de Fuca, a Greek pilot in the employ of the Spanish king, in Venice in 1596. Lok said he was told of a voyage Fuca had made to find the Strait of Anian, finding instead his Strait of Juan de Fuca, into which he had sailed for twenty days until he came to the North Sea (presumably the Atlantic Ocean)—clearly a fabrication. It seems likely that Lok concocted the entire tale to further his own ends. Nonetheless, the story gave rise to maps of inland seas (see page 22) and to Barkley's naming of the real strait in 1787.

MAP 34 (*right, top*).
Fur trader John Meares, who had Charles Barkley's papers in his possession, showed *John de Fuca's Straits* on a map in his 1790 book.

MAP 35 (*right*).
It was left to fellow British fur trader Charles Duncan in 1788 to draw the first published map of the Strait of Juan de Fuca. This is Duncan's map, complete with a view of the entrance at bottom, which shows a pillar on the southern side, tantalizingly similar to pillars that Fuca—or Lok—had described, but on the *northern* side.

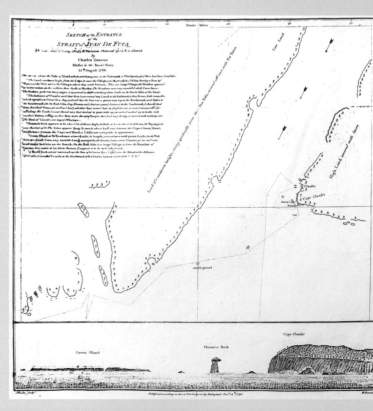

THE ENGLISH CLAIM

The original British (at first English) claim to what is today British Columbia stems from the 1579 visit to the West Coast by Sir Francis Drake during his circumnavigation of the world between 1577 and 1580. The fact that he landed somewhere to repair his ship is documented, but the precise position of the landing is not. A debate has raged for many years as to this location, with coves and bays in California and Oregon as well as British Columbia becoming the objects of theorists' delight. In fact the most probable location of Drake's landing is at Drakes Bay, in the lee of Point Reyes, California, just north of San Francisco. Despite claims to the contrary, there is currently no evidence that Drake reached British Columbia.

Especially insidious is the notion that because a coastline on an old map appears to vaguely match part of what we see on a map today, this somehow constitutes a proof of Drake's visit. Nothing could be further from the truth, as there are a myriad of scribbled coastlines on hundreds of old maps that seem to represent modern coastlines that are no more than cartographic good luck. Mapmakers had to sell maps and did not like to let their customers know they had no knowledge of unexplored regions and so, as we have seen, filled in the gaps any way they could.

No one knows how far north Drake sailed. He could have reached British Columbia; indeed, since he appears to have searched for a western entrance to a Northwest Passage to get home to England quickly (Drake thought he was being followed by the Spanish, whose Manila galleon he had plundered), it is entirely possible he entered the Strait of Juan de Fuca looking for the passage. Those who believe that his information was suppressed to fool the Spanish seem to have overlooked the fact that the English presence—with a Northwest Passage—was noted on several maps published at the time (MAP 37, *right*).

But Drake's voyage did establish—in English minds at least—an English claim to the Northwest Coast, and this was used in later boundary negotiations; certainly because of it the world came to recognize that it was fair game—under the rules of the day—for British Columbia to be British. For Drake's New Albion (New England) would migrate north.

MAP 36 (*above*).
Part of the world map of the French mapmaker Nicola van Sype, published in 1583, showing the West Coast part of Drake's voyage. Note how the termination of his track north is carefully hidden by a drawing of his ship.

MAP 37 (*right*).
This map, the first to record Drake's voyage, was published in 1582 by the English promoter of the Northwest Passage concept, Michael Lok. A grossly enlarged Baja California extends north to about 45°, where a Northwest Passage opens up across the north of the continent. At that point is written *Anglorum 1580* (English, 1580; Lok got his date wrong by a year, substituting the date of Drake's return to England).

MAP 39 (*below, left*), with *inset*. Another world map, this one by Jodocus Hondius, produced in 1589. Here the northward track of Drake appears, some claim, to have been erased, as though information it once contained about how far north he sailed was suppressed in some way. The *inset*, which is shown as such on the larger map, shows the bay where Drake careened his ship for necessary repairs to its hull—and claimed the land for England. The location of *Portus Novæ Albion*—Port New England—is the subject of ongoing controversy. Will we ever know if it is Drakes Bay, California, or Boundary Bay, British Columbia?

MAP 40 (*below, right*).
A favourite of those who would attribute a vaguely correct coastline shape to Drake is part of a globe created in 1592 by English globe maker Emery Molyneux, supposedly suppressed, and then reissued in Amsterdam in 1603. The indentation in the coastline by the *A* of *Nova Albion* is similar to that of the Strait of Juan de Fuca and is conveniently at about 49°N. It is possible but unconvincing evidence.

MAP 38 (*left*).
Part of a map published in a 1647 atlas by Robert Dudley, whose father had been involved with financing Drake's voyage and was therefore possibly privy to some inside knowledge of it. The map shows what has been interpreted as the coast of Washington, with Grays Harbor and Cape Flattery, before petering out farther north. If Dudley had inside knowledge of Drake's voyage, this is surely evidence that the great navigator did not reach British Columbia.

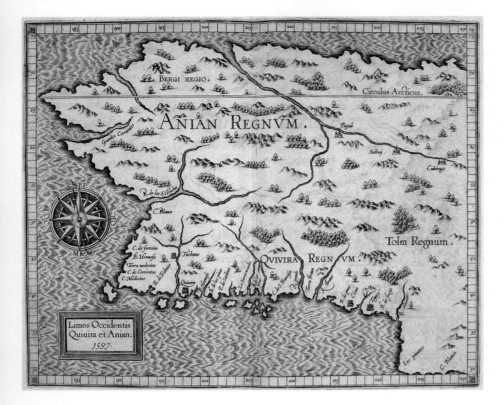

MAP 41 (*left*).
Published in the first atlas to cover all of North America in regional sheets, this map of the Northwest by Cornelius Wytfliet appeared in 1597. It was in reality an imaginary map of *Anian*, the land to the east referred to in the works of Marco Polo. It just happened to look vaguely like Alaska and the West Coast.

MAP 42 (*below*).
This map, also imaginary, is considered to be the first regional map of the Northwest. Beautifully drawn and illustrated, it was contained in a 1593 atlas created by Dutch mapmaker Cornelis de Jode. *El Streto de Anian*—the Strait of Anian or Northwest Passage—stretches across the north of the continent. To its south is Marco Polo's *Anian*. Farther south still is *Quivira Regnu* (also shown on MAP 41), a mythical city of gold contained in a story told to Spanish explorer Francisco Vásquez de Coronado in 1540. Here British Columbia is nowhere to be seen, but its position would have been about where the unicorn sea monster is drawn.

The farthest north that the Spanish reached before the eighteenth century was around Cape Mendocino in California; this was the voyage of Sebastián Vizcaíno in 1602, sent north to try to find harbours that could be used by the annual galleon that sailed from the Philippines to New Spain and was often in distress nearing the end of its voyage. The galleon followed a northerly path eastward across the Pacific in order to utilize the prevailing winds but, nearing the coast of North America, turned southward towards New Spain. One of Vizcaíno's ships, captained by Martín de Aguilar, became separated in a storm and was pushed farther north, where he found an opening that was considered as a possible Strait of Anian, a Northwest Passage then thought to link the Pacific with the Atlantic. It may have been the mouth of the Columbia or even the Strait of Juan de Fuca, and it appeared as the "Entrada de Aguilar" or something similar on many early maps.

A search for the Strait of Anian or Northwest Passage was something that interested the Spanish but not enough to motivate them until they thought others might find it first (see page 24).

Indeed, it was fur trader Mikhail Gvozdev who first sighted the Alaska coast in 1732, and voyages by Vitus Bering and Aleksei Chirikov in 1841 found points on the Alaska Panhandle, the former at Cape St. Elias, which Bering named, and the latter on Prince of Wales Island. The pair had been dispatched by Tsar Peter the Great as part of a decades-long effort to find a Great Land to the east—North America—but the effort was not followed up until much later (eventually leading to the chartering of Hudson's Bay Company rival the Russian America Company in 1799) but did serve to motivate the Spanish to explore north and find out just what those Russians were up to.

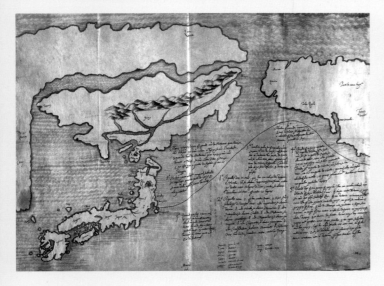

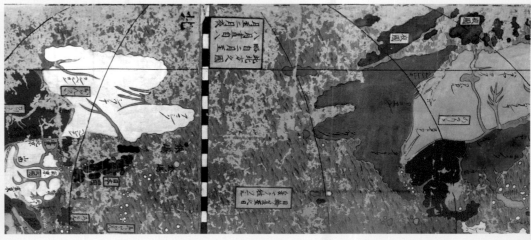

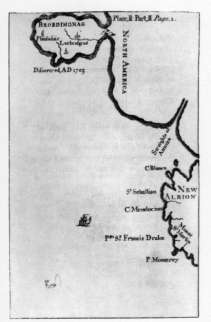

A selection of maps from pre-European exploration of British Columbia. Map 43 (*above, left*) is a map of the Pacific drawn in 1621 by Girolamo de Angelis, a Jesuit missionary in Japan, which, including Hokkaido to the north, is greatly exaggerated in size. On the eastern shore of the ocean, Baja California is all that is identifiable. Map 44 (*above, right*) is from a Japanese map of the world created in 1645; again only Baja, found by Hernán Cortes in 1533, is recognizable, though British Columbia is there, somewhere to the north. Map 45 (*left*) is Jonathan Swift's fictitious interpretation of British Columbia as *Brobdingnag*, his land of giants in *Gulliver's Travels*. Nothing was known of the Northwest, so where better to locate a tale of fiction? This map was from the 1726 edition. Map 46 (*right*) is a typical eighteenth-century map of North America, this one by French mapmaker Jacques-Nicolas Bellin from 1743. A *Fleuve de l'Ouest* (River of the West) flows to—or from—the Northwest. The idea of this river derives from Aboriginal tales of rivers flowing into Lake Winnipeg as told to early French fur traders. The river is also shown on Map 47 (*right*), from a 1778 edition of Jonathan Carver's *Travels*; here the position of the strait *Discovered by Juan de Fuca* is remarkably accurate and connects with the opening *Discovered by [Martín de] Aguilar* as well as the *River of the West*.

The position that would be British Columbia is here a *Western Sea*. Note Drake's *New Albion*, still appearing on maps two hundred years after his visit. Below it (Map 50) is a more honest, less speculative map dated 1758, also by Bellin. Here only the supposed positions of actual west coast discoveries have been noted, though the tentative position of the *Mer de l'Ouest* (Sea of the West; see next page) is noted about where British Columbia would be. Map 48

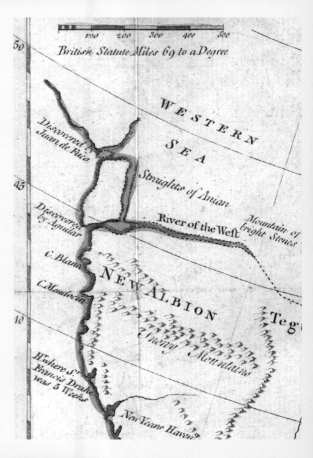

(*left*) is even more honest. Drawn in 1731 by Joseph-Nicolas De L'Isle, it was given to Vitus Bering as a supposed guide to what he should discover; the entire Northwest was correctly shown as a vast blank. Finally we have Map 49 (*below*), which is a map of the 1741 North American landfalls of Bering and Aleksei Chirikov, sailing from Kamchatka in two ships. Bering's is the northernmost one, with a landfall at Cape St. Elias, and Chirikov's is to the south, at Baker Island, off Prince of Wales Island—only 100 km or so from what is now British Columbia. The double coastline shown is due to errors of dead reckoning, with the position being worked back from the eastern land in case of the return voyage.

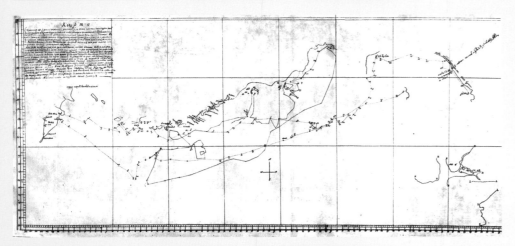

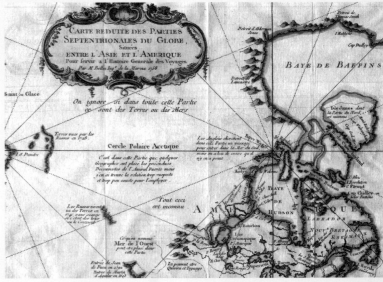

Water, Water, Everywhere

Seas and straits abound in the Northwest in this selection of bizarre maps, produced in response to accounts of the supposed voyages of Juan de Fuca, Bartholomew de Fonte, and others. Many of these maps leave one scratching one's head—where is British Columbia?

The absurd configurations of the coastlines of the Northwest shown on these pages have their origins in the interpretations,

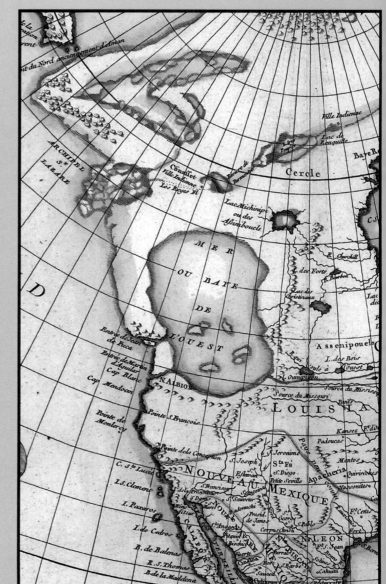

MAP 51 (*above*), enlarged detail, and whole map, *inset*. Perhaps the finest and at the same time most ridiculous depiction of the West Coast was on this 1748 map produced for the Hudson's Bay Company and then—all but six copies—destroyed when it was recognized as widely inaccurate even in terms of geographical knowledge at the time. *De Gama's Land* juts out across the Pacific. *Straights* and *Anian* are approximately in the position of the Strait of Juan de Fuca. Note *De Fonte Track* offshore to the *Straights & Isles of St. Lazarus*, an archipelago that also features in the de Fonte hoax story, but again positioned where islands—those of the Alaska Panhandle—actually exist. Hudson Bay and the Great Lakes are depicted too far west; *Upper Lake or Superior* almost reaches British Columbia.

mostly by French mapmakers, of the mythical, or at least unproven, stories of João da Gama (*De Gama's Land* on MAP 51), Juan de Fuca, Martín de Aguilar, and Bartholomew de Fonte. Those of Fuca (page 18) and Aguilar (page 20) have been noted, but the most nonsensical story was that of de Fonte, which grew into a major hoax. The story was that a Spanish admiral, de Fonte, had in 1640 sailed north along the Northwest Coast and found channels—many of which are shown on the maps—that led to Hudson Bay. This story was published as an anonymous letter in a magazine in Britain in 1708, then receiving virtually no attention. However, in 1744 Arthur Dobbs, a member of the British parliament and a firm believer in the existence of a Northwest Passage, found the original letter and promoted its ideas, which were in turn picked up and publicized by others, culminating in the map interpretations of the story created by influential French cartographers Joseph-Nicolas De L'Isle and Philippe Buache, whose maps were in turn copied and embellished by others. The story remained in currency until Jacinto Caamaño in 1792 (see MAP 85, *page 33*) and George Vancouver in 1794 (see page 36) finally proved the falsehood of the bizarre tale.

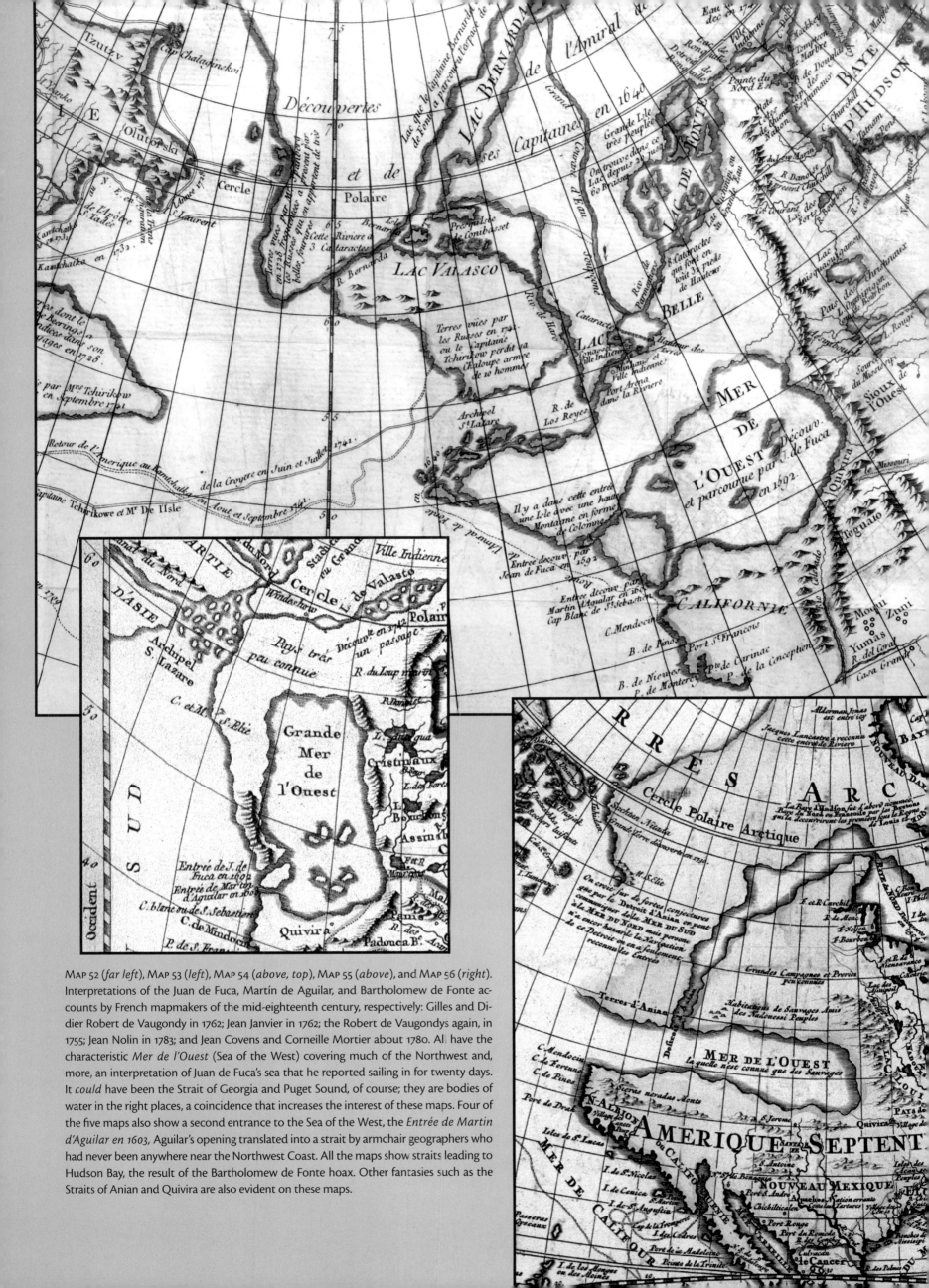

Map 52 (*far left*), Map 53 (*left*), Map 54 (*above, top*), Map 55 (*above*), and Map 56 (*right*). Interpretations of the Juan de Fuca, Martín de Aguilar, and Bartholomew de Fonte accounts by French mapmakers of the mid-eighteenth century, respectively: Gilles and Didier Robert de Vaugondy in 1762; Jean Janvier in 1762; the Robert de Vaugondys again, in 1755; Jean Nolin in 1783; and Jean Covens and Corneille Mortier about 1780. All have the characteristic *Mer de l'Ouest* (Sea of the West) covering much of the Northwest and, more, an interpretation of Juan de Fuca's sea that he reported sailing in for twenty days. It *could* have been the Strait of Georgia and Puget Sound, of course; they are bodies of water in the right places, a coincidence that increases the interest of these maps. Four of the five maps also show a second entrance to the Sea of the West, the *Entrée de Martin d'Aguilar en 1603*, Aguilar's opening translated into a strait by armchair geographers who had never been anywhere near the Northwest Coast. All the maps show straits leading to Hudson Bay, the result of the Bartholomew de Fonte hoax. Other fantasies such as the Straits of Anian and Quivira are also evident on these maps.

THE SPANISH MOVE NORTH

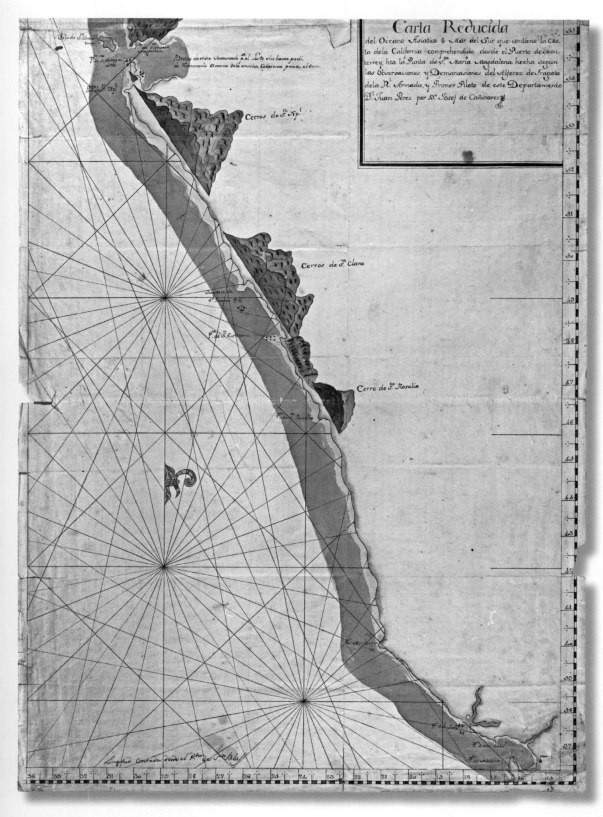

Spain's policy in North America underwent a change in 1765 with the appointment of José de Gálvez as visitador general (inspector general) to New Spain and, in 1776, secretary of the Indies. He was an expansionist determined to protect Spanish interests in the New World. And Spain became increasingly concerned about possible Russian and British incursions.

The voyage of Vitus Bering and Aleksei Chirikov to the West Coast in 1741 seems to have set off Spanish imaginations of Russians everywhere, a concern bolstered by the 1759 publication of a book, *I Moscoviti nella California* (Muscovites in California). In addition, several more books warned of impending British discovery of the Northwest Passage. British commodore George Anson sacked a city in Peru in 1741, and in 1764 Commodore John Byron was dispatched to the Pacific to search for the passage; he got nowhere near it, but this was unknown to the Spanish.

A 1767 move by the Spanish king to expel Jesuits began a grand scheme by Gálvez to colonize Alta California and move the replacement Franciscans there. The so-called Sacred Expedition of 1769 led to the settlement of San Diego the following year and San Francisco the next. Then the king ordered several voyages far to the north to seek out any Russian trespassers and perform acts of possession, believing them to secure ownership.

In 1774 the viceroy of New Spain dispatched a senior naval officer, Juan Pérez, with instructions to sail to 60°N. It was during this voyage that Europeans first discovered British Columbia. Pérez made it as far north as Langara Island, at the tip of Haida Gwaii, which he named Punta de Santa Margarita, shown *below*. It was here that the first known contact between Aboriginal people and Europeans took place, on 19 July 1774, when Haida in canoes

MAP 57 (*above*), with enlarged detail (*right*).
The first map of British Columbia. This is the map of Juan Pérez's 1774 voyage, drawn by José de Cañizares, a pilot, early the following year. *Pta. de Sta. Margarita* is at the top, *Surgidero de Sn. Lorenzo* (Nootka Sound) and *Pta. de Sn. Estevan* (Estevan Point) are named on the British Columbia coast; California's San Francisco Bay and Monterey Bay are at the bottom.

Below.
Point of contact. This is St. Margaret Point—Punta de Santa Margarita—probably looking today much as it would have looked to Pérez in 1774. The point is also shown (*right*) on Pérez's map, enlarged from the map above. Langara Island is shown attached to the mainland of Haida Gwaii.

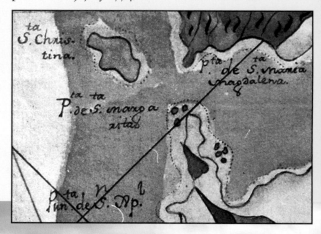

came out to investigate Pérez's ship. Pérez was wary of landing because of what had happened to Chirikov in 1741: the latter had sent two boats to land and neither returned.

On his return voyage Pérez kept close to the coast, anchoring at one point off Nootka Sound, and this enabled him to map the coast with limited accuracy, producing the first map of any part of British Columbia drawn from exploration rather than imagination (MAP 57, *left*).

Apart from the significance of the first contact, Pérez's voyage was also important because it established a Spanish claim to the coast as far as 54°40′N, a latitude that was to figure large in later boundary negotiations (see page 54); it is today the southern tip of the Alaska Panhandle.

The viceroy was not pleased with Pérez, for he had not reached as far north as he had been instructed, and so the next year another expedition was dispatched under Bruno de Hezeta; Pérez was demoted to first officer. Two ships sailed north, one under Hezeta and the other, a shallow draft vessel intended for inshore use and only about 11 m long, under an ambitious young naval officer, Juan Francisco de la Bodega y Quadra. His ship became separated from Hezeta's, likely deliberately, for Hezeta, whose crew was suffering from scurvy, wished to return home. Hezeta returned along the coast, finding a bay that later turned out to be the mouth of the Columbia River. Pérez died on this voyage. Bodega y Quadra determined to follow his instructions and continued north, braving heavy seas in a voyage that can only be described as epic, reaching 58°N and making a landfall in the vicinity of Sitka. But he found no Russians.

Bodega y Quadra, who seventeen years later as a senior Spanish officer would negotiate with George Vancouver at Friendly Cove, produced several maps of his voyage, but because he was forced to sail offshore most of the time, they show detail of only the Washington and Alaska coasts, with that of British Columbia simply drawn as a connecting line (MAP 59, *right*).

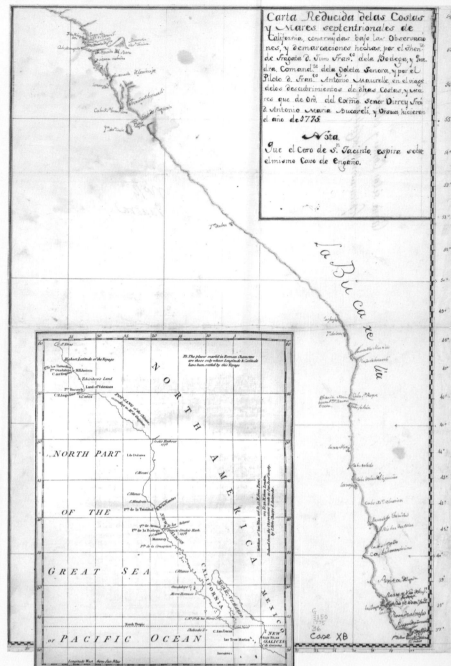

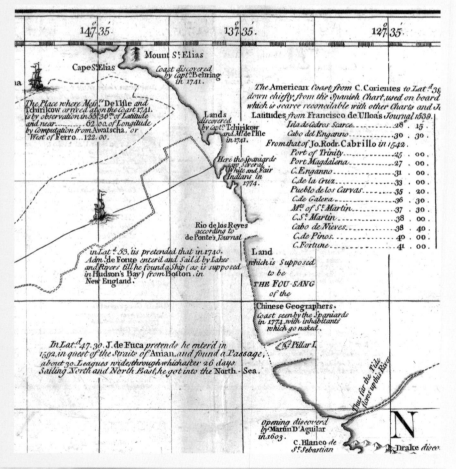

MAP 58 (*left*).
The Spanish were typically very secretive with their maps, and so it is somewhat of a surprise that information about the Pérez voyage turned up in Thomas Jefferys's *American Atlas* the year after. The map shows, in British Columbia, the notation *Here the Spaniards saw Several White and Fair Indians in 1774* and, farther south, a *Coast seen by the Spaniards in 1774 with inhabitants which go naked.* The dates make it clear that these refer to the Pérez voyage. Note also the notations regarding *Captn. Bearing* and *Captn. Tchirikov* (Bering and Chirikov) in 1741, another about the Strait of Juan de Fuca, and yet another designating British Columbia as *Land which is Supposed to be the Fou Sang of the Chinese Geographers* (see page 18). The coastal configuration suggests a map made from a written description rather than copied from another map.

MAP 59 (*above, main map*).
One of the maps of the voyages of Bodega y Quadra and Hezeta, drawn in 1775. Detail of the Alaskan coast is at top, and detail of the Washington coast southward begins halfway down the map, but the coast of British Columbia is depicted as a nearly straight line connecting the two sets of information, the result of Bodega y Quadra's offshore sailing.

MAP 60 (*above, inset*).
Also remarkable was the publishing in 1780 of an English translation of the entire journal of Francisco Antonio Mourelle de la Rúa, Bodega y Quadra's second-in-command. It contained this map. Again British Columbia is a straight line designated *Fou-Sang of the Chinese*. It also shows *Cook's Harbour 1778*, an early reference to James Cook's third voyage (see next page).

THE WORLD FINDS BRITISH COLUMBIA

BRITAIN'S GREAT NAVIGATOR

If the Spanish were secretive, the British were not. They had less need to be, perhaps, for their navy dominated the oceans of the world by the late eighteenth century. And it was a British explorer, James Cook, who was the next to arrive on the British Columbia coast.

Overlooking Victoria's Inner Harbour stands a statue of Britain's most famous navigator, still considered by many to be the founder of *British* British Columbia. Certainly he is responsible for—quite literally—putting British Columbia on the map, for the Northwest then found its way onto the printed map of the world with its position correctly delineated. (The confusion beforehand is illustrated by MAP 914, *page 368.*) For Cook carried with him some of the very first chronometers, which would revolutionize location finding and mapmaking because they allowed navigators to determine longitude without having to go through laborious land-based astronomical observations. Although maps after Cook would now show the correct position of the coast, it took a while for the interior features to be mapped in their correct relative positions, for explorers could not carry chronometers in their canoes.

The revelation by Cook's crew and those who read their accounts that the coast held a fortune in "soft gold"—the prized fur of the sea otter, soon to be hunted almost to extinction—led to a series of commercially motivated voyages to the Northwest that would begin to build up geographical knowledge of the British Columbia coast.

Cook was dispatched in 1777 on his third voyage of exploration, with orders to search for the western entrance of the Northwest Passage. He arrived off the Oregon Coast on 7 March 1778, and after being then driven both south and offshore once more, he battled his way north through stormy conditions until he found refuge in Nootka Sound at the end of the month. Much has been made of the fact that Cook missed the Strait of Juan de Fuca, but he had been instructed to search for a strait north of 65°N and so was not especially looking for one at that point.

Cook left Nootka at the end of April to continue his explorations north, and his mapping of much of the coast of Alaska right up to the wall of the Arctic ice in Bering Strait added more detail to the world map than his charting of the coast of British Columbia. But the trend of the coast, and the important position of the coast, was now defined—and that knowledge would soon be released to the world.

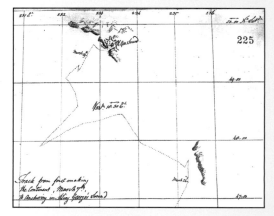

MAP 61 (*above*).
The track of Cook's ships off the West Coast from Oregon to Nootka Sound, here *K. Geo. Sound* (King George's Sound), is shown on a page from the journal of James Burney, one of Cook's officers. It shows the ships to be far offshore as they passed the Strait of Juan de Fuca.

MAP 62 (*far left*).
One of the first maps to show Cook's third voyage was this one, published in 1781. Details of the voyage were leaked by many members of Cook's crew years before the official book was published in 1784. The map shows clearly Cook's track from Hawaii, which he discovered and mapped earlier in 1778.

MAP 63 (*left*).
Another page from Burney's journal maps Nootka Sound. *Ship Cove*, today Resolution Cove, is the place where Cook anchored his ships, *Resolution* and *Discovery*, in April 1778. One of the midshipmen on the latter ship was none other than George Vancouver (see page 36). *Below* is Resolution Cove today, likely looking much as it did to Cook in 1778. *Top, left.* James Cook in a painting by Nathaniel Dance.

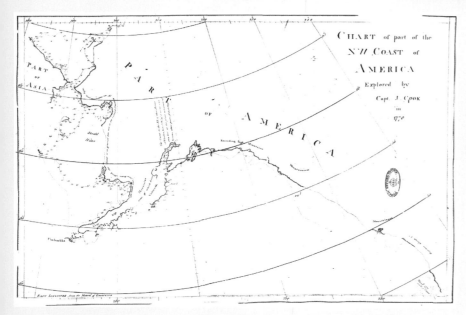

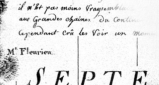

MAP 64 *(above).*
Cook's original manuscript summary map of the West Coast was contained in a letter to Philip Stephens, Secretary of the Admiralty, dated 20 September 1778. It was given to Russian fur trader Gerasim Izmailov at Unalaska for delivery overland through Russia. Explorers often tried to use multiple routes to transmit information about their voyages as insurance in case they did not return. Cook's letter and map arrived safely in London on 6 March 1780, thirteen months after his death in Hawaii. Nootka Sound is the only detail of the British Columbia coast shown, as Cook, sailing north, again was forced by the weather to stay offshore.

MAP 65 *(right, top).*
One of James Cook's contributions to science was that he finally placed the Northwest Coast, correctly positioned, on the map of the world. He had commissioned one of his officers, Henry Roberts, to add all new information from all three of his voyages to a world map. In 1784 Roberts published it, and it was at the time the most complete and up-to-date map of the world available. Here it shows *Nootka or K. Georges S^d.* and an accurate depiction of the shape of Alaska. Interior information from the explorations of Samuel Hearne in 1771–72 is also shown (see *Coppermine River* at top), but because Hearne did not correctly measure his longitude, it is placed too far west. The British government knew of Hearne's explorations and knew that they essentially precluded the possibility of a Northwest Passage existing below that latitude—provided he had measured the latitude correctly—yet would still send out George Vancouver nearly twenty years later just to make sure.

MAP 66 *(below).*
One of the maps that La Pérouse dropped off in Australia was this hand-drawn survey of the West Coast, drawn in 1786 by La Pérouse's surveyor, Gérault-Sébastien Bernizet, and his astronomer, Joseph Lapaute Dagelet. La Pérouse kept generally offshore from British Columbia, and the only location shown in detail is *Nootka* Sound.

A FRENCH RESPONSE

Rivalry between France and Britain led to a round-the-world scientific voyage by France to counterbalance the perceived loss of prestige in the wake of James Cook. A highly regarded naval officer, Jean-François de Galaup, comte de la Pérouse, led the expedition, which carried with it several of the still rare chronometers.

La Pérouse arrived on the West Coast in the summer of 1786 and was supposed to survey the entire coast in three months, a clearly impossible task. He sailed from Maui to Alaska and then southward down the coast. His officers made a good running survey, positionally correct, thanks to chronometers, but lacking in detail in some locations. La Pérouse was wrecked in the South Pacific two years later, but not before handing off some of his maps to British authorities at Botany Bay, Australia. The French Revolution intervened, La Pérouse, an aristocrat, fell out of favour, and his work was not published until 1797, just in time to be eclipsed by the meticulously surveyed and detailed maps of George Vancouver published the following year (see page 36).

In Search of the Sea Otter

Cook's crew had traded for sea otter pelts from the coastal peoples, intending simply to keep themselves warm during the upcoming Arctic part of their voyage. When they reached Canton (now Guangzhou), China, on the way home, however, they found the fine pelts were eagerly sought after and worth a small fortune. It did not take long for commercial ventures to be put together, and a trade developed between the Northwest Coast and China. In the process, many traders made maps that began to fill out details of the complex coast of the region. And the poor sea otter was hunted almost to extinction over the next few decades.

Accounts of Cook's third voyage leaked out as early as 1781 (see Map 62, *page 26*). The first fur trader seeking sea otter pelts, James Hanna, arrived on the British Columbia coast in 1785. Before long American traders got into the act, with Robert Gray, discoverer of the Columbia River, arriving on his first voyage in 1788. The increased activity on the Northwest Coast of course concerned the Spanish, who initiated a further set of voyages ostensibly to protect their territory—or what they saw as their territory—from claim by other powers.

Fur traders drew maps principally for their own use, since they wanted to be able to find a fur-rich location again, though a few clearly had their eye on fortune or fame from the production of a book. Some of the maps of the British Columbia coast produced by the fur traders are reproduced here. It is apparent that despite their intermittent coverage, they added enormously to the sparse knowledge of the coast and set the stage for the detailed mapping that Britain's George Vancouver would carry out beginning in 1792 (see page 36). For the first time a world long oblivious to British Columbia was now *interested* in it.

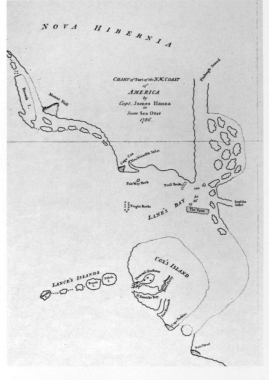

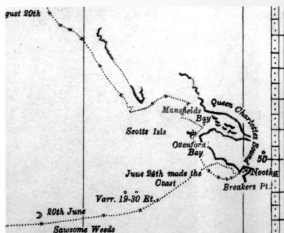

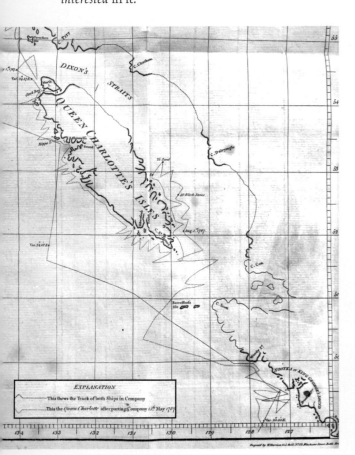

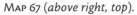

Map 67 (above right, top).
Sponsored by private commercial interests with connections to the British East India Company, James Hanna was the first to visit British Columbia in search of sea otter pelts. He arrived at Nootka Sound in August 1785—in the appropriately named *Sea Otter*—and collected 560 pelts. On a second voyage the following year he was much less successful, collecting only 50 pelts, because another trader, James Strange, had visited just before his arrival. This map is of the northern tip of Vancouver Island, here *Cox's Island*, with Quatsino Sound to the south. *Lance's Islands* are now the Scott Islands, and *St. Patrick's Bay* is San Josef Bay. *Lane's Bay* is Smith Sound on the mainland coast.

Map 68 (above, right).
A map produced by James Strange, who arrived in 1786 on a voyage privately sponsored by David Scott but supported by the East India Company. It shows much the same area as that covered in Hanna's map, the north part of Vancouver Island and the adjacent mainland. The *Scott Isls* and Cape Scott, the northern tip of Vancouver Island, were named by Strange for his sponsor. *Queen Charlottes Sound*, named for the wife of King George III, is today Queen Charlotte Strait.

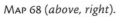

Map 69 (above).
George Dixon and Nathaniel Portlock, both Cook's ex-crew, came to the Northwest Coast in 1786 and 1787, overwintering in Hawaii. In June 1787 Dixon found many furs on some islands he named after his ship: *Queen Charlotte's Isles*. Dixon sailed all around the islands, proving their insularity. Only recently has the name Haida Gwaii been adopted as the official name for the Queen Charlotte Islands. In 1789 Dixon published a book, *A Voyage Round the World; But More Particularly to the North-West Coast of America*, which included this map. *Dixon's Straits* (now Dixon Entrance) was named by famous botanist Joseph Banks after being shown Dixon's map and invited to name some features.

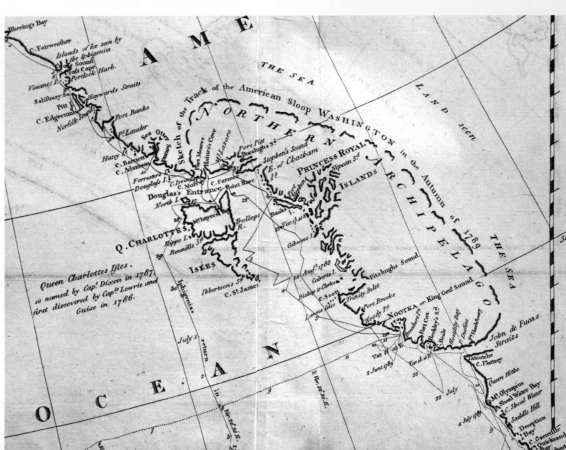

Charles William Barkley was the next to reach the coast, arriving in 1787 with his wife, Frances, in the *Imperial Eagle,* flying Austrian East India Company colours to circumvent monopolies by the British East India Company and South Sea Company. Barkley Sound was named by Barkley himself. After a stay at Nootka Barkley sailed into a large opening he thought was the same as Juan de Fuca's strait and promptly named it after him (see MAP 35, page 18). John Meares, who was one of the owners of the ship, acquired Barkley's journal and maps after the ship was seized following legal action by the East India Company, and Meares shamelessly used Barkley's information without credit when he published a book about his own travels in 1790.

MAP 70 (*left, bottom*), MAP 71 (*above, left*), and MAP 72 (*above, right*).
John Meares had arrived in Alaska in 1786 but was trapped in Prince William Sound over the winter, being rescued by Dixon and Portlock the following May. In 1788 Meares returned to the coast from Macao flying Portuguese colours and established a camp at Friendly Cove in Nootka Sound where a small vessel, the *North West America,* was built—the first European ship to be built in the Northwest. In 1790 Meares published a book, *Voyages Made in the Years 1788 and 1789, from China to the North West Coast of America*, in which this map appeared. Meares had fabricated details of a voyage by Robert Gray, which purported to show that Vancouver Island and a massive chunk of the mainland were in fact an island. Meares gained a reputation as a liar, which outshone his actual achievements in the exploration of the coast, including the production of this map (MAP 71) of part of Clayoquot Sound: *Low Wood Lands* is the Tofino peninsula, and the *Village* is Opitsaht, opposite today's Tofino. Russian traders produced their own, even more exaggerated, version of Meares's ridiculous map a couple of years after his book was published (MAP 72).

MAP 73 (*below, left*).
Charles Duncan arrived on the coast in 1787 with James Colnett (see next page). He found and named the *Princess Royal Islands*, shown on a map of the world published by up-and-coming British mapmaker Aaron Arrowsmith in 1790. The map now shows an *Entrance of Juan de Fuca* leading to a hint of an inland sea. The northern part of Vancouver Island is recognizable.

MAP 74 (*below, right*).
British geographical bon vivant, Northwest Passage believer, and hydrographer to the East India Company Alexander Dalrymple published this map in 1790 as a summary of all that was known of the emerging shape of the west coast of North America. The entrance to the Strait of Juan de Fuca, the northern part of Vancouver Island, and Dixon's Queen Charlotte Islands—here named *Nova Hibernia* (New Ireland)—and other features from fur-trader maps are shown on his outline of the coast.

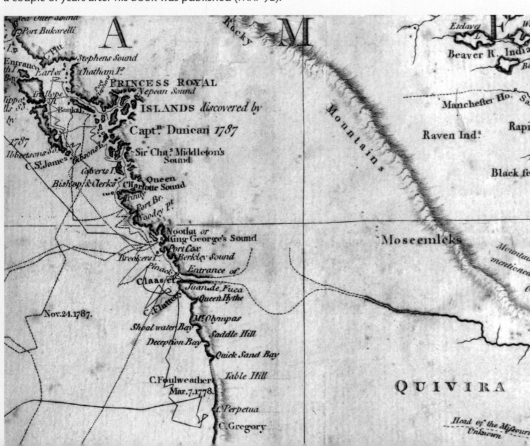

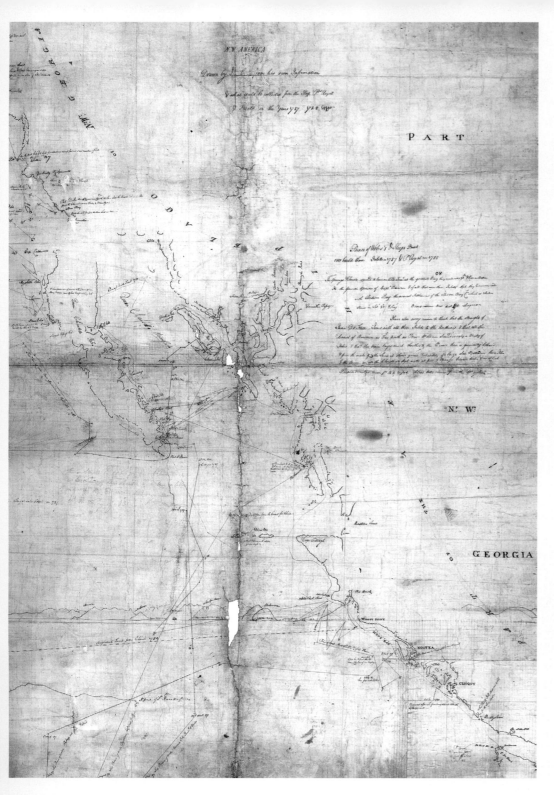

Map 75 (left).

Perhaps the most historically significant fur trader to arrive on the British Columbia coast between the voyages of James Cook and George Vancouver was James Colnett, who not only mapped the coast but also nearly precipitated a war. Colnett first arrived in 1787, in company with Charles Duncan. His officers included James Johnstone and botanist Archibald Menzies, both of whom would later sail with Vancouver. Returning to Canton in late 1788, he met John Meares, who joined with him and others to establish a permanent settlement as a base for fur-trading operations, but when he arrived at Nootka in July 1789, he was arrested by Esteban José Martínez, in charge of a new Spanish attempt to establish sovereignty over the Northwest Coast (see page 32). The Spanish also confiscated the *North West America*, the ship Meares built (see previous page). This was the Nootka Sound Incident, fanned by Meares into a crisis that saw Britain threatening war on Spain unless it was resolved. One of the missions of George Vancouver was to meet the Spanish at Nootka as part of the resolution process (see page 36). This map was drawn by Colnett during his 1787–88 voyage and shows his route in detail. Vancouver Island and Nootka are at bottom right, and Haida Gwaii and the adjacent coast at the top. Colnett correctly surmised that the coast consisted of numerous islands.

Map 76 (below) and Map 77 (below, inset).

"Boston men"—Americans—soon arrived on the coast to share in the fur bonanza they had heard about. Robert Gray and John Kendrick arrived at Nootka in September 1788 and overwintered there. They wanted to establish a trading base, and Kendrick purchased land from Nootka (now Mowachaht) chief Maquinna and built his *Fort Washington* (named on Map 77) in Marvinas Bay, about halfway between Friendly Cove and today's Tahsis. Kendrick returned to the coast in 1791 and was there for the summers of 1793 and 1794. He tried without success to enlist the support of the American government in a colony on Vancouver Island, and this is alluded to in the areas shown as "deeded" on these maps, drawn by Oregon settlement booster Hall Jackson Kelley in 1839 in support of American claims to the coast. These are two versions of similar maps. Map 76 shows *Quadra and Kendrick's I.* rather than "Quadra and Vancouver's Island," the name bestowed by George Vancouver.

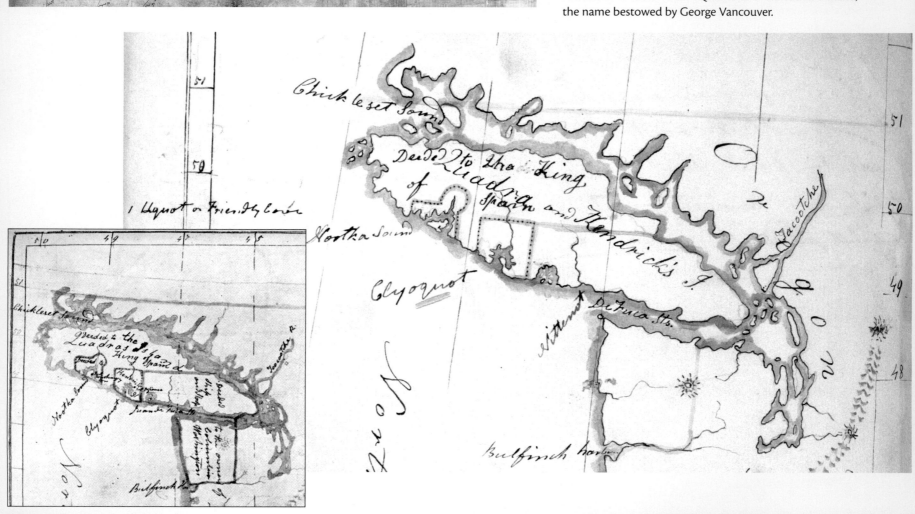

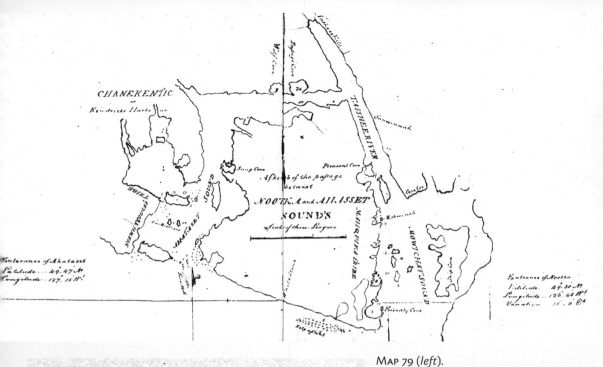

MAP 79 (left).

Robert Haswell was first an officer under Kendrick and later under Gray before being appointed captain of a small sloop, the *Adventure*, built at Clayoquot Sound in early 1792. He visited Haida Gwaii in 1791 and 1792. He drew maps and sketches of the places he visited, including this one in 1792 of the head of Masset Sound. The *Village* is the important Haida settlement of Masset.

MAP 80 (right) and MAP 82 (right, bottom).

These two unusual maps were drawn by American trader Joseph Ingraham, who first came to the coast with Kendrick in 1788 and circumnavigated the world with Gray in 1789–90. The maps feature *Washingtons Isle*s—Haida Gwaii—and *Quadra's Isles*—Vancouver Island, the latter with north more or less to the left. The maps were drawn in 1792, when he was back on the coast in his own ship, *Hope*.

MAP 81 (below).

William Goldson was one of numerous so-called armchair geographers who proliferated before the world was fully known. In 1793 (before any information from Vancouver's voyage had reached Britain) Goldson tried to reconcile the supposed geography of the Northwest as delineated by previous hoaxes and false accounts such as those that led to the supposition of a vast inland sea and a Northwest Passage in temperate latitudes (see page 22). They included one by Lorenzo Ferrer Maldonado, referred to in the title of Goldson's map, who claimed in a 1609 memorial to the king of Spain to have discovered the Northwest Passage in 1588. Goldson dexterously weaves the emerging knowledge of the Northwest Coast as derived from the published accounts of the fur traders—and especially the fictions of John Meares—with the supposed previous knowledge. Soon to be run out of town by George Vancouver's maps, Goldson's rendering lives on to amuse us.

MAP 78 (left).

Kendrick's compatriot Robert Gray is well known for his 1792 discovery of the Columbia River, which he named after his ship the *Columbia Rediviva*. Gray traded on the British Columbia coast in 1789 after overwintering at Nootka with Kendrick, and it was then that he drew this map of Nootka Island and its surrounding waterways. In 1789 he sailed to Haida Gwaii, naming it Washington's Island (as on MAP 82, *below*). Leaving later that year he returned to Boston westward becoming the first American circumnavigator of the world in the process. In 1791 he was back on the coast and, with Kendrick, built Fort Defiance in Clayoquot Sound as a base. The remains of this structure have recently been located by archaeologists. Early the following year Gray discovered what he thought was a plot by the local people to seize his ship and in retaliation burned down the village of Opitsaht, opposite today's Tofino. On 28 April, as he was heading south, he met George Vancouver he found the Columbia River, which Vancouver had missed, thirteen days later.

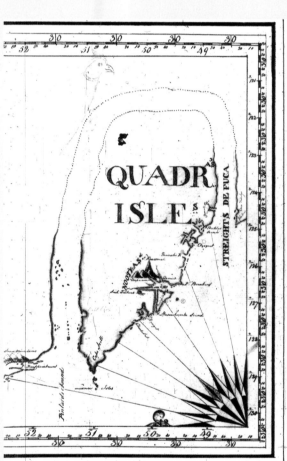

MAP 83 (*right*).
A map of Manuel Quimper's exploration of the Strait of Juan de Fuca, drawn in 1790.

Left. The Spanish fort built in Friendly Cove.

THE SPANISH RETURN

The Spanish never got over their fear that Russians were encroaching on their territory in the North Pacific, and every little rumour fuelled their paranoia. In 1788 Esteban José Martínez, who had been with Pérez on the first voyage to British Columbia in 1774 (see page 24), and Gonzalo López de Haro were dispatched to Alaska, where indeed, they did meet Russian traders, but Martínez was more interested in drinking with them than driving them away. And he did learn of an impending Russian expedition that intended to occupy Nootka. However, unknown to Martínez, the expedition had been cancelled when war broke out between Russia and Turkey in 1787.

Nevertheless, the Spanish authorities were sufficiently alarmed to order Martínez to occupy Nootka the following year; this is when he arrested British trader James Colnett (see page 30) and set off the so-called Nootka Sound Incident, nearly causing a war between Spain and Britain.

The next year senior officer Francisco de Eliza led a contingent of Spanish military men, this time including soldiers under Pedro de Alberni. This contingent stayed at Nootka for two winters, building a fort in Friendly Cove, which acted as a base for a number of forays into the nearby channels to determine if

what would have been a strategically important western entrance to the Northwest Passage lurked in these waters.

In the three years between 1790 and 1792 the Spanish carried out exploratory voyages in British Columbia waters that defined Vancouver Island and a great deal of the coast, and they even shared information with Britain's George Vancouver.

In 1790 Eliza dispatched Salvador Fidalgo north to Alaska and Manuel Quimper south to the Strait of Juan de Fuca. López de Haro accompanied Quimper as pilot. They sailed the length of the strait but failed to penetrate either south into Puget Sound or north through the San Juan Islands (MAP 83, *above*). Notably, however, Quimper did locate and map Puerto de Córdova—Esquimalt Harbour, later to be used as a British naval base.

The following year Eliza himself led a voyage into the Strait of Juan de Fuca with two ships, the *San Carlos* and the small schooner *Santa Saturnina,* the latter commanded by José María Narváez. Basing himself first at Esquimalt and then at Port Discovery on the south side of the strait, Eliza sent one of his pilots, José

Continued on page 35.

MAP 84 (*below*).
Undoubtedly the most important map of the Northwest Coast produced by the Spanish was the *Carta que comprehende* of 1791. It was compiled by Eliza and Narváez, doubtless with the aid of their pilots, and sent to New Spain, where several fair copies

were made. The map summarized the explorations of 1791 and includes the first mapping of the Strait of Georgia and the first mapping of any part of what is today Metro Vancouver.

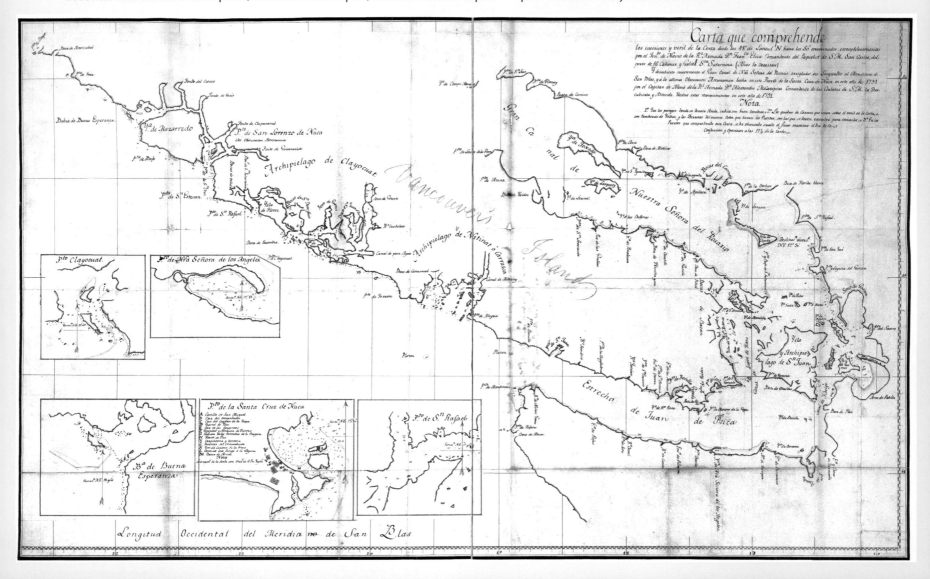

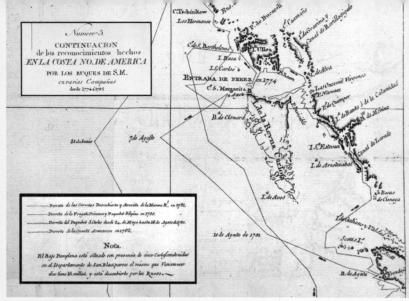

Map 85 (left).

In 1792 Juan Francisco de la Bodega y Quadra dispatched one of his officers, Jacinto Caamaño, northward in his ship *Aránzazu* to attempt to find the entrance to the Northwest Passage noted in the supposed 1640 account of Bartholomew de Fonte (see page 22). His voyage proved to the satisfaction of the Spanish that de Fonte's account was a fabrication. While sailing north, Caamaño's crew jettisoned an olive jar near Langara Island, which was recovered in 1985 in a fishing net and is now displayed at the Masset Museum (photo, *left*). The map shows the *Entrada de Perez en 1774* (Dixon Entrance) and the track of the *Aránzazu* (note the key), a somewhat erratic path, as Caamaño battled adverse winds. The map is from the *Relación* of 1802 (see page 35).

Map 86 (left).

This chart of Clayoquot Sound was made by Narváez with the help of Eliza's pilot, Juan Pantoja y Arriaga, in May 1791 on their way eastward into the Strait of Juan de Fuca. The detail is also shown on the *Carta que comprehende* (Map 84, *far left, bottom*). Eliza met local chief Wickaninnish at Clayoquot. Today's Tofino is at bottom, centre, at the end of the long peninsula. Sandbanks and shoals have been coloured orange.

Map 87 (left, bottom).

A Spanish map of Friendly Cove—*Cala de los Amigos*—originally drawn in 1791 and reproduced in the 1802 *Relación*. The photo shows Friendly Cove—or Yuquot—as it is today. It can be compared with the 1791 engraving, *far left, top*. A lighthouse has replaced the Spanish fort on the islets that conveniently block the Pacific swells from entering the anchorage.

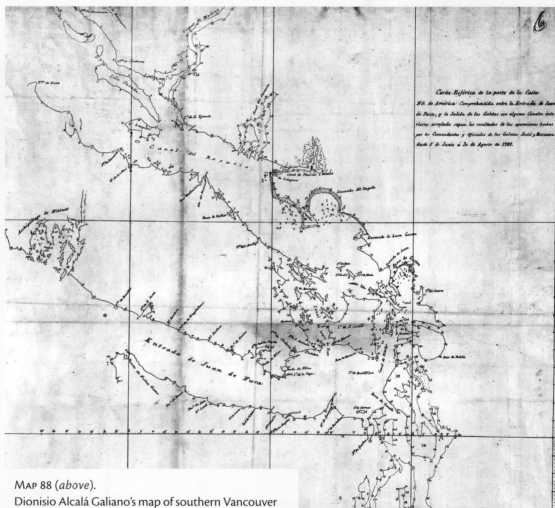

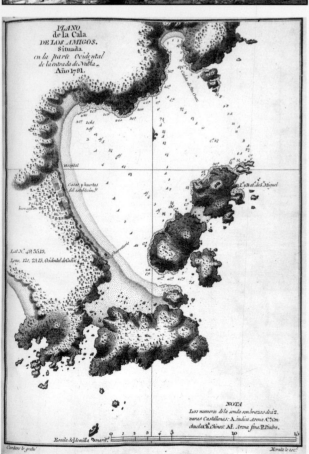

Map 88 (above).

Dionisio Alcalá Galiano's map of southern Vancouver Island and the lower mainland of British Columbia, with information from George Vancouver added, notably the configuration of Puget Sound. Point Roberts now appears as a peninsula rather than an island. Burrard Inlet and Indian Arm, together with Howe Sound, have now been explored to their heads. Detail of what is now Downtown Vancouver, the West End, and Stanley Park is also evident, shown as an island.

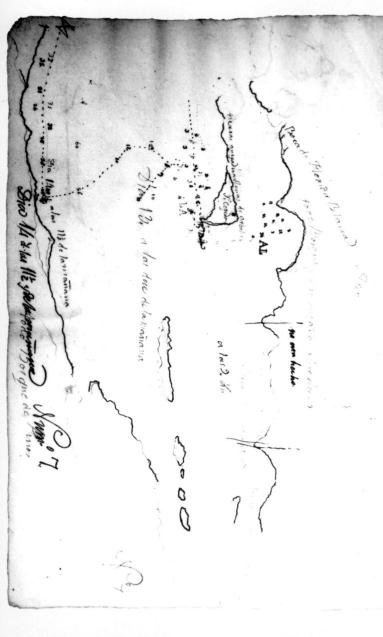

MAP 89 (*left*) and MAP 90 (*below*).

These are two of Galiano's *borradores*—on-the-spot rough drafts—drawn during his sojourn in the southern Strait of Georgia in June 1792. It may be that the base was partly copied from the Narváez map (MAP 84, *page 32*), to which Galiano added his track. MAP 89 shows Boundary Bay with *Y*ˢ *Cepeda*, Point Roberts, shown as an island, and the coast of what is now Galiano Island and the other Gulf Islands to the west. MAP 90 depicts his probing of *Pᵗᵒ. ó Bocas de Porlier* (Porlier Pass), between Galiano Island to the southeast (right) and Valdes Island to the northwest (left), found by Narváez the previous year. Both islands were named in 1859 by British hydrographic surveyor George Henry Richards. What may be a later hand has labelled *Boca del Carmelo* (Howe Sound) across the Strait of Georgia to the north (top). Galiano's track on these maps is northwestward, being noted from off Point Roberts across to Galiano Island and then northward along the eastern shore of the Gulf Islands.

MAP 91 (*below, right*), MAP 92 (*below, centre*), and MAP 93 (*below, left*).

These three maps, the British Columbia section of larger maps of the West Coast, were all drawn by Bodega y Quadra and represent, respectively, the geographical knowledge of the Spanish at the end of 1790, 1791, and 1792. The map sequence ably demonstrates the emergence of the British Columbia coast—and especially Vancouver Island as an island—through Spanish eyes.

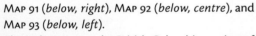

MAP 91 shows the Strait of Juan de Fuca as a closed-end basin; MAP 92 reflects the explorations of Narváez during 1791; and MAP 93 includes the circumnavigation of Vancouver Island by Galiano and Valdes in 1792, together with information from Caamaño's explorations to the north. On the latter map Vancouver Island is now named *Yᵃ. de Quadra y Vancouver*, the name bestowed by Bodega y Quadra and Vancouver following their negotiations over the Nootka Treaty in 1792.

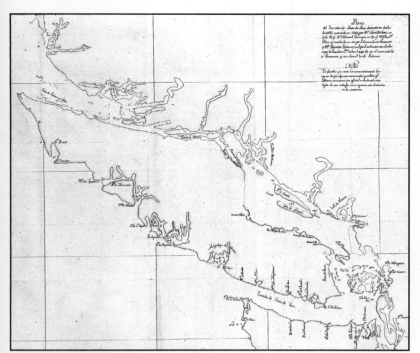

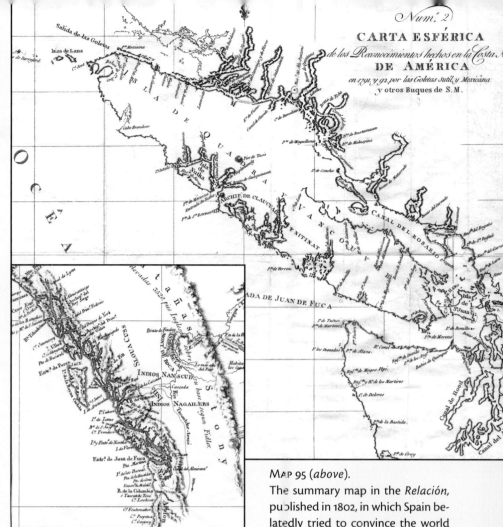

MAP 94 (above).
The co-operation between Galiano and George Vancouver is depicted on this map by Bodega y Quadra. Coasts mapped by the Spanish are in brown, while those mapped by Vancouver are in red. Vancouver drew a comparable map (MAP 100, overleaf).

Verdía, in a longboat to explore the channels leading into the Strait of Georgia. Menaced by the local inhabitants, Verdía soon returned, whereupon Eliza dispatched the longboat again, with another pilot, Juan Pantoja y Arriaga, this time covered by the guns of the schooner. After they returned with news of a larger sea to the north, Eliza sent the *Santa Saturnina* out once more, this time accompanied by Juan Carrasco in the longboat. Narváez took twenty-two days to explore the Strait of Georgia north beyond Texada Island, including the area that is now Metro Vancouver, which was mapped for the first time. The results of the explorations of 1791 were recorded on the landmark *Carta que comprehende* (MAP 84, page 32), the first map of southwestern British Columbia.

The *Santa Saturnina* never returned to Nootka, instead being forced by adverse winds south to New Spain. Here Carrasco, who had been placed in charge when Eliza took Narváez on board his ship to work on the charts, briefed Alejandro Malaspina, Spain's round-the-world scientific voyager and its answer to James Cook; Malaspina and the viceroy decided as a result to dispatch another expedition to probe the new-found sea farther in one last hope that it might harbour the western entrance of the Northwest Passage. Two experienced officers, Dionisio Alcalá Galiano and Cayetano Valdés y Flores, were detached from Malaspina's command for this purpose.

In 1792 Galiano and Valdés sailed north in the *Sutil* and *Mexicana*, arriving at Nootka Sound in May. From there they began a voyage that made them the first to circumnavigate Vancouver Island, meeting George Vancouver off Punta de Langara—Point Grey and nearby Spanish Banks—on 21 June and exchanging information with him. Galiano's longboats checked Burrard Inlet before they continued north, probing, like Vancouver, most of the long inlets to ensure they were dead ends. Rounding the northern tip of Vancouver Island, they headed back to Nootka, which they reached on 31 August. Here their new commander—Juan Francisco de la Bodega y Quadra, who had sailed with Pérez in 1774 and had now replaced Eliza—waited. They were able to give him valuable information before he met with George Vancouver to hand over lands specified in the Nootka Convention of 1790 (see next page).

Galiano's account of his voyage was suppressed until 1802 because Malaspina fell to political intrigue in Spain. That year a full account was published in an important book called the *Relación del Viaje hecho por las Goletas Sutil y Mexicana en el año 1792*, together with accounts of previous other voyages until then also suppressed by a Spanish policy of secrecy only relaxed because the publication of George Vancouver's account in 1798 was deemed threatening to Spanish sovereignty in the Northwest. By that time, however, Spain's influence was on the wane; what is probably the last Spanish map of British Columbia was drawn in 1817 (MAP 912, page 355). The Transcontinental Treaty of 1819 transferred all Spanish claims to the Northwest Coast to the United States.

MAP 95 (above).
The summary map in the *Relación*, published in 1802, in which Spain belatedly tried to convince the world of her claim to the Pacific Northwest through prior exploration. Accounts of earlier voyages were also included.

MAP 96 (left, centre).
Also published in 1802, this Spanish map retains the Spanish names but includes British information derived from both George Vancouver and the 1793 overland explorations of Alexander Mackenzie (see page 38).

MAP 97 (below).
Spanish maps after 1792 may have sometimes been prettier, but they contained little further information unless it was gleaned from other sources. This delightful illustrated map was drawn in 1808, possibly as a decorative copy, as many known details, especially islands, are omitted.

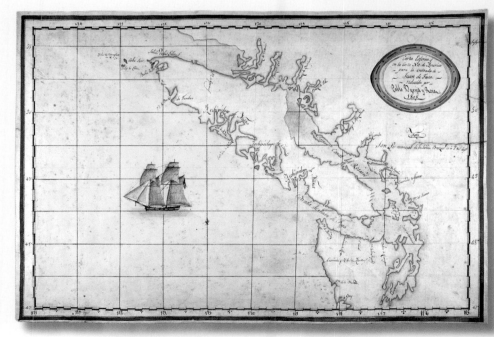

British Explorers

George Vancouver

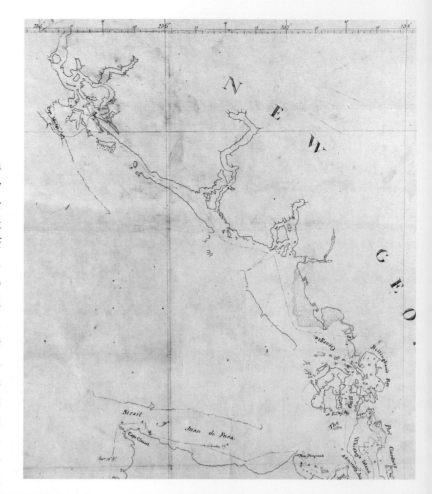

The Spanish arrest of James Colnett and Spain's assertion of its territorial rights in Nootka Sound led John Meares, one of the syndicate members who had supposedly lost land at Nootka, to lobby the British government for redress. Prime Minister William Pitt was pressured to assert the doctrine of freedom of the seas, and he took the opportunity to force Spain to abandon its claim to the North Pacific by right of first discovery; from now on prior occupation was to be the criterion for sovereignty. Spain, without its ally France because of the revolution in that country, was forced to sign the (first) Nootka Convention in 1790; this gave Meares his compensation, and Spain also agreed to return land and buildings supposedly taken from him at Nootka.

To determine what land should be returned, Britain dispatched George Vancouver to meet with the Spanish representative—Bodega y Quadra—at Nootka, where, not surprisingly, they could not resolve the issue, since Meares had falsified his claims. Since Britain now had as much right to the Northwest as Spain, Vancouver was instructed to survey the coast and in particular find the elusive western entrance to the Northwest Passage—or at least prove it was not there, which, of course, is what he did.

So began one of the most comprehensive maritime surveys ever undertaken. Vancouver arrived on the coast in April 1792 in his ship, *Discovery*, accompanied by *Chatham*, with William Broughton as captain. Without visiting Nootka, they sailed directly into the Strait of Juan de Fuca. Vancouver soon realized that the survey of the intricate coastline would have to be carried out in small boats, and so the ships would anchor for days or weeks at a time while longboats probed the fiords and inlets of the Washington, British Columbia, and Alaska coasts. Vancouver soon found the previously unmapped Admiralty Inlet and Puget Sound—and named the latter after his officer Peter Puget. In June Vancouver himself surveyed Burrard Inlet and the shores of the city that now bears his name. Here he met the Spaniards Galiano and Valdés and agreed to share surveying tasks and maps.

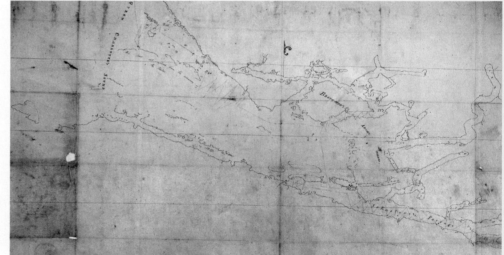

Vancouver surveyed as far north as Burke Channel (near Bella Coola) that first year before visiting Nootka to meet with Bodega y Quadra. After overwintering in Hawaii, he continued his mapping north into Alaska in 1793 and completed it in Alaska the following year. His monumental survey was published, posthumously, in 1798, three years after he had returned to Britain; his remarkable survey remained the standard for maps of the coast for fifty years.

MAP 98 (*right, top*).
George Vancouver's preliminary chart of southwest British Columbia, including the area now occupied by the City of Vancouver. It was drawn by his officer Joseph Baker on board *Discovery* in June 1792. Vancouver left out Vancouver Island because he concentrated on what he termed the continental shore, where a Northwest Passage could have been lurking.

MAP 99 (*right, centre*).
Vancouver's initial chart of *Johnstone's Passage*, surveyed by James Johnstone, master of *Chatham*, in July 1792. This was the key waterway that separated the mainland from Vancouver Island and allowed its circumnavigation. Vancouver later gave it the name Johnstone's Straits (now Johnstone Strait).

MAP 100 (*right, bottom*).
A preliminary summary chart carried by William Broughton to Britain from Monterey at the end of the 1792 surveying season showing the results of that year's work. The map shows the coast surveyed by Vancouver in black and that surveyed by the Spanish in red. Compare this map with its Spanish equivalent, MAP 94, *previous page*.

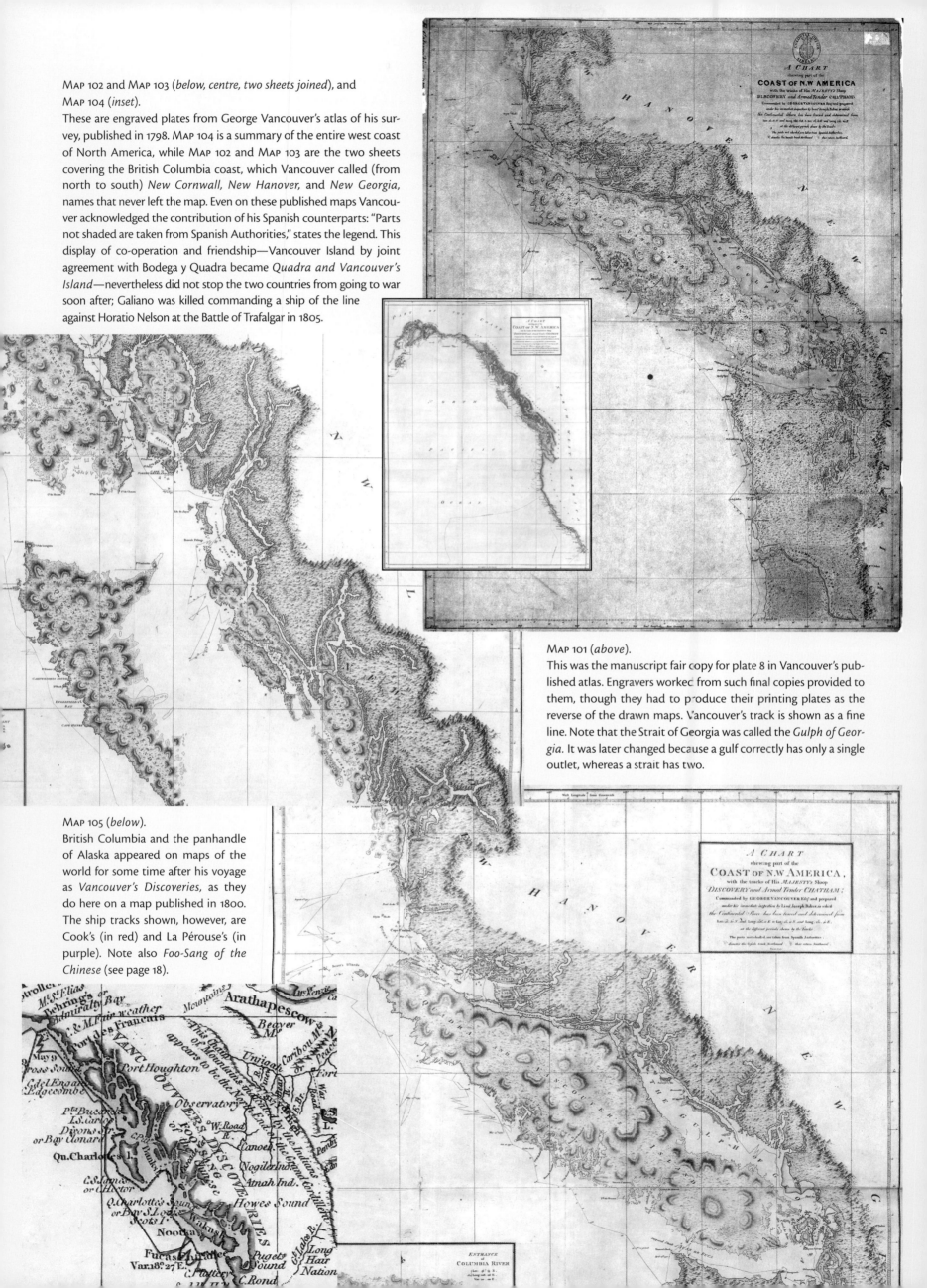

Map 102 and Map 103 (*below, centre, two sheets joined*), and Map 104 (*inset*).

These are engraved plates from George Vancouver's atlas of his survey, published in 1798. Map 104 is a summary of the entire west coast of North America, while Map 102 and Map 103 are the two sheets covering the British Columbia coast, which Vancouver called (from north to south) *New Cornwall, New Hanover,* and *New Georgia,* names that never left the map. Even on these published maps Vancouver acknowledged the contribution of his Spanish counterparts: "Parts not shaded are taken from Spanish Authorities," states the legend. This display of co-operation and friendship—Vancouver Island by joint agreement with Bodega y Quadra became *Quadra and Vancouver's Island*—nevertheless did not stop the two countries from going to war soon after; Galiano was killed commanding a ship of the line against Horatio Nelson at the Battle of Trafalgar in 1805.

Map 101 (*above*).

This was the manuscript fair copy for plate 8 in Vancouver's published atlas. Engravers worked from such final copies provided to them, though they had to produce their printing plates as the reverse of the drawn maps. Vancouver's track is shown as a fine line. Note that the Strait of Georgia was called the *Gulph of Georgia.* It was later changed because a gulf correctly has only a single outlet, whereas a strait has two.

Map 105 (*below*).

British Columbia and the panhandle of Alaska appeared on maps of the world for some time after his voyage as *Vancouver's Discoveries,* as they do here on a map published in 1800. The ship tracks shown, however, are Cook's (in red) and La Pérouse's (in purple). Note also *Foo-Sang of the Chinese* (see page 18).

OVERLAND FROM CANADA

MAP 106 (*below*) and MAP 107 (*below, bottom*).
These two maps from North West Company partner Peter Pond show how he mapped the river and lake system of the Northwest and then became convinced that the Pacific was nearby. MAP 107 is the map Pond drew for Mackenzie to present to Catherine, empress of Russia. The rivers and lakes are relatively accurately drawn but inaccurately positioned longitudinally. Pond has placed the coastline, the position of which was taken from James Cook's book (see page 26), in the right position, but the juxtaposition of it with the rivers and lakes is wrong. The Pacific looks tantalizingly close. Pond and Mackenzie overwintered in 1788–89 on the river just south of *Lake of the Hills* (Lake Athabasca). MAP 106 was a later development of Pond's theory, published in a British magazine in 1790. Here he has taken Cook's report of *Cook's River* in Alaska—actually Cook Inlet—and connected it with a fictitious river flowing out of (Great) *Slave Lake*—enough to motivate Alexander Mackenzie to try to find it.

At the same time that George Vancouver was surveying the British Columbia coast, another explorer was planning an expedition to reach the coast overland. His reasons were different: he wanted to find a route to ship interior furs out to the new markets in China.

Alexander Mackenzie, a partner in the North West Company, had overwintered in 1788–89 with Peter Pond, another company partner who had for years meticulously mapped the interior rivers and lakes used by the fur traders as their highways. With the position of the coast mapped by James Cook, Pond became convinced he could find a short route to the Pacific. What he didn't realize was that he had his longitudes all wrong; the distance to Cook's coast was in fact much greater than he thought. In 1789 Mackenzie decided to try to reach the Pacific, even taking with him a map prepared by Pond to present to the empress of Russia (MAP 107, *below*), but he took a wrong turn at the western end of Lake Athabasca and ended up

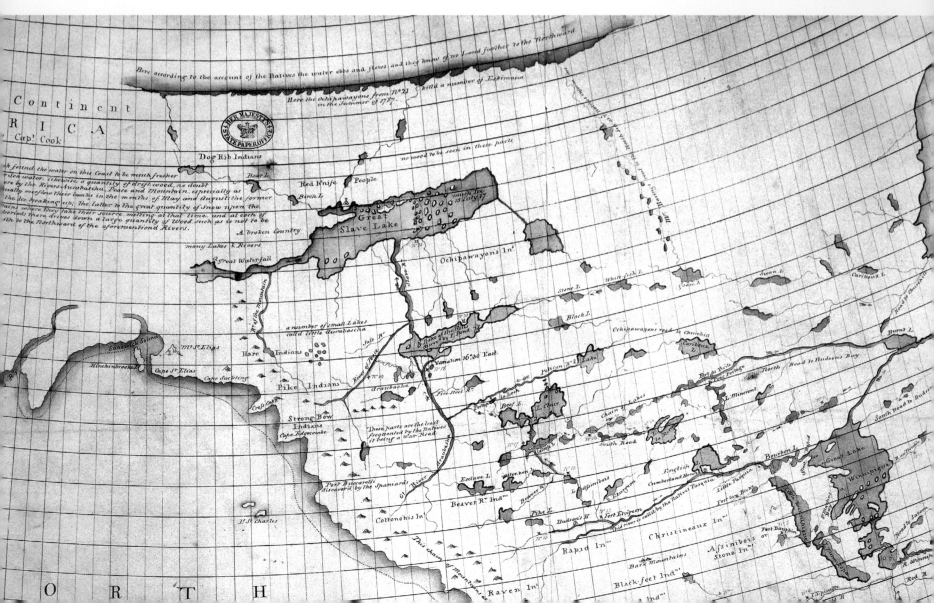

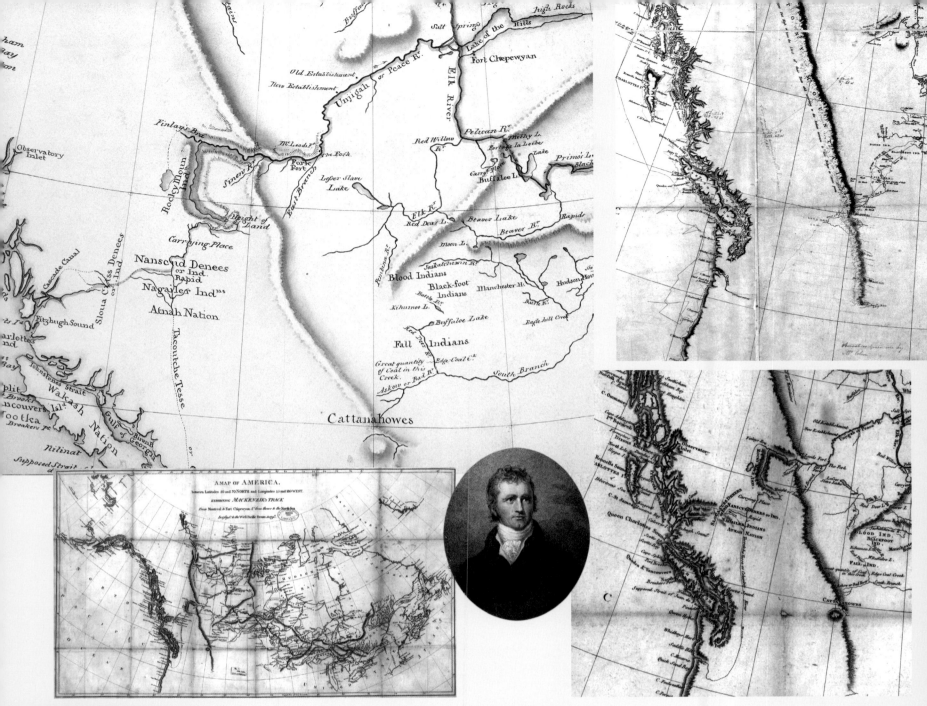

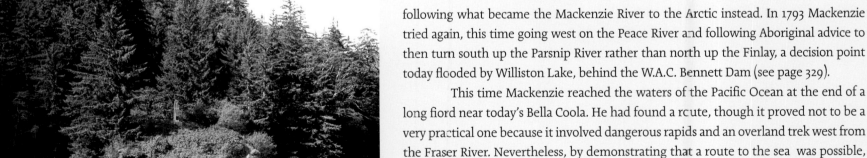

Map 108 (*above, top left*), and Map 109 (*above*) and detail (*above, right*).

The map that was reproduced in Alexander Mackenzie's 1801 book *Voyages from Montreal* (Map 109) provided the first published geographical detail from exploration of any part of interior British Columbia. Map 108 is the manuscript map prepared for the engraver. Mackenzie followed the Parsnip River south when the *Unjigah or Peace R^r.* branched, reaching a *Height of Land* (the Continental Divide) and then fearsome rapids (the James River; Mackenzie called it the Bad River) leading to the Fraser. On local advice he left the Fraser at the West River and trekked overland, arriving at the Nuxalk village of Qomq'-ts (see Map 212, *page 78*), where he borrowed a canoe and made it to a rocky ledge in Dean Channel now called Mackenzie's Rock (photo, *below, left*). Here with vermilion he wrote on a rock: "Alex Mackenzie from Canada by land 22d July 1793." The engraving of Alexander Mackenzie (*above*), after a painting by Thomas Lawrence, is from Mackenzie's book. Note the tentative connection between the Fraser and the Columbia, not disproved until Simon Fraser's expedition in 1808 (see next page).

Map 110 (*above, top right*).

The earliest mapping of Mackenzie's expedition to the Pacific is this 1796 map by Aaron Arrowsmith, on which he has noted in red ink the point on the coast Mackenzie reached as *52°22´N, 129°15´W*. This information came from Joseph Colen, then factor at the Hudson's Bay Company's York Factory, on Hudson Bay. How a rival fur trader would have acquired this information has not been determined.

following what became the Mackenzie River to the Arctic instead. In 1793 Mackenzie tried again, this time going west on the Peace River and following Aboriginal advice to then turn south up the Parsnip River rather than north up the Finlay, a decision point today flooded by Williston Lake, behind the W.A.C. Bennett Dam (see page 329).

This time Mackenzie reached the waters of the Pacific Ocean at the end of a long fiord near today's Bella Coola. He had found a route, though it proved not to be a very practical one because it involved dangerous rapids and an overland trek west from the Fraser River. Nevertheless, by demonstrating that a route to the sea was possible, Mackenzie began the process by which the territory west of the Rocky Mountains was opened up to interior fur traders. Within fifteen years they would establish trading posts in what is today British Columbia (see page 44). And Mackenzie's achievements became embedded in every map of the West for several decades.

A River Not the Columbia

Mackenzie had mixed up the Fraser and the Columbia, surmising that they were one and the same river. In 1808 the North West Company's Simon Fraser set out to follow the Fraser River to the sea and thus settle the issue.

Fraser had arrived west of the mountains in 1805, finding an easier route than the one Mackenzie had taken (Map 112, *below*). The company set about building a trading network. Fort George had been built in 1807 where the Nechako met the Fraser River, and it was from here that Fraser set out on 28 May 1808 with nineteen men in four canoes down the river that was soon to bear his name.

The river proved more difficult than anything Fraser had ever encountered—unrelenting rapids, impassable waters, and near impossible portages around them. His journal speaks of "tumultuous waves" and his "narrow escape from perdition." On 15 June, just below Lillooet, some local people drew a map for him that showed "another large river which runs parallel to this to the sea." This was the Columbia, and the map was the first hint that the river he was descending might not be that river.

On 20 June Fraser passed the Thompson River, naming it for his fellow partner, David Thompson, whom he supposed had built a post, Kootenae House, farther upstream. It fact it was on the Columbia (see page 42). Thompson would later return the compliment, naming the Fraser after his friend on his map (Map 113, *far right*).

In the Fraser Canyon, the going got even more difficult, and Fraser and his men had to traverse Aboriginal-made wooden steps, like ladders strung from tree boughs along the canyon edge. "We had to pass where no human being should venture," he wrote.

Fraser arrived at the sea on 2 July, at a village "called by the Natives Misquiame (Musqueam)," but its inhabitants were unfriendly and intimidating. He knew he was at the Strait of Georgia from Vancouver's map and wanted to continue to what he called "the main ocean" but, feeling threatened, decided against it. He confirmed his latitude as being "49° nearly," and, knowing that the Columbia's mouth was at 46° 20´N, he sadly concluded: "this River, therefore, is not the Columbia." The North West Company would have to find a different route to the Pacific.

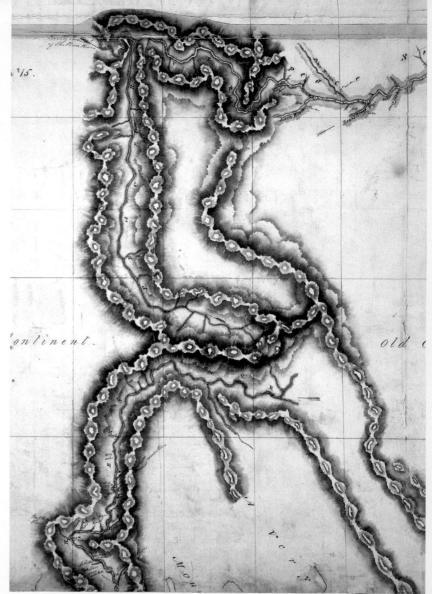

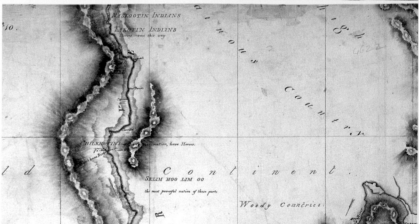

Map 111 (*above*).
The first part of Fraser's journey down the river that would bear his name is shown on this detail of a large map by North West Company surveyor David Thompson. By the second *R* of Fraser's River is written *NW Coy* (North West Company), which refers to Fort George (now Prince George), established in 1807, where Fraser began his journey in 1808. Just before the gap is *Quesnels River,* which Fraser named after one of his clerks, Jules Quesnel.

Map 112 (*left*).
This 1898 map shows the route Simon Fraser took when he first arrived in British Columbia in 1805. It also shows the route Mackenzie took, which makes an interesting comparison. Both ascended the *Parsnip River.* Mackenzie stayed with it, reaching the "Height of Land" shown on his map (previous page), here *Portage 817 paces* and *Portage 175 paces,* to his Bad River that led to the *N. Fork Fraser R.* Fraser, however, found the *Pack R.* flowing out of *McLeod L,* and he followed this route, establishing the first continuously inhabited European settlement west of the Rockies on the shores of this lake, Trout Lake Post (later Fort McLeod), in late 1805. Note the route in red used later: through Pine Pass across the Rockies, through *Ft. McLeod* to *Ft. St. James.*

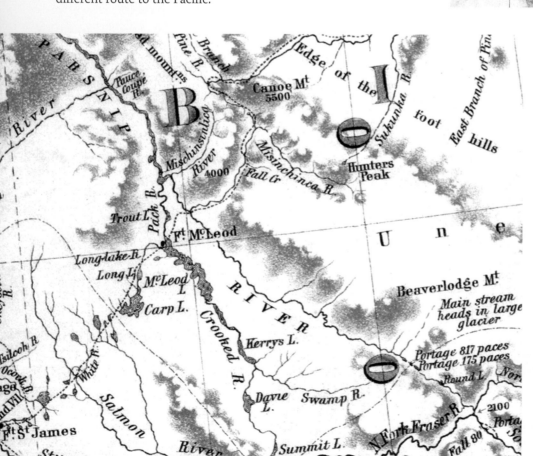

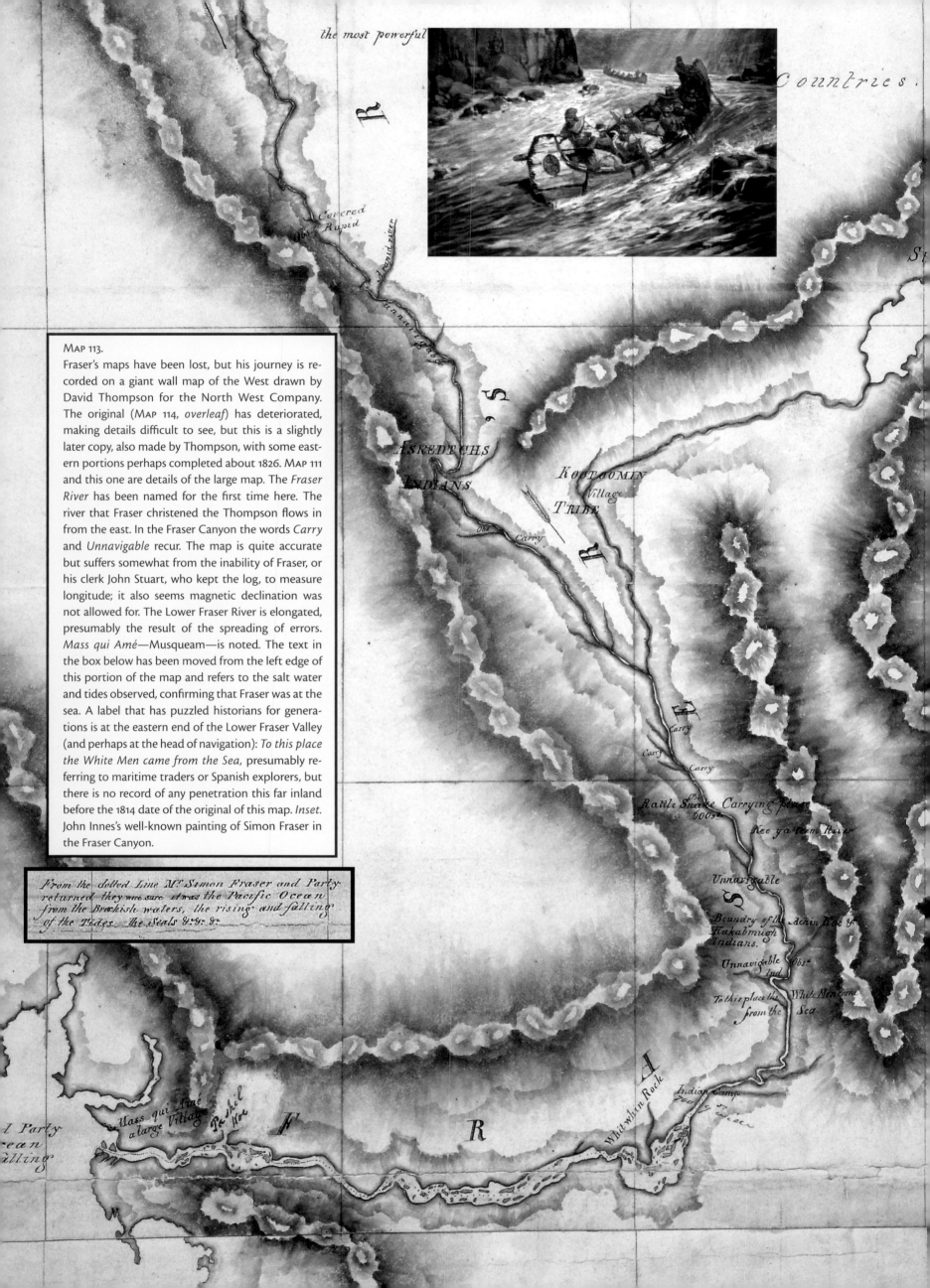

the most powerful

Countries.

R

Covered Rapid
Obs.

Fraser river

Unnavigable

ASKEDECHS
INDIANS

Obs.
Carry

KOOTOOMIN
Village
TRIBE

R

MAP 113.

Fraser's maps have been lost, but his journey is recorded on a giant wall map of the West drawn by David Thompson for the North West Company. The original (MAP 114, *overleaf*) has deteriorated, making details difficult to see, but this is a slightly later copy, also made by Thompson, with some eastern portions perhaps completed about 1826. MAP 111 and this one are details of the large map. The *Fraser River* has been named for the first time here. The river that Fraser christened the Thompson flows in from the east. In the Fraser Canyon the words *Carry* and *Unnavigable* recur. The map is quite accurate but suffers somewhat from the inability of Fraser, or his clerk John Stuart, who kept the log, to measure longitude; it also seems magnetic declination was not allowed for. The Lower Fraser River is elongated, presumably the result of the spreading of errors. *Mass qui Amé*—Musqueam—is noted. The text in the box below has been moved from the left edge of this portion of the map and refers to the salt water and tides observed, confirming that Fraser was at the sea. A label that has puzzled historians for generations is at the eastern end of the Lower Fraser Valley (and perhaps at the head of navigation): *To this place the White Men came from the Sea*, presumably referring to maritime traders or Spanish explorers, but there is no record of any penetration this far inland before the 1814 date of the original of this map. *Inset.* John Innes's well-known painting of Simon Fraser in the Fraser Canyon.

Carry

Carry

Carry
Carry

Rattle Snake Carrying place
600 yd.

Kee ya toom River

Unnavigable

S

Boundry of the Achin Roe &
Hakabmuoh
Indians.

Unnavigable
Obs.
Ind.

To this place the   White Men came
from the   Sea

From the dotted Line Mr Simon Fraser and Party
returned they were sure it was the Pacific Ocean
from the Brackish waters, the rising and falling
of the Tides the Seals &c. &c. &c.

A

R

Whulwhun Rock

Indian Camp
Peaty River

Party
cean
alling

Mass qui Amé
a large Village
Peshil Roe

R

R

THE GREAT MAPMAKER

With the knowledge that the Fraser River was not practically navigable, the North West Company needed to find an alternative route to the sea. The route was obvious: the real Columbia River. William Broughton, with George Vancouver's survey in 1792, had mapped the lower part of the river, downstream of today's Portland, but east of that its location was unknown. The American explorers Meriwether Lewis and William Clark reached the mouth of the Columbia overland in 1805, but their route was unknown to all but a few fur traders before the publication of their book in 1814.

No urgency was felt, however, until 1810, when the North West Company heard that a competitor, the American John Jacob Astor's Pacific Fur Company, intended to establish a fort at the mouth of the Columbia. Company surveyor and astronomer David Thompson was dispatched to follow the Columbia to the ocean and set up a North West Company post there.

Thompson was a meticulous mapmaker, taking time-consuming astronomical observations almost every night, which thus fixed his position numerous times and allowed him in 1814 to draw what would remain the most accurate and comprehensive map of the West for decades (MAP 114, *right*).

Thompson had first crossed the Rockies in 1807, building Kootenae House (MAP 118, *far right, bottom*), a base from which he explored much of the Upper Columbia and Kootenay Rivers in the following two years. In the middle of the winter of 1810–11 he crossed the Rockies through the Athabasca Pass and spent the rest of the winter at Boat Encampment, on the Big Bend of the Columbia. In the spring, initially following routes he knew, he went up the Columbia, crossing at Canal Flats to the Kootenay, and then to the Pend Oreille River, the Spokane River, and the Colville River back to the Columbia. On 15 July 1811 Thompson reached the mouth of the Columbia only to find that the Americans had already built their Fort Astoria there. Nevertheless, he had found what Mackenzie had sought—a navigable path to the sea, and a route by which furs could be shipped to China.

The War of 1812 allowed the North West Company to gain control of Fort Astoria in October 1813. It was taken over by the Hudson's Bay Company in 1821 and renamed Fort George. In 1825 it was abandoned for what was considered a more suitable location. Where the Willamette entered the Columbia the company built Fort Vancouver, which would remain as the headquarters of the Columbia Department, the area west of the Rockies, until 1849. Every year, fur "brigades" would converge on the fort laden with their pelts for export (see page 50).

MAP 114 (*below*).
This is the British Columbia part of David Thompson's 1814 great map of the West. It is said to have hung for many years on the wall at Fort William, the North West Company's major post at the western end of Lake Superior, though this may have been a different copy. The map is now on display at the Archives of Ontario in Toronto, where a curtain shelters it from light unless being viewed. Because Thompson used ink made from oak galls, the map has deteriorated to the point where much detail is very hard to see. Most of the information on the map came from Thompson's own extensive travels, but he also added information from other sources, notably from his fellow partner, Simon Fraser; the course of *Fraser's River* can be seen at left. Information from Mackenzie, the *Peace River*, is at top.

Map 115 (*below, inset*) and **Map 116** (*right*).

For many years British commercial mapmaker Aaron Arrowsmith published updated versions of his 1795 map of North America, "with all the latest discoveries." **Map 115** is the 1802 edition of the map; it shows only Mackenzie's path to the sea. **Map 116** is the 1814 edition of the same map and now incorporates much new information. Virtually all the new British Columbia information (and most of that in Washington and Oregon) is derived from David Thompson's surveying work between 1807 and 1812. The difference between the two maps illustrates well the immense contribution that Thompson made to geographical knowledge west of the Rockies. Mackenzie's tentative link between the Fraser and the Columbia Rivers has gone, and the rather circuitous path of the Columbia and all its numerous tributaries have now been added. Fraser's river, however, is not yet shown.

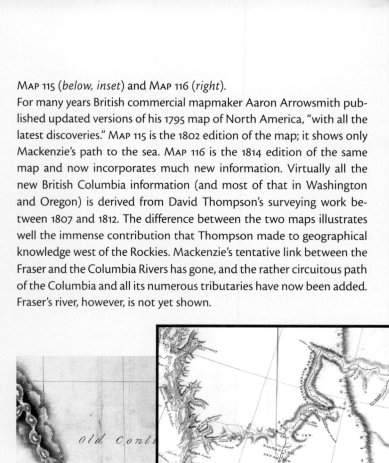

Map 117 (*left*).

This is the southeastern part of the British Columbia section of Thompson's later copy of his 1814 map of the West. It is by far the most comprehensive map of the British Columbia interior to that date. The part to the west is **Map 113**, *page 41*. Kootenae House is the *NW Co* marked just above the *N* in *Mount Nelson*, a 3,283-m-high peak in the Purcell Mountains. Another *NW Co*, Boat Encampment, is marked at the Big Bend of the Upper *Columbia River. Okenawkane Lakes* is Okanagan Lake; *McGillivray's Lake* is Kootenay Lake, and the Upper Arrow Lake is marked but not named on the south-flowing part of the *Columbia River.*

Map 118 (*below*).

The site of Thompson's Kootenae House, built in 1807 and in use until 1812, is shown by the red square on this 1920s map; the photo is of the site of the post, a National Historic Site, as it is today.

43

Right. The fur warehouse at Fort St. James, today a National Historic Site, on Stuart Lake. *Above.* The North West Company seal.

The North West Company, looking for new fields to conquer and thus compete with its archrival the Hudson's Bay Company, edged ever westward as the eighteenth century came to a close. Posts were established on the Peace River in the late eighteenth century. The company decided to expand west of the Rockies in 1805, and late that year Simon Fraser followed Alexander Mackenzie's route up the Peace and Parsnip and found the latter's tributary, the Pack River, which led to McLeod Lake (at first called Trout Lake). Here were friendly Sekani, and so he built Trout Lake Post (the *fort* on Map 121, *below, right*). The post was abandoned soon after following dissension between the men left there over the winter.

This inauspicious start did not derail the North West Company's plan. In early 1806 James McDougall, at first sent to investigate the abandoned post, reconnoitred overland and found the large Stuart Lake, which was surrounded by friendly Carrier. The trade possibilities looked good, and so Fraser decided to establish his main post here. In three canoes, twelve men including Fraser and his clerk John Stuart found a water route to the lake, to the Fraser River, up the Nechako, and then up the Stuart River. The post that became Fort St. James was built that summer. Another post was built at Fraser Lake, some distance to the west; this was Fort Fraser.

The following year a post was built at the confluence of the Nechako with the Fraser. This became Fort George, at the site of present-day Prince George, and was the starting point for Fraser's epic 1808 journey down

Map 119 (*below*).
The boundaries of the North West Company's territory have been drawn in red on this map. This was the area, constrained by the Russian American Company territory to the north and Spanish territory to the south, that included the Oregon Country and the Athabasca district, all outside the boundaries of the Hudson's Bay Company's Rupert's Land—the land draining into Hudson Bay.

Map 120 (*left*).
The North West Company's region west of the Rocky Mountains became known as New Caledonia, derived from the old name for Scotland, the homeland of many of the fur traders. The name first appeared in 1808 in the journal of fur trader Daniel Harmon. This is Harmon's map from his 1820 book. The information on the map comes principally from one published by Aaron Arrowsmith in 1819. The Pack River is shown but not named. *Stuarts L.* is named. Some lakes are shown with outlets to the Pacific via the *Naccotain R.*, possibly the Skeena. The map incorporates traders' interpretations of Aboriginal reports; the grossly exaggerated *Great Bear Lake*, emptying into *Finlays Bra.*, is presumably Thutade Lake, the source of the Finlay River (and the Peace River system) and was not found by traders until 1824.

Map 121 (*right*).
James Wyld's 1823 map showed an enormous *Stuart's Lake* as well as a huge *Great Bear's Lake*. At *Frazer's Lake* is a *Settlemt. establ. 1806*, and a *Fort* is shown on *McLeod's Lake.* The geography of New Caledonia as displayed to the world was clearly still rather confused.

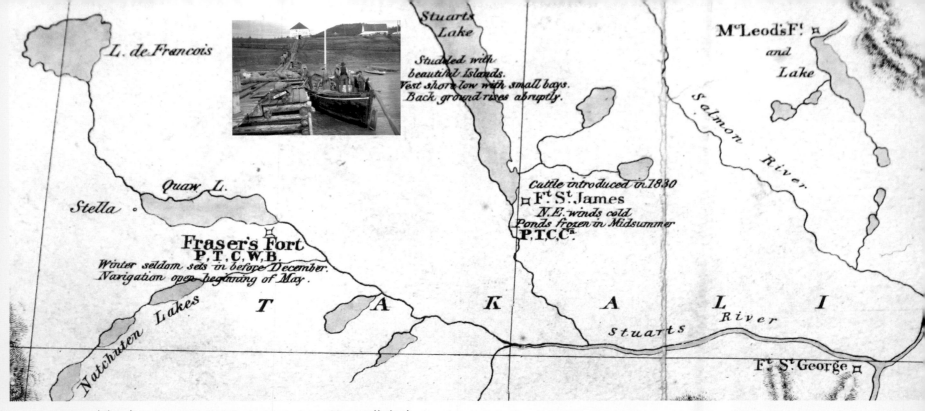

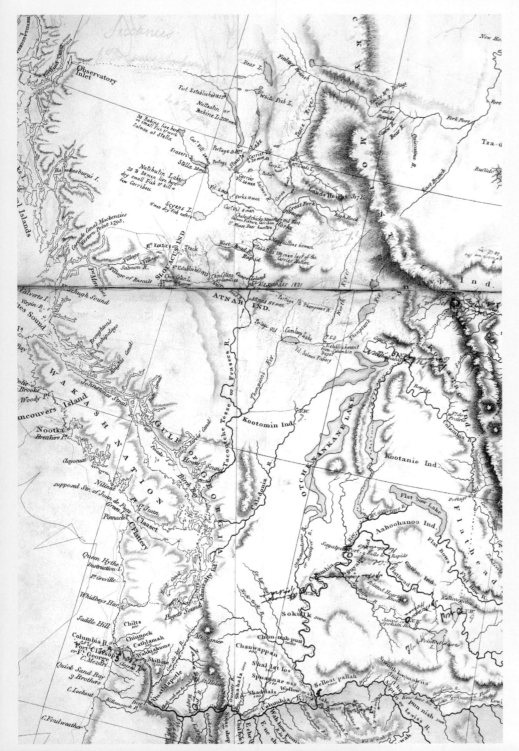

Map 122 (*above*).

The early fur-trading posts that the North West Company founded are shown on this map: *McLeod's F^t.* in 1805; *F^t. S^t. James* and *Fraser's Fort* (Fort Fraser) in 1806; and *F^t. S^t. George* (Fort George) in 1807. *Stuarts River* is the Nechako. Stuart River connects *Stuarts Lake* with the Nechako and is the route Fraser and his men took in 1806.

Map 123 (*below*).

The 1824 map of North America published by the British mapmaking firm of Arrowsmith (by that time Aaron's sons Aaron and Samuel and nephew John) has been added to on this copy with new information in New Caledonia from the Hudson's Bay Company, which had merged with the North West Company three years earlier. The new information gives employee strength at each post and some indication of the size of local villages. *Bear Lake* is now correctly much reduced in size, and it and *Nattcotin or Babine Lake* are shown, again correctly, as connecting to a westward-flowing river—the Skeena (though, as Simpson's River, it was for a long time confused with the Nass). The Skeena never seems to have been considered as a Pacific outlet for shipping furs until much later, though steamboats would in a few decades ply the lower part of the river. The southern branch of the Peace River has been labelled *Parsnip* in pencil. The *Trout R.* (Pack River) leads to *McLeod's L.* The route of the fur brigades from New Caledonia to the Columbia is evident on this map. A *Portage to Thompson's R.* runs from *Ft. Alexandria 1821* on the Fraser to the *North (Thompson) River*, and at its confluence with *Thompson's River* is *P.S.O. Establishment from Columbia Depot*—seemingly an error meaning the Pacific Fur Company—which built a post here in 1812 that the North West Company took over the next year, combining it with its own Fort Shuswap (**Map 125**, *overleaf*). The fictitious *Caledonia R.*, with another post at *N.W.*—again representing Kamloops—is also shown on this map (see page 55). This map, from Hudson's Bay Company archives, is unique in that it also indicates the route of Governor George Simpson in 1824; this is the red dotted line. Simpson crossed the Rockies via the North Saskatchewan River and ended his speedy visit at *Fort Clatsop or F^t. George.*

the river to the sea (see page 40). John Stuart, his trusty lieutenant, took overall charge of New Caledonia operations the following year, a post he held until 1824.

In 1813 Stuart put together a practicable route to transport furs between the Fraser and the Columbia. This left the Fraser at Alexandria, where a new fort was built in 1821, went overland to Kamloops, and then south through the Okanagan Valley to the Okanagan River, a tributary of the Columbia. At Kamloops a post built in 1812 by the Pacific Fur Company was taken over in 1813 as part of the forced purchase of the company by the Nor'westers at Fort Astoria in 1813 during the War of 1812.

The North West Company was merged with the Hudson's Bay Company in 1821 following years of acrimonious competition; that company's fur empire now extended right across today's Canada. George Simpson, its energetic North American governor, travelled to the West Coast in 1824, covering the distance from York Factory, on Hudson Bay, to Fort George (previously Fort Astoria), at the mouth of the Columbia, in eighty-four days. Here he made the decision to move the company headquarters to a new post farther

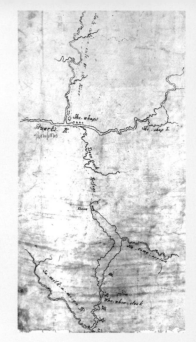

MAP 125 (*right*).
Part of a large map of much of the Columbia Basin drawn in 1821 and added to in 1849 by fur trader Alexander Ross. This shows *She-Whaps* (Shuswap), the Pacific Fur Company and after 1813 first North West Company post at Kamloops, together with *Portage,* the fur brigade route to Okanagan Lake. Note the pencil alteration changing *Stuart's R.* to *Thompsons*.

up the Columbia, and Fort Vancouver was constructed later that year. He also developed a plan for combatting Russian competition on the coast, and the decision to establish Fort Langley was part of this strategy. Fort Langley was built in 1827 after an 1824 reconnaissance of the Lower Fraser to find the right location (see page 49).

In 1828 Simpson, not convinced of the unnavigability of the Fraser River despite Simon Fraser's harrowing experience twenty years before, decided to see for himself. Accompanied by chief traders James Yale and Archibald McDonald, Simpson managed to scare himself so badly that he gave up, once and for all, the notion of using the Fraser as a means of getting furs to the sea. Descending the Fraser, he declared, would result in "certain death nine times out of ten."

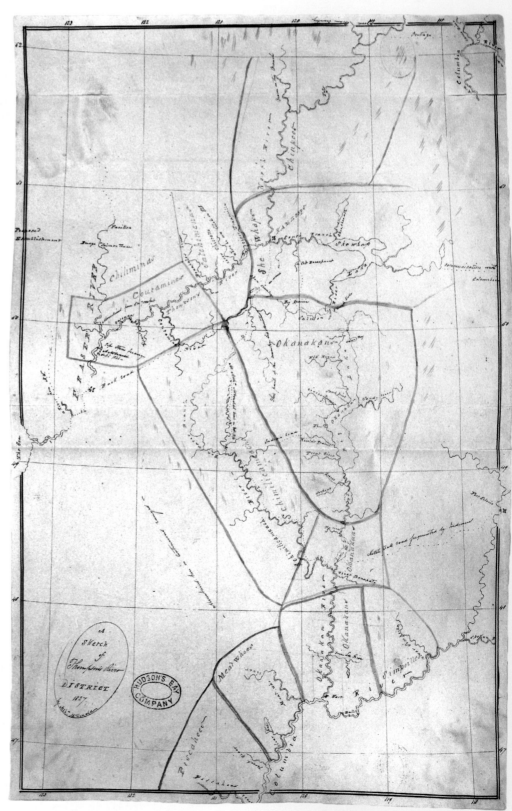

MAP 124 (*above*).
The *Department of Columbia*—the Hudson's Bay Company's dominions west of the Rocky Mountains—is shown along with the company posts in red ink additions to an Arrowsmith map of 1824. The additions also show some results of northern explorations (as on MAP 127, *far right*), but this information appears to be slightly earlier here in that it is less complete. *Thompson's Riv Ft.* is at Kamloops. *Kootanee Ft.* is David Thompson's post on the Upper *Columbia R.* (see page 42), though this was abandoned in 1812. Babine Lake (with *Babine Fort N.W.,* built in 1822) and *Bear L.* are (correctly) shown draining to the Pacific coast by *Simpson's R.,* actually the Skeena headwaters combined with the lower reaches of the Nass River. *Chilcotin* has been written next to *Fort NW* at the confluence of the Chilko and Chilcotin Rivers, the latter incorrectly labelled *West Road R.;* Alexander Mackenzie's route west is shown correctly farther north. Fort Chilcotin was established in 1822, information shown on the version of this map on the previous page (MAP 123).

MAP 126 (*right*).
The first regional map of the Okanagan and southern interior of British Columbia is this map showing rivers, lakes, and Aboriginal group areas drawn in 1827 by Archibald McDonald. He had been in charge of the post at *Kamloops,* then called Thompson's River Post. The year before, accompanied by Okanagan chief Nicola, he had explored *Thompson's River* to where it met the *Fraser River* and had fixed the position of the confluence. Details from this expedition were included in his map, drawn to illustrate his district report. At bottom, at the confluence of the *Okanagan River* with the *Columbia River,* is a *Fort.* This was Fort Okanagan, built in 1811 by the Pacific Fur Company and taken over by the North West Company in 1813. The *Schimilicameach River* is the Similkameen. *The Sea* (Strait of Georgia) is off the margin at left.

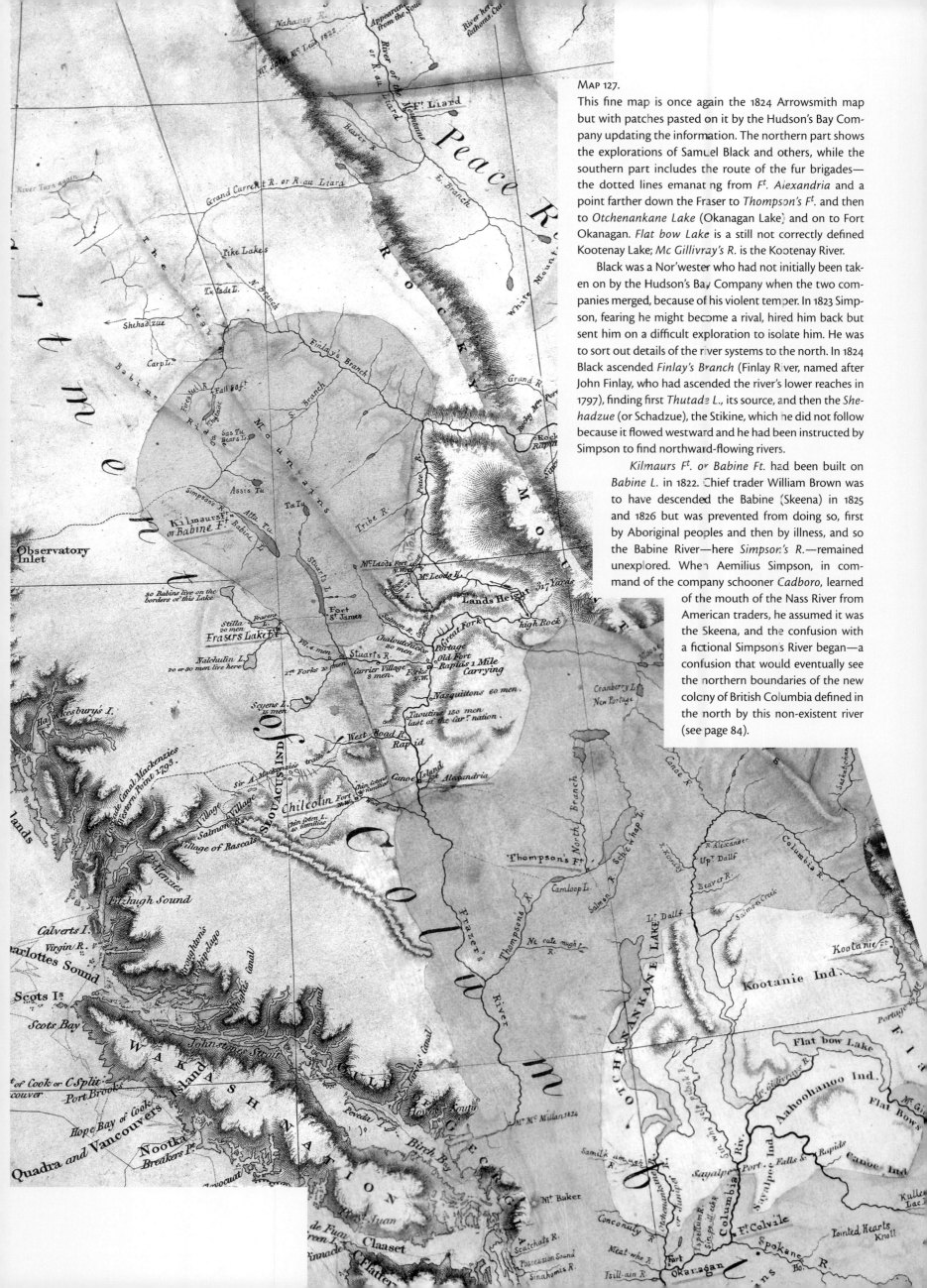

MAP 127.

This fine map is once again the 1824 Arrowsmith map but with patches pasted on it by the Hudson's Bay Company updating the information. The northern part shows the explorations of Samuel Black and others, while the southern part includes the route of the fur brigades—the dotted lines emanating from *Ft. Alexandria* and a point farther down the Fraser to *Thompson's Ft.* and then to *Otchenankane Lake* (Okanagan Lake) and on to Fort Okanagan. *Flat bow Lake* is a still not correctly defined Kootenay Lake; *Mc Gillivray's R.* is the Kootenay River.

Black was a Nor'wester who had not initially been taken on by the Hudson's Bay Company when the two companies merged, because of his violent temper. In 1823 Simpson, fearing he might become a rival, hired him back but sent him on a difficult exploration to isolate him. He was to sort out details of the river systems to the north. In 1824 Black ascended *Finlay's Branch* (Finlay River, named after John Finlay, who had ascended the river's lower reaches in 1797), finding first *Thutade L.*, its source, and then the *Shehadzue* (or Schadzue), the Stikine, which he did not follow because it flowed westward and he had been instructed by Simpson to find northward-flowing rivers.

Kilmaurs Ft. or *Babine Ft.* had been built on *Babine L.* in 1822. Chief trader William Brown was to have descended the Babine (Skeena) in 1825 and 1826 but was prevented from doing so, first by Aboriginal peoples and then by illness, and so the Babine River—here *Simpson's R.*—remained unexplored. When Aemilius Simpson, in command of the company schooner *Cadboro*, learned of the mouth of the Nass River from American traders, he assumed it was the Skeena, and the confusion with a fictional Simpson's River began—a confusion that would eventually see the northern boundaries of the new colony of British Columbia defined in the north by this non-existent river (see page 84).

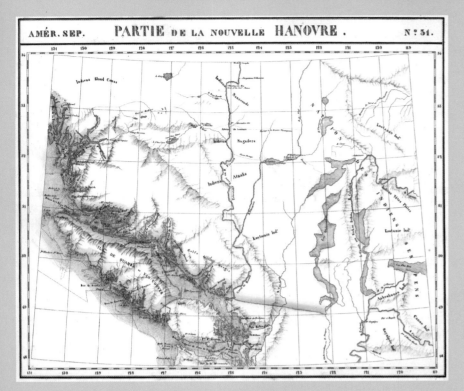

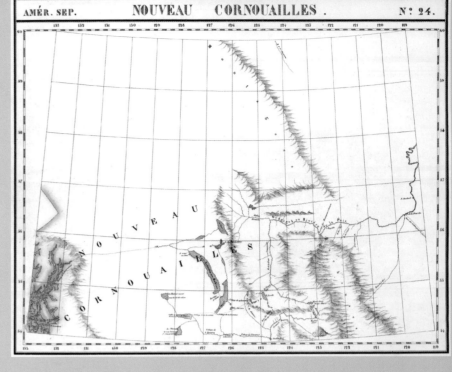

Map 128 (*above, left*) and Map 129 (*above, right*).
In 1827 Belgian mapmaker Philippe Vandermaelen published a notable atlas. His massive *Atlas universel* contained 380 sheets drawn on a conical projection that allowed them to be fitted together to make a globe. Such a globe would be about 7.75 m in diameter, and it is thought that one was constructed in Brussels soon after the atlas was published. Today, Princeton University Library has put together a virtual one, available online. The atlas also contained forty pages of statistical tables, was the first to show the world at a consistent scale (about 1 cm to 16.5 km), and was the first to be printed using the then-new process of lithography. The two maps shown here cover the area that would become British Columbia. Although by no means totally inaccurate, the maps are not as correct or up-to-date as others shown in this section because Vandermaelen did not have the benefit of

updates provided by the Hudson's Bay Company. Interestingly, the area of the United States is shown to include southern Vancouver Island, along the lines of some of the American boundary proposals (see page 54), where the forty-ninth parallel, agreed by a convention in 1818 as a boundary east of the Rockies, is simply extended west right to the Pacific Ocean. The Fraser River, here still called the *Tchoutché Tessé ou R. Frasers*, is shown erroneously flowing to the *Canal de Burrard* (Burrard Inlet). A fur trade post (Fort Kamloops) is correctly shown at the confluence of the *R. de Nord* (North Thompson) and *R. Thompsons*, east of *L. Camloop*. The fur brigade trail from *Ft Alexandre 1821* is shown, though there is confusion about its continuance to *Lac Otchenankan*, Okanagan Lake. *Lac Chatth* is Upper and Lower Arrow Lakes, part of the course of the *R. Columbia*.

Simpson was keen on expanding the company's territory, and this was done in the south by overtrapping on the Snake River in Oregon, which was intended to combat competitive American fur traders—the famous "mountain men." In the northern interior there was no competition, but the rivers were unknown, and so Simpson ordered a series of explorations to open up the territory to fur-gathering operations. This effort would culminate in the mapping of the Upper Yukon in 1851 by Robert Campbell and reveal the complex paths of the northward-flowing rivers.

Samuel Black was sent up the Finlay River in 1824 and found Thutade Lake, which is the river's source and also that of the entire Mackenzie River system. Continuing north, he found a westward-flowing river he called the Schadzue. This was the Stikine and would have proved useful to the company had he followed it to the sea, though it would have flowed through Russian territory. But Black had been instructed to search to the north. Following his instructions, Black was disappointed to find an eastward-flowing river; since he decided to give up his exploration at this point, he named it the Turnagain River. It was a tributary of the Kechika, itself a tributary of the Liard, which flows into the Mackenzie.

Part of Governor Simpson's plans included developing the coastal fur trade. In 1825 the company ship *William & Ann* under captain Henry Hanwell carried out a reconnaissance of the whole coast, but here there was American and, farther north, Russian competition. Hanwell was considered too timid for the task of trading, so Simpson hired his cousin Aemilius, formerly a lieutenant in the Royal Navy. In 1829 Aemilius Simpson (who had helped found Fort Langley—see opposite page) visited the mouth of the Nass River in the company's ship *Cadboro*, trading fine furs. Two years later he and Peter Skene Ogden built a new post here, which, after Simpson died in September 1831, was named Fort Simpson in his honour. It was moved in 1834 to a better trading location on the Tsimshian peninsula and became the headquarters of the coasting and land-based trade of the entire Northwest Coast (Map 143, *page 51*). In 1836 the company added another coastal trade advantage: the steam paddlewheeler *Beaver*, the first steam vessel in the entire Pacific Ocean. The ship was able to penetrate deep into inlets inaccessible to sailing ships.

Map 130 (*below*).
The North West Company had established posts in the Mackenzie Valley early on in the nineteenth century, and this base could be expanded upon by exploring westward. At the same time Black was exploring northward, company clerk Murdoch McPherson was instructed to explore the Liard River west from his post at *Fort Simpson*, at the confluence of the Liard with the Mackenzie River, at top. *Fort de Liard*, established in 1805, is on the Liard, while on the *East Branch* (the Fort Nelson River) is *Fort Brullé*, the second Fort Nelson, first built in 1806 or 1807, is at bottom right.

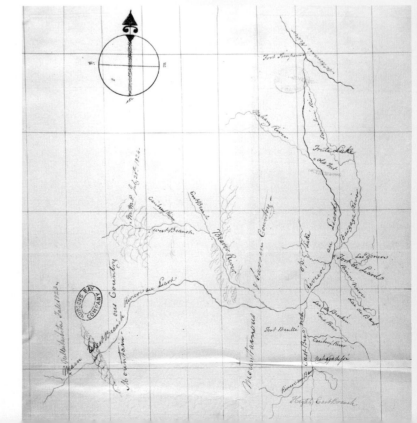

Fort Langley

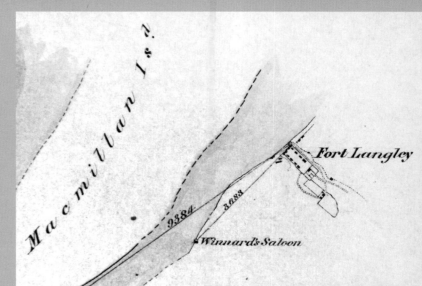

MAP 131 (*right*).
Fort Langley in 1862, at the location it was moved to in October 1838, the same as the National Historic Site today. The new fort burned down only eighteen months after it was completed but was rebuilt at the same location. Opposite, *Macmillan Is^d.* was named after fort founder James McMillan.

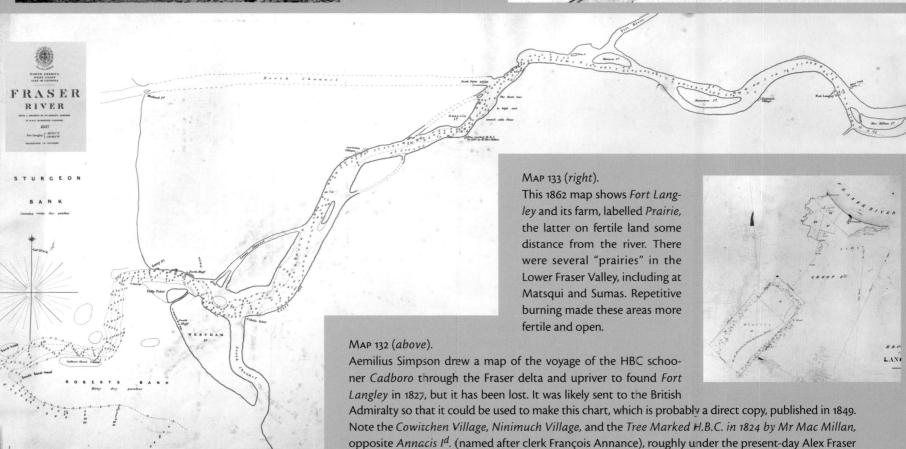

MAP 133 (*right*).
This 1862 map shows *Fort Langley* and its farm, labelled *Prairie*, the latter on fertile land some distance from the river. There were several "prairies" in the Lower Fraser Valley, including at Matsqui and Sumas. Repetitive burning made these areas more fertile and open.

MAP 132 (*above*).
Aemilius Simpson drew a map of the voyage of the HBC schooner *Cadboro* through the Fraser delta and upriver to found *Fort Langley* in 1827, but it has been lost. It was likely sent to the British Admiralty so that it could be used to make this chart, which is probably a direct copy, published in 1849. Note the *Cowitchen Village, Ninimuch Village*, and the *Tree Marked H.B.C. in 1824 by Mr Mac Millan*, opposite *Annacis I^d.* (named after clerk François Annance), roughly under the present-day Alex Fraser Bridge. Fort Langley is shown at its first location, north of the Salmon River, not where it was in 1849.

Since the location of a boundary between British and American territory west of the Rocky Mountains had not yet been determined, in 1824 Governor George Simpson decided to consider a post at the mouth of the Fraser that could, if necessary, become the company headquarters. He did not yet believe in Simon Fraser's assessment of the Fraser as unnavigable.

In November 1824 James McMillan was sent to reconnoitre the Lower Fraser. He avoided the river openings in the delta in favour of a route up the Nicomekl River and a portage to the Salmon River. Near here was the site he chose for the post (MAP 135, *below*).

On 22 June 1827 the company's schooner *Cadboro*, commanded by Aemilius Simpson, entered the Fraser with McMillan aboard, and on 31 July they began to build the new post, named Fort Langley after Thomas Langley, a company director. The *Cadboro* departed on 18 September after a secure structure had been built and provisioned.

The following year George Simpson himself descended the Fraser and soon concurred with Simon Fraser's assessment as to its unsuitability for navigation. As a result the plans for Fort Langley were scaled back. It only survived as a company post because it could supply agricultural produce to other posts and, after 1839, to Russian American Company posts (see next page).

MAP 134 (*right, next to bottom*) and MAP 135 (*right, bottom*).
Aaron's Arrowsmith map, 1824 edition, and with an updating patch pasted on by the HBC. The update shows James McMillan's reconnaissance that year in which he determined the site for Fort Langley, built three years later. Both the Nicomekl and Salmon Rivers, used by McMillan to first access the Fraser River by canoe, are shown but not named. McMillan canoed to the mouth of the Fraser before returning via the two rivers, and the updated version shows the channels of the Fraser delta defined, delineating Richmond's Lulu Island for the first time.

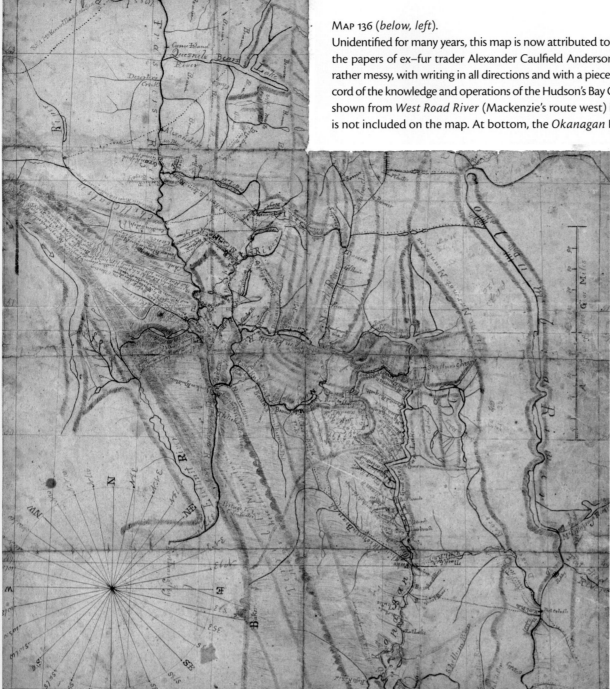

MAP 136 (*below, left*).
Unidentified for many years, this map is now attributed to Samuel Black. The 1835 map was found in 1915 among the papers of ex–fur trader Alexander Caulfield Anderson, mapmaker extraordinaire (see page 108). Although rather messy, with writing in all directions and with a piece missing, the map is full of detail and is an excellent record of the knowledge and operations of the Hudson's Bay Company at its prime in British Columbia. *Fraser's R.* is shown from *West Road River* (Mackenzie's route west) in the north to *Fort Langley* in the south. The coast is not included on the map. At bottom, the *Okanagan* River runs into the *Columbia River.*

MAP 137 (*above*).
In 1845 HBC fur trader Alexander Caulfield Anderson explored several routes to *Fort Langley* from Fort Kamloops as an alternate path to the sea after New Caledonia was cut off from the Lower Columbia by the border. This was one route, which was not used, from *Fraser's River* to *Harrison's R.* Several other routes were used for periods of time.

There was a problem farther north in that the Alaska Panhandle cut the company off from the interior. Although the 1825 agreement with Russia had given Britain navigation rights on this coast, the Russians prevented Ogden from building a fort at the mouth of the Stikine River in 1834. This problem was solved in 1839 with a ten-year lease given to the Hudson's Bay Company that allowed it to build posts, hunt, and trade furs. In return, the company was to supply Russian American Company posts with food and supplies.

In 1840 James Douglas, with the *Beaver,* set up two new posts, Fort Taku—officially Fort Durham, at Taku Harbor (just south of today's Juneau)—and Fort Highfield, at the mouth of the Stikine River. Two years earlier Robert Campbell had dangerously entered a great Tlingit trading rendezvous farther up the Stikine, and it was known to be a rich fur-trading ground. Fort Highfield, also known as Fort Stikine, had previously been the Russian Redoubt St. Dionysius, built in 1834 to keep the Hudson's Bay Company out, but the latter took it over as part of the 1839 lease (MAP 138, *below, left*). Taku was closed in 1843, but Highfield lasted until the termination of the lease in 1849 and for a decade allowed the company easy access to the western part of its northern interior territory.

The route by which most interior furs were taken out was through Fort Vancouver, accessible to ships of some size. Furs were compacted in large presses and made up into packs of about 40 kg. They were then transported using canoes, where feasible, or pack horses, where not. Several routes were developed, but the principal one was from Fort Alexandria, on the Fraser, overland to the North Thompson and Fort Kamloops, then overland again to the Okanagan Valley to Fort Okanagan and on down the Columbia.

In 1846 this route was truncated by the new border (see page 54) and alternate routes had to be found (MAP 137, *above*), but they were never as satisfactory as the original. But the fur trade era was drawing to a close in any case; although a downsized fur trade continued, fashions had changed and volumes were much smaller. In 1870 the Hudson's Bay Company sold its vast trading territory of Rupert's Land to the new Canadian government.

In less than seventy years the fur traders had explored much of British Columbia. In their quest for ever more furs they had expanded geographical knowledge dramatically and produced many maps, laying the groundwork for the Royal Engineers who would come to British Columbia following the 1858 gold rush and the proclamation of the new colony.

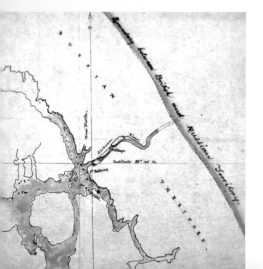

MAP 138 (*left*).
Highfield marks the location of Fort Highfield at the tip of Wrangell Island near the mouth of the *Stikeen River.* The fort was built in 1840 by James Douglas as part of the Hudson's Bay Company's 1839 agreement with the Russian American Company.

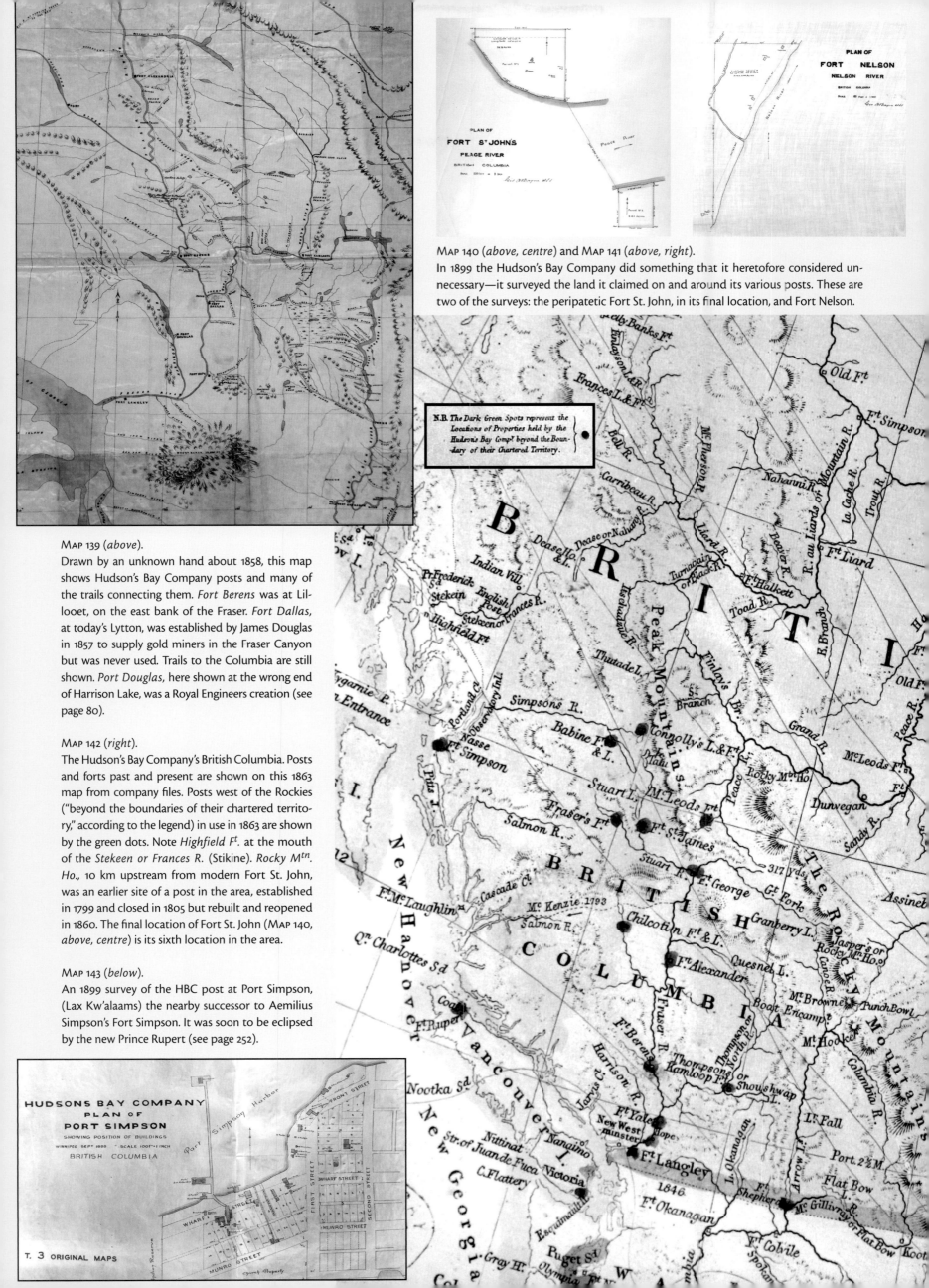

MAP 140 (*above, centre*) and **MAP 141** (*above, right*).
In 1899 the Hudson's Bay Company did something that it heretofore considered unnecessary—it surveyed the land it claimed on and around its various posts. These are two of the surveys: the peripatetic Fort St. John, in its final location, and Fort Nelson.

PLAN OF
FORT St. JOHN'S
PEACE RIVER
BRITISH COLUMBIA

PLAN OF
FORT NELSON
NELSON RIVER
BRITISH COLUMBIA

MAP 139 (*above*).
Drawn by an unknown hand about 1858, this map shows Hudson's Bay Company posts and many of the trails connecting them. *Fort Berens* was at Lillooet, on the east bank of the Fraser. *Fort Dallas*, at today's Lytton, was established by James Douglas in 1857 to supply gold miners in the Fraser Canyon but was never used. Trails to the Columbia are still shown. *Port Douglas*, here shown at the wrong end of Harrison Lake, was a Royal Engineers creation (see page 80).

MAP 142 (*right*).
The Hudson's Bay Company's British Columbia. Posts and forts past and present are shown on this 1863 map from company files. Posts west of the Rockies ("beyond the boundaries of their chartered territory," according to the legend) in use in 1863 are shown by the green dots. Note *Highfield Ft.* at the mouth of the *Stekeen or Frances R.* (Stikine). *Rocky Mtn. Ho.*, 10 km upstream from modern Fort St. John, was an earlier site of a post in the area, established in 1799 and closed in 1805 but rebuilt and reopened in 1860. The final location of Fort St. John (**MAP 140**, *above, centre*) is its sixth location in the area.

MAP 143 (*below*).
An 1899 survey of the HBC post at Port Simpson, (Lax Kw'alaams) the nearby successor to Aemilius Simpson's Fort Simpson. It was soon to be eclipsed by the new Prince Rupert (see page 252).

N.B. The Dark Green Spots represent the Locations of Properties held by the Hudson's Bay Comp.y beyond the Boundary of their Chartered Territory.

HUDSONS BAY COMPANY
PLAN OF
PORT SIMPSON
SHOWING POSITION OF BUILDINGS
WINNIPEG SEPT 1899   SCALE 100F=1 INCH
BRITISH COLUMBIA

T. 3 ORIGINAL MAPS

FORT VICTORIA

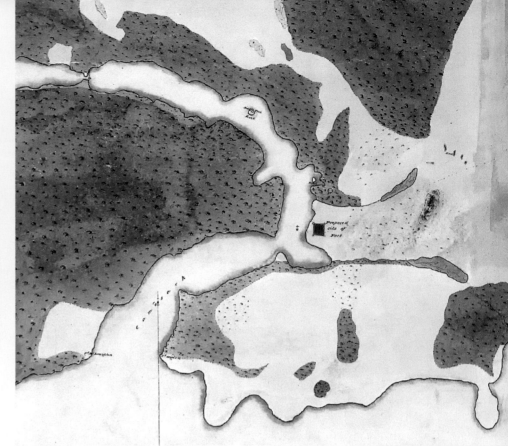

As American settlers began to stream into the Willamette Valley to the south, the Hudson's Bay Company, with its headquarters for all of Columbia and New Caledonia Districts at Fort Vancouver, could see the writing on the wall. In 1842 company governor George Simpson instructed John McLoughlin, chief factor at Fort Vancouver, to locate a new headquarters on Vancouver Island, which he hoped would be salvageable from American territorial ambitions. And, as history would show, he was right.

McLoughlin dispatched his deputy, James Douglas, to find the right location. Douglas found the site "for the proposed new Establishment" in the "Port of Camosack" (Camosun)—Victoria Harbour. Here was timber in abundance, land suitable for agriculture, and potential tidal water power.

MAP 144 (right, top).
This is a detail of James Douglas and his assistant Adolphus Lee Lewes's 1842 map locating the *Proposed Site of Fort*—Fort Victoria. It is a land use map; dark green shows forested land and the light green "plains" suitable for agriculture. Yellow shows wet marshes; brown, rocks and hills.

MAP 145 (right).
An 1850 map of Fort Victoria by Walter Colquhoun Grant, the Colony of Vancouver Island's first settler. The colour and number keys are included.

MAP 146 (below).
An 1850 company map of the Victoria area. The squares are square miles.
Above. A painting of Fort Victoria in 1845 by Henry Warre.
Below, bottom. The first photograph of Fort Victoria, and likely the first of anywhere in British Columbia, taken in 1857 by Lieutenant Richard Roche, an officer on the British naval ship *Satellite.*

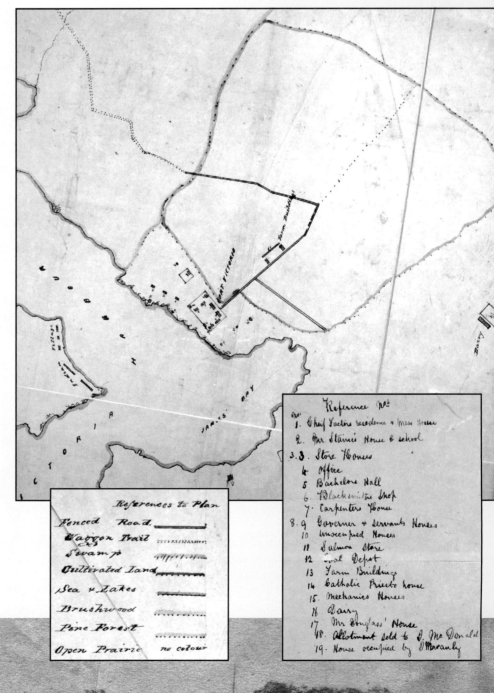

References to Plan

Fenced Road
Waggon Trail
Swamp
Cultivated Land
Sea & Lakes
Brushwood
Pine Forest
Open Prairie   no colour

Reference No.
1. Chief Factors residence & Mess House
2. Mr Staines House & school
3.3. Store Houses
4 Office
5 Bachelors Hall
6 Blacksmith Shop
7 Carpenters House
8.9 Governer & Servants Houses
10 Unoccupied Houses
11 Salmon Store
12 Coal Depot
13 Farm Buildings
14 Catholic Priests house
15 Mechanics Houses
16 Dairy
17 Mr Douglass' House
18 Allotment sold to J. Mc Donald
19 House occupied by J. McAulay

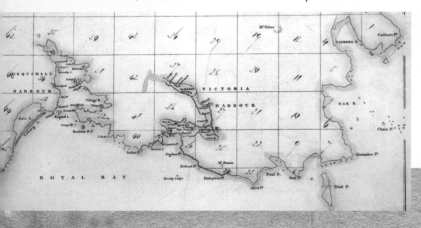

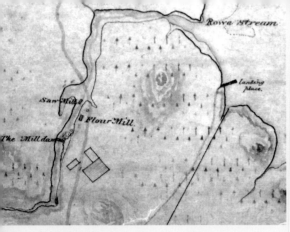

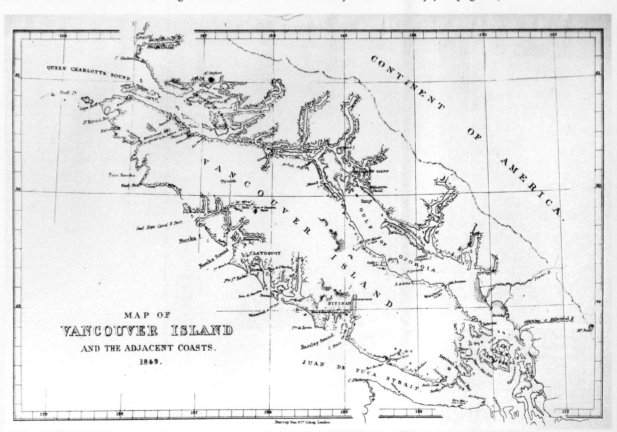

MAP 147 (in two parts, *above* and *right*).
Drawn in 1853 by the company's surveyor, Joseph Despard Pemberton, this map shows *Victoria*, *The Fort*, *Government House*, and the *Songé Village* and (*above*) a *Mill dam*, *Saw Mill*, and *Flour Mill* located at the head of Esquimalt Harbour. *Rowe Stream* is now Millstream. The Hudson's Bay Company's saw mill was built in 1846, the flour mill the year after.

The company, in search of efficiencies, was also revising its trading methods. Existing coastal forts were to close down and the *Beaver*, operating out of the new fort, was to substitute for them.

On 13 March 1843 Douglas and fifteen men arrived and set to work building a new fort, offering the local Aboriginal people, the Songhees, who,

Douglas reported, were pleased that the company was to locate there, a blanket for every fifty pickets they supplied. Indeed, it seems that the Songhees were pleased, as they moved their village from Cadboro Bay to the Inner Harbour. In June the Northern Department directors decided on a name for their new post: it would be called Fort Victoria, after the new British queen.

SEVEN SHILLINGS A YEAR

There had been proposals to colonize Vancouver Island before—one notable one in the 1790s, by creating a prison colony à la Botany Bay—but nothing had found the support of the British colonial office. But with the settling of the boundary issue (see next page) came a desire to place a loyal British population on the northern side of the border as a bulwark against possible future American expansionist plans.

On 13 January 1849 the British government issued a Charter of Grant of Vancouver's Island to the Hudson's Bay Company on the strict condition that it establish a settlement or settlements of colonists within five years. All the colonists were to come from "Great Britain, Ireland, or the Dominions"; no Americans allowed. Control of Vancouver Island was given to the company for an annual rent of seven shillings. A barrister, Richard Blanshard, was appointed governor, but with little or no support from company men, he lasted only a few months. After that, the British government bowed to the inevitable and appointed James Douglas, Hudson's Bay Company chief factor, who then held both jobs.

MAP 148 (*right*).
This rather sketchy and poorly printed map of the new Colony of Vancouver Island appeared in the 1849 booklet *Colonization of Vancouver's Island*, printed and circulated in Britain, to little effect. The coastline is largely based on George Vancouver's map, and the interior of the island is a blank. *Victoria* is marked, and also *Wentuhuysen Inlet*, where the discovery of coal led to the founding of Nanaimo (see page 224). George Vancouver's "Quadra and Vancouver's Island" has now become Vancouver Island, matching the name of the colony.

Attempts at colonization were half-hearted and ill-founded. By 1854 there were only 112 people on the island not working for or dependent on the company. The colony's first settler was Walter Colquhoun Grant, a once-wealthy British army officer who had fallen on hard times and decided to start anew in "the colonies." When he arrived in 1849 he famously shot a cow thinking it was a buffalo.

The Colony of Vancouver Island, which the company much preferred as its own private fiefdom uncluttered by settlers who might get in its way, was essentially a failure as a colony until 1858, when the influx of gold seekers created an entirely different story (see page 62).

BRITISH BRITISH COLUMBIA

It was by no means certain that the area that is now British Columbia would be British at all. No one could show how far up the coast Francis Drake had sailed in 1579. The Russians had a good case in that their navigators Vitus Bering and Aleksei Chirikov had found the coast—in two places independently—in 1741. The original grant to the Hudson's Bay Company was for Rupert's Land, the region drained by rivers flowing into Hudson Bay, and excluded the lands west of the mountains. The latter was leased to the company after 1821 by the British government. Alexander Mackenzie had arrived overland in 1793, Lewis and Clark not until 1805, the year the British North West Company began to trade in the region following the arrival of Simon Fraser. The Americans had beaten David Thompson to the mouth of the Columbia in 1811. In 1818, when the United States and Britain first tried to decide where to draw a boundary, they had decided to call it a draw and made the entire Oregon Country, as it became known, a region of joint occupancy. And so it sat until the mid-1840s, by which time the United States had far more of its citizens living in the Pacific Northwest than Britain.

The Russian claim to British Columbia, rooted in Bering and Chirikov, was becoming stronger by the end of the eighteenth century. Russian fur traders progressively hopped along the islands of the Aleutian chain, and in 1799 one company, the Russian American Company, was granted a charter by the tsar that gave it a monopoly of the fur trade. A headquarters was established, first at Kodiak Island and then, in 1804, at Sitka. In 1821 the tsar was persuaded to issue a ukase, or imperial edict, claiming Russian sovereignty south to 51°N (Map 152, *below, left*), which produced formal complaints from both Britain and the United States. The Russian claim was negotiated back to 54°40′N by treaty with the United States and with Britain the following year (Map 153, *below, right*).

Spain, whose claims to the Northwest were legitimate but unenforceable by that declining power, formally ceded them to the United States in the Transcontinental Treaty of 1819. The United States thus inherited from the Spanish its claim north to Juan Pérez's

Map 149 (*right, top*).
A typical British map dated 1814 shows "British Dominions" extending right to the Pacific Coast, where the names George Vancouver bestowed are in evidence: *New Georgia, New Hanover,* and *New Cornwall.*

Map 150 (*right, centre top*).
This 1787 map by Russian fur trader and later founder of the Russian American Company Grigorii Shelikov is based on James Cook's final map (Map 65, *page 27*) but displays a Russian southern boundary reaching the northern American one at about 45°N, completely shutting out the British from the Northwest Coast. The boundary portrays Russian ambitions at the time.

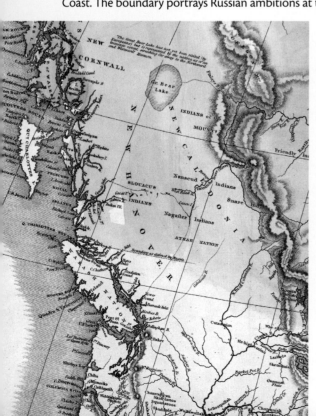

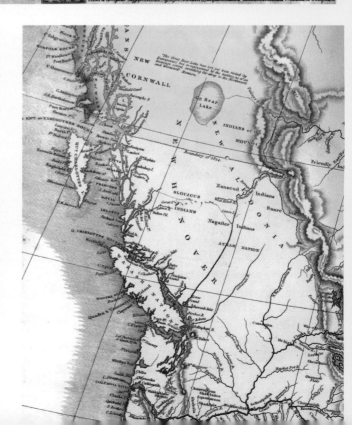

Map 151 (*above*).
This 1821 Russian map was drawn by Vasili Berkh, who reached Sitka in 1804 with Uri Lisianski on the first Russian circumnavigation. The map shows no Russian boundary to its Alaskan possessions, which stretch across North America to Hudson Bay, the latter depicted too far west.

Map 152 (*left*) and **Map 153** (*right*).
These maps, from two editions of the same atlas, published in 1822 and 1825, respectively, by American mapmaker Henry Tanner, attempt to show the claimed Russian boundary at 51°N (Map 152), well into today's British Columbia, and back at 54°40′N (Map 153), as agreed by treaty in 1824 and 1825. The 1821 claim was the catalyst for the Monroe Doctrine, issued in 1823, an American policy that North America was not to be further colonized by European powers.

54°40′N (see page 25). From that point on there were really only two contenders for the Oregon Country: Britain and the United States.

In the lead-up to the negotiations in 1818 the North West Company, trying to protect its fur trade, produced a map with a deliberate falsehood in the form of a phantom river, the Caledonia. This was an extension of the Thompson River south to Puget Sound, where it broke into two streams (Map 154, *right*).

Boundary negotiations between the United States and Britain took place again in 1823 and 1826–27, and the joint occupancy agreement was maintained, although the British offered a boundary following the most northeasterly branch of the Columbia, the Kootenay River (Kootenai in the United States), then known as McGillivray's River, and then following the Columbia to the sea, not that they had any kind of exact idea where that river might be. This boundary and variants of it showed up on maps for many years (Map 160, *overleaf*). Another variant, which appears to have originated

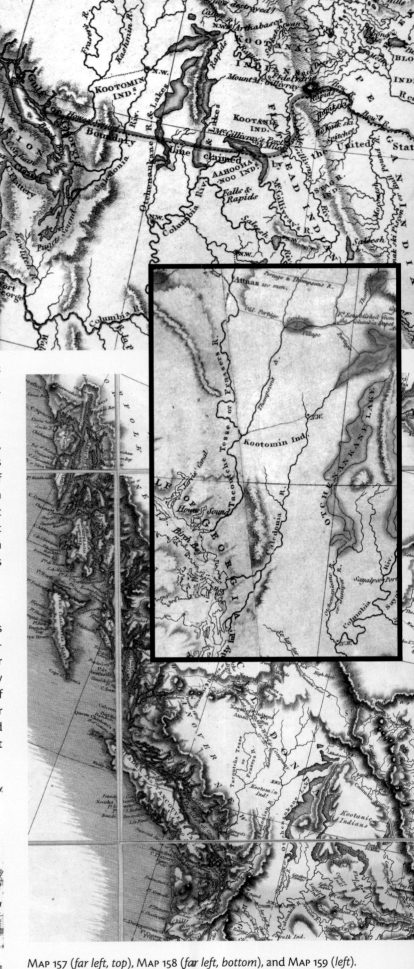

Map 154 (*right, top*).
The North West Company's map of 1817, almost certainly produced by Simon McGillivray, a partner in the firm, showing a fictitious *Caledonia Riv.* emptying into *Pugets Sound.* His intention was to demonstrate how a possible forty-ninth-parallel boundary would cut off trade. The *Boundary Line claimed by the United States* shown here is, however, not drawn on that parallel. The red dot marked *N.W.* is the company's Fort Kamloops. A boundary that would have been drawn south of this river was offered by the American negotiator Albert Gallatin to the British representative, Lord Castlereagh, in 1818; had it been accepted, British Columbia would have stretched south to today's Everett, Washington. Twenty-eight years later Britain would settle for a boundary 130 km farther north.

Map 155 (*right*) and **Map 156** (*right, centre inset*).
True to form, McGillivray's phantom river was promptly copied by other mapmakers. This is not really surprising, because the North West Company was the definitive source for the geography of the Northwest at the time, having considerably more knowledge than any other source. The *Caledonia R.* is shown on these maps, both published in 1823. Map 155 is by James Wyld, while Map 156 was published by Aaron Arrowsmith, both British mapmakers of some repute. Wyld's map is particularly interesting, for it shows a flowing international boundary (the dashed red line) ending precisely where McGillivray wanted it to: south of the mouth of his fictitious river.

Map 157 (*far left, top*), **Map 158** (*far left, bottom*), and **Map 159** (*left*).
Three maps, dated 1815, 1815, and 1819, respectively, show interpretations of the boundary between the United States and British territory, and, on Map 159, with Russian territory. All of these maps were published by Adrien-Hubert Brué, a French commercial mapmaker. On the 1815 maps British territory is shown as far south as the Snake River in the east and the Columbia on the west, while on the 1819 one it is restricted to a relatively narrow band and appears to allocate Vancouver Island to the Americans.

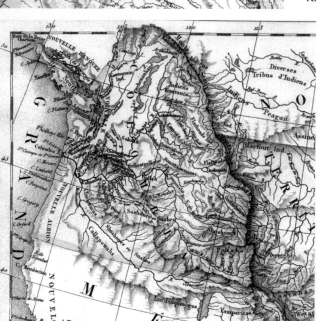

Map 160 (*left*).
Here the proposed international boundary follows the Columbia and the *R. Coohamie or McGilli-vray* (Kootenay/Kootenai River)—insofar as the latter's course was known—before joining the forty-ninth parallel at the crest of the Rockies. The map was published in 1834 by the wonderfully named Society for the Diffusion of Useful Knowledge.

Map 161 (*left*).
The northern boundary of *Columbia*—presumably Oregon—follows a huge arc from the *Fraser R.* to the *Monts Stony* (Rocky Mountains) on this 1825 map from French map-maker Louis Vivien de St. Martin. Note the *Caledonia* River. The large *Lac Otchenankane* is Okanagan Lake.

Map 162 (*above*).
For many years the Oregon Country was depicted on maps like this, typically with American sources such as this one, joined to the United States and shown as an extension of its territory. The Oregon Country covered the entire region west of the Rockies between 54°40´N and 42°N; the latter is today the boundary between Oregon and California and was the demarcation between Spanish and American territory agreed between those two countries in 1819.

Map 163 (*right*).
Much the same configuration is shown on this delightful—but hopelessly out-of-date—map of unknown Dutch origin. British Columbia is here part of *Oregon Distrikt*. The map was drawn about 1838.

Map 164 (*left*).
This influential American map, compiled by U.S. Army topographical engineer Washington Hood and published in 1838, accompanied a bill introduced by Senator Lewis Linn to authorize the American occupation of Oregon by military force. His bill did not pass, but the same effect was achieved by immigration of American settlers to the Willamette Valley, at first a trickle but by 1845 a flood. This map shows what the British, in the form of the Hudson's Bay Company, were trying to avoid at all costs: an international boundary west of the Rocky Mountains following the 49th parallel all the way to the Pacific—including through Vancouver Island.

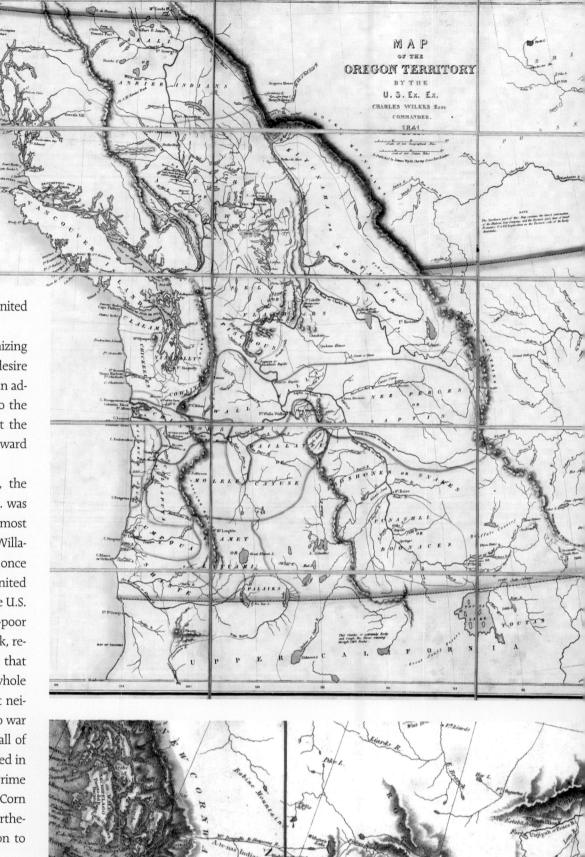

with Hudson's Bay Company director George Simpson, was to drop the forty-ninth-parallel boundary to 46°20´N at the Rockies and then run it to the Snake River, which it would then follow to the Columbia and thence to the Pacific (MAP 157, *page 55*). But American negotiators maintained their contention that the whole of the Oregon Country belonged to the United States, and this claim was reflected in their maps.

During the 1826–27 negotiations Britain, recognizing that the United States' position was partly due to a desire for the deep-water harbours of Puget Sound, offered an additional enclave of territory roughly corresponding to the Olympic Peninsula. This offer was turned down, but the proposal showed up on maps for many years afterward (MAPS 167 and 168, *overleaf*).

Expansionist advocates in the United States, the new cry of "Manifest Destiny"—the idea that the U.S. was destined to cover the whole of North America—and, most importantly, the increasing tide of immigration to the Willamette Valley brought the boundary matter to the fore once more in the 1840s. In addition, the 1841 visit of the United States Exploring Expedition led to a realization that the U.S. needed Puget Sound as a safe harbour on a harbour-poor coast. In 1845 a new expansionist president, James Polk, rescinded the joint occupancy agreement, maintaining that the U.S. had a "clear and unquestionable right" to the whole of the Oregon Country. The stage was set for war, but neither side wanted it: the Americans were about to go to war with Mexico—a war that would result in its gaining all of the modern Southwest—and the British were embroiled in political controversy due to a famine in Ireland and Prime Minister Robert Peel's determination to repeal the Corn Laws, which maintained import duties on corn. Nevertheless, Britain dispatched two military officers to Oregon to report on the American defences.

Peel had no time for Oregon. He instructed his foreign minister, the Earl of Aberdeen, to settle the troublesome Oregon problem. Compromise was reached in a proposal to

Continued on page 60.

MAP 165 (*above, right*).
This is James Wyld's version of the United States Exploring Expedition map of the Oregon Country, published by its leader, Charles Wilkes, in 1844. The expedition visited the region in 1841 and lost a ship trying to cross the sandbar at the mouth of the Columbia River. This led Wilkes to the conclusion that the United States ought to possess Puget Sound, with the only sizeable safe harbours on the West Coast north of San Francisco. His report was enormously influential in cementing American demands for a boundary settlement that included Puget Sound.

MAP 166 (*right*).
The 1845 edition of James Wyld's map (MAP 155, *page 55*) still indicates a British claim south to the mouth of the Columbia by the use of red colouring along the coast. Geographical details are improved, and the Caledonia River has now vanished.

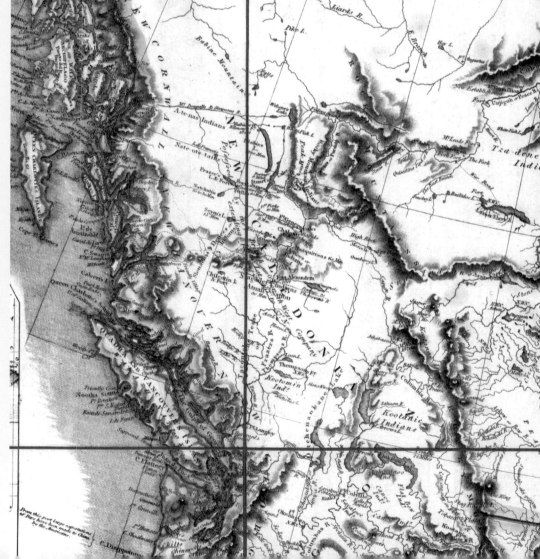

MAP 168 (*above*).
James Wyld drew this map in 1845 to show the history of claims to the Oregon Country, with an accompanying leaflet that was supportive of the British point of view. His British bias leads him to describe the United States' "claims," comparing them with British "rights." The purple line at right is the never-well-defined boundary of Louisiana, inherited by the United States—shown as excluding Oregon, of course. The green line is the *Line Proposed by the Americans in 1824 & 1825*, and the red line is that *Proposed by English in 1826*.

MAP 167 (*above*).
A good map of the various boundary proposals is this map by Eugène Duflot de Mofras in 1844. Proposed boundaries are in red. There is a *Limite proposée par les Anglais en 1826* along the forty-ninth parallel and then down the Columbia, and including the Olympic Peninsula cutoff, and a *Ligne proposée par les Américains en 1824 et 1826* that follows the Upper Columbia to the forty-ninth parallel and then directly to the ocean, right across Vancouver Island. Duflot de Mofras was a trade official with the French legation in Mexico City who travelled the West Coast in 1841 looking for trade opportunities and in the process collected information that resulted in the most accurate map of the West for its time, one copied by many others. Other red lines indicate routes to the West Coast; some depict those of immigrant settlers, but Alexander Mackenzie's track in 1793 and Lewis and Clark's route in 1805 are also shown. The northern boundary line (also in red) is that of the Russia–United States treaty of 1824.

MAP 169 (*below, left*) and MAP 170 (*below, right*).
These maps are two editions of Philadelphia mapmaker Samuel Augustus Mitchell's 1846 *New Map of Oregon, Texas and California*—all disputed areas in 1846. MAP 169 was published while negotiations were ongoing, while MAP 170 was published after the Treaty of Washington had been signed in June. The first shows several boundary settlement possibilities, including the one straight through Vancouver Island (still shown here as *Quadra & Vancouver I.*), while the second shows the boundary as decided—although the status of the San Juan Islands, even on this map by an American—is still noticeably ill-defined. Did he intend the islands to be within or without the red line? The hopelessly inaccurate configuration of southern Vancouver Island shown on this map demonstrates why there was a problem with the San Juan boundary in the first place.

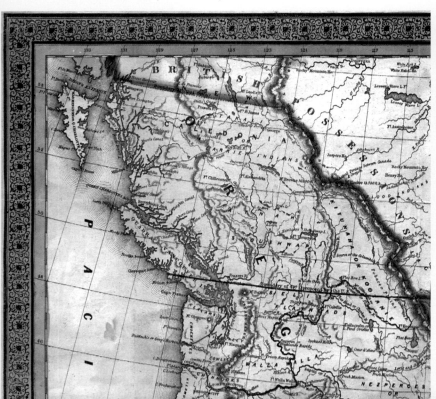

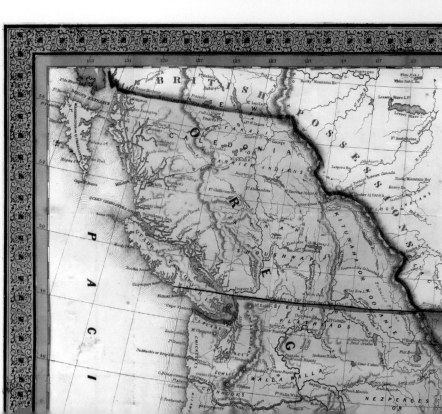

Map 171 (*main map, above*) and **Map 171A** (*inset, right*).

This is the British Foreign Office copy of a commercial map published by W. & A.K. Johnston, dated 1 January 1846, on which there are many manuscript additions. The forty-ninth-parallel boundary line has the notation *Boundary settled by Treaty of 1846*, and its previous extension through Vancouver Island has been erased, leaving a white band in its place. The map as published, part of which is shown as **Map 171A**, *inset*, shows this line as the *Boundary proposed by the United States 1824–26*. The boundary Britain wanted at this time is shown by the red line following the *Columbia R*. The Olympic Peninsula is the *Detached Territory Offered by G*. *Britain 1826*. The locations of the Hudson's Bay Company posts are shown, which someone has underlined in black. The map was based on that of Duflot de Mofras (**Map 167**, *previous page*).

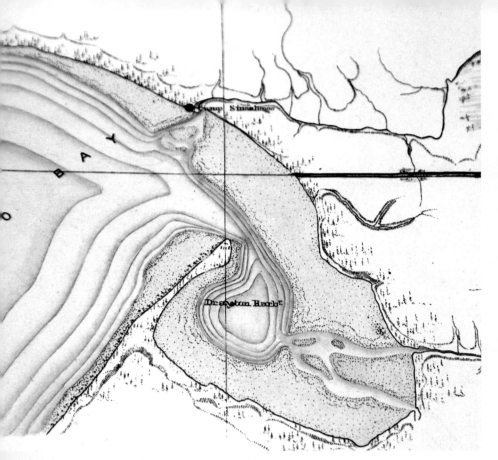

MAP 172 (*above*) and MAP 173 (*below*).
Two parts of a map of the boundary survey dated 1869 but published in 1899. MAP 172 shows the international boundary at Semiahmoo Bay, with *Camp Simiahmoo*, the British surveyors' camp at the mouth of the Campbell River, and MAP 173 shows it on *Point Roberts*, adjacent to today's Tsawwassen.
Inset is the boundary *obelisk* shown on the map; this was placed in 1862. The left side of the obelisk is in British Columbia, the right in Washington State.

extend the forty-ninth-parallel boundary to the sea but include the whole of Vancouver Island in British territory. This was begrudgingly satisfactory to the Hudson's Bay Company, which had feared worse, since it spared its new post on the southern tip of the island, Fort Victoria (see page 52). The agreement was negotiated in Washington, D.C., by British ambassador Richard Pakenham and American foreign secretary James Buchanan, and the Treaty between Her Majesty and the United States of America, for the Settlement of the Oregon Boundary, popularly known in Britain as the Treaty of Washington and in the United States as the Oregon Treaty, was signed on 15 June 1846.

The area that would be British Columbia was now indisputably British, and that should have been the end of the matter, but it was not, largely because of incomplete geographical knowledge of the islands between the mainland and Vancouver Island. The treaty merely stated that the boundary should be in the "middle of the channel," but through the San Juan Islands there were multiple channels. This thorny issue again nearly led to a war but was finally resolved in 1872 by arbitration (see page 101), and the definition of British Columbia was, apart from some adjustments along the border with Alaska (see page 234), complete.

MARKING THE BOUNDARY

There seemed no reason and certainly no rush to actually define the new boundary on the ground. All this changed with the discovery of gold on the sandbars of the Thompson in 1857, and the two countries then agreed that each would field a commission to survey and mark the boundary. Thence began a surveying saga.

The American Boundary Commission, led by Archibald Campbell, began work later that year, and it was joined by the British Boundary Commission—Royal Engineers led by Captain John S. Hawkins—the following summer. The two agreed in August 1858 that the boundary should be marked at all points where trails, rivers, and other permanent natural features crossed it, and this methodology produced a series of maps like MAP 178, *right,* which shows where the boundary crosses the Chilliwack River. The Americans had already located the beginning point for the boundary—on the west side of Point Roberts, where they erected a stone cairn, replaced in 1862 with a stone obelisk (*left*).

MAP 174 (*below*) and MAP 175 (*right, bottom*).
Two sheets showing between them the entire southern boundary of British Columbia as surveyed by the British Boundary Commission and drawn in 1862. Trails are shown in red. Note that the two sheets overlap; Kootenay Lake and the *Salmon River* are shown on both maps.

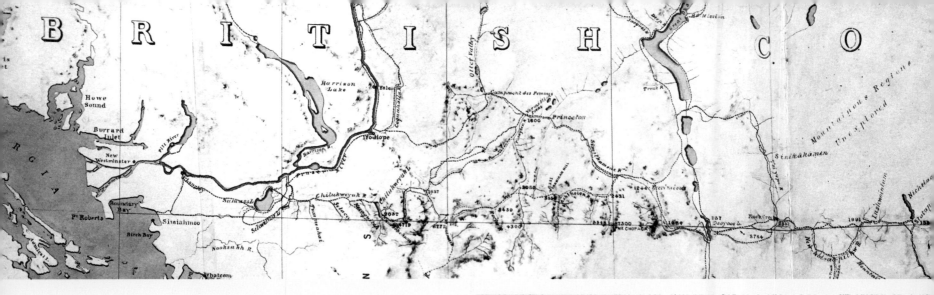

The boundary survey was completed in 1862, but the Americans refused to issue a joint report, and so both countries each produced their own. They did get together to produce final maps, however, only to find there were some discrepancies between the position of the boundary marked by each side. This problem was solved by simply splitting the difference, but, since these boundaries were also marked on the ground, this led to further issues in more settled areas. And then—unbelievably—the original reports of *both* countries went missing. The British one eventually turned up at the Greenwich Observatory in England and was finally published—in 1899. It was dated 7 May 1869 and was signed by both Hawkins and Campbell. A part of the American report was found a year later.

Both governments now decided to resurvey the boundary, a task carried out between 1901 and 1907. Yet more issues emerged. The boundary line in some places was not on the forty-ninth parallel but, as in the case of the Blaine area map shown here (Map 177, *right*), about 300 m north of it. It took another Treaty of Washington, signed in 1908, for both sides to accept the boundary as surveyed, even where it was inaccurate.

MAP 176 (*above, top*).
This was the map published by the Boundary Commission in 1869 showing the line of the boundary across the western half of British Columbia.

MAP 177 (*right, centre top*).
Part of the map of the revised survey of 1903–07, published in 1913. The positions of smaller marker obelisks are noted; the photos (*inset*) show these markers along 0 Avenue in Surrey (*left* to *right*): marker *6, 7,* and *8,* as shown on the map. The plaque on *7* is *inset* on that photo. Note the boundary is north of the forty-ninth parallel (the thin black line close to the boundary line).

MAP 178 (*right*).
The international boundary at *Chiloweyuck Lake* (Chilliwack Lake). This is from the atlas of maps dated 1869 but lost and not published until 1899. Note the boundary surveyors' *Camp Chiloweyuck.*

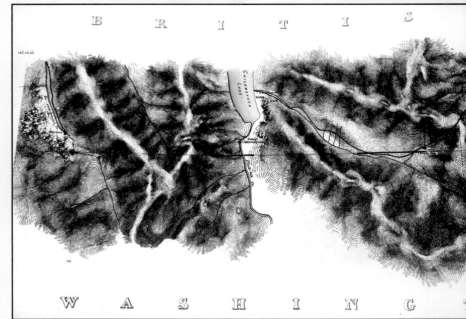

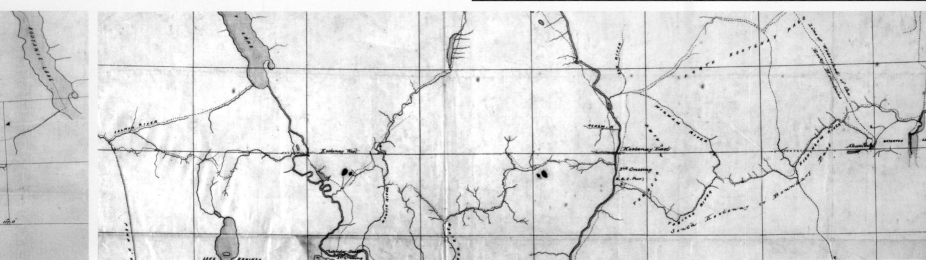

THE GOLDEN COLONY

Gold created British Columbia. It was the reason for the mainland colony and for the beginning of a process that dragged it into the twentieth century. Before gold, geographical knowledge of British Columbia was sparse and spotty; after it, roads were built, population levels surged, and suddenly the world knew of British Columbia's existence.

There were gold rushes before—to the Queen Charlotte Islands (Haida Gwaii) in 1850—and there would be many others after—to the Cariboo, Omineca, East Kootenay, Big Bend, Vancouver Island, and more—but none had the lasting impact of the Fraser River Gold Rush of 1858. Its impact is all the more amazing since the rush itself was of such limited duration.

The Hudson's Bay Company had traded gold retrieved from the Fraser or Thompson in 1856, and in 1857 American miners had migrated north in search of gold streams and had found some in the Kamloops area. That year the company shipped a substantial amount of gold to the mint in San Francisco, thus revealing the secret to the world. The rush was on. Gold seekers, mainly Americans, flooded in, first to Victoria, then to the mainland, ascending the Fraser by whatever means they could find. By April 1858, James Douglas reported seventy or eighty, the following month a thousand more, and by July reported that "this country and Fraser's River have gained an increase of 10,000 inhabitants within the last six weeks." Indeed, before the end of the year upwards of 25,000 had joined the rush for gold.

The massive influx of Americans concerned the British government, which was not about to see British Columbia fall to the United States the way Oregon had because of settlers to the Willamette Valley in the 1840s, where occupation had proved to be the harbinger of sovereignty. Resources, therefore, were quickly marshalled to ensure a firm British grip was sustained.

Above. One of a number of books hurried into print to guide gold seekers was this one by William Hazlitt. Portions of the map referred to on the cover are Map 10, *page 11,* and Map 187, *page 67.*

Map 179 (*below*).
One of the first maps to show British Columbia and its gold fields, published by Kinahan Cornwallis, one of a handful of authors who had actually visited the region. It appeared in his 1858 book *The New El Dorado. Gold* is also noted on *Queen Charlottes Isl^ds.,* as is *coal* on *Vancouver Island.*

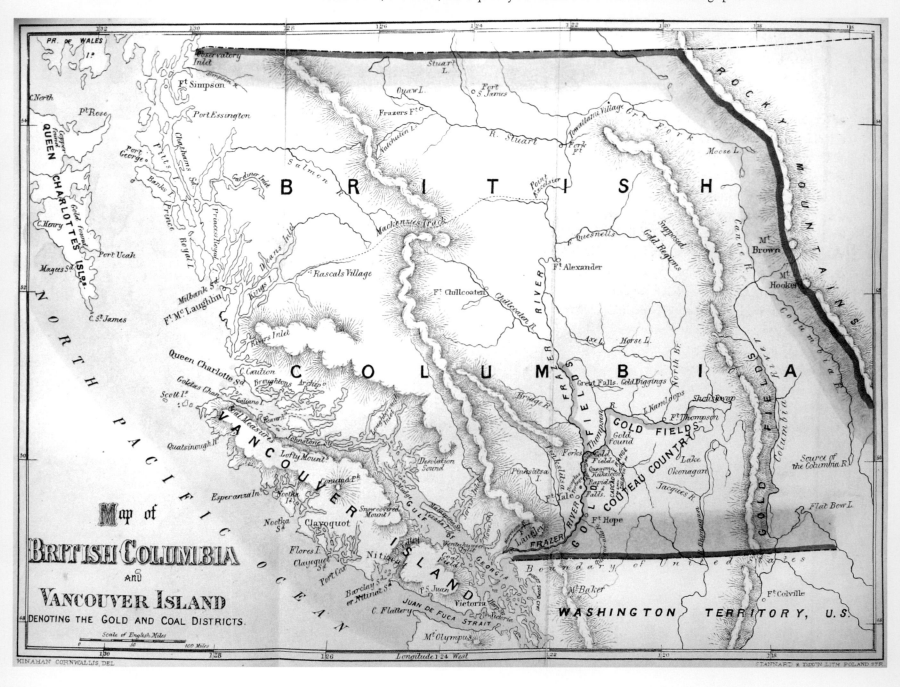

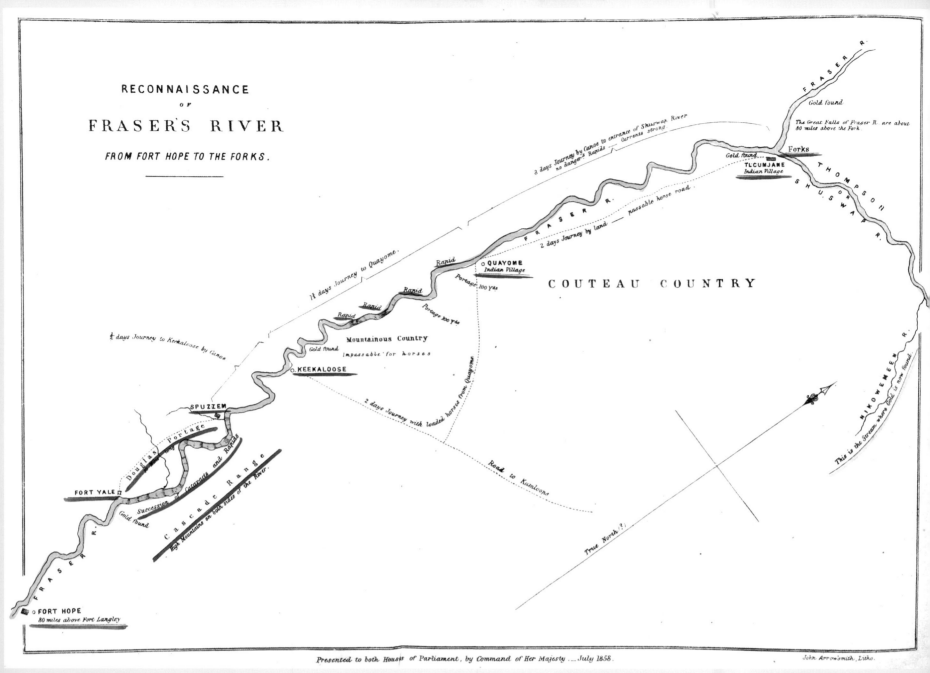

RECONNAISSANCE
OF
FRASER'S RIVER

FROM FORT HOPE TO THE FORKS.

Presented to both Houses of Parliament, by Command of Her Majesty .— July 1858.

John Arrowsmith, Litho.

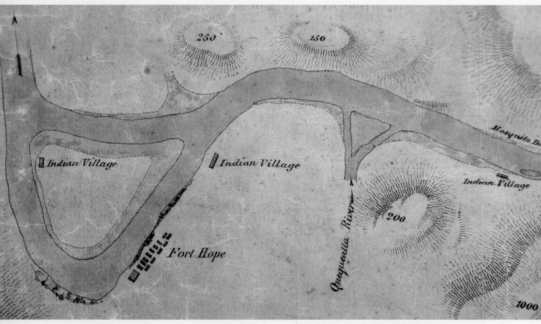

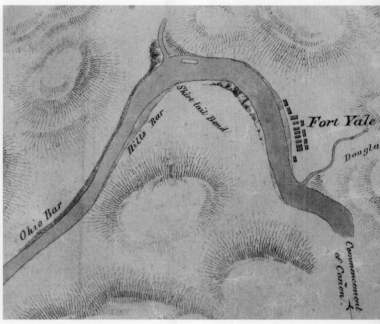

The government's agent on the ground, of course, was the redoubtable James Douglas, chief factor of the Columbia District of the Hudson's Bay Company and governor of the Colony of Vancouver Island, who initially exceeded his legal powers to ensure that order—British order—prevailed on the mainland.

In August 1858 the British parliament passed *An Act to Provide for the Government of British Columbia*—a name decided upon by Queen Victoria herself. And on 19 November 1858 Douglas presided over a

MAP 180 (*above, top*).
Douglas sent a steady stream of "despatches" to the British government in 1858, reporting on the discovery of gold on the mainland. Enclosed with one dated 6 April was this map showing, in a barebones fashion, the location of gold finds in the Fraser Canyon between *Fort Hope* and *Tlcumjame*, at the location of the later townsite of Lytton. Engraved by mapmaker John Arrowsmith, the map was presented to the British parliament in July 1858. The Aboriginal peoples of the area above the canyon referred to themselves at this time as "Nicoutameen," which fur traders corrupted to *couteau*, French for knife. Hence the term *Couteau Country* on the map.

MAP 181 (*above, in two parts*).
Details of *Fort Hope* and *Fort Yale* and *Indian Villages* on an 1858 map. The *Quequealla River* is the Coquihalla.

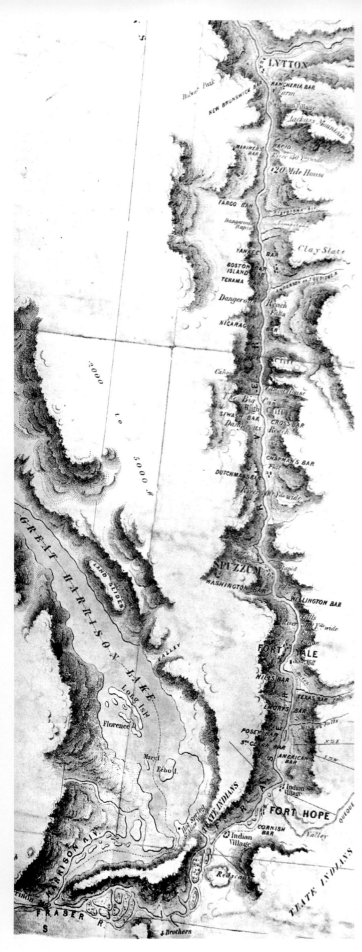

MAP 182 (*above*).
A detailed map of the gold workings on the *Fraser R.* between *Fort Hope* and *Lytton,* surveyed and drawn by naval officer Lieutenant Richard C. Mayne.

MAP 183 (*right*).
Possibly one of the finest maps ever to depict British Columbia, this splendid bird's-eye view was rushed into print in September 1858 to illustrate the gold rush to the British public. The *Frazer River* is dominant here. Alas, the map is woefully inaccurate in many details, including the shape of Vancouver Island, the islands in the *Gulf of Georgia, Pᵗ Roberts, Fᵗ. Yale* (shown on an extensive plain), and even an apparently vast *Victoria Harbour.* But it did not matter, for the map's potential audience likely had no notion of what the real British Columbia looked like anyway.

A COMPLETE VIEW OF THE NEWLY DISCOVERED GOLD F

ROCKY MOUNTAINS

DE FRANÇOIS
STUARTS L.
FT. ST JAMES

QUAW LAKE
FT FRAZER

GREAT FORK

MT BROWN

PRINE PORTAGE
BETWEEN R. FRAZER & OREGON

MT HOOKER

FRAZERS BLUFF

MACKENZIES TRACK

MT STEPHENS

FORK T

GOLD REGIONS

GREGANS BLUFF

CARRIER INDIANS

FT. EXCELSIOR

GOLD FIELDS

TISH COLUMBIA

GOLD WASHINGS

R. QUESNELLS

FT. CHILCOATEN

FT. ALEXANDER

GOLD COUNTRY

R. CHILCOATEN

FRAZER R.

HORSE L.

NORTH R.

KAMLOOPS L.

THOMPSON R.

FRAZER R.

TITCUMJANE

COUTEAU COUNTRY

RAPIDS

FT. YALE

GOLD
DIGGINGS & WASH.

FRAZER RIVER

LANGLEY

BOUNDRY

BOUNDRY OF THE UNITED STATES

BOUNDRY

PUGET SOUND

WASHINGTON TERRITORY

MT BAKER

UNITED STATES

GEORGIA

FLATHEAD INDIANS

ESQUIMALT HARBOR

BELLINGHAM BAY

CAPE FLATTERY

JAUN DE FUCA STRAIT

R. F. CALIFORNIA & OREGON

London. Pub.d Sep.t 13th 1858. by Read & Co. 10 Johnsons C.t Fleet St.

READ & C.o IMP

HE NEW EL DORADO.

N BRITISH NORTH AMERICA, WITH VANCOUVER ISLAND AND THE WHOLE OF THE SEA-BORD FROM CAPE FLATTERY TO PRINCE OF WALES ISLAND.

Compiled from Authentic Views Plans & Charts in the possession of
LIEU.T CHARLES BARWELL. UNITED STATES NAVY.

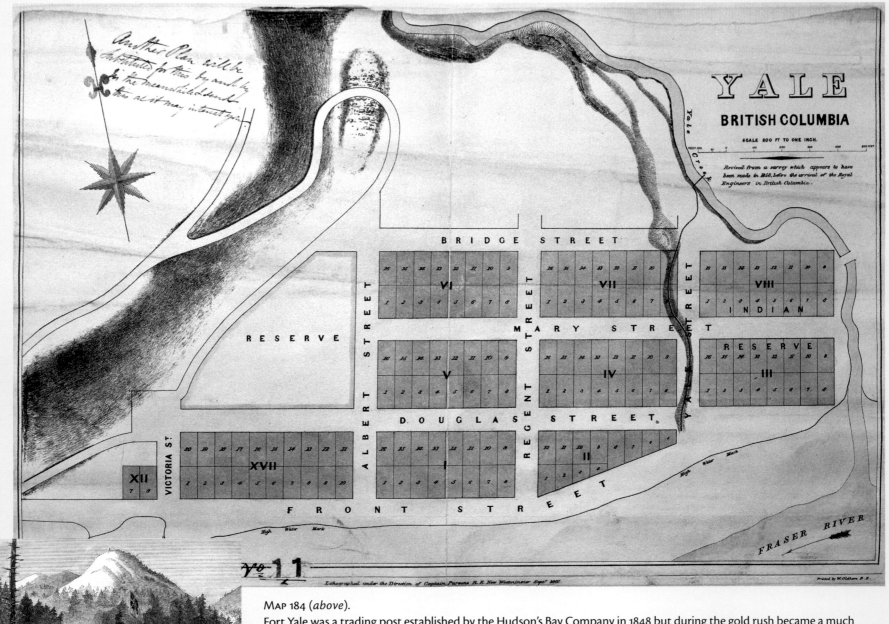

MAP 184 (above).
Fort Yale was a trading post established by the Hudson's Bay Company in 1848 but during the gold rush became a much larger settlement owing to its position at the head of steamboat navigation on the Fraser. In 1859 the Royal Engineers laid out a formal townsite, shown here. See also MAP 220, *page 81.* The engraving, *left,* was published in *Harper's Weekly* in 1858 and shows gold seekers camped along the riverbank at Fort Yale. The artist clearly had not been to Yale, for the fort was in reality but a single log structure, with no stockade.

ceremony at Fort Langley proclaiming the creation of the mainland Colony of British Columbia—still distinct from Vancouver Island—with himself as first governor. The British government had decided—probably quite correctly—that Douglas, who had reigned over New Caledonia, in a manner of speaking—was the best person to take the reins of its new colony. He was, however, required to resign from the Hudson's Bay Company. This was to be a British colony in a way distinct even from the Colony of Vancouver Island, not a company fiefdom.

The new British regard for its northwestern possessions engendered one overridingly important decision, made in this case by colonial secretary Edward Bulwer-Lytton (famous for his novel that began, "It was a dark and stormy night" and the quotation "The pen is mightier than the sword"). He anticipated a request for soldiers by Douglas and sent the Royal Engineers to survey land and build roads and bridges. It was their collective work in creating a transportation network infrastructure that began what might be referred to as the opening up of British Columbia (see page 70). A no doubt grateful Douglas renamed a company post, Fort Dallas, after him (MAP 139, *page 51*).

MAP 185 (right).
In contrast to the fancier maps produced by commercial mapmakers is this miner's map, crudely drawn in C.O. Phillips's notebook in November 1858. There is more information on the Harrison Lake–Lillooet route than on that through the Fraser Canyon. A trail to *Ft. Langley* from Semiahmoo Bay and the *Whatcom Trail* are shown; the latter was sometimes used as a back door by American miners trying to get to the gold fields without paying $5 for a licence.

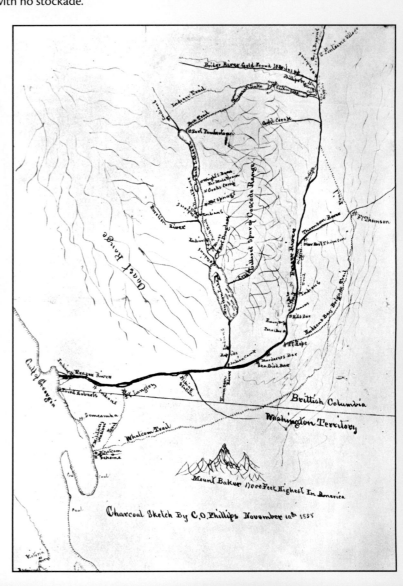

MAP 186 (above).

Reasonably enough, publishers of maps of gold fields liked to use yellow to mark the areas where gold could be found. Gaps in the supposed gold fields would have been hard to note accurately, and the profusion of yellow helped give the illusion that gold was everywhere—so buy the map to find it! This effort is by James Wyld, published in 1858. Note that the boundary of British territory is drawn to include the San Juan Islands. *Vancouver Island* is separate from *New Caledonia*, the latter soon to be British Columbia. Wyld published a new edition of this map later in the year in which the words *British Columbia* were superimposed over *New Caledonia*.

MAP 187 (left).

Gold Found is repeated many times on this part of the map accompanying William Hazlitt's 1858 guide to the gold fields, illustrated on page 62.

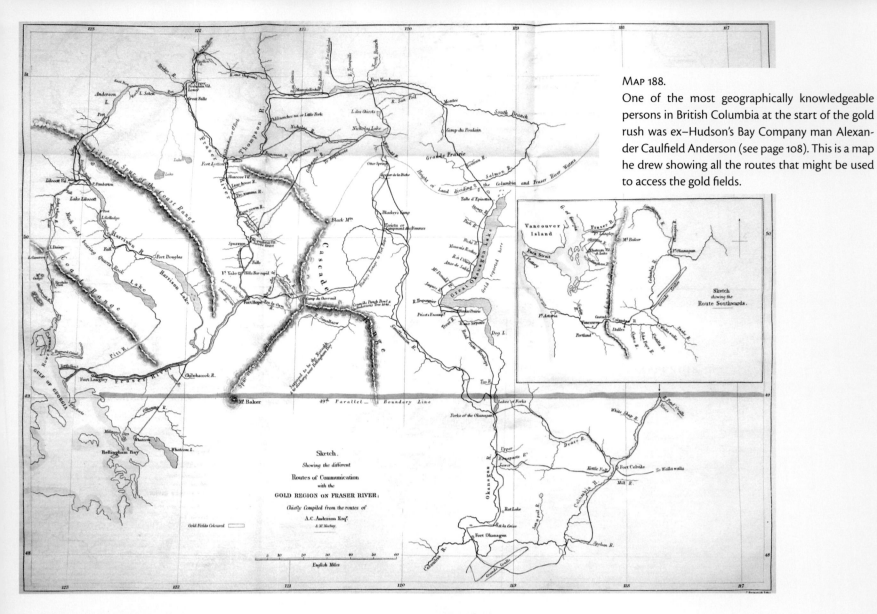

Map 188.

One of the most geographically knowledgeable persons in British Columbia at the start of the gold rush was ex–Hudson's Bay Company man Alexander Caulfield Anderson (see page 108). This is a map he drew showing all the routes that might be used to access the gold fields.

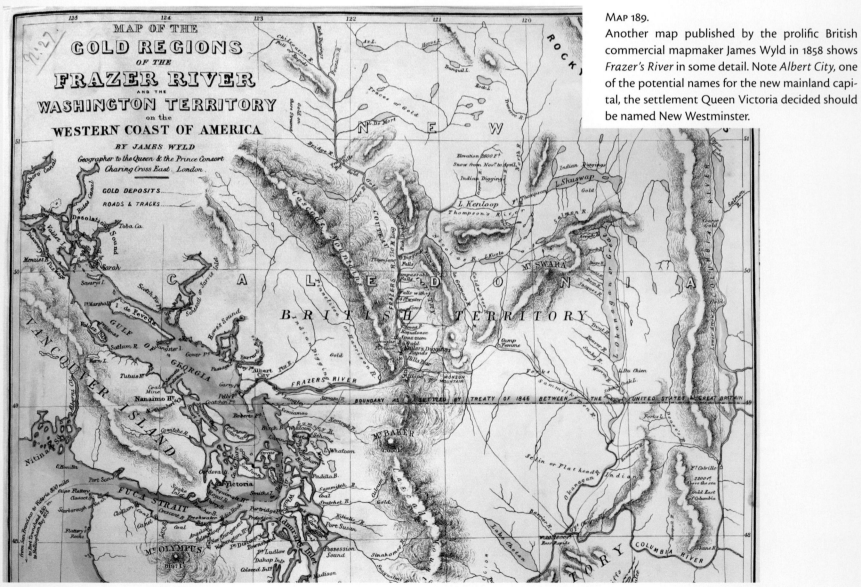

Map 189.

Another map published by the prolific British commercial mapmaker James Wyld in 1858 shows *Frazer's River* in some detail. Note *Albert City*, one of the potential names for the new mainland capital, the settlement Queen Victoria decided should be named New Westminster.

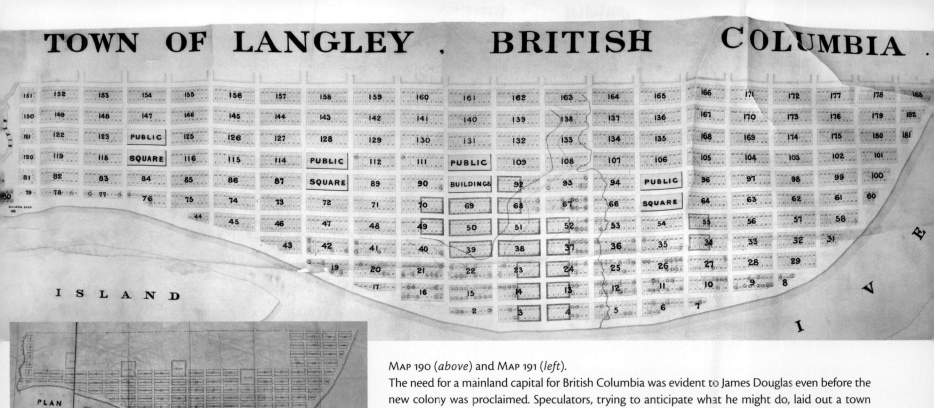

PLAN
TOWN
LANGLEY

Map 190 (*above*) and Map 191 (*left*).

The need for a mainland capital for British Columbia was evident to James Douglas even before the new colony was proclaimed. Speculators, trying to anticipate what he might do, laid out a town on the Fraser just below Fort Langley. Douglas was not amused but liked the site and so ordered Joseph Despard Pemberton, the Vancouver Island colonial surveyor, to continue to draw up plans for the new town. The result was the town of Langley. Map 190 is a very large map, about 5 m long, used to sell lots to Victoria investors. Map 191 is a slightly different version. On 25 November 1858 nearly four hundred lots were sold, all over a reserve price of $100 each, with 10 per cent paid immediately. Yet two days before the sale, Captain John Grant, with the first contingent of Royal Engineers to arrive from Britain, warned Douglas that the site was poor because, being on the south side of the river and also on flat land, it was indefensible against a possible American attack. This view was endorsed by the Royal Engineers commanding officer Colonel Richard Moody when he arrived on 25 December. Early in 1859 new surveys were made to find a better site; Mary Hill (now in Port Coquitlam) was considered, but today's site of New Westminster chosen. Douglas cleverly quieted the clamouring of the Victoria investors by arranging for amounts paid on Langley lots to be transferred to New Westminster sales.

Map 192 (*above, centre*).

German geographer Augustus Petermann published this illustrated map of the Fraser up to *Fort Yale* in his magazine *Petermanns Geographische Mitteilungen* in 1859.

Map 193 (*left*).

Alfred Waddington, later a proponent of a road or railway to connect British Columbia to eastern Canada, drew up this colourful map in 1858 relating the geology, poorly known at that time, to the gold fields. Judging by his map, gold "formations" seem to cover half of British Columbia as well as much of Washington Territory. Note *Albert Town*, soon to be New Westminster, is also shown on this map.

69

THE ROYAL ENGINEERS

The elite Royal Engineers of the British Army were sent to British Columbia to ensure it remained British but also to build an elementary infrastructure to allow the new colony to function. They forged trails, surveyed and built roads, surveyed and subdivided land, and laid out townsites—the latter two so that land might be sold or pre-empted—encouraging permanent settlement and creating a revenue stream for the cash-strapped colonial government. The force, wrote colonial secretary Edward Bulwer-Lytton to James Douglas, was "sent for scientific and practical purposes and not solely for military objects." A considerable number of surveys and maps are one of the legacies of the Royal Engineers.

After many years of neglect and indifference to its western North American possessions, Britain, with a new government under Lord Derby, suddenly had awoken to its value.

The Columbia Detachment of Royal Engineers arrived in British Columbia in four groups between 29 October 1858 and 27 June 1859, with their commander, Colonel Richard Clement Moody, arriving on Christmas Day 1858. They stayed until 1863, and even then, many took discharges offered to them with grants of land that enabled them to stay in British Columbia, forming an invaluable backbone of both settlers and surveyors.

Moody first had to assess what needed to be done, and his rough survey of the area around the mouth of the Fraser (MAP 194, *below*) shows he quickly determined that the capital should be moved from *Langley Town*, south of the river, to the site of New Westminster. *High ground* to the east, Mary Hill, was also considered as a potential site. He was far-sighted, as evidenced by the location of a *probable line of railway* close to where the Canadian Pacific would arrive twenty-six years later.

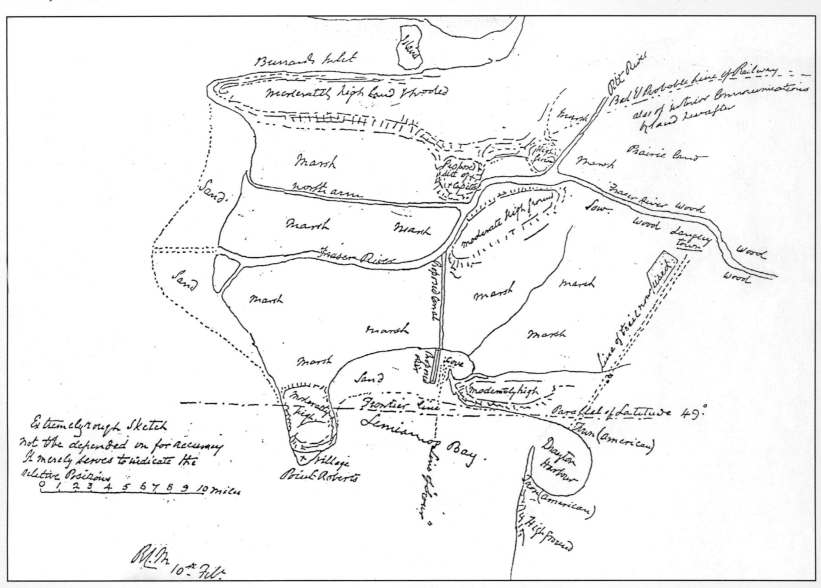

MAP 194 (*above*).
Colonel Moody's initial reconnaissance sketch of the Lower Mainland dated 10 February (1859) showing the *Proposed site of the Capital* and a *Proposed Canal* from the Fraser to Boundary Bay, an idea that was intended to bypass the treacherous channels at the river's mouth.

MAP 195 (*right*).
A survey of Sea Island noting the vegetation, part of a survey commissioned by Moody for civilian surveyor Joseph Trutch and carried out in 1859. The map is from his field notebook. Trutch began by surveying the Coast Meridian (now along 160th Street in Surrey) north from the intersection of the international boundary with Semiahmoo Bay and then subdividing what is now Richmond, North Surrey, and an area around the Pitt River.

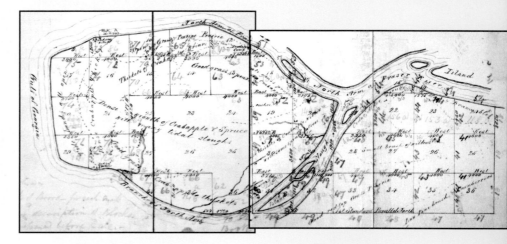

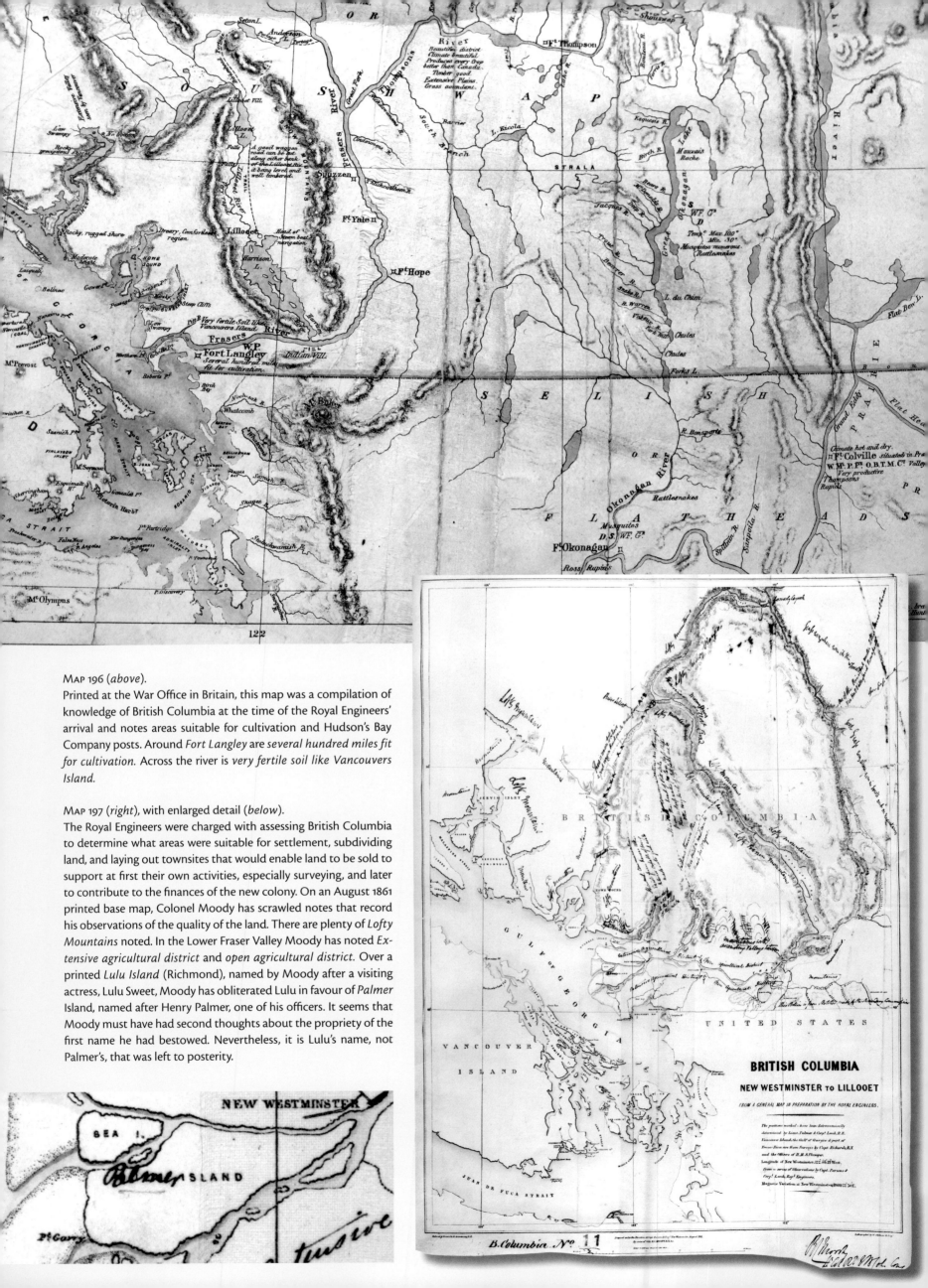

MAP 196 (*above*).

Printed at the War Office in Britain, this map was a compilation of knowledge of British Columbia at the time of the Royal Engineers' arrival and notes areas suitable for cultivation and Hudson's Bay Company posts. Around *Fort Langley* are *several hundred miles fit for cultivation*. Across the river is *very fertile soil like Vancouvers Island*.

MAP 197 (*right*), with enlarged detail (*below*).

The Royal Engineers were charged with assessing British Columbia to determine what areas were suitable for settlement, subdividing land, and laying out townsites that would enable land to be sold to support at first their own activities, especially surveying, and later to contribute to the finances of the new colony. On an August 1861 printed base map, Colonel Moody has scrawled notes that record his observations of the quality of the land. There are plenty of *Lofty Mountains* noted. In the Lower Fraser Valley Moody has noted *Extensive agricultural district* and *open agricultural district*. Over a printed *Lulu Island* (Richmond), named by Moody after a visiting actress, Lulu Sweet, Moody has obliterated Lulu in favour of *Palmer* Island, named after Henry Palmer, one of his officers. It seems that Moody must have had second thoughts about the propriety of the first name he had bestowed. Nevertheless, it is Lulu's name, not Palmer's, that was left to posterity.

BRITISH COLUMBIA

NEW WESTMINSTER TO LILLOOET

FROM A GENERAL MAP IN PREPARATION BY THE ROYAL ENGINEERS.

B. Columbia No. 11

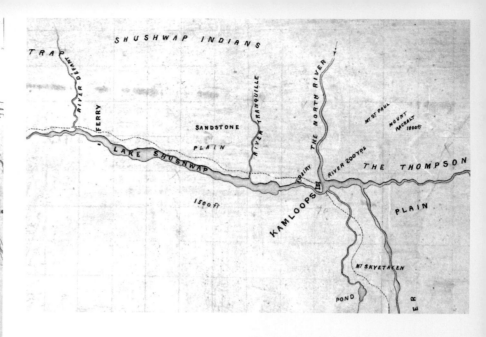

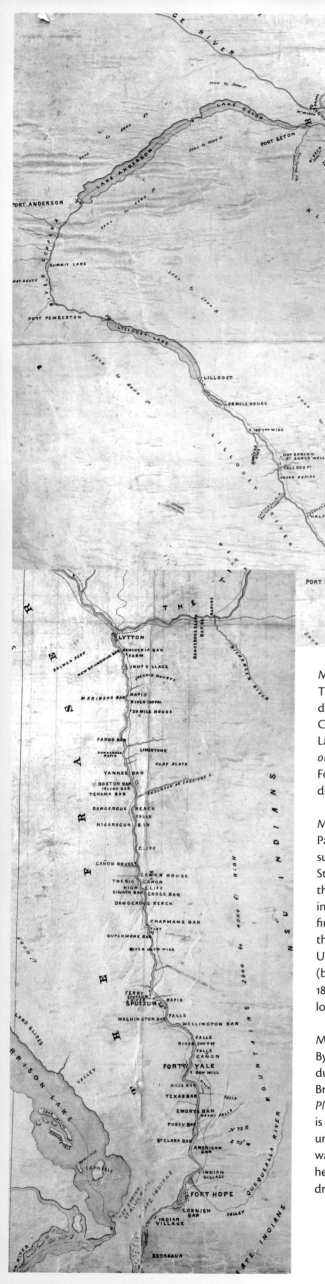

The Royal Engineers and British navy were already engaged in surveying the international boundary (see page 60), and the ships took on the task of hydrographic mapping to find and map safe harbours and channels for the considerably increased volume of shipping that was now arriving in the colonies. HMS *Plumper,* a surveying ship commanded by Captain George Henry Richards, was used to draw a number of important hydrographic charts in this period. One of its officers, Lieutenant Richard Charles Mayne (after whom Mayne Island is named) was seconded to interior exploration and mapping, about which he wrote a book, *Four Years in British Columbia and Vancouver Island,* now a significant source of historical information. Maps drawn up by the men of the *Plumper* were later used in the negotiations to settle the international boundary line through the San Juan Islands (see page 101).

MAP 198 (*above* and *left*).
Three details of a large map produced by naval officer Richard C. Mayne: the route via Harrison Lake to the *Fraser River* at *Cayoush* (Lillooet), the area around Fort *Kamloops,* and the gold diggings of the Fraser Canyon.

MAP 199 (*right, centre*).
Part of a large hydrographic survey map of the southern Strait of Georgia published by the United States Coast Survey in 1858. It shows *Active Passage,* first named on this map after the American surveying ship USS *Active,* which steamed by (but not through) the pass in 1855 en route to Nanaimo to load coal.

MAP 200 (*right*).
By contrast is this map produced by the officers of the British surveying ship HMS *Plumper* in 1858; Active Pass is named *Plumper Pass,* but, unusually, the American name was the one that prevailed here. This is a preliminary hand-drawn sketch.

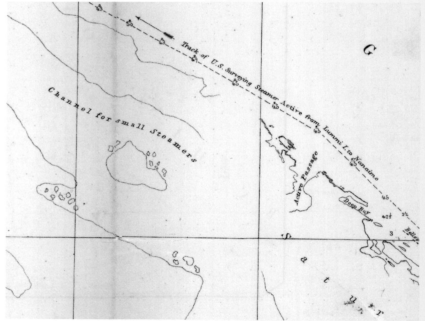

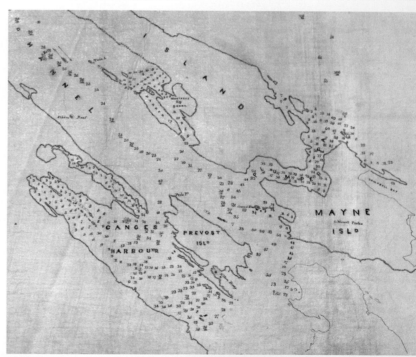

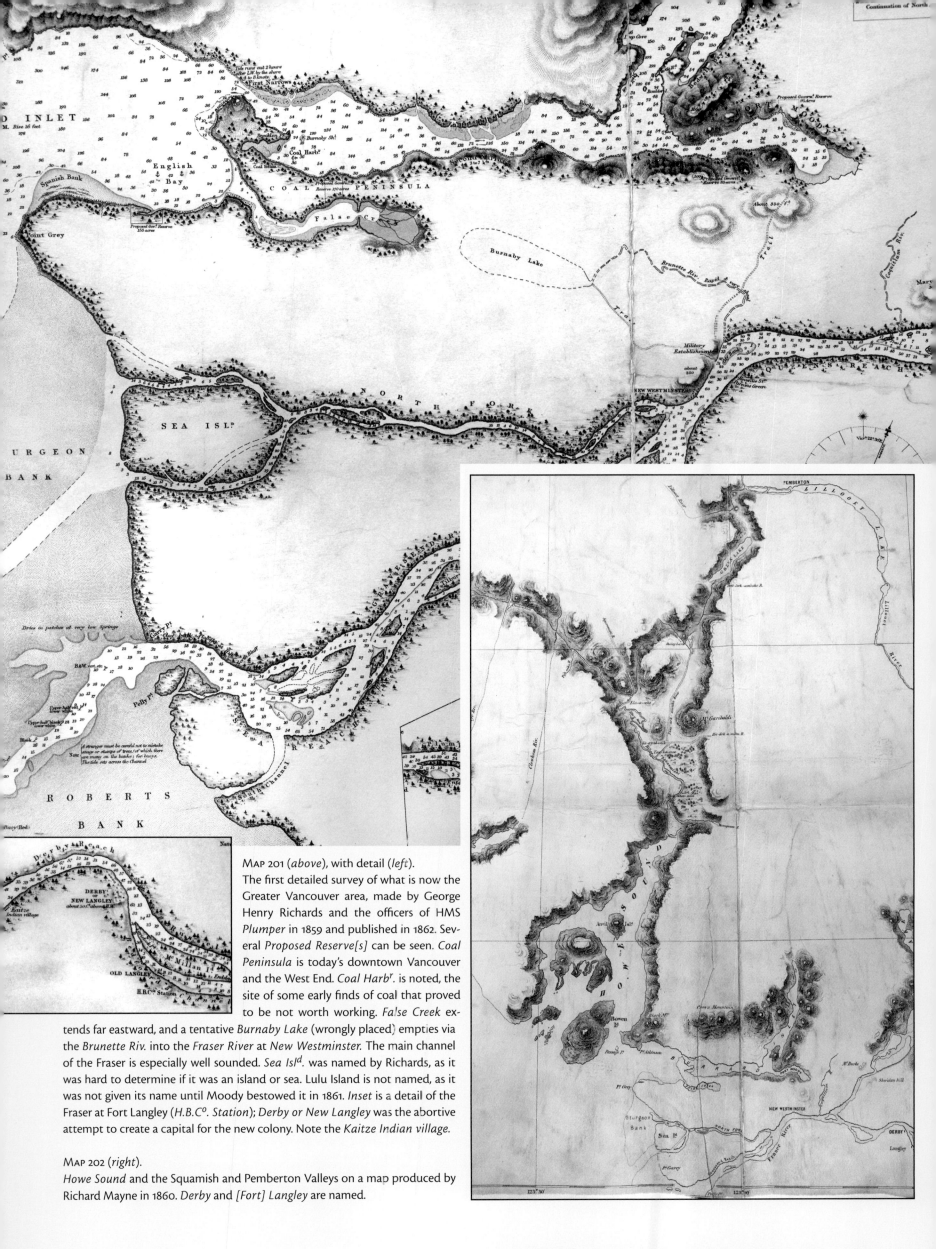

Map 201 (*above*), with detail (*left*).
The first detailed survey of what is now the Greater Vancouver area, made by George Henry Richards and the officers of HMS *Plumper* in 1859 and published in 1862. Several *Proposed Reserve[s]* can be seen. *Coal Peninsula* is today's downtown Vancouver and the West End. *Coal Harb*[r]. is noted, the site of some early finds of coal that proved to be not worth working. *False Creek* extends far eastward, and a tentative *Burnaby Lake* (wrongly placed) empties via the *Brunette Riv.* into the *Fraser River* at *New Westminster*. The main channel of the Fraser is especially well sounded. *Sea Isl*[d]. was named by Richards, as it was hard to determine if it was an island or sea. Lulu Island is not named, as it was not given its name until Moody bestowed it in 1861. *Inset* is a detail of the Fraser at Fort Langley (*H.B.C*[o]. *Station*); *Derby* or *New Langley* was the abortive attempt to create a capital for the new colony. Note the *Kaitze Indian village*.

Map 202 (*right*).
Howe Sound and the Squamish and Pemberton Valleys on a map produced by Richard Mayne in 1860. *Derby* and *[Fort] Langley* are named.

The Royal Engineers were occasionally called on to assist in maintaining order. In July 1858, boundary survey engineers assisted Douglas in quelling a riot in Victoria, and in August 1858 a group of Royal Engineers and marines formed a bodyguard for Douglas when he made a tour of the gold fields. Then there was "Ned McGowan's War," in December 1858 and January 1859, when two groups of American miners at Yale and Hill's Bar, antagonistic to each other, came to blows. A man who had been jailed was set free by the opposing group, and Edward McGowan, after whom the fracas was named, was seen sailing up the Fraser flying the American flag, a scene guaranteed to give Governor Douglas cause for pause. Douglas sent about twenty-six Royal Engineers and marines to Yale along

with newly arrived judge Matthew Begbie, while more marines were sent to Fort Langley to guard against possible intervention by the American boundary surveyors. In the end, Begbie's diplomatic skills were brought to bear, and the situation was resolved without a shot being fired.

MAP 203 (*below, left*).
The beginnings of the Cariboo Road (see page 90) are visible on this map of the roads (solid red line) and trails (dashed red line) of the Thompson Valley drawn in 1862. *Savonas Ferry* crosses the river at the western end of *Lake Kamloops*. The pencil notation notes a "fine flat spoken of by [Royal Engineer] Mr. [George] Turner" and notes that "Francois Savenna" (Francis Saveneux) pre-empted land here in August 1862. Saveneux had established his ferry here in 1859, and the location became the place where stagecoach riders transferred to lake steamboats to continue eastward. *[Fort] Kamloops* is at right, on the north side of the river.

MAP 204 (*below*).
This map shows completed and proposed roads in 1862 or 1863. Alfred Waddington's road from *Bute Inlet* (see page 79) and Henry Palmer's survey from *Bentinck Arm* (see page 78) are shown as proposed roads. North of Harrison Lake it can be seen how roads were constructed to portage from one lake to the next.

MAP 205 (*right, in two parts*).
Details of two of these portage roads, between *Lake Lillooet* and *Lake Anderson* (top) and between *Lake Harrison* and *Lake Lillooet* (bottom).

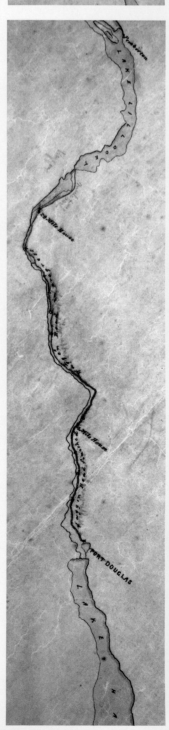

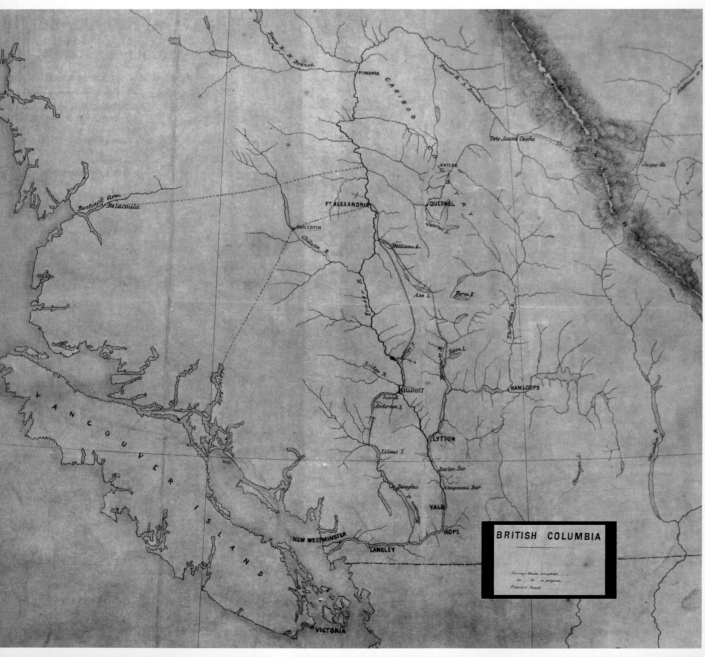

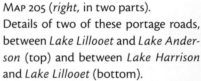

Gold Field Justice

Matthew Baillie Begbie was a judge with a sense of geography. Travelling all over the new colony to bring justice to transgressor and victim alike, he made maps of his journeys that were so useful the Royal Engineers incorporated his information into their own maps.

Begbie, a London lawyer of independent mind, had been selected by the British government in August 1858 to become the chief justice of the new Colony of British Columbia. He arrived in Victoria on 16 November and a few days later travelled with James Douglas to Fort Langley, where Douglas proclaimed the new colony and was sworn into office by Begbie, who in turn was sworn in by Douglas.

Begbie became a trusted advisor to Douglas and was responsible for much of the colony's legislation passed before 1866, including the 1860 Pre-emption Act.

It was not long before Begbie realized that he would have to take the law to the miners, for they were unlikely to come to him. In January 1859 he helped Douglas resolve a miners' dispute and the following month walked from New Westminster

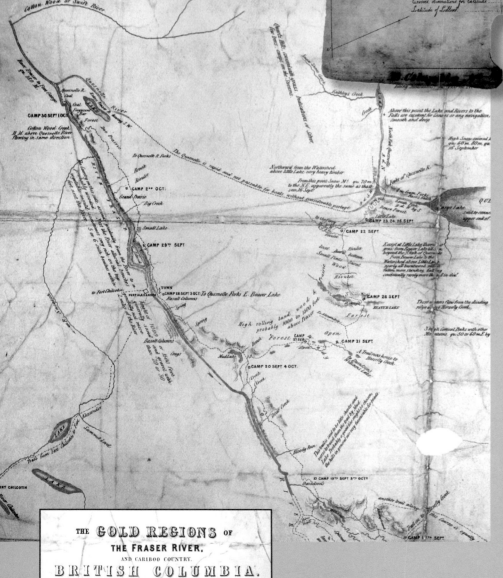

to Lillooet and back to familiarize himself with the country. He took to riding his horse "on circuit" to virtually every part of the colony where there was settlement, often setting up his court in a tent and by the power of his personality brought what has been called "the majesty of British justice" to a young and rough country. He was known to be fair to Aboriginal people, who called him "Big Chief."

Begbie went on to become chief justice of the united colony of British Columbia in 1870 and the province in 1871. When he died in 1894 Victorians gave him a massive funeral second only to that given Douglas himself.

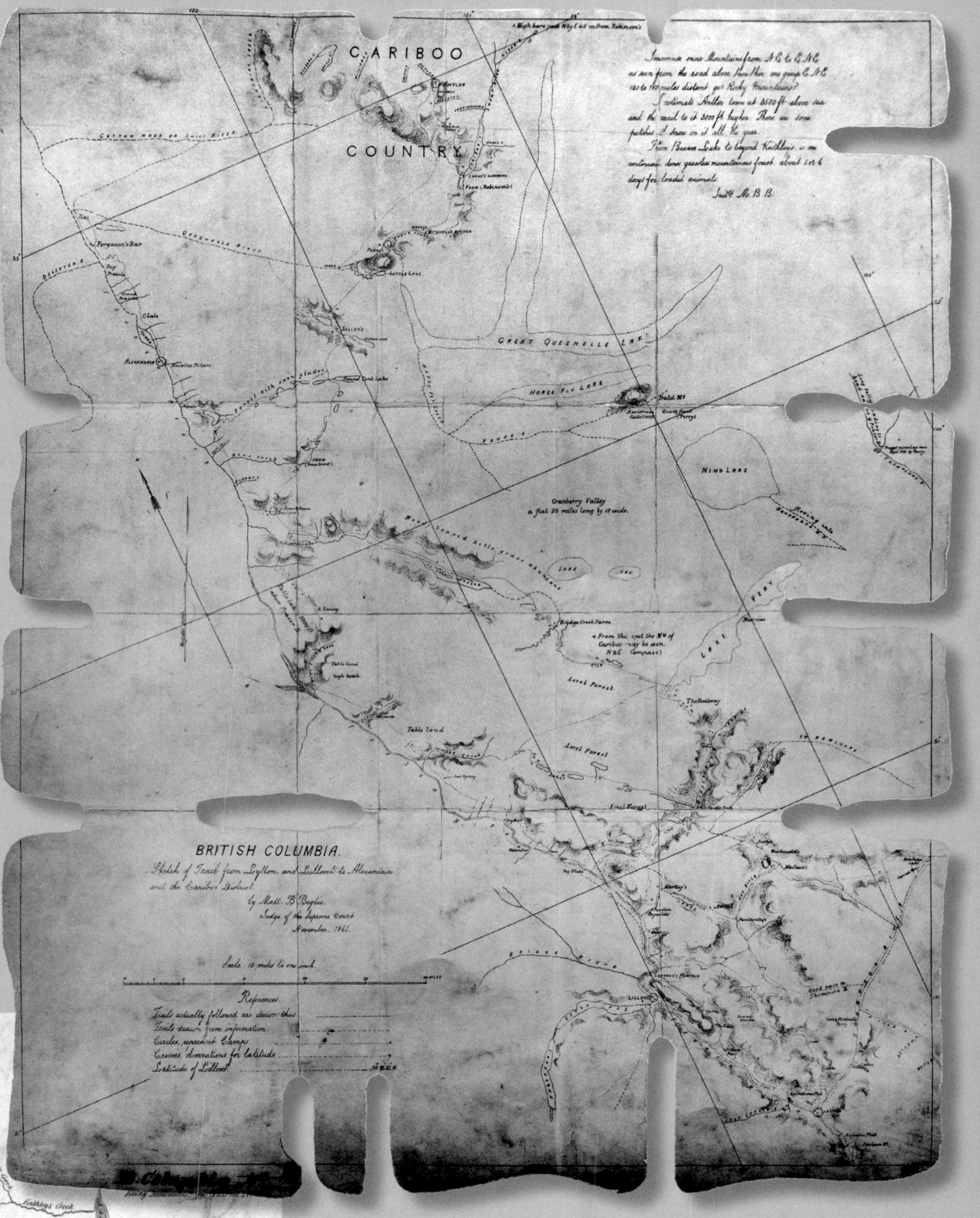

MAP 206 (*above*).
This is Matthew Begbie's own meticulously detailed map, drawn in November 1861 at the start of the Cariboo Gold Rush, of the trails between *Lytton* (at bottom) and *Lillooet* and *Fort Alexandria* and the *Cariboo Country* (at top). Begbie travelled this area in the fall of 1861 (see MAP 207, *left*). Crosses mark the position of places where Begbie checked the latitude in order to make his map as accurate as possible.

MAP 207 (*left*).
The Royal Engineers used Begbie's information in the compilation of their own maps. This is a map showing part of his itinerary in the fall of 1861 after which his own map (MAP 206, *above*) was drawn.

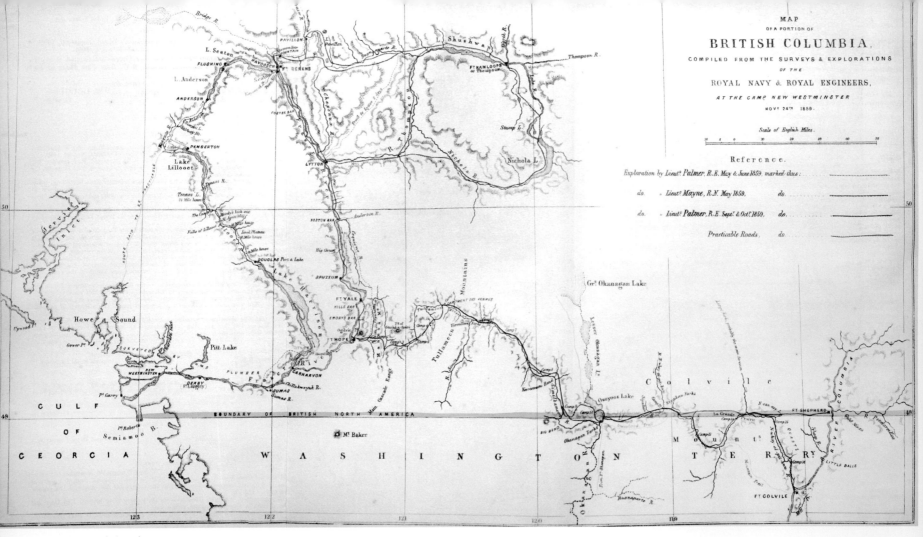

MAP 208 (above).

A compilation of the early explorations and surveys of Lieutenant Henry Spencer Palmer of the Royal Engineers and Lieutenant Richard Charles Mayne of the Royal Navy. Mayne's maps are shown on pages 72–73. In September and October 1859 Palmer explored from *F^t. Hope* to *F^t. Colvile* on the *River Columbia*. A trail was built by Edgar Dewdney the following year over the section from Hope to the valley of the *Similkameen R.*; this was the first part of the Dewdney Trail (see page 104).

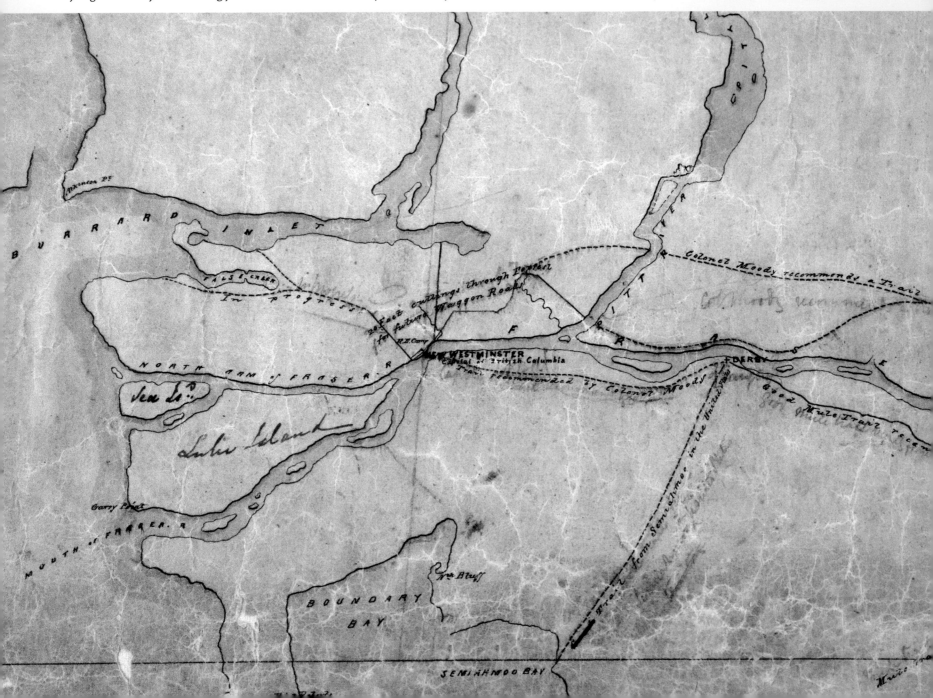

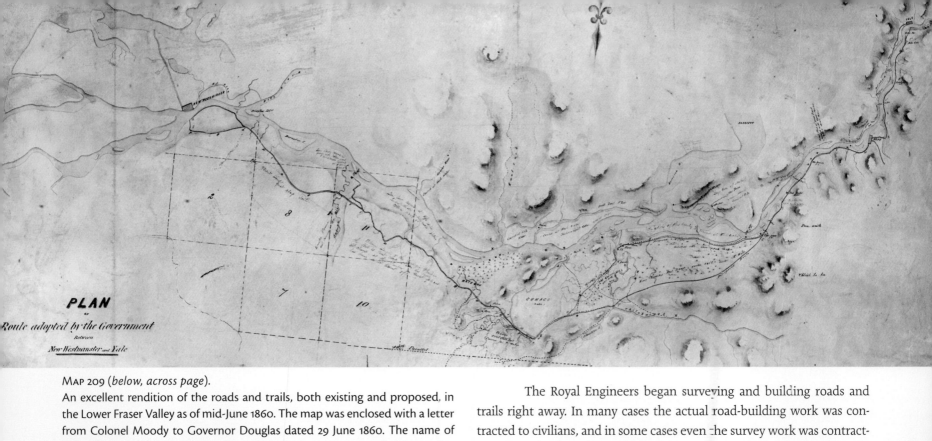

Map 209 (*below, across page*).
An excellent rendition of the roads and trails, both existing and proposed, in the Lower Fraser Valley as of mid-June 1860. The map was enclosed with a letter from Colonel Moody to Governor Douglas dated 29 June 1860. The name of *Lulu Island* may have been added later. Note the large *Sumass Lake* obstructing the building of roads on the south side of the river in the eastern valley. The *R.E. Camp* is at *New Westminster*. The trail to *False Creek* and English Bay is shown *In progress*. North Road (not named) is shown due north to the eastern end of *Burrard Inlet*. A notation that appears to refer to this trail is written across it: *20 Feet cuttings through Forest for Future Waggon Roads*. This trail appears not to have been used very much, if at all. On the north side of the river *Colonel Moody recommends a Trail be formed on this side of the River*—a more militarily strategic location. Moody said that he always kept the military part of his job foremost in his mind. Note *Derby* (town of Langley) and the short-lived *Carnarvon*. *Alice's Well (Hot Spring)* is today Harrison Hot Springs.

Map 210 (*above*).
But it was the road on the south side of the river that was the first usable one, as shown on this map dated about 1860. This became the line also followed by the telegraph in 1866; as part of the agreement with Western Union the government was to maintain a 20-foot cutting so as to avoid tree branches that might fall onto the telegraph line (see page 98). The Douglas Road to Second Narrows and the *N.E. Road* (Pitt River Road) are also shown on this map.

The Royal Engineers began surveying and building roads and trails right away. In many cases the actual road-building work was contracted to civilians, and in some cases even the survey work was contracted out, as it was to Edgar Dewdney for the first part of the Dewdney Trail east from Hope in 1860. The first roads in the Lower Mainland, not surprisingly, radiated from New Westminster, where the Royal Engineers had their camp. A trail—North Road—was hacked to Burrard Inlet to provide a possible "back door" for New Westminster should the city be besieged or, more likely, the river freeze, though this appears to have not been used at the time. Another trail was built to False Creek and English Bay at Jerry's Cove (Jericho Beach); this is the trail that eventually became

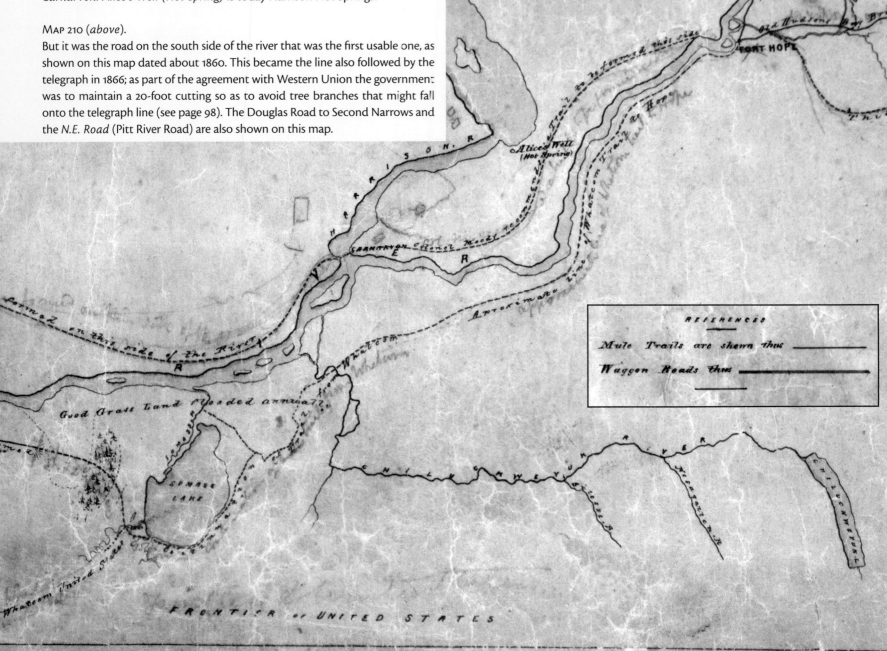

Kingsway in Vancouver and Burnaby. Another accessed the Pitt River. The Douglas Road to the Second Narrows, where Hastings Townsite would grow, was begun by the Royal Engineers but finished after they left in 1863 (see MAP 256, *page 94*).

Settlers, anxious for a means to transport their produce to market, also built roads, notably the Kennedy Trail south from opposite Annacis Island, and the North Arm Road, which followed most of the route that is today Vancouver's Southeast and Southwest Marine Drive.

Many of the first roads were portage roads, built between lakes or rivers, on which travel was easier. MAP 198, *page 72,* shows the string of such roads between Harrison Lake and Lillooet. Seton Portage, between Anderson and Seton Lakes, on this route, was the site of British Columbia's first railway, a wooden-railed, mule-drawn and gravity affair built by a civilian, Carl Dozier, in 1861. It fell into disuse following the completion of the Cariboo Road.

In 1861 gold was discovered in the Cariboo (see page 85), and in 1862 construction of the Cariboo Road began (see page 90). A possible alternative and shorter route was surveyed that year by Lieutenant Henry Spencer Palmer from North Bentinck Arm, at today's Bella Coola, to Alexandria, on the Fraser 45 km south of Quesnel. His detailed mapping is shown here. The route, however, proved less than ideal and not a good route to the Cariboo after all.

MAP 211 (*right*), MAP 212 (*right, centre*), and MAP 213 (*below*).
Royal Engineer Lieutenant Henry Spencer Palmer explored a possible route to the Cariboo gold fields in 1862 east from the head of North Bentinck Arm, the location of today's Bella Coola and the place where Alexander Mackenzie first reached Pacific tidewater (see page 38). The Bella Coola River and its tributaries occupy a deep canyon and require a precipitous climb to reach the plateau lands of the Chilcotin. Today the road takes many hairpin bends to navigate this cliff. It was only completed in 1953 after the citizens of the valley took it into their own hands and did the work themselves; the highways department had refused to construct the road owing to its difficulty and expense. MAP 213 shows *The Precipice* where Palmer made his climb out of the canyon; this is where today's road also begins its ascent. This map also shows the contrast between the Bella Coola River canyon and the Chilcotin Plateau to good effect. The straight red line is the *Desirable direction for a Coast Route* and the heading Palmer would have been aiming for to reach *Ft. Alexandria,* shown at right. The other red line is his actual route. Aboriginal groups are indicated: *Bella Coola* (Nuxalk) near the coast and *Atnayo Carrier* on the plateau. MAP 212 is Palmer's map of the mouth of the *Nookhalk (Bella Coola) River,* with Nuxalk villages *Ko-om-ko-otz* and *Soonoohlim.* Note the *Salmon Dam* at the latter location, a weir designed to trap salmon, the staple food of these people. (See also enlarged detail, MAP 12, *page 11.*) Two *Settler's Shanties* are also noted, showing that the valley already had some EuroCanadian settlement. There is also a *Site available for a Settlement* noted; if this was to become a route to the Cariboo, a town would certainly be required here. MAP 211 likewise shows a *Town Site,* the location of Bella Coola today, also shown in the aerial photograph (*inset*).

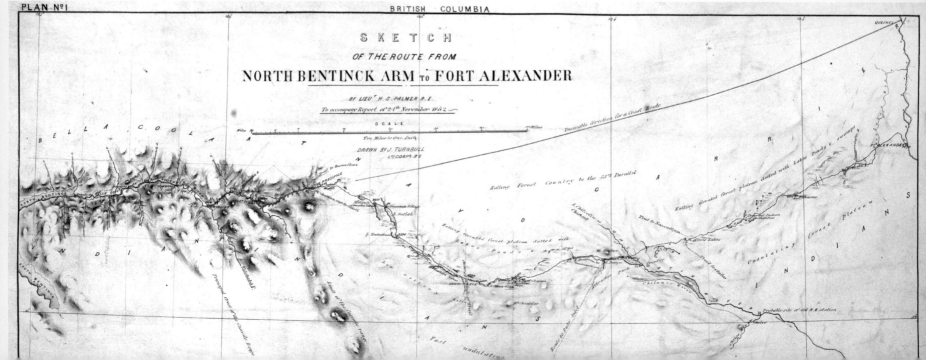

PLAN Nº1   BRITISH COLUMBIA

SKETCH
OF THE ROUTE FROM
NORTH BENTINCK ARM TO FORT ALEXANDER
BY LIEUᵗ H. S. PALMER R. E.
To accompany Report of 24ᵗʰ November 1862
SCALE
Ten Miles to one Inch
DRAWN BY J. TURNBULL
Lᵗ·CORPS R·E·

The Chilcotin Massacre

Alfred Waddington, a Victoria businessman, thought he knew a shorter way to the Cariboo following a route from Bute Inlet—a route he would later advocate using for a transcontinental railway (see page 114). Waddington had visited Bute Inlet in 1861 and was beguiled by the apparently easy approach to the interior. But the Homathko River runs from the Coast Range in a nearly perpendicular canyon fed by glacial torrents. In 1861 and 1862 he financed surveys; the first penetrated only a short distance, and the second, by surveyor Hermann Tiedemann, resulted in Tiedemann's near death from starvation. Nevertheless, Waddington pressed ahead. In 1862 he obtained a charter to build a wagon road and collect tolls, and construction began.

On 29 April 1864, hungry Tsilhqot'in (then known as Chilcotin), angered by deaths from smallpox they thought had been introduced by British or Americans, demanded food from a ferryman on the Homathko and killed him when he refused. Provisions intended for the construction works were taken. The next morning the Tsilhqot'in fell on the construction camp at dawn, killing nine workers; three escaped. Then they located William Brewster, the foreman, and three others farther up the trail and killed them too. Later they ambushed a pack train and killed more people.

These events sparked fears of a general uprising, and Governor Frederick Seymour organized two parties to track down the killers. Hudson's Bay trader Donald McLean was killed while leading a scouting party, but William Cox, a gold commissioner, made contact and persuaded the Tsilhqot'in leaders to surrender. The chiefs thought they had been given immunity, but they had not; Chief Klatsassin and four other chiefs were arrested and in September 1864 tried by Judge Begbie at Quesnelmouth (Quesnel). They were found guilty and hanged. Chief Klatsassin, in his defence, said that the Tsilhqot'in had been waging war, not committing murder.

Waddington died in 1872 of the disease that had indirectly led to the massacre while building his road—smallpox. More than a century later, in 1993, recognizing a miscarriage of justice, the British Columbia attorney general apologized for the hanging of the Tsilhqot'in chiefs.

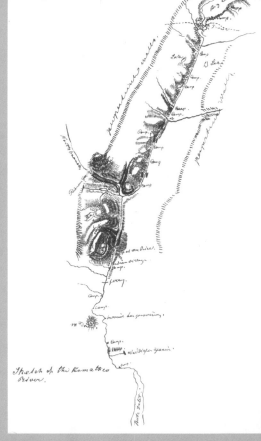

MAP 214 (*above*).
This is surveyor Hermann Otto Tiedemann's map of the Homathko Valley surveyed for Alfred Waddington in 1862. *Bute Inlet* is at bottom. Tiedemann and three others took a month (24 May–25 June) to reach Fort Alexandria, where he arrived "reduced to a skeleton, unable to walk." They had lost their provisions while crossing a river.

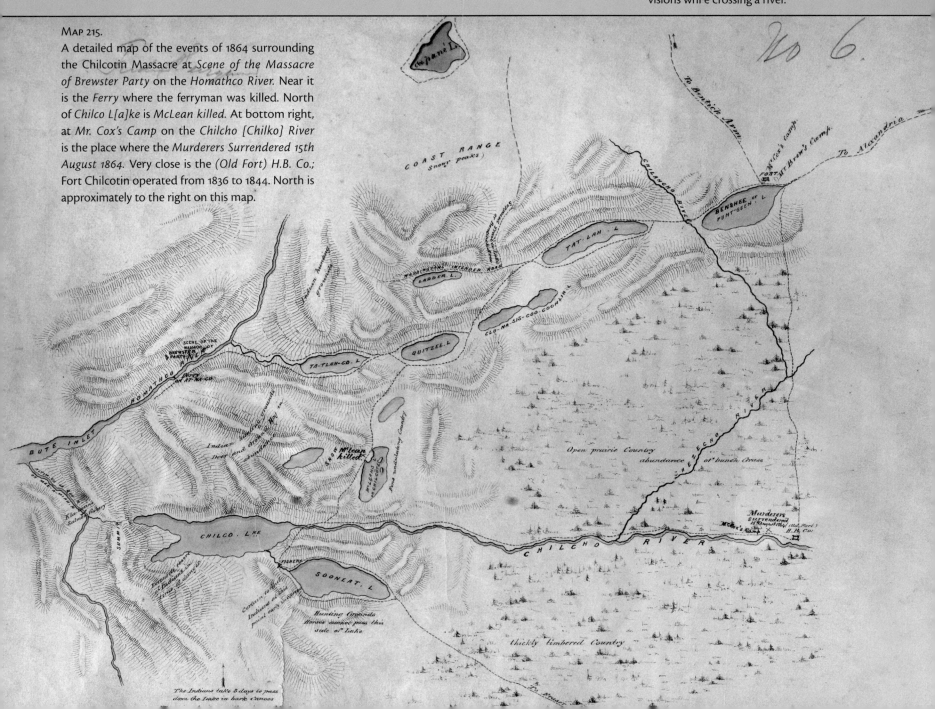

MAP 215.
A detailed map of the events of 1864 surrounding the Chilcotin Massacre at *Scene of the Massacre of Brewster Party* on the *Homathco River*. Near it is the *Ferry* where the ferryman was killed. North of *Chilco L[a]ke* is *McLean killed*. At bottom right, at *Mr. Cox's Camp* on the *Chilcho [Chilko] River* is the place where the *Murderers Surrendered 15th August 1864*. Very close is the *(Old Fort) H.B. Co.*; Fort Chilcotin operated from 1836 to 1844. North is approximately to the right on this map.

The primary responsibility of the Royal Engineers was to survey land and lay out townsites, which by the sale of lots were intended to defray their own costs and those of the colonial government. The Pre-emption Act of 1860 gave them a lot of work in this regard.

New Westminster, as the capital, was the first townsite to be laid out, and it was created in a style befitting a capital expected to one day be filled with grand government buildings (Map 226 and Map 228, *page 83*).

The townsites of Yale, Hope, and Douglas (at first Port Douglas) were next. The Royal Engineers laid out townsites wherever they judged there might be a demand: at the beginnings or ends of portages between lakes, at the confluences of rivers, or in a controlling position on a route. Some of their carefully surveyed and often beautifully coloured plans are shown on these pages.

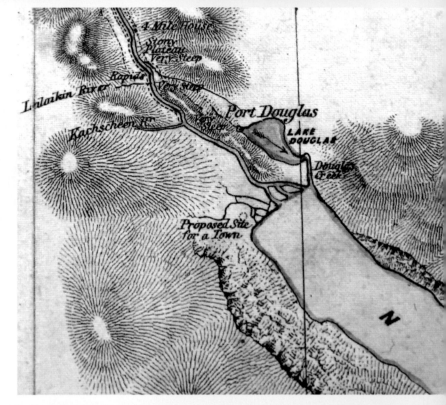

Map 216 (*below*), Map 217 (*right, top*), and Map 218 (*right*).
Douglas, at first Port Douglas, at the northern end of Harrison Lake, was the first stop on the Harrison–Lillooet trail. Map 216 also shows *Carnarvon* where the Harrison River meets the Fraser, a townsite that was proclaimed but never used. Map 217 shows the *Proposed Site for a Town*; Douglas was not a good townsite, as it was surrounded by swampy land and backed by a difficult hill. On the detailed plan, Map 218, Colonel Moody himself crossed out and initialled the word *Port*, creating just *Douglas*. A working plan, it carries the notation *This Plan Cancelled* from a later date. In June 1859 fifteen lots were auctioned, selling at an average of $115 each. Douglas had over two hundred inhabitants at its peak, but by 1874 this was reduced to just one. All three of these maps are dated 1859.

Map 219 (*below*).
Yale, shown on a map from about 1861. Two years later St. John the Divine church (*inset, left*) was built (on Lot 6 on *Douglas Street* between *Albert St.* and *Victoria St.* on this map). The subdivision of lots has expanded compared with the earlier map (Map 184, *page 66*).

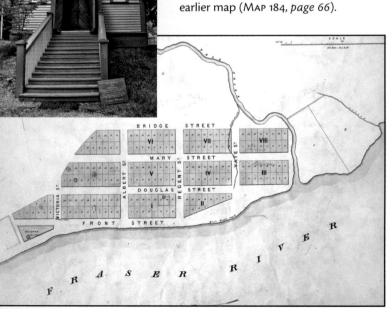

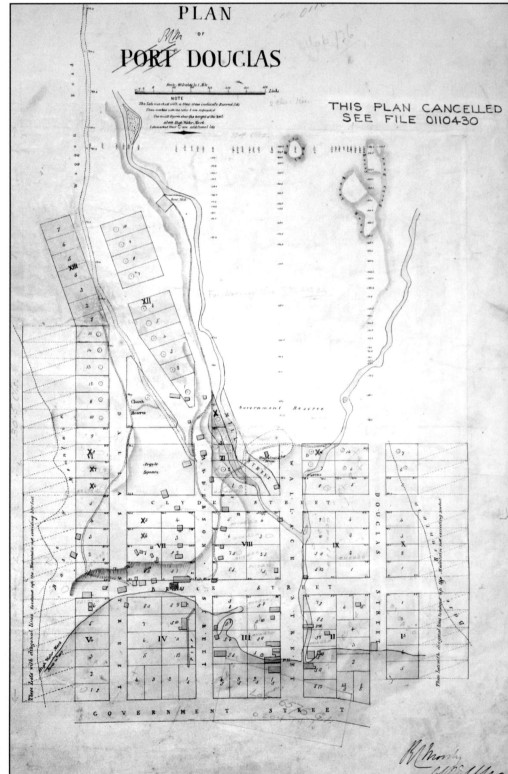

Map 220 (*right*).

This fine map of Yale and vicinity was produced by the Royal Engineers in 1862 and was drawn by James Launders. In addition to the townsite shown on other maps, this map includes a survey of larger lots, in this case "suburban" lots, as was done around New Westminster, Hope, and a few other townsites. In New Westminster's case there were also even larger "country" lots beyond the suburban ones. Purchasers of suburban and country lots were expected to farm their land, whereas town lots were for merchants and other town residents.

The neat and tidy map does not tell the real story of Yale, which at the time of the gold rush had a particularly bad reputation. "A worse set of cut-throats and all-round scoundrels than those who flocked to Yale from all parts of the world never assembled anywhere," complained one contemporary writer.

Map 221 (*below*).

Another version of the Royal Engineers' townsite plan of Douglas. Lots that appear to be underwater were on swampy ground that was expected to be reclaimed.

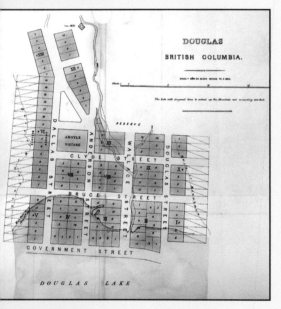

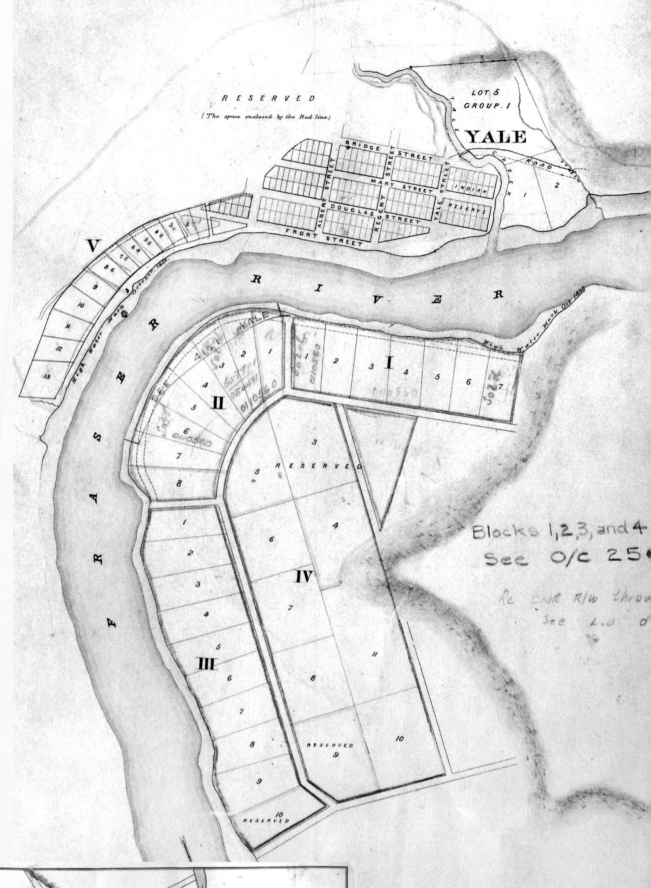

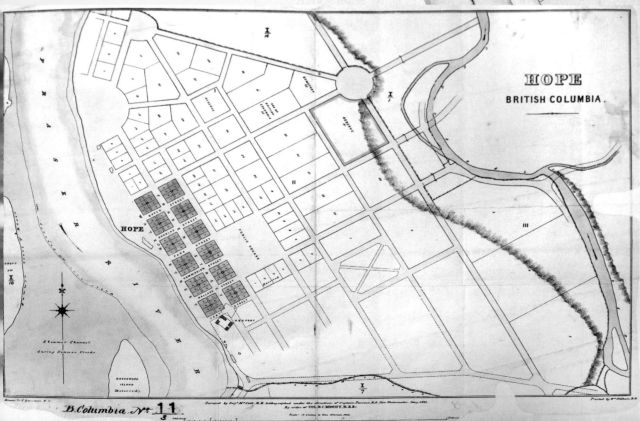

Map 222 (*left*).

Hope, on a curve of the Fraser (at left) at the point it meets the Coquihalla River (at right), was the recipient of a particularly elegant plan, shown here with its circles and radial boulevards. Here again there are larger "suburban" lots farther from the town centre. The radial streets remain to this day, though the circles seem never to have been constructed. Fort Hope was built in 1848–49 after a trail had been found through the mountains that the Hudson's Bay Company *hoped* would prove to be a feasible all-British route.

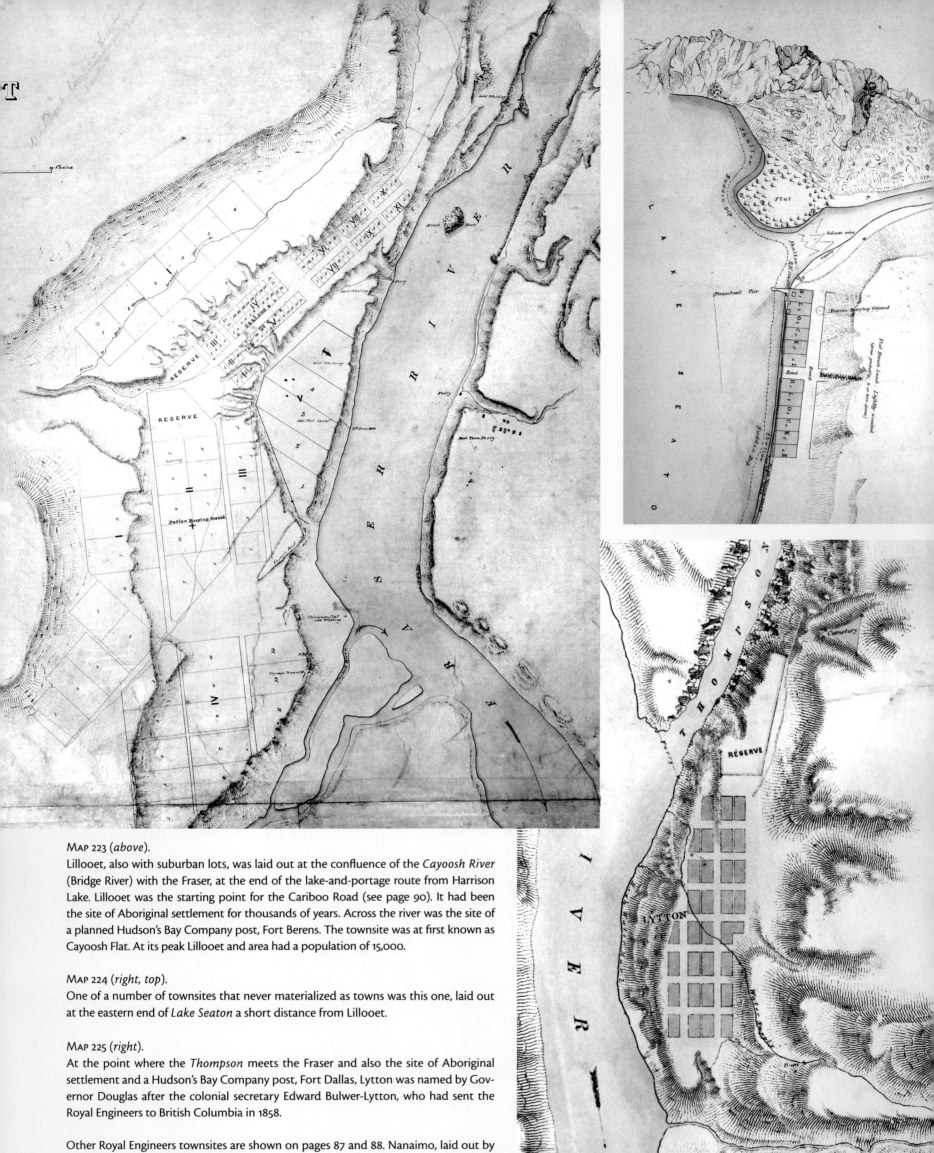

MAP 223 (*above*).
Lillooet, also with suburban lots, was laid out at the confluence of the *Cayoosh River* (Bridge River) with the Fraser, at the end of the lake-and-portage route from Harrison Lake. Lillooet was the starting point for the Cariboo Road (see page 90). It had been the site of Aboriginal settlement for thousands of years. Across the river was the site of a planned Hudson's Bay Company post, Fort Berens. The townsite was at first known as Cayoosh Flat. At its peak Lillooet and area had a population of 15,000.

MAP 224 (*right, top*).
One of a number of townsites that never materialized as towns was this one, laid out at the eastern end of *Lake Seaton* a short distance from Lillooet.

MAP 225 (*right*).
At the point where the *Thompson* meets the Fraser and also the site of Aboriginal settlement and a Hudson's Bay Company post, Fort Dallas, Lytton was named by Governor Douglas after the colonial secretary Edward Bulwer-Lytton, who had sent the Royal Engineers to British Columbia in 1858.

Other Royal Engineers townsites are shown on pages 87 and 88. Nanaimo, laid out by the Vancouver Coal Mining & Land Company in 1862 (see page 224), is the only formal townsite laid out during this period that was not created by the Royal Engineers, whose mandate was the mainland colony.

Map 226 (*below*) and Map 228 (*below, bottom*).
The first townsite laid out by the Royal Engineers was Queensborough, which, once selected as the capital for the mainland colony, became New Westminster, a name selected by Queen Victoria along with the name *British Columbia*. Sales of lots were brisk at the new capital, aided by the transfer of titles for purchasers of lots at Langley (Map 190, *page 69*). The colourful grand plan, with its boulevards, circles, and vistas, belies the reality of a few wooden cabins among the stumps of the forest. By the time the engraving of the city appeared in *Harper's Weekly* in August 1865 (*inset*), it was looking a bit more like an emerging metropolis. Map 226 is dated 1861; Map 228, covering a larger area and showing municipal, government, and hospital reserves, is dated 1862.

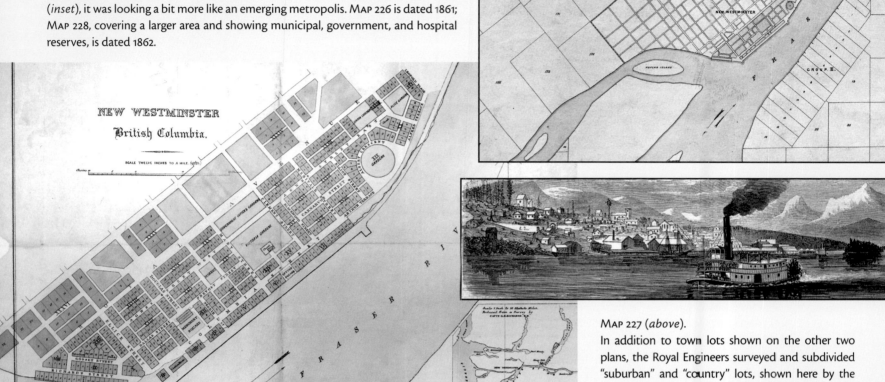

Map 227 (*above*).
In addition to town lots shown on the other two plans, the Royal Engineers surveyed and subdivided "suburban" and "country" lots, shown here by the two densities of subdivision outside of the (darker-coloured) town. These maps were printed by the Royal Engineers at their camp and then hand coloured.

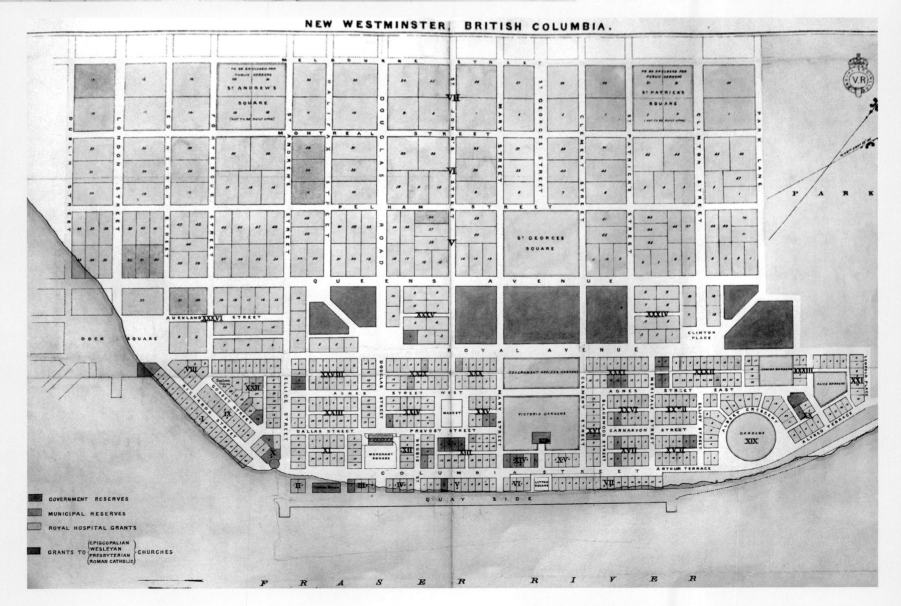

A Northern Boundary

When the mainland Colony of British Columbia was brought into being by an act of the British parliament in August 1858 (see MAP 5, *page 8*), it was necessary to define the area it was to include. The southern, eastern, and western boundaries were all relatively clear and definable, but the northern boundary was defined as "Simpson's River, and the Finlay Branch of the Peace River." This last boundary, the northern one, confused mapmakers because Simpson's River did not exist. Its mouth, at least, was actually the Nass River, explored in 1831 by the Hudson's Bay Company's Aemilius Simpson (hence its name) but shown on maps as flowing from the east when it actually flows from the north.

Thus we have various interpretations of the northern boundary of British Columbia shown on the first maps. The situation was resolved a few years later after gold was found on the Stikine River farther north (see page 210). In July 1862 the British government created a separate Stickeen Territories between the northern boundary of British Columbia and 62°N, but the gold rush proved short lived, and in July 1863 the area was incorporated into British Columbia.

In 1898, with the creation of Yukon Territory following the Klondike Gold Rush, the boundary was set at 60°N, where it remains today.

MAP 229 (*above*).
The British government asked mapmaker James Wyld to suggest boundaries for the new colony in July 1858, and this is his map, with the *Proposed Boundary* at the Stikine River—a real possible boundary. The fictitious *Simpsons R.* is also shown, as is *Finlays Br.*

MAP 230 (*left*).
This map by James Wyld, dated 10 September 1858, shows the northern boundary of British Columbia as a nearly straight line following *McDougalls* or *Simpson's R.* but ignoring the northern part of the course of *Finlay's Br.* (River), also specified as the boundary, even though the river is shown, albeit inaccurately, on the map.

MAP 231 (*left, bottom*).
John Arrowsmith's map, dated 1 June 1859, has a more accurate northern boundary, following the *Finlay River*, but it also follows the non-existent *Simpson or Babine R.* The *Queen Charlotte Islands* were also made part of British Columbia in the 1858 act.

MAP 232 (*right*).
New Columbia reaches from *British Columbia* to *Alaska* on this 1867 map. The name seems to have been a figment of the mapmaker's imagination.

MAP 233 (*below*).
Stekin Territories on an 1872 map, ten years after it had been incorporated into *British Columbia*.

Gold in the Cariboo

As miners found it harder to find gold on the gravel bars of the Fraser Canyon, some migrated farther north, searching for an as-yet-untouched bonanza. And they found it, though ultimately most of the gold was deep underground, and recovering it required lots of capital to purchase water wheels, pumps, hoisting gear, and the like. When the value of the gold taken out hit $4.5 million by 1862, Governor James Douglas authorized the construction of the Cariboo Road, itself requiring the borrowing of a lot of money to finance (see page 90).

Gold appears to have first been found in the Cariboo—a misspelled name for the region bestowed by Douglas in 1861—on the Horsefly River in 1859. Substantial amounts of gold were found in the river gravel on Keithley Creek in 1960 by "Doc" Keithley and his friends. They then followed the creek upstream and crossed over to Antler Creek, where they found even richer deposits. But winter was closing in, and so they returned to Keithley for supplies. Other miners working there got wind of their discovery, and quite a few set out for Antler in mid-winter, over a metre or two of snow, to stake claims.

Map 234 (right, top).
The gold creeks have been hand coloured in yellow on this 1862 map from naval officer Richard C. Mayne's book, published in 1862. Note *Keithley Cr.*, *Antler Cr.*, *Lightning Cr.*, and *Williams Cr.*

Map 235 (below).
One of the best maps to show the creeks and settlements of the Cariboo gold fields is this map published in 1882; all the easy gold had been found by then.

Map 236 (right).
One of the earliest maps of the Cariboo gold fields was this one drawn up by the Royal Engineers in August 1862. The trails and gold areas, the latter for the Cariboo shown only as a single large area, have been drawn in on a printed map. A notation at the bottom states that "this map will be reproduced when further information is obtained."

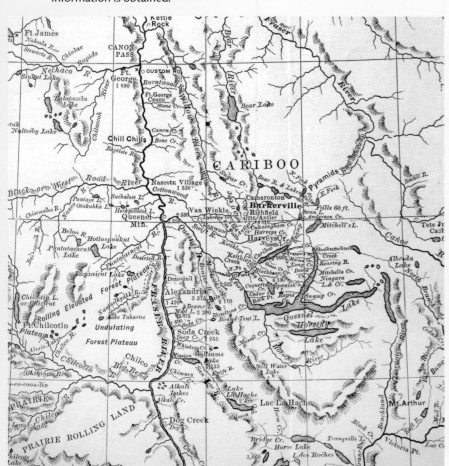

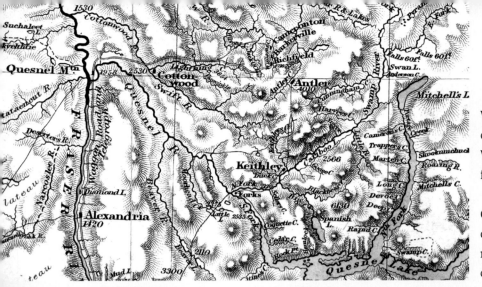

At the same time another party of miners, which included one William Dietz—known as Dutch Bill—began a search for other, less crowded creeks and found an exceptionally rich one that they dubbed Williams Creek. At least, it was found to be rich later, for Dietz's initial finds were worthwhile but not exceptional deposits.

By the spring of 1861 word of the new finds reached beyond the Cariboo, and the rush was on. As it happened, almost all of the creeks of this region were gold bearing to some degree or other, and so many made their fortunes. The gold finds were publicized in Britain and Canada and by the following year resulted in yet more men flooding in to find their fortunes.

On one creek, Lowhee, miner Richard Willoughby, who had found it, took out 3,037 ounces of gold, helped by four to seven men in a six-week period between the end of July and beginning of September 1861. It was worth $50,000, a fabulous sum. Lightning Creek was reported to have yielded 900 ounces in one day, $100,000 worth in three months. On Williams Creek, the returns had been more moderate at first—poor enough to be given the moniker Humbug Creek—and only six claims had produced gold. But a probe dug about 5 m deep in the summer of 1861 found a stratum of gold-bearing gravel that proved to be stunningly rich, producing 50 ounces in a couple of days. The discovery produced another minor rush, with claims being staked for nearly 10 km along the creek (MAP 238, below and left). This was to be the future of the Cariboo gold

MAP 237 (left, top).
The Cariboo mining district as depicted in 1871 on a large Lands and Works map of that date (MAP 913, pages 358–59).

MAP 238 (left and below, bottom).
This rare find is a detailed map of all the mining claims on Williams Creek in 1862 and 1863, together with a list of the claim owners. Many of the claims are now worked by companies, reflecting the increased capital required to buy equipment as the gold deposits were found deeper and deeper. In the middle of the claims is Richfield Town. The map title is dated September 1862, but the list has an 1863 date despite being all on the same sheet of paper. The map appears to have been drawn up officially, perhaps by the gold commissioner; it seems unlikely that a miner would take the time to plot his rivals' claims.

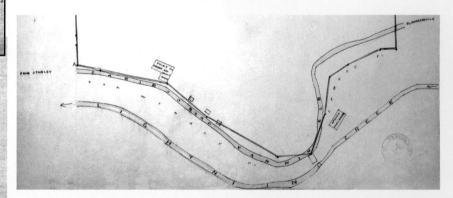

PLAN
OF
BARKERVILLE

WILLIAMS CREEK

SCALE

Five Feet to One Mile

DRAWN BY J TURNBULL R.E. 1st AUG. 1869

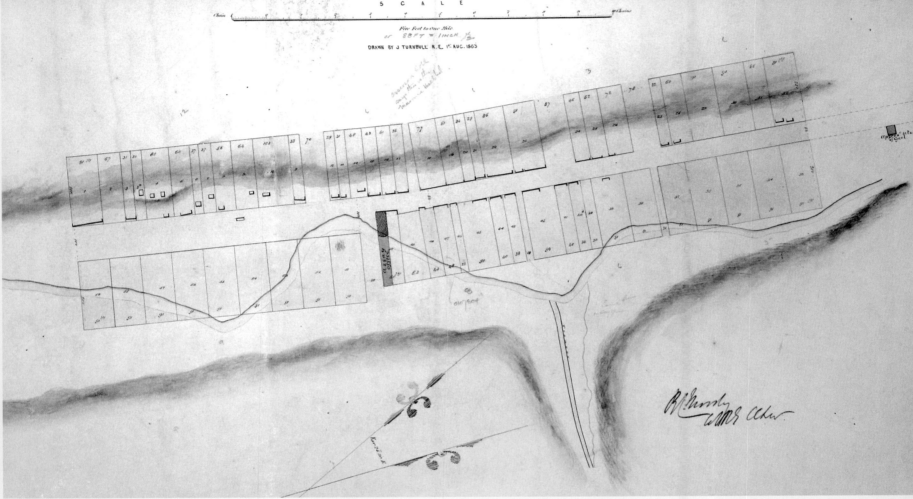

Map 239 (left).

Two company shaft houses are located on the *Cariboo Road Van Winkle* between *Stanley* and *Barkerville* on *Lightniing (sic) Creek*. The map is dated 1874.

Map 240 (above), with two photos.

This is Barkerville, which sprang up on Williams Creek in 1862 and, with five thousand people, became the largest western town north of San Francisco. It remains today, restored, as a National Historic Site. The town was at the centre of the Williams Creek mining district, with Richfield and Cameronton (see next page) on either side of it. Barkerville was named after Billy Barker, who hit pay dirt here at a depth of 52 feet, eventually removing 37,500 ounces of gold. The town became the northern terminus of the Cariboo Road. The two photos are from 1864 (left) and about 1865 (right). Some buildings were constructed right over Williams Creek, providing an easy way to dispose of household wastes in those less environmentally concerned times. Barkerville burned down in 1868 but was promptly rebuilt.

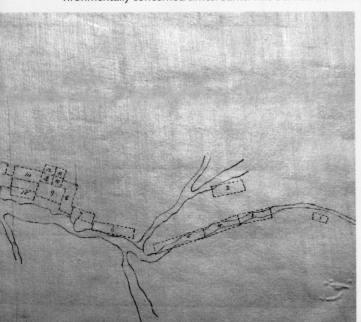

Map 241 (right).

An elegant 1893 plan of the townsite of Quesnelle Forks, a supply centre established in 1860. It is at the confluence of the north and south forks of the Quesnel River, 100 km upriver from Quesnel. The town began to die when it was bypassed by the Cariboo Road in 1865. A few buildings remain today.

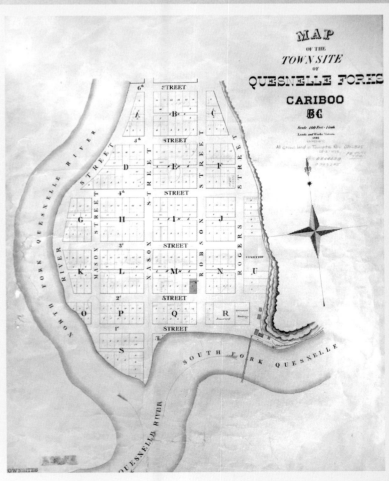

MAP
OF THE
TOWN SITE
OF
QUESNELLE FORKS
CARIBOO
BC

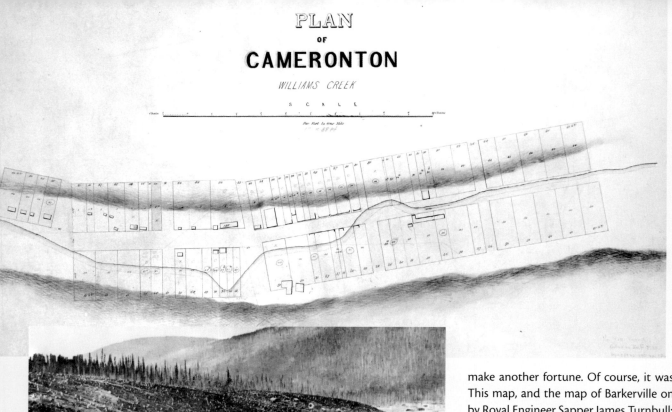

PLAN
OF
CAMERONTON
WILLIAMS CREEK

SCALE

Map 242 (left).
Cameronton was on Williams Creek downstream from Barkerville and grew up around the claim of John A. "Cariboo" Cameron, another exceptionally rich claim on Williams Creek. The photo, *inset*, is of Cameronton about 1863 and shows a perhaps even more ramshackle collection of cabins than those at Barkerville. Again the creek runs though the settlement, its advantages outweighing its disadvantages. Although Cameron returned to his native Ontario with a considerable sum of money, he lost it all on subsequent investments and returned to the Cariboo to make another fortune. Of course, it was not to be, and he died here in 1888. This map, and the map of Barkerville on the previous page, were both drawn by Royal Engineer Sapper James Turnbull; they are dated 1 and 5 August, respectively, and both are countersigned in the margin by Colonel Moody.

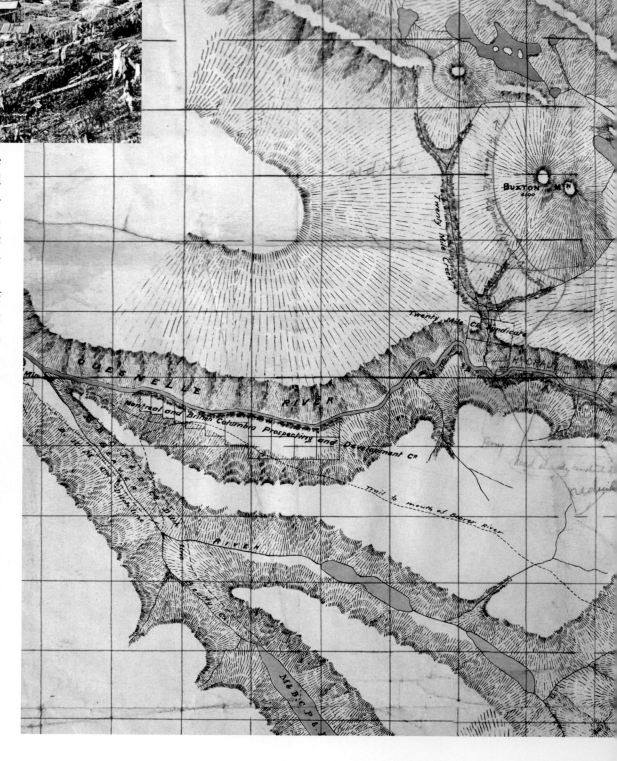

fields: most of the richest deposits lay at some depth and required more intensively capitalized operations to dig shafts and pits, causing enormous destruction of the creek beds in the process. Later maps show the proliferation of hydraulic mining methods, which in washing away entire mountainsides were even more destructive.

In 1861 some $2.67 million worth of gold was mined from the Cariboo, the equivalent of perhaps $68 million today. Some $2.66 million worth was mined the following year. In 1863, the peak year, $3.91 million worth of gold was extracted from the Cariboo's creeks.

Map 243 (right).
Hydraulic extraction of gold by mining companies is very much in evidence on this 1896 map of the area around *Quesnelle Forks*, with whole mountainsides now allocated to these companies. Hydraulic mining used high-pressure water to flush out the gold and was by its very nature extremely destructive. The practice was used in the province until about 1920 but is not permitted today. This map very effectively uses hachures—straight lines drawn sloping downhill—to portray the topography. *Inset*. Hydraulic mining on Cunningham Creek about 1923.

Map 244 (far right, top).
Using unusual contour shading to depict the mountains and valleys, this 1890s map shows *Williams Creek*, with *Richfield* and *Barkerville*. Cameronton, downstream of Barkerville, has disappeared. *Lowhee Creek* is also shown.

The Cariboo Road

The discovery of gold in the Cariboo was the catalyst for one of the important developments in British Columbia history—the building of the Cariboo Road. This was to be a wagon road rather than a pack trail, which was all that had existed previously. The non-surface nature of much of Cariboo mining required equipment and supplies whose transport was more suited to wagons than pack animals, and since the mining required capital, men were willing to work as employees on road building. Nevertheless the road builders often had trouble keeping their workforce as news of lucky strikes passed them by every week. And, of course, the revenues to the colony emerging from all that gold allowed James Douglas to borrow money to finance the road.

The Cariboo Road was built in sections. Captain John Grant and the Royal Engineers, with some civilian labour, built the first—and most difficult—section for 10 km north of Yale (MAP 247, *far right, bottom*). Surveyor Joseph Trutch was responsible for the next section, from Chapman's Bar to Boston Bar. This included the first Alexandra Suspension Bridge, which was completed in September 1863. Thomas Spence constructed the section from Boston Bar to Lytton. Later, in 1865, Spence was contracted to replace Mortimer Cook's ferry at the Nicola River with a bridge—Spence's Bridge.

The section from Lytton north was built by contractors, except for a 14-km difficult section near Spences Bridge that was again built by the Royal Engineers, this time under

MAP 245 (*right*).
The *Cariboo Road* to *Quesnelle* and *Barkerville* is clearly shown and labelled on this 1897 map.
Above. A very fine 1883 painting by Edward Roper of the Cariboo Road, complete with oxen-drawn wagon.
Below. Wagons on the Cariboo Road north of Cache Creek, probably about 1890.
Below, right. By 1913, the date of this ad, a tri-weekly bus service of sorts ran between Ashcroft and Quesnel.

MAP 246 (*right*).
The Cariboo Road, clearly marked but not named, is shown on this 1884 map. The road runs from *Lilloet* to *Barkerville* via *Quesnellemouth* (now Quesnel) and joins the road from *Ashcroft* at *Clinton*.

Left. The easy way to get to the gold fields—by road steamer. Several steamers were purchased from Scotland by F.J. Barnard and J.C. Beedy. The first left Yale in the spring of 1871, pulling six tons of freight. It got as far as Jackass Mountain before proving unable to cope with the grade. Needless to say, the experiment was short lived.

Lieutenant Henry Palmer. Another road, from Lillooet to Clinton, was built by Royal Engineers under Sergeant John McMurphy. At Clinton the road connected with the Cariboo Road (MAP 246, *right*).

The Cariboo Road was completed in September 1863 as far as Soda Creek, where a steamboat on the Fraser took over, but by 1865 it was completed through to Barkerville.

Although widely regarded as an engineering triumph, the Cariboo Road was seen by the British government as yet another colonial expense, and this was one of the reasons the Royal Engineers were recalled to Britain in July 1863.

MAP 247 (*below*).
This 1862 Royal Engineers map by Sergeant William McColl locates the start of the proposed Cariboo Road around the first rock bluff north of *Yale*. The line of road had *the trees . . . blazed and marked . . . with Red paint*. There is an existing *Mule Trail* (dashed line), which was the track taken before the road was built. The Cariboo Road was a wagon road, considerably wider than a mere trail. *Black line proposed road* is higher than the mule trail. This first difficult section of the road and another short section along the Thompson were built by the Royal Engineers themselves, though much of the rest of it was contracted out.

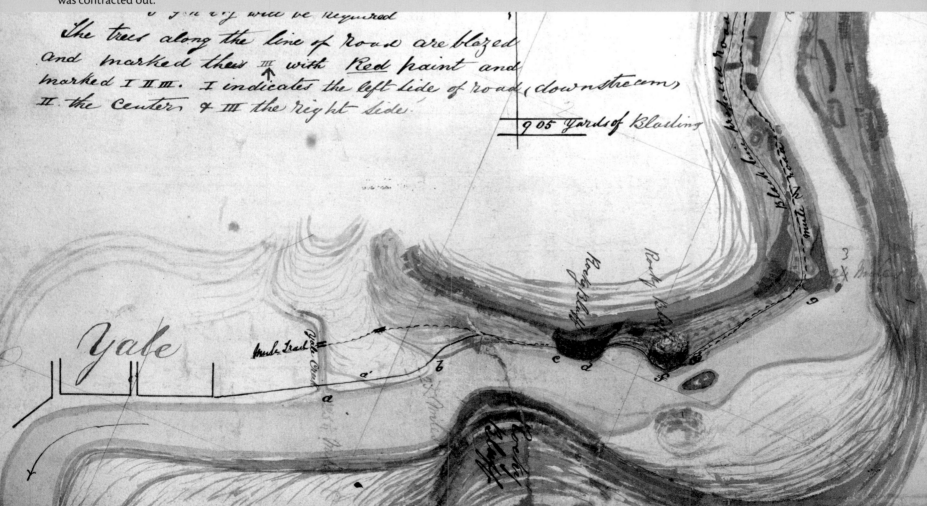

SETTLEMENT ON BURRARD INLET

The first European settlements in mainland southwestern British Columbia, as with New Westminster, lined the Fraser River, the gateway to the gold fields of the interior. The Royal Engineers had built roads to Burrard Inlet for strategic reasons, but settlement on the inlet began near sawmills established to exploit the forest resources near at hand.

At Moodyville, on the North Shore, land was pre-empted by three entrepreneurs in 1862 for their sawmill, called Pioneer Mills. The business did not thrive, however, and was sold twice before in 1865 ending up in the hands of American businessman Sewell Prescott Moody, who made the mill a success. It continued under his ownership until 1875, when he was drowned on a ship bound for San Francisco that sank off Cape Flattery.

The outstanding quality of the trees lining Burrard Inlet and their initial easy access encouraged another entrepreneur, Captain Edward Stamp, to build another sawmill, this time on the south side of the inlet. Stamp had at first wanted to build his sawmill in what is now Stanley Park; if he had, the appearance of today's Vancouver would doubtless have been very different. The current at this site was not really suitable for log booming, in any case. Stamp instead built farther east at what is today Centennial Pier. From 1865 the mill processed spars cut by Jeremiah Rogers at Jerry's Cove—Jericho. In 1867 the mill was expanded, but none-theless, like its North Shore counterpart, it was not a success and closed down. It was reopened in 1870 as Hastings Mill after being purchased by Edward Heatley and George Campbell.

A settlement grew up around the mill, and nearby, in September 1867, John "Gassy Jack" Deighton built his Globe Saloon. In 1870 the provincial government laid out a townsite, and Deighton's premises ended up in the middle of what was technically a road (MAP 259, *page 95*)—for Deighton was but a squatter. At a government auction Deighton bought a surveyed lot and built a new hotel. This remained in business until being wiped out by the 1886 fire, though it was run by others after 1875, when Deighton died. The name of the neighbourhood—Gastown—remains to commemorate its beginnings.

A few kilometres east another townsite was laid out by the provincial government in 1869, confusingly also called Hastings. This was at the north end of a town reserve created where Douglas Road reached the inlet. Two hotels had already been constructed there to serve the residents of New Westminster who had begun to frequent the location in the summer to escape the mosquitoes of the Fraser Valley. The resort was known as New Brighton, as it had much the same relationship as the British seaside resort Brighton did to its capital, London.

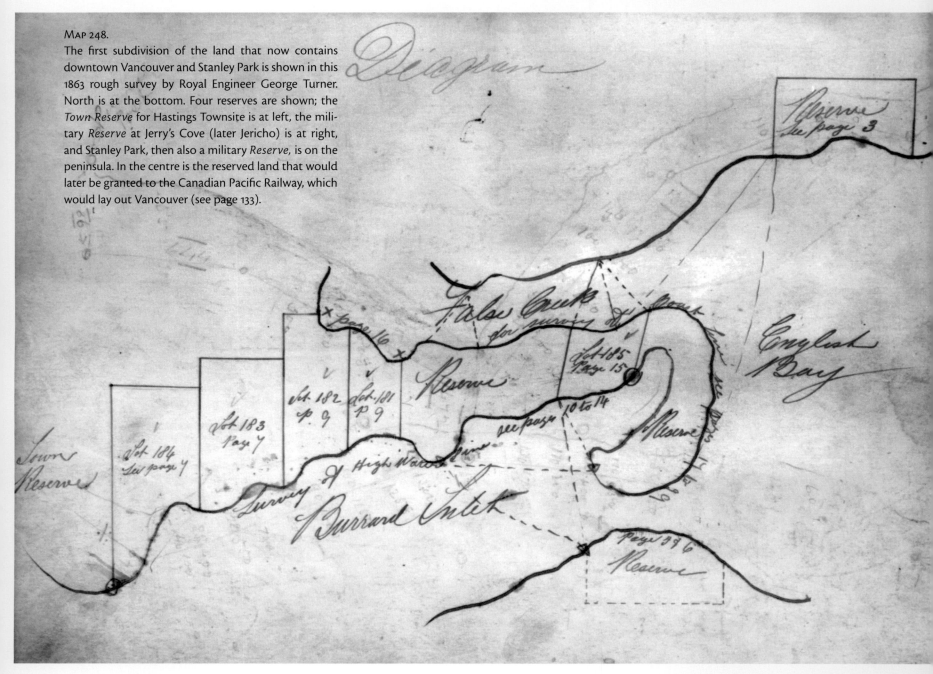

MAP 248.
The first subdivision of the land that now contains downtown Vancouver and Stanley Park is shown in this 1863 rough survey by Royal Engineer George Turner. North is at the bottom. Four reserves are shown; the *Town Reserve* for Hastings Townsite is at left, the military *Reserve* at Jerry's Cove (later Jericho) is at right, and Stanley Park, then also a military *Reserve,* is on the peninsula. In the centre is the reserved land that would later be granted to the Canadian Pacific Railway, which would lay out Vancouver (see page 133).

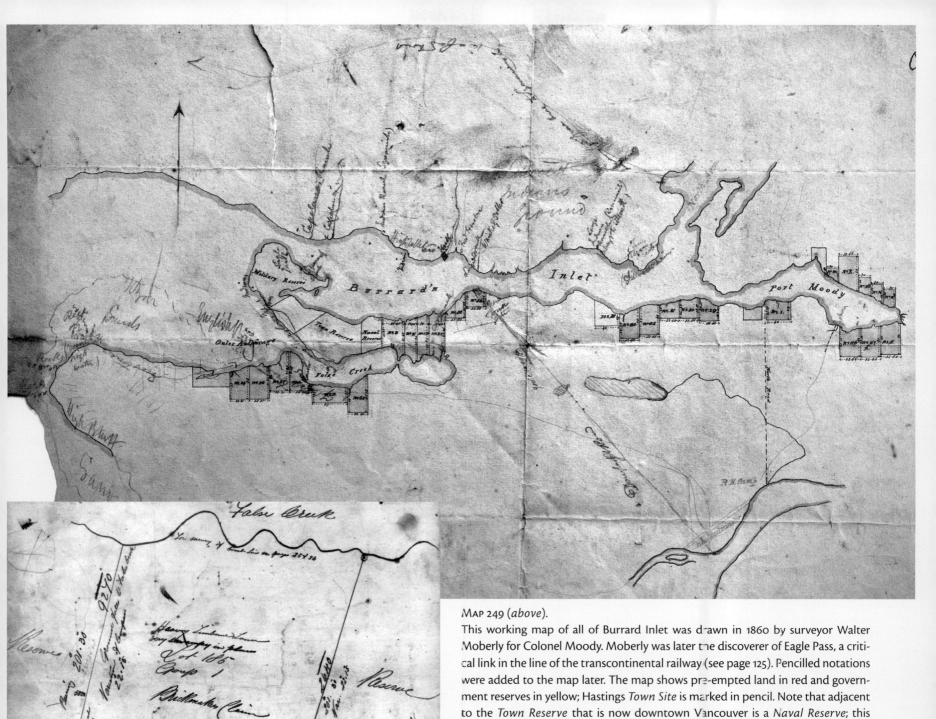

MAP 249 (*above*).
This working map of all of Burrard Inlet was drawn in 1860 by surveyor Walter Moberly for Colonel Moody. Moberly was later the discoverer of Eagle Pass, a critical link in the line of the transcontinental railway (see page 125). Pencilled notations were added to the map later. The map shows pre-empted land in red and government reserves in yellow; Hastings *Town Site* is marked in pencil. Note that adjacent to the *Town Reserve* that is now downtown Vancouver is a *Naval Reserve*; this was the land on which the townsite of Granville was laid out in 1870. On the North Shore, *Mill* denotes the later location of Moodyville.

MAP 250 (*above*).
The 218 ha (540 acres) between the reserves that became Stanley Park and downtown Vancouver were pre-empted in 1862 by John Morton, a brick maker, his cousin Samuel Brighouse, and a friend, William Hailstone. Morton was interested in the possible clay associated with the coal seams of Coal Harbour, and the three took turns living on the property to establish their claim. The then apparent uselessness of the claim led to their being dubbed the "Three Greenhorn Englishmen." This map of the claim was drawn by George Turner in 1863. The three had the last laugh, however, selling the land in 1882 to a syndicate that laid out a private townsite, Liverpool, hoping to cash in on the arrival of the transcontinental railway (see MAP 315, *page 122*). Today this *Heavy Timbered Land very swampy in places* is Vancouver's West End.

MAP 251 (*right*).
This obviously well-used and taped-up map dated 1865 shows Edward Stamp's *Saw Mill*, later called Hastings Mill, with a proposed *Flume*, which was to bring water from Trout Lake for the boilers. South of the *Mill Site* is *Additional Land required by Sawmill Co*—land with the needed timber.

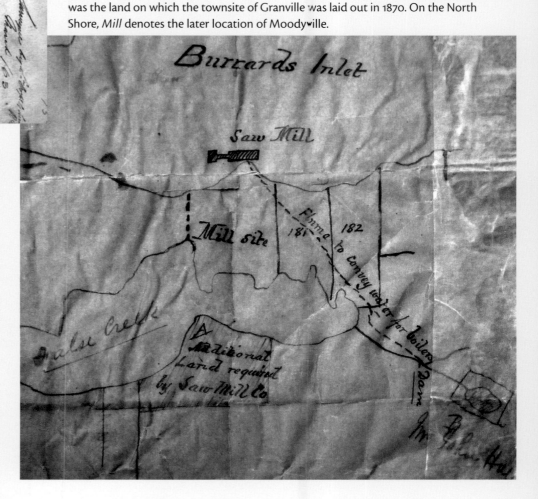

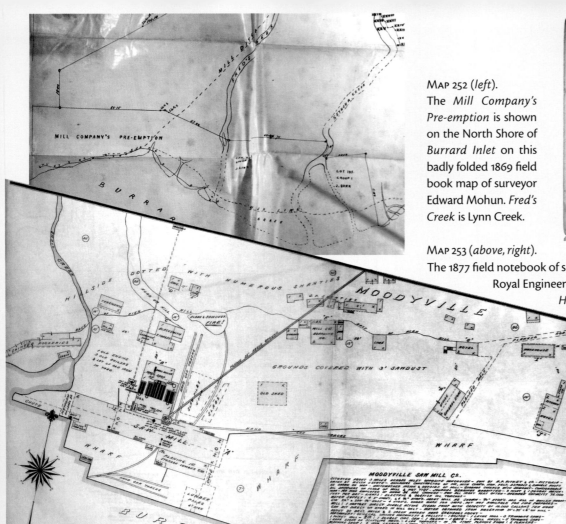

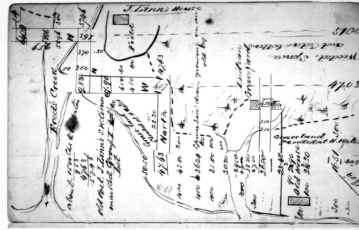

Map 252 (*left*).
The *Mill Company's Pre-emption* is shown on the North Shore of *Burrard Inlet* on this badly folded 1869 field book map of surveyor Edward Mohun. *Fred's Creek* is Lynn Creek.

Map 253 (*above, right*).
The 1877 field notebook of surveyor John Jane shows the North Shore pre-emption of ex–Royal Engineer John Linn, after whom the Lynn Valley is named, and *J. Linn's House*. Linn pre-empted his land in 1871 on the east bank of the river that would be renamed after him.

Map 254 (*left*).
This 1889 fire insurance plan of Moodyville shows in detail the layout of the mill site. *Ground covered with 3' sawdust* to deal with mud would not today be considered anything but a fire hazard. Behind the mill is a *Hillside dotted with numerous shanties*. To the west of the mill is a building labelled *Chinese Rookeries*—presumably segregated accommodation for Chinese workers, and a *Tenement. Hand car tracks* lead to and from the mill, including one set to the wharf. At centre are *Slabs & Sawdust, Fire!* The grain elevators of the Saskatchewan Wheat Pool now occupy this site.

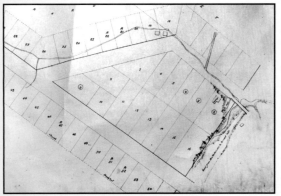

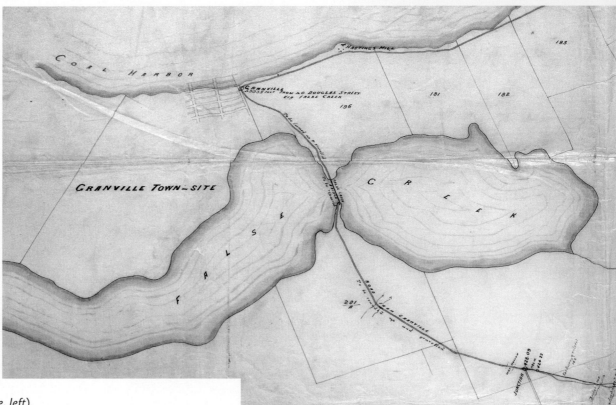

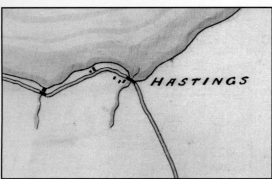

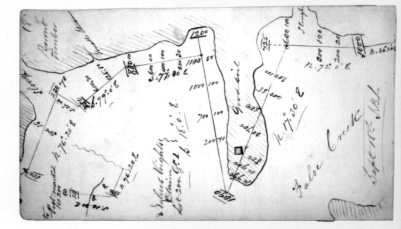

Map 255 (*above left, centre*) and Map 256 (*above, left*).
Map 255 shows the partially laid-out townsite of Hastings drawn for a sale of lots held on 10 July 1869; only seven lots were sold. The *Road from New Westminster* is at right. Map 256 shows more realistically the actual size of the community and its position on Douglas Road.

Map 257 (*above, right*).
Drawn in 1883, this map shows the *Granville Town-site*, the part actually laid out (now Gastown), *Hastings Mill*, and the *Road from Granville* to New Westminster. The comments along the road reveal the map's source: from a contract for road improvement. The map shows clearly the former eastward extension of False Creek.

Map 258 (*right*).
Another page of a surveyor's field notebook, this 1869 one from that of ex–Royal Engineer James Benjamin Launders, shows the peninsula, complete with *Good Soil* noted, that divided the two parts of *False Creek*, now Main Street north of Terminal Avenue.

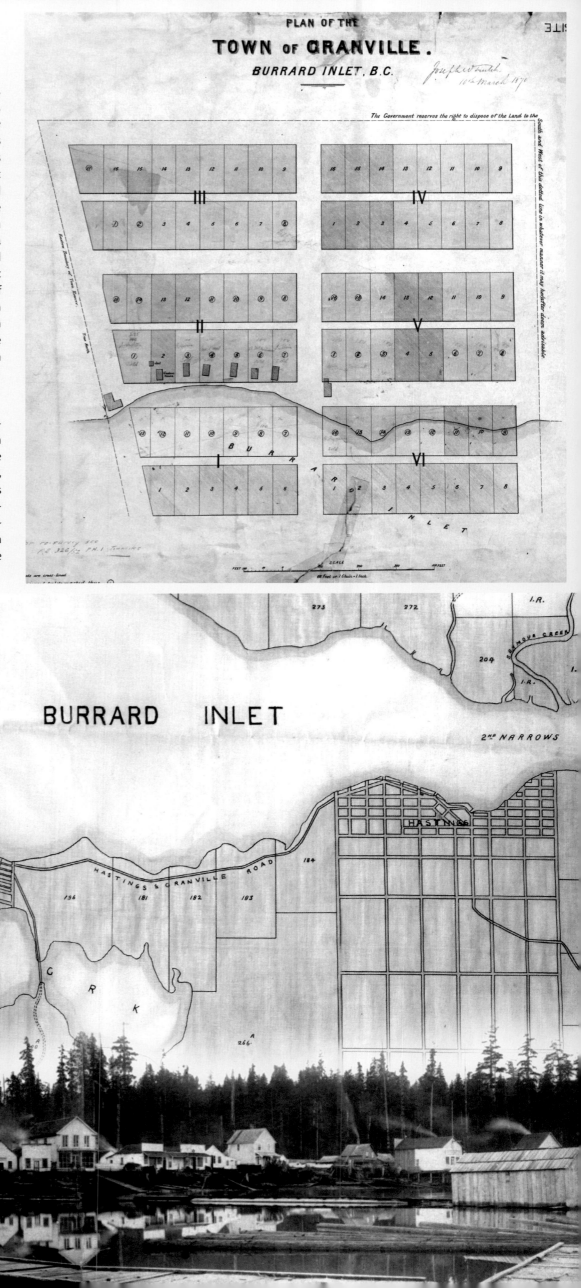

Map 259 (*right*).

A historically very important map of the townsite of Granville, laid out by the colonial government near Hastings Mill in 1870, the year before British Columbia joined Confederation. The survey was carried out by ex–Royal Engineer James Launders and signed by Joseph Trutch, then commissioner of Lands and Works but a year later British Columbia's first lieutenant governor. The existing structures shown on the map were all technically squatters on a government reserve, and hence some do not fall within surveyed lots. Notable in this regard is John Deighton's Globe Saloon, which sits right in the middle of a surveyed road at left, at what is now the intersection of Carrall and Water Streets. Deighton purchased a lot at the auction that prompted this survey and rebuilt his saloon "off-road." Some of the surveyed lots are over water; some structures were built on piles, as can be seen in the photo *below*, a view of Granville in late 1884 or before August 1885. The left-most building in the photo is the Sunnyside Hotel addition; this hotel was built on Lot 12 of Block 1, which can be located, over water, on the map.

Map 260 (*below*).

The street pattern of the 1870 survey can be seen in the northeast corner of the streets shown on this map, drawn seven years later. The mapmaker has added streets spilling onto the rest of the town reserve owned by the provincial government, but these were non-existent except on paper. The plan for this area would await the coming of Lauchlan Hamilton, the Canadian Pacific surveyor eight years hence. The *Hastings & Granville Road* connects Granville with Hastings, likewise shown with a road network that did not exist at that date. A bridge crosses False Creek at its narrowest part.

Vancouver Island Gold

After the gold discoveries of the Fraser Canyon and the Cariboo, Victoria was not growing at the frenetic pace of its earlier years. When a new colonial governor, Arthur Kennedy, arrived in 1864, he offered to partially finance an expedition that would explore the southern part of Vancouver Island, ostensibly to fill in the still-extensive gaps in the map but in reality to search for gold that, it was hoped, would revive the flagging ambitions of Victoria. Indeed, small amounts of gold had been found close to the city the year before at the appropriately named Goldstream.

An expedition was assembled that would be led by Robert Brown, a Briton on Vancouver Island to collect botanical specimens, and included two ex–Royal Engineers, Peter Leech and John Meade. The artist Frederick Whymper also joined what was named the Vancouver Island Exploring Expedition.

The expedition explored in small groups, covering a lot of ground. Many areas of Vancouver Island that were then essentially unknown were mapped, and the expedition collected much significant information on the agricultural and mineral potential of the island. The expedition made two major discoveries: coal in the Co-

mox Valley (see page 229) and placer gold on the upper reaches of the Sooke River. The latter discovery led to a minor gold rush, raising the hopes of Victoria to new highs. An instant town sprang up, named Leechtown after the discoverer of the gold there. At its peak, there were 1,200 mining operations in progress and a workforce of 4,000; the area reputedly produced $100,000 worth of gold that first year.

As is often the case with gold rushes, the excitement was over within a couple of years as the easily found gold was taken away. Digging to find the source of the gold proved very difficult. Yet gold continued to be found in small quantities, and mining briefly began again in earnest during the Depression of the 1930s, and the area around Leechtown still produces small amounts of gold, a century and a half after its first discovery.

Map 261 (below).
The field notebook of ex–Royal Engineer and member of the Vancouver Island Exploring Expedition John Meade. The map on the right shows the Sooke River between *Sooke Lake* and *Sooke Inlet.* Leechtown sprang up at the notation *gold found 200 yards from Camp 18.* This location, the confluence of the two rivers, is at the bottom of the left-hand map, which shows the Leech River.

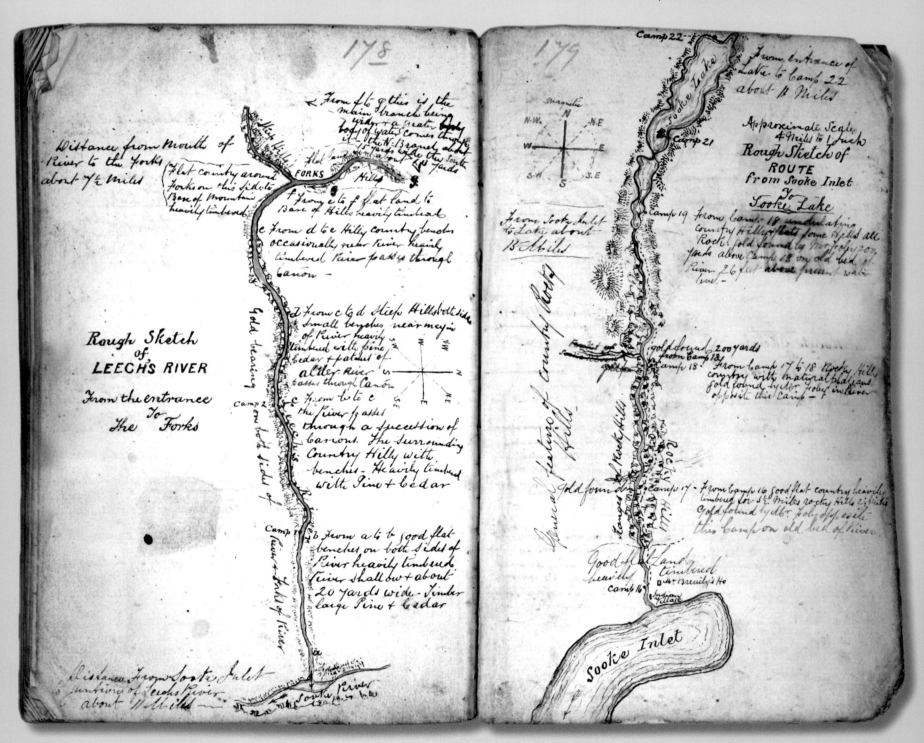

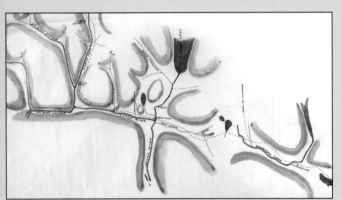

MAP 262 (*left*).
Another page of Meade's notebook shows part of the Leech River with *gold found* in several locations.

MAP 263 (*right*).
Expedition member John Buttle made this map of the area around Kennedy Lake, now abreast of the road to Long Beach, the unnamed coast to the left. The lake was named after expedition sponsor Governor Arthur Kennedy.

MAP 264 (*below*).
An 1881 map with *Leech Town* at the junction of the *Sooke River* and *Leech River*. *Goldstream Creek*, where gold had been found in 1863, is at right, with *Langfords Lake* at bottom right. The map was drawn by surveyor William Ralph, who prospected the area himself in 1864 and returned years later when he met a man who told him he had a stolen map showing the location of a hidden gold stash. Ralph helped the man dig for it, to no avail. But there is still a legend that the area holds hidden gold.

MAP 265 (*below*).
The considerably increased knowledge of the interior of Vancouver Island, especially the southern half, is displayed on this excellent map. It was based on the work of Robert Brown's Vancouver Island Exploring Expedition and was published in 1869 by German geographer Augustus Petermann in his long-running magazine, *Petermanns Geographische Mitteilungen*. The topography has been shown quite effectively using hachures, short lines drawn up and down the direction of slope.

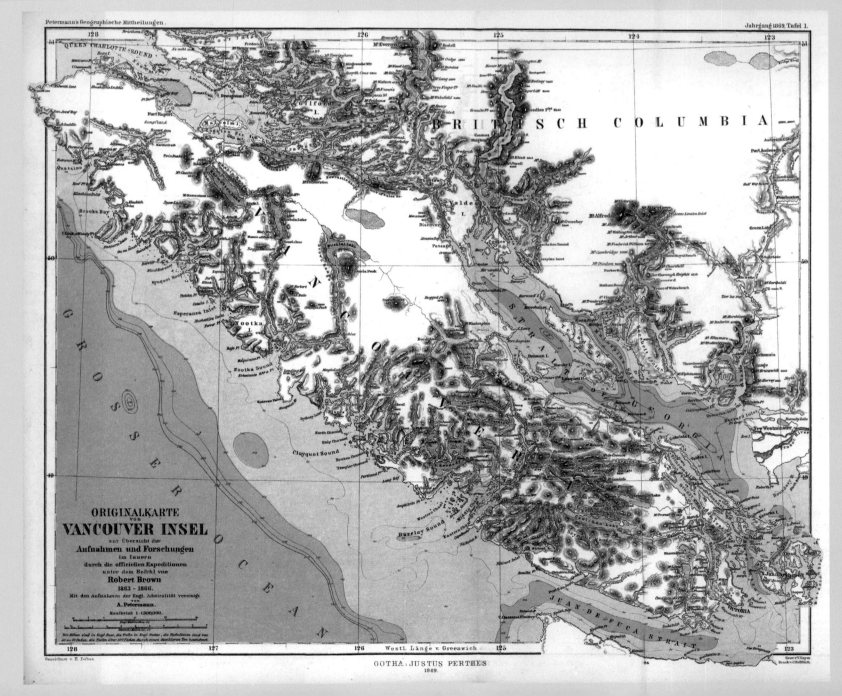

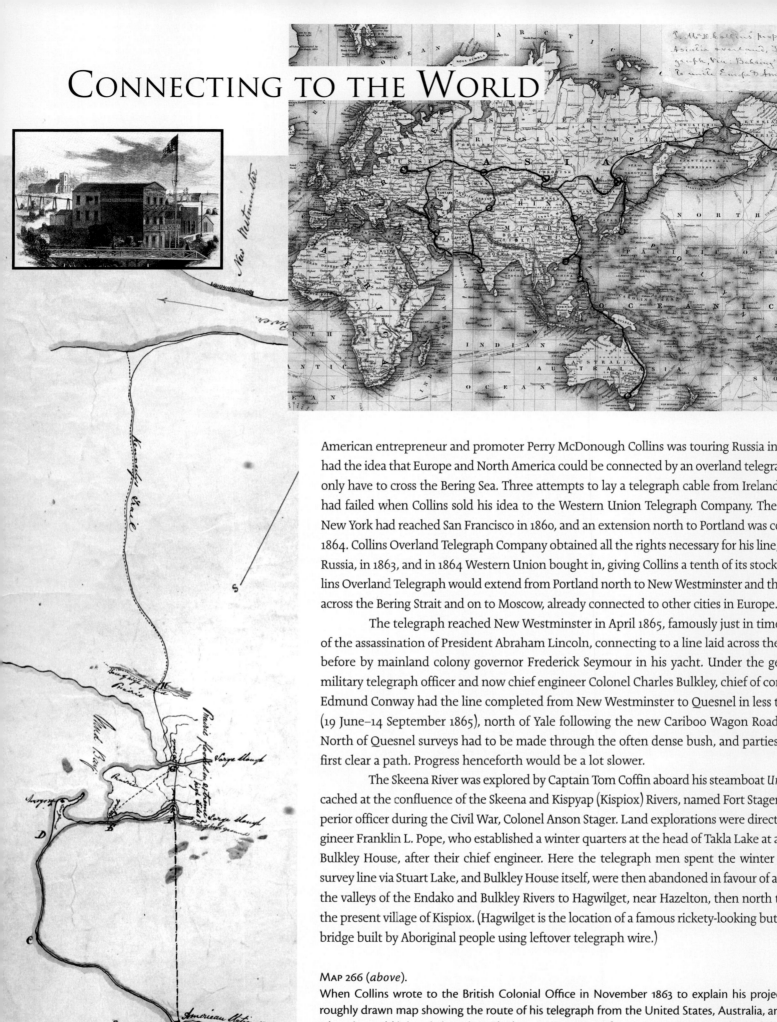

Connecting to the World

American entrepreneur and promoter Perry McDonough Collins was touring Russia in 1856 when he first had the idea that Europe and North America could be connected by an overland telegraph line that would only have to cross the Bering Sea. Three attempts to lay a telegraph cable from Ireland to Newfoundland had failed when Collins sold his idea to the Western Union Telegraph Company. The latter's cable from New York had reached San Francisco in 1860, and an extension north to Portland was completed in March 1864. Collins Overland Telegraph Company obtained all the rights necessary for his line, including those in Russia, in 1863, and in 1864 Western Union bought in, giving Collins a tenth of its stock in return. The Collins Overland Telegraph would extend from Portland north to New Westminster and then north to Alaska, across the Bering Strait and on to Moscow, already connected to other cities in Europe.

The telegraph reached New Westminster in April 1865, famously just in time to relay the news of the assassination of President Abraham Lincoln, connecting to a line laid across the Fraser the month before by mainland colony governor Frederick Seymour in his yacht. Under the general direction of military telegraph officer and now chief engineer Colonel Charles Bulkley, chief of construction Captain Edmund Conway had the line completed from New Westminster to Quesnel in less than three months (19 June–14 September 1865), north of Yale following the new Cariboo Wagon Road as far as possible. North of Quesnel surveys had to be made through the often dense bush, and parties of axemen had to first clear a path. Progress henceforth would be a lot slower.

The Skeena River was explored by Captain Tom Coffin aboard his steamboat *Union*. Supplies were cached at the confluence of the Skeena and Kispyap (Kispiox) Rivers, named Fort Stager, after Bulkley's superior officer during the Civil War, Colonel Anson Stager. Land explorations were directed by electrical engineer Franklin L. Pope, who established a winter quarters at the head of Takla Lake at a place they named Bulkley House, after their chief engineer. Here the telegraph men spent the winter of 1866–67. Pope's survey line via Stuart Lake, and Bulkley House itself, were then abandoned in favour of another route using the valleys of the Endako and Bulkley Rivers to Hagwilget, near Hazelton, then north to Fort Stager, near the present village of Kispiox. (Hagwilget is the location of a famous rickety-looking but structurally sound bridge built by Aboriginal people using leftover telegraph wire.)

Map 266 (*above*).
When Collins wrote to the British Colonial Office in November 1863 to explain his project, he enclosed this roughly drawn map showing the route of his telegraph from the United States, Australia, and India to Moscow, where it would join existing wires. The base map was torn from a contemporary atlas and the map's original title pasted over with his own. The atlas map was centred on North America, and hence the telegraph line extended off the left-hand side and continued on the right; here part of the right side has been placed next to the left to better display the telegraph's route. Note the three alternative crossings of the Bering Sea.

Map 267 (*left*).
Drawn in 1865, this rough sketch shows the line the telegraph would take from the United States north to New Westminster. The *British Station HQ* and *American Station HQ* refer to the locations of the boundary survey camps. The boundary was surveyed between 1857 and 1862. The *Kennedy Trail*, which continues north to the Fraser, was used for the northern part, while parts of an existing trail, now called Telegraph Trail, were followed across the Semiahmoo Peninsula north of Campbell *River*. Inset is the telegraph office established in New Westminster.

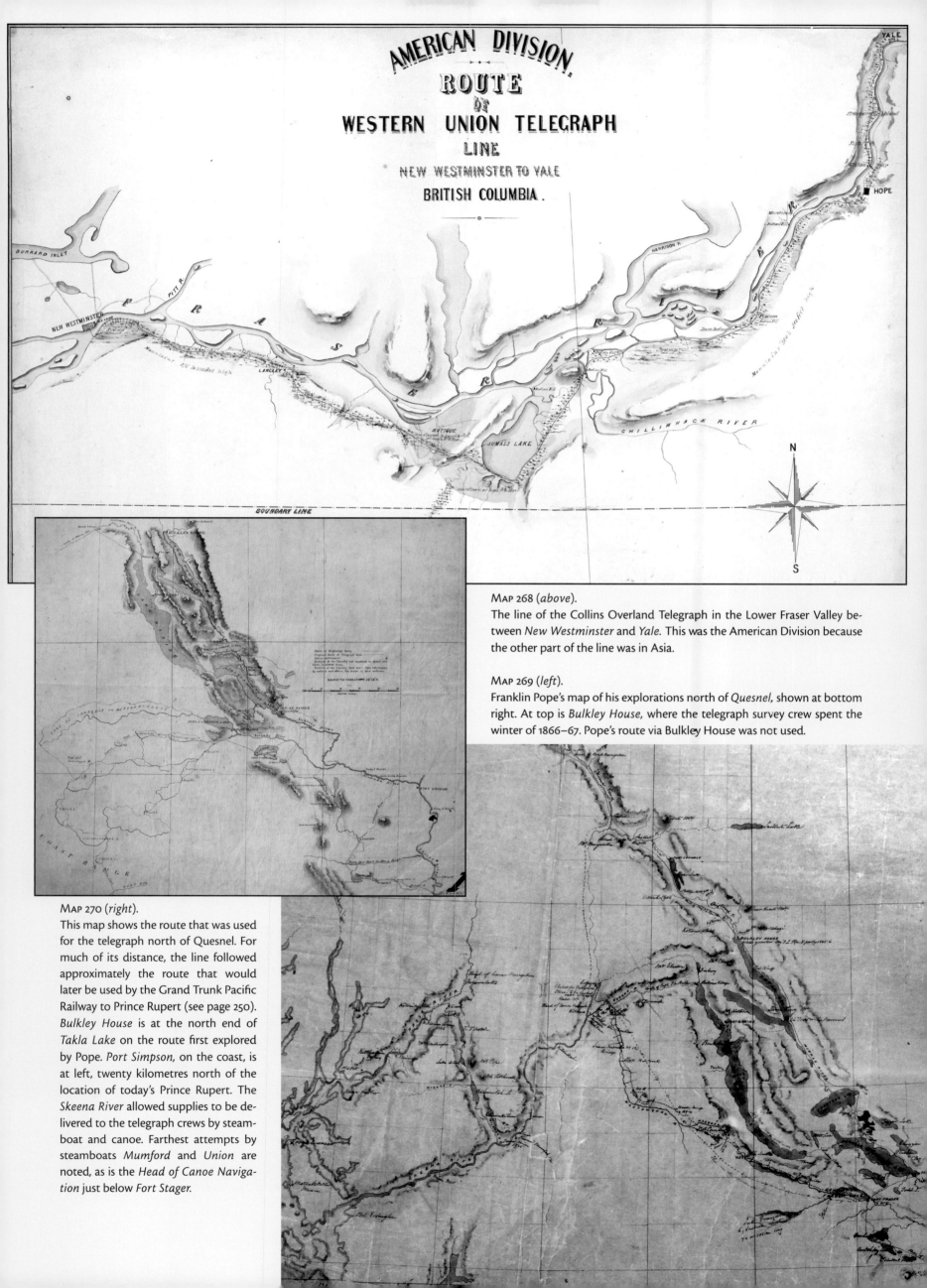

AMERICAN DIVISION.
ROUTE
OF
WESTERN UNION TELEGRAPH
LINE
NEW WESTMINSTER TO YALE
BRITISH COLUMBIA.

MAP 268 (*above*).
The line of the Collins Overland Telegraph in the Lower Fraser Valley between *New Westminster* and *Yale.* This was the American Division because the other part of the line was in Asia.

MAP 269 (*left*).
Franklin Pope's map of his explorations north of *Quesnel,* shown at bottom right. At top is *Bulkley House,* where the telegraph survey crew spent the winter of 1866–67. Pope's route via Bulkley House was not used.

MAP 270 (*right*).
This map shows the route that was used for the telegraph north of Quesnel. For much of its distance, the line followed approximately the route that would later be used by the Grand Trunk Pacific Railway to Prince Rupert (see page 250). *Bulkley House* is at the north end of *Takla Lake* on the route first explored by Pope. *Port Simpson,* on the coast, is at left, twenty kilometres north of the location of today's Prince Rupert. The *Skeena River* allowed supplies to be delivered to the telegraph crews by steamboat and canoe. Farthest attempts by steamboats *Mumford* and *Union* are noted, as is the *Head of Canoe Navigation* just below *Fort Stager.*

In the summer of 1866 news arrived of the completion of the transatlantic cable on the fifth attempt to lay it, though it was discounted at the time because some previous attempts had at first succeeded, only to break soon after. And so work continued. But Western Union slowly came to the realization that this time the undersea line was permanent, and in 1867 orders where sent out to cease work on the overland telegraph.

The line had reached 40 km up the Kispiox River north of Fort Stager but went no farther. The working portion between Quesnel and Fort Stager was abandoned, but the portion south to New Westminster continued in commercial operation. In 1868 the line was extended east to Barkerville, in the Cariboo gold fields. Three years later the British Columbia colonial government obtained a perpetual lease on all the company's lines within its boundaries, and later that year, when British Columbia joined Confederation, the federal government took over the lease. In 1880 the federal government purchased the line outright.

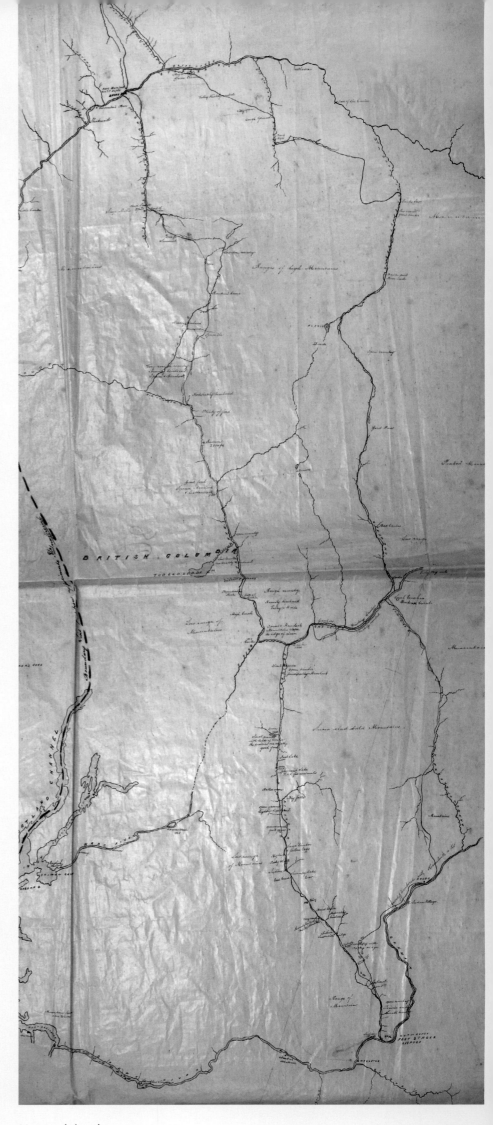

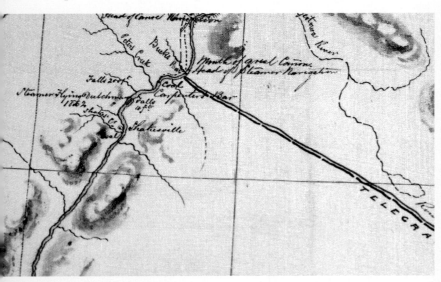

Map 271 (*above*).
The surveyed line of the telegraph as it approached the Stikine River. The *Head of Steamer Navigation* is shown, at *Buck's Bar*. The settlement that grew up here became known as Telegraph Creek.

Map 272 (*below*).
The *Yukon Telegraph*, shown here on a 1924 map as a black line with black dots, finally made it to *Telegraph Creek* and on to the Klondike. Built in 1899 by the federal government, it followed the intended route of the Collins Overland Telegraph.

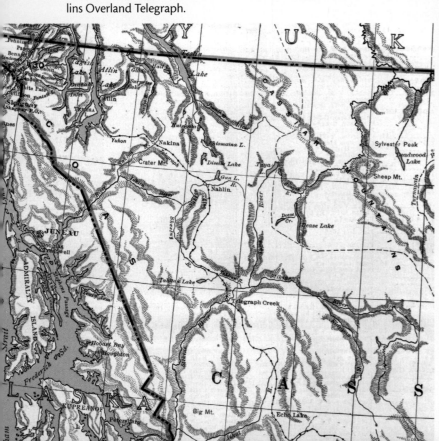

Map 273 (*above*).
The proposed route north of *Fort Stager* of the Western Union telegraph line as far as *Buck's Bar* (Telegraph Creek) on the *Stikeen River*. On the river, *Fort Mumford W.U.T.C. Depot* is seen. This was named after Western Union's steamboat, the *Mumford*. The dashed red line at left is the Alaska boundary.

THE SAN JUAN BOUNDARY DISPUTE

When the international boundary was agreed upon in 1846 (see page 58), the British were glad enough to retain all of Vancouver Island, as they were then seeking. The treaty establishing the new boundary line simply stated that the forty-ninth-parallel line was to be extended to the "middle of the channel" between Vancouver Island and the mainland, and then to follow it southward. At the time it seems that there was not a very good understanding of the geography of the Strait of Georgia—all of George Vancouver's early work counted for naught (he did not map the San Juan Islands in detail). The same lack of understanding on both sides allowed the creation of Point Roberts as an isolated peninsula, an island of American territory, but both sides wanted the boundary issue resolved and were not concerned about details; these could be worked out later. However, this "working out" nearly precipitated a war. For islands inhabited the southern strait, and it was not clear where the "middle of the channel" lay.

The British may have at first thought that Haro Strait was a fair interpretation of the "middle," but it soon became clear that nearby San Juan Island could threaten the new settlement of Victoria, and for Britain, it became imperative that the island be in British hands.

In 1856 Britain and the United States agreed to carry out a joint boundary survey. While the survey was taking place, a dispute erupted over the ownership of San Juan Island. In June 1859 an American settler shot a pig that had wandered onto his land—a pig that belonged to the Hudson's Bay Company, which had a farm there. The settler then refused to pay the $100 in damages demanded by the company. General William S. Harney, aggressive and anti-British military commander of the Department of Oregon, ordered troops to the island, and their commanding

MAP 274 (right).
This British map, with key, inset, is dated 3 December 1857. It was copied from an American naval map, and the potential boundary lines through the San Juan Islands were added. The map was signed by George Henry Richards, captain of HMS *Plumper*, and James Prevost, captain of HMS *Satellite*, both British survey ships. The black line passes through *Rosario Strait*, the red through the *Canal de Haro* (Haro Strait and Boundary Pass), and the green through the middle channel, which would have given *Belle Vue or San Juan I.* to the British but the rest of the San Juan Islands to the United States. The blue line, according to the key, "passes through the centre of the whole space between the Continent and Vancouver's Island" at the expense of bisecting *Orcas Island* and *Shaws I.* Some of the San Juans are noted by the names given to them by Charles Wilkes of the 1841 U.S. Exploring Expedition.

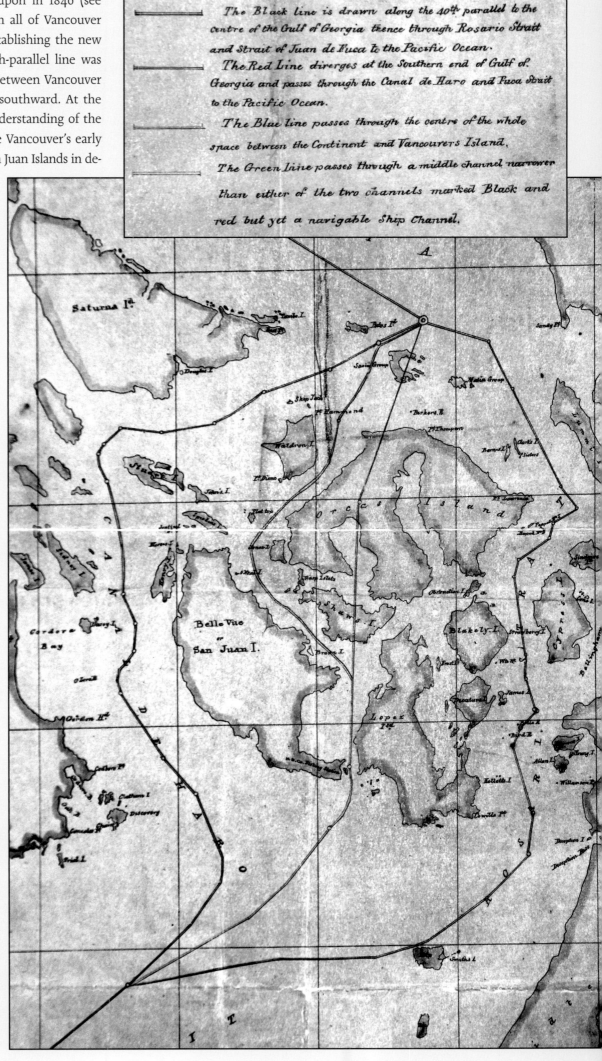

The Black line is drawn along the 40th parallel to the centre of the Gulf of Georgia thence through Rosario Strait and Strait of Juan de Fuca to the Pacific Ocean.

The Red Line diverges at the Southern end of Gulf of Georgia and passes through the Canal de Haro and Fuca Strait to the Pacific Ocean.

The Blue line passes through the centre of the whole space between the Continent and Vancouvers Island.

The Green line passes through a middle channel narrower than either of the two channels marked Black and red but yet a navigable Ship Channel.

Petermann's Geographische Mittheilungen.

Jahrgang 1859 Tafel 19.

KARTE
DES
SAN JUAN- od. HARO- ARCHIPELS.
Nach den Aufnahmen der Engl. Admiralität
unter KELLETT, RICHARDS &c.
1847, 1858 und 1859.
Von A. Petermann.
Maassstab 1 : 500000.

Deutsche Meilen (15 = 1°)

Die Zahlen im Lande bezeichnen die Höhe in Engl. Fuss,
diejenigen im Meere dessen Tiefe in Englischen Faden.

☐ Englisches Gebiet.
☐ Gebiet der Verein. Staaten.
☐ Strelliges Gebiet.

Abkürzungen.
B. . Bai.   Mt. . Mount.
B⁹. . Bank.   P. . Pass.
C. . Cap.   Pass. . Passage.
Ch. . Channel (Canal). Pen. . Peninsula.
Dr. . Dorf.   Pk. . Peak.
H. . Hill (Berg).   Pt. . Point (Spitze).
H.B.C. Hudsons B. Comp. R. . River (Fluss).
Hd. . Head.   Rge. . Range (Kette).
Hr. Harbour (Hafen). Rf. . Reef (Riff).
I. . Island (Insel). Rk. . Rock (Felsen).
L. Lake (See).   Sd. . Sound.

GOTHA: JUSTUS PERTHES
1859.

L. Printerichsen del.

MAP 275 (left).
A well-produced map of the San Juan boundary issue published by the influential German geographer Augustus Petermann in his magazine *Petermanns Geographische Mitteilungen* in 1859. This particular copy found its way into American government files. Here the options are but two: Rosario Strait or Haro Strait.

MAP 276 (below).
Published in 1871, the year before arbitration, this map depicts the San Juan Islands from both countries' perspectives, circled by the British and American claim lines.

copied vast numbers of maps to bolster their respective cases, each trying to show that the other had intended the boundary they wanted.

The kaiser handed down his decision in October 1872, and it was one entirely unfavourable to the British. He selected as the boundary the middle of what is the widest continuous channel—through Haro Strait and its northern connection with the Strait of Georgia; this became known as Boundary Pass. The San Juan Islands were lost to British Columbia and to Canada.

---

officer, Captain George E. Pickett (later famous for Pickett's Charge in the American Civil War Battle of Gettysburg in July 1863) issued a proclamation claiming the island as United States territory.

In response, Governor James Douglas dispatched three ships and detachments of Marines and Royal Engineers. The resulting standoff was dubbed the Pig War. The British ensconced themselves at the northern end of the island, and the Americans established a camp at the southern end. The U.S. government was not pleased with Harney's actions and sent General Winfield Scott, veteran of the war with Mexico, to deal with the matter. Joint occupancy returned to the Oregon Country when Scott and Douglas agreed that a company of each of their forces would both occupy the island until it could be determined to which nation it belonged. From 1860 to 1872 this is how the situation remained.

Finally, in 1871 Britain and the United States agreed to submit their dispute to the German kaiser for arbitration, and minds such as the great chancellor Otto von Bismarck were brought to bear on the issue. Both sides

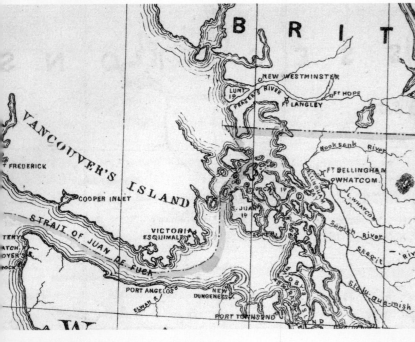

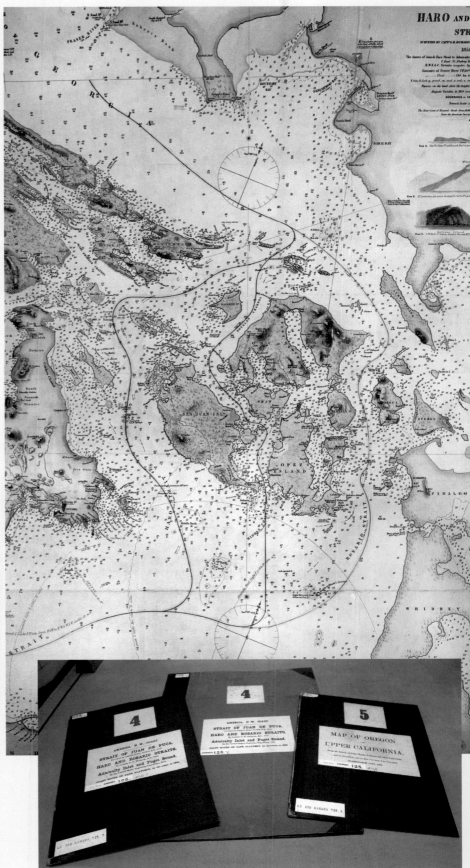

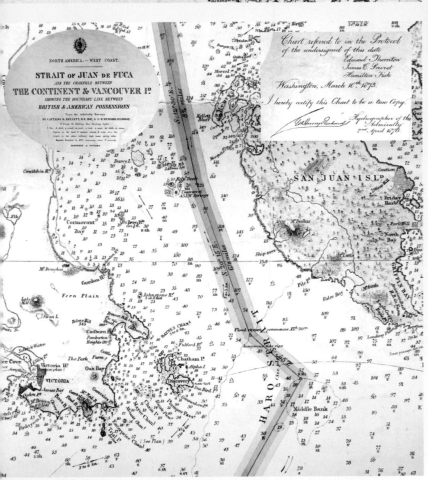

MAP 277 (*left, top*) and **MAP 278** (*left, centre*).
An 1866 American map and an 1858 British map, both produced during the period of conflicting claims to the San Juans. Each shows its respective nation's position as if it were a *fait accompli.* Note the American military presence at *Ft. Bellingham* on **MAP 277**.

MAP 279 (*above*).
From the British case presented to the emperor of Germany for arbitration in 1871–72, this colourful map also depicts the three possible boundaries he was to consider. One can perhaps see why the British were confident they would retain at least *San Juan Island*, since an obvious compromise would have been the middle line.
Inset are some of the many folders containing maps used to present the British case. They are now in the British National Archives.

MAP 280 (*left*).
Every boundary settlement needs its documentation. This is a small section of the map used in the signed protocol of March 1873 to exactly define the new arbitrated boundary between *San Juan Isl^d.* and Vancouver Island down the middle of the now-defined channel: *Haro Strait.* Note the *H.B. Co. Farm* at *American Bay* on San Juan Island.

GOLD ON WILD HORSE CREEK

In nineteenth-century British Columbia gold rushes came and went, often without major consequence, but one rush that did leave a legacy was that of Wild Horse Creek (today River), an east bank tributary of the Kootenay River in the East Kootenays. Here, because the easiest access was north from the United States, Governor Frederick Seymour was concerned, as Douglas was in 1858, that there would be an uncontrollable influx of American miners that might yet lead to a claim to the region—or other parts of southern British Columbia—by the United States. In any case, the lack of a road meant a loss of trade for Victoria or New Westminster merchants. Hence Seymour ordered the extension of the Dewdney Trail all the way east to Wild Horse Creek.

British engineer and surveyor Edgar Dewdney had constructed a pack trail from Hope to the Similkameen Valley in 1860 and 1861. In April 1865 he was contracted to extend this trail east. The 700-km-long trail he built became the first highway linking eastern British Columbia with the coast and was the predecessor of much of today's difficult Highway 3.

Jack Fisher discovered gold in the canyon of Wild Horse Creek in 1863, setting off a rush that for a short time saw a thousand miners prospecting in the area. Fisherville, the town that sprang up to service the miners, for a time had five thousand inhabitants. In 1865 the town was moved, and renamed Kootenai or Wild Horse, after it became apparent that a rich gold deposit lay right under it. The town's location, labelled *Old Town,* can been seen on MAP 284, *far right.*

Dewdney hacked his trail eastward, reaching Wild Horse Creek in September 1865. Yet by the following year most of the miners had disappeared, leaving for the promise of easier gold elsewhere, so coast merchants failed to achieve the supply business they coveted. However, the Dewdney Trail, despite its rough nature, proved to be a vital link in the early communication routes of the province.

MAP 281 (*below*).
This three-dimensional map of the Dewdney Trail is in the Fort Steele Museum. *Fort Hope* is at left; *Galbraith's Ferry* and *Wild Horse* are at right.

MAP 283 (*above, right*).
The course of *Wild Horse Creek,* flowing into the *Kootenay River,* is shown on this 1865 map. *Dromedaryville* lies just to the north, suggesting the presence of dromedaries—Arabian camels—perhaps? They briefly plied the Cariboo Road, but it is not known that any reached here.

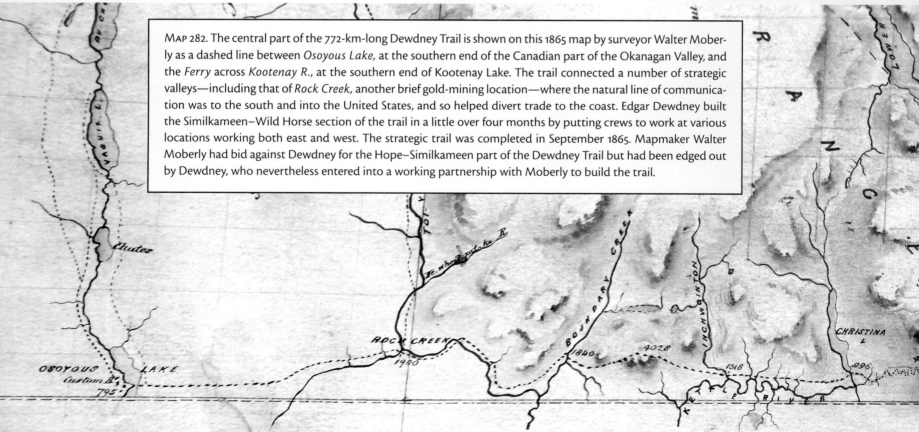

MAP 282. The central part of the 772-km-long Dewdney Trail is shown on this 1865 map by surveyor Walter Moberly as a dashed line between *Osoyous Lake,* at the southern end of the Canadian part of the Okanagan Valley, and the *Ferry* across *Kootenay R.,* at the southern end of Kootenay Lake. The trail connected a number of strategic valleys—including that of *Rock Creek,* another brief gold-mining location—where the natural line of communication was to the south and into the United States, and so helped divert trade to the coast. Edgar Dewdney built the Similkameen–Wild Horse section of the trail in a little over four months by putting crews to work at various locations working both east and west. The strategic trail was completed in September 1865. Mapmaker Walter Moberly had bid against Dewdney for the Hope–Similkameen part of the Dewdney Trail but had been edged out by Dewdney, who nevertheless entered into a working partnership with Moberly to build the trail.

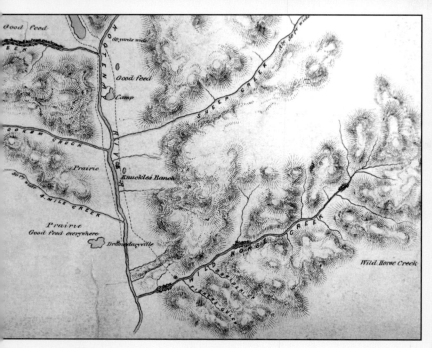

Map 284 (*right*).

An 1897 blueprint map of the mineral claims on *Wild Horse River* and vicinity. *Old Town* marks the location of Kootenai or Wild Horse, previously Fisherville, on the *Wagon Road* to *Fort Steele*.

Map 285 (*below*).

Details of the Trutch map of 1871 (Map 913, pages 358–59) showed Galbraith's *Ferry*, established in 1864 to transport miners across the Kootenay to *Wild Horse Ck.* Galbraith's original cabin (*inset, right*) is preserved at Fort Steele. *Josephs Prairie* is the location of today's Cranbrook.

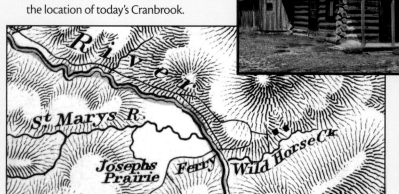

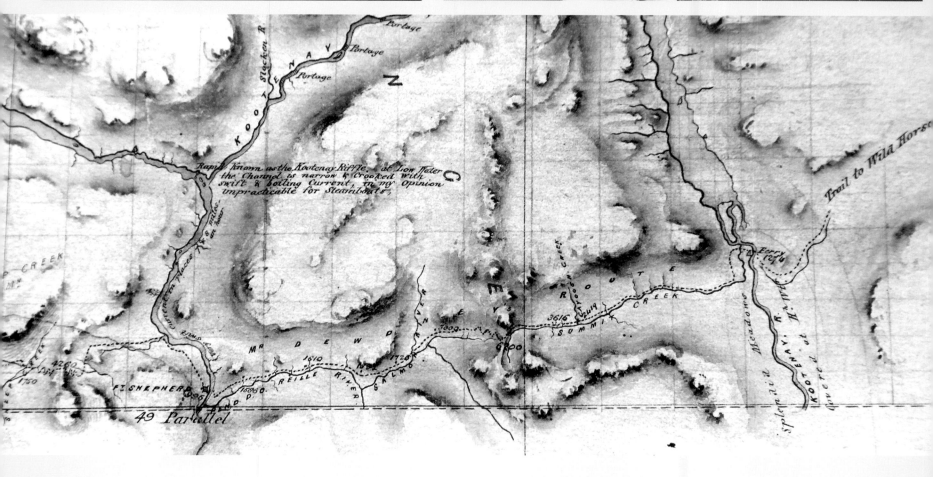

Sam Steele's Fort

John Galbraith set up his ferry across the Kootenay in 1864, and a small settlement had grown up around it. In 1887 a contingent of North-West Mounted Police arrived in response to a request from the local magistrate, Colonel James Baker, who was anticipating an uprising by the Kootenay people of the region led by Chief Isadore. The Mounties were led by the redoubtable Major Samuel Benfield Steele, who secured a temporary lease from Galbraith to build his camp. The police travelled, as with the North-West Rebellion of Louis Riel three years earlier, by the new Canadian Pacific Railway. Steele arrived at Golden and marched up the Columbia Valley and down the Kootenay after seeing a steamboat onto which his supplies were loaded sink right before his eyes.

As it happened, the police force proved hardly necessary, for Steele was an excellent diplomat and negotiated a peace with Chief Isadore, and an accused murderer whom Steele found no evidence against was released. Steele and his men left in 1888, but not before Galbraith had renamed his town Fort Steele in honour of his guest. Steele went on to a continued distinguished career, commanding forces in both the Boer War and World War I in Europe.

MAP 286 (right).
Included in Sam Steele's report was this map showing the route from *Golden City*, patrols while at Fort Steele, and *The Fort*, complete with a large red flag marking the spot. *Col. Baker's* is the location of Colonel James Baker's lands, the place where, with the arrival of the railway, the city of Cranbrook was to be established, eclipsing Fort Steele as a regional centre.

MAP 287 (below).
A plan of the Fort Steele Townsite originally filed in 1897; this is a 1941 certified copy. *Below, right* is a street in Fort Steele today. The entire town is now a National Historic Site.

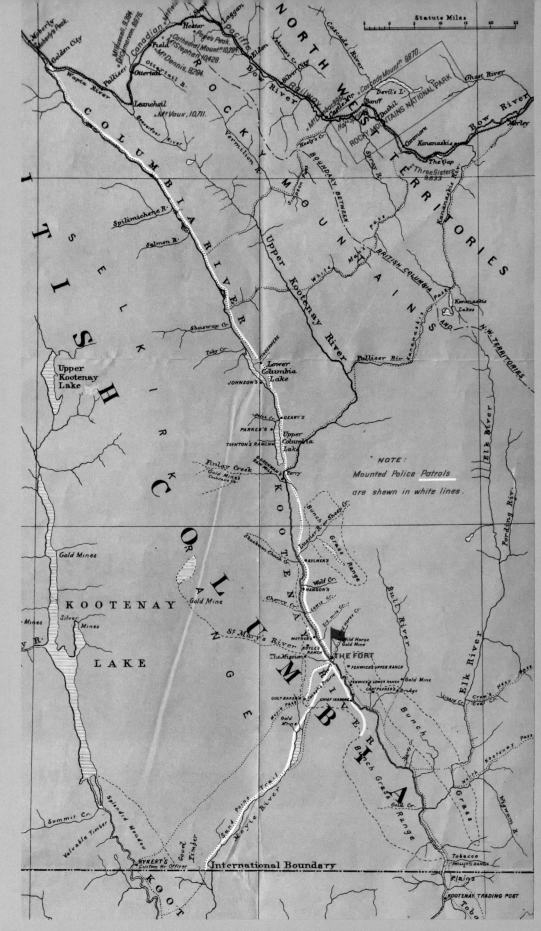

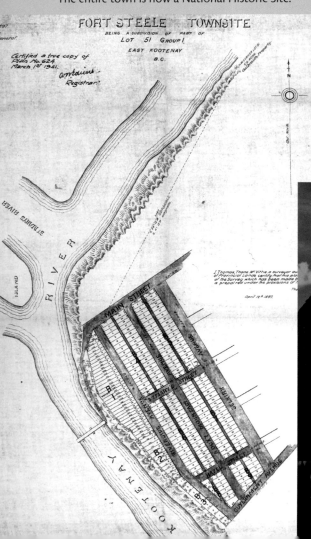

Omineca Gold

When the Wild Horse Creek gold rush subsided, many of the gold seekers left for the Big Bend area of the Columbia, where more gold finds had been reported. That proved to be less than lucrative for most, but then the miners were able to move even farther north, to the Omineca region, named for the Omineca River, which flowed into the Finlay, the northern tributary of the Peace River (it now flows into Williston Lake, behind the W.A.C. Bennett Dam).

Gold had been found in the area in 1861, but more substantial finds in 1869 set off a rush. Despite the region's difficult accessibility, by June 1870 some four hundred miners had arrived, including an American, James Germansen. He found an as-yet-unknown creek from which vast amounts of gold were recovered. Subsequently it was named Germansen Creek. One group of five miners reportedly found 390 ounces of gold there in thirteen days the following year.

By 1871 1,200 miners were working on Omineca creeks. Some 31,000 ounces of gold were reportedly extracted in 1871, and 29,000 the following year. However, like most gold rushes, this one was over quite quickly, and in 1873 news of the next strike, this time in Cassiar (see page 210), induced many miners to move on. The gold rush's main legacy was that knowledge of this previously little-known and remote region was considerably improved.

MAP 289 (below).
William Patterson's 1870 map was the first to show the Omineca gold fields, and his map also showed the Cariboo fields, both brightly coloured yellow. The map was produced to show gold seekers how to reach this latest promised bonanza. It correctly shows *Germansen C[reek]* flowing into the *Omineca R.*

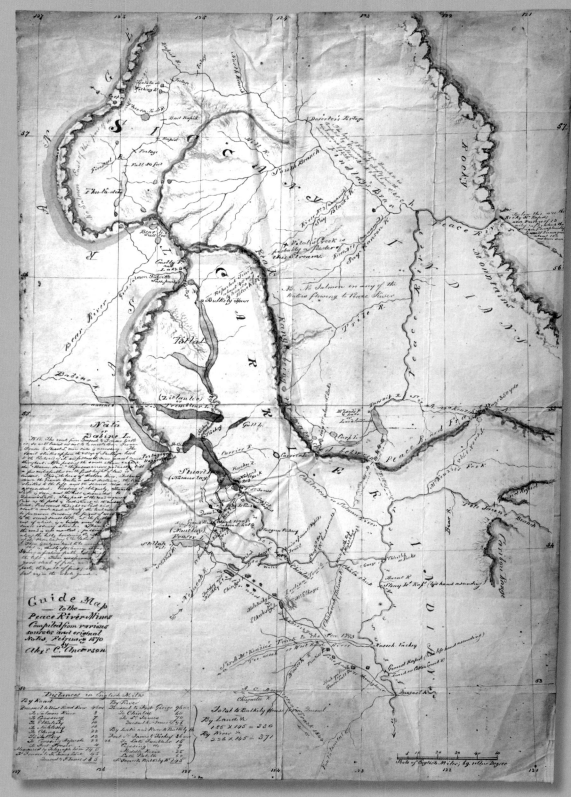

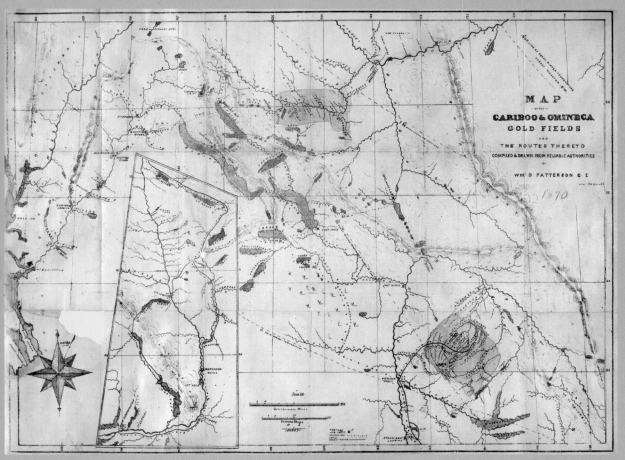

MAP 288 (above).
This map of the Omineca was drawn by ex–Hudson's Bay Company man Alexander Caulfield Anderson, who produced a number of detailed maps of early British Columbia (see next page). This one, drawn in 1870, is full of detailed notes and includes the names of Aboriginal groups written across their locales. It is unclear whether this map had much influence at the time, however, as it is a single manuscript rather than, like Patterson's map, printed for bulk distribution. *Findlay's Branch* is the Finlay River, and the *Peace River* is the Parsnip south of its confluence with the Finlay, where it becomes the Peace. This area is today submerged under Williston Lake. The river against which Anderson has written *Say Manson R.* is more likely the Omineca itself and is the river into which Germansen Creek flowed. The notation about *Vital's Creek* in red above it refers to the location of the earliest substantial gold find by Vital LaForce, previously employed by the Overland Telegraph Company (see page 98).

Alexander's Amazing Map

One of the most detailed, artistically excellent, and historically detailed maps of British Columbia is the one that Alexander Caulfield Anderson drew in 1867. Anderson spent twenty-two years, from 1832 to 1854, in the Northwest in the employ of the Hudson's Bay Company and during that time conducted a number of expeditions and lived in many different locations, all the time keeping meticulous records and drawing maps. In 1867 he put all this knowledge together to compile one enormous map, his *Map of a portion of the Colony of British Columbia*, which has survived and is currently held in the British Columbia Archives, where it has recently been restored.

Because of Anderson's record keeping and his careful map drawing, this map is a unique historical document of British Columbia's Hudson's Bay Company years. Even when it was drawn, the map was intended as a historical document as well as a record of geographical minutiae for a region still then not fully explored.

Anderson joined the Hudson's Bay Company in 1831 and arrived at Fort Vancouver the following year. In 1834 he was with Peter Skene Ogden when the company attempted to establish a fort on the Stikine River but was blocked by the Russians (see page 50). The following year he was placed in charge of the New Caledonia District and was stationed at Fraser Lake and Fort George.

In 1846 and 1847 Anderson led expeditions that attempted to find a route between Fort Kamloops and Fort Langley, the latter being seen as a possible replacement for Fort Vancouver, after 1846 in American territory. Between 1849 and 1860 the company did utilize one of the routes Anderson found, via the Coquihalla and Tulameen Rivers.

After stints at Fort Colville and Fort Vancouver Anderson retired from the Hudson's Bay Company in 1854 at the early age of forty. He then began a series of jobs ranging from postmaster at Victoria, collector of customs, Dominion inspector of fisheries, commissioner on Indian lands, and part-owner of the Victoria Steam Navigation Company. It was during these years that Anderson compiled a number of his detailed maps, including the one shown here.

While Anderson was collector of customs, his assistant had embezzled government money, and Anderson somehow—and unjustly—became involved, losing his job in the process. At the same time his steamship interests failed. Anderson got into considerable debt and in 1867 was unable to pay the taxes and arrears (amounting to about $1,000) on a farm he owned in North Saanich. He offered his newly drawn map to the government as payment, an offer that was accepted (though Anderson did not in the end receive a tax credit). This is how the map came to be in the British Columbia Archives.

Anderson was the author of several books about British Columbia, including his 1858 *Hand-book and Map to the Gold Region of Frazer's and Thompson's Rivers*. Other maps by Anderson include MAP 137, *page 50*; MAP 188, *page 68*; and MAP 288, *previous page*.

MAP 290 (*below*).
Alexander Caulfield Anderson's 1867 *Map of a portion of the Colony of British Columbia*. The map is huge, measuring 1.26 x 1.77 m. It contains much information from Anderson's own travels and includes some Aboriginal population distribution information, the line of the telegraph, most of the trails used by the Hudson's Bay Company, and all of its posts.

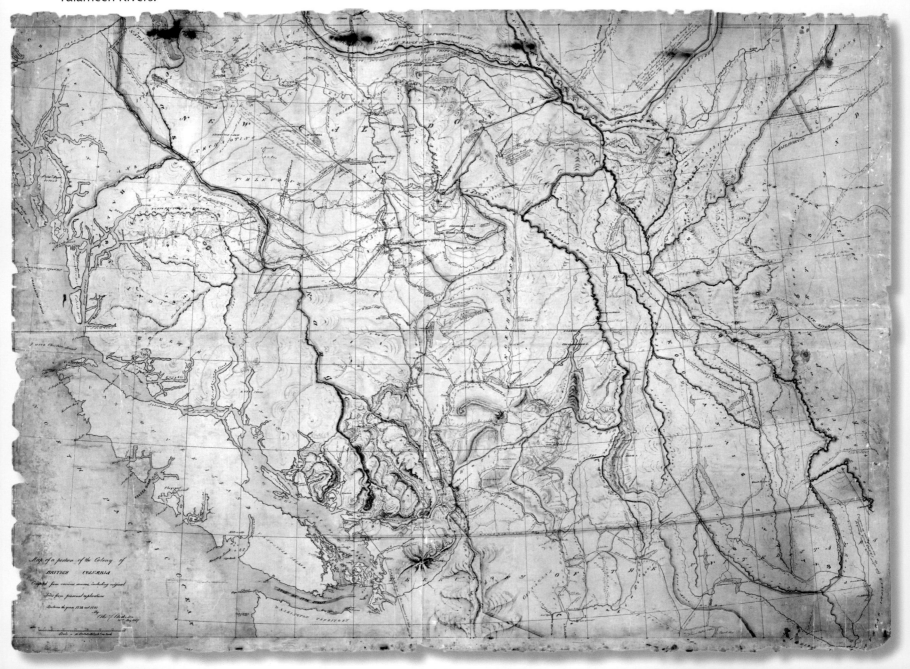

Map 291 (*above*) and Map 292 (*below*).

Details of Anderson's map. Map 291 shows the region between the *Fort* at *Kamloops* and *Fort Langley* with the routes he explored in 1846–47 to try to find a new all-British route to the Pacific following the 1846 boundary settlement. His routes included one through *Lillooett Town (sic),* at the time called Cayoosh Flat, and then down *Harrison Lake,* and another that approximately follows the path of today's Coquihalla Highway. Map 292 shows part of New Caledonia. Anderson was at the Hudson's Bay Company post at *Fraser's Lake* between 1836 and 1839 and at *F^t. George,* at the *Forks,* in 1839–40. The location of Aboriginal villages was of importance to the company, so it is of no surprise that Anderson noted all their locations. The map reveals its 1867 date with the addition of features and names beyond those of the New Caledonia era.

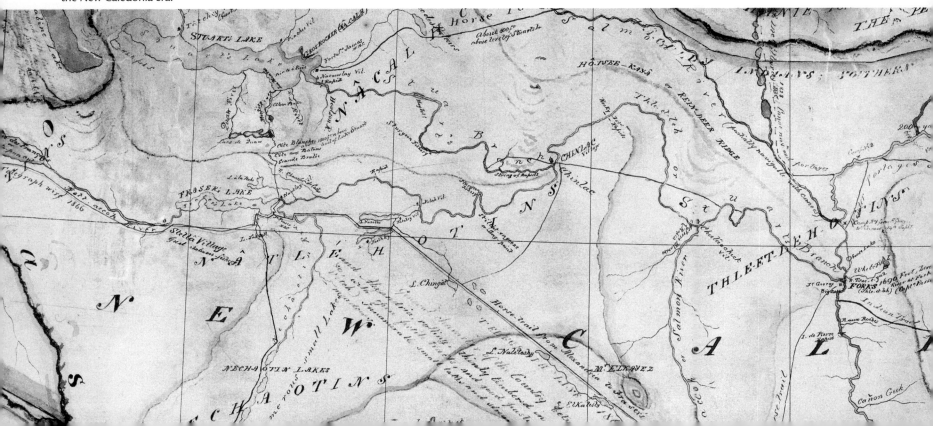

Metlakatla

The British Church Missionary Society believed that the British Empire provided an excellent means for the dissemination of its form of evangelical Christianity. According to the *Church Missionary Intelligencer,* "The Heathen cry, and they cry to us—to us Englishmen of the nineteenth century."

One of the strongest proponents of the society's principles was a missionary named William Duncan, who was sent to the northern coast in 1857 to convert the Tsimshian. He settled at Port Simpson (which has today reverted to its Aboriginal name of Lax Kw'alaams) but by 1862 had resolved to rid his Aboriginal followers of what he considered the bad influence of the Hudson's Bay Company and also, he hoped, the scourge of smallpox prevalent at Port Simpson. And so Duncan moved his flock to Metlakatla, on Digby Island to the south, about 10 km from where Prince Rupert would later be established. Metlakatla had been the site of Tsimshian settlement before its inhabitants had moved to Port Simpson in 1832 in order to be close to the Hudson's Bay post.

Duncan's goal was to found a model Christian village, and this he did, building a distinctly European-style settlement with a very large church, a sawmill, a cooperage, a tannery, a cannery, and a school. This self-proclaimed Christian utopia saw its population grow to about 1,200 persons at its peak.

Duncan had his own form of religion and a devoted band of followers, but he increasingly disagreed with the Church Missionary Society and in particular with an Anglican bishop it sent to reside in Metlakatla in 1879, William Ridley. By 1887 the fissure had become a chasm, and Duncan had resolved to repeat his earlier act: he moved with all his followers—most of the Tsimshian supported him and only a few Bishop Ridley—north to Annette Island, Alaska, where he founded New Metlakatla. There he reproduced all the facilities of the first Metlakatla. The British Columbia Metlakatla still exists but has only about a hundred inhabitants, the remnant of an imperial religious experiment that blossomed and flourished and then withered.

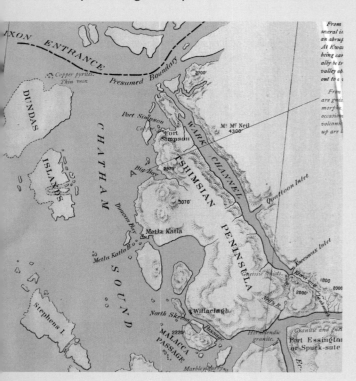

MAP 293 (*right, top*).
Four years after Metlakatla was founded, this map showed *Metlakatla Mission* on the Tsimshian Peninsula.

MAP 294 (*right*).
The *Stations of the Church Missionary Society,* including *Metlakatla,* are located on this 1879 atlas map.

MAP 295 (*left*).
An 1880 report from geologist George Mercer Dawson showed *Metla Katla* on the *Tsimshian Peninsula.*

MAP 296 (*above, right*).
A plan of Metlakatla in 1895, together with a listing of the names of Indian lot owners, drawn by the Department of Indian Affairs and signed by *C. Todd, Indian Agent.* It appears to be a reduced copy of a larger plan that Todd drew. The map approximately represents the view shown in the photo; in 1901 the community experienced a devastating fire but much was rebuilt. Note the industrial school behind the rows of houses. The church, at the time the largest in British Columbia, is on the *Church Lot* behind the *Mission Reserve.*

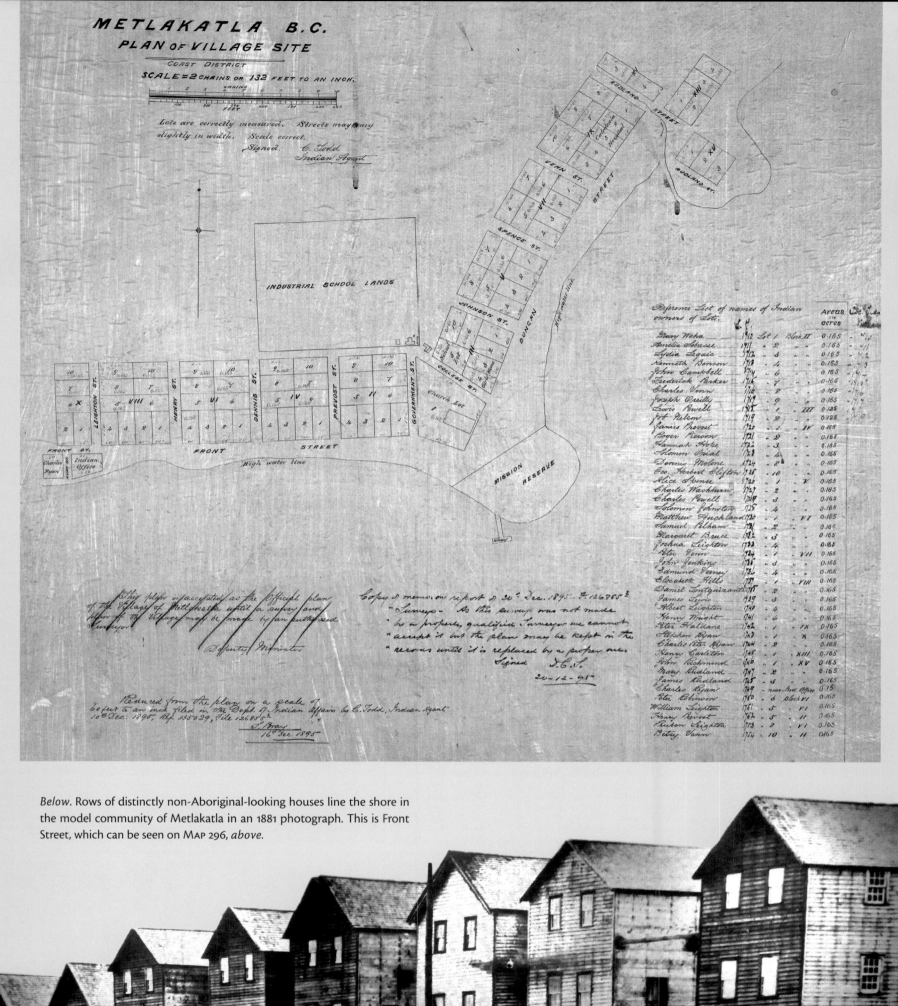

METLAKATLA B.C.
PLAN OF VILLAGE SITE
COAST DISTRICT
SCALE = 2 CHAINS OR 132 FEET TO AN INCH.

Below. Rows of distinctly non-Aboriginal-looking houses line the shore in the model community of Metlakatla in an 1881 photograph. This is Front Street, which can be seen on MAP 296, *above.*

Baillie-Grohman's Canal Scheme

In 1882 William Adolph Baillie-Grohman, a British sportsman and author, had a vision. He thought that the low-lying land at the southern end of Kootenay Lake—the Kootenay Flats (west of today's Creston)—could be reclaimed, not by diking but by lowering the level of the lake itself. At first he proposed to do this by widening the constricted outlet of the lake, principally at what is today called Grohman Narrows, west of Nelson, but the following year, while on a hunting trip, he came up with a better idea: also divert the waters of the Kootenay River into the Columbia River by diverting them across the narrow neck of land between the two—the place that David Thompson had called McGillivray's Portage—north into Columbia Lake, the source of the Columbia. And how to do it? Cut a shallow canal across the 2,000-metre-wide portage.

Baillie-Grohman was not the first to conceive the idea of diverting the Kootenay, which side-swipes the Upper Columbia in the East Kootenay before meeting it farther west. In 1862 a group of gold miners thought they could divert the Kootenay entirely, leaving its bed and—they hoped—placer gold exposed. They could not afford such a magnificent project, but Baillie-Grohman could.

First he negotiated with the provincial government for about 19,400 ha of the Kootenay Flats. In this he had to fight George Ainsworth, who also wanted the land granted for his proposed Columbia & Kootenay Railway. Baillie-Grohman won this fight, obtaining a ten-year lease on the flats—and the first right of purchase—in December 1883. As part of the agreement he obtained the right to dig a "ditch" on Crown lands anywhere.

Baillie-Grohman then formed the Kootenay Lake Syndicate and began raising money for his project in Britain. Then he returned to Victoria with a proposal whereby he would be granted partially free land in the upper Kootenay valley if his "Upper Kootenay River Canal" was completed. There were conditions—one of which was that he begin steamboat service on the Upper Columbia to Golden. Baillie-Grohman imported a little steamboat, the *Midge*, even brilliantly circumventing the payment of customs duties on it by declaring it to be an agricultural implement—it was to pull a steam plow on the flooded land he was to reclaim! The boat had to be manhandled over part of the route from Sandpoint, the nearest railway, to Bonner's Ferry, Idaho.

The construction of the Kootenay Canal was approved by the provincial government in September 1885, although technically it also needed federal government approval. Some Golden residents interceded, worried that the Upper Columbia would be flooded, but Baillie-Grohman gave proper assurances, a lock appeared in the canal scheme, the provincial government agreed to a 30,000-acre land grant in the Upper Kootenay valley, and a new agreement was signed in October 1886.

Baillie-Grohman set to work. He formed yet another company to finance the work, the Kootenay Valleys Company, and a force of horses and scrapers and Chinese workers descended on the portage. By July 1889 the canal was complete and the land grant was awarded. Then work began on the Kootenay Flats with yet another company, the Alberta and British Columbia Exploration Company, which also widened the lake outlet at Grohman Narrows. At this point Baillie-Grohman lost control, a result of complex legal manoeuvring, documents that went astray, and his lawyer absconding with his money. Baillie-Grohman gave up and left for London. The Alberta and British Columbia Exploration Company by 1894 reclaimed a large section of the Kootenay Flats by diking—which Baillie-Grohman had opposed—and received a 7,000-acre (2,830-ha) land grant in 1894.

Later excavation of the lake outlet—270,000 m^3 of gravel and rock was removed by West Kootenay Power & Light—and further diking on the Kootenay Flats itself did ultimately reclaim the area for farming. At the portage the canal is no more, but a small community established there remains, appropriately named Canal Flats.

MAP 297 (*right*).
This map appeared in a pamphlet produced for potential investors by the Kootenay Valleys Company in 1886. The *Site of Canal & Kootenay City* is at the southern end of *Upper Columbia Lakes,* at the point that the *Kootenay R.* comes close to the Columbia. At the southern end of *Kootenay Lake* a shaded area shows the *Kootenay C°. Land* proposed for reclamation. In the west arm of the lake, *Outlet* and *Rapids* mark the locations of proposed widening.

MAP 298 (*below*).
This 1914 survey for the Kootenay Central Railway is the only detailed plan of the canal known. The canal is marked by the *Right-of-Way of Abandoned Canal–Crown Land.* The Kootenay River is at left.

MAP 299 (*far right*).
This fine map was published by the provincial Department of Lands and Works in 1885 to illustrate the Baillie-Grohman scheme, which can be confusing, since it required work at one location to reclaim land at another. Note that the *Kootenay River* descends into the United States before re-entering Canada, flowing into Kootenay Lake and joining the *Columbia* via the West Arm of that lake. Kootenay Flats are the *Splendid Meadows 45,000 Acres.* An inset map (which has here been enlarged) shows details of the proposed canal and the *Site of Kootenay City*—today the community of Canal Flats, a name given to it in 1913. Note the many proposed railways shown on this map.

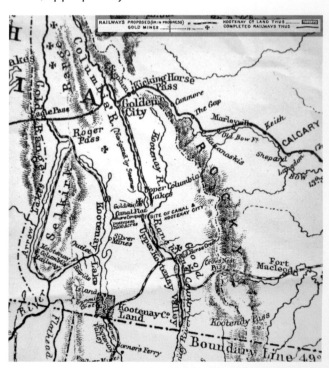

Left.
Chinese workers excavating Baillie-Grohman's canal in 1887.

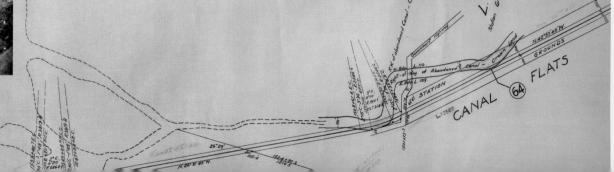

MAP OF THE KOOTENAY DISTRICT IN BRITISH COLUMBIA

From the specially prepared map compiled in the Department of Lands and Works, Victoria B.C. by order of Honble Wm Smite Chief Coms of Lands and Works, 1885, with additions taken from the Report of the Geological and Natural History Survey of Canada under Director Alfred R.C. Selwyn L.L.D., F.R.S., F.G.S. upon the Rocky Mountains beween Lat. 49° and 51°30' N. by G.M. Dawson D.S., F.G.S., Associate Royal School of Mines

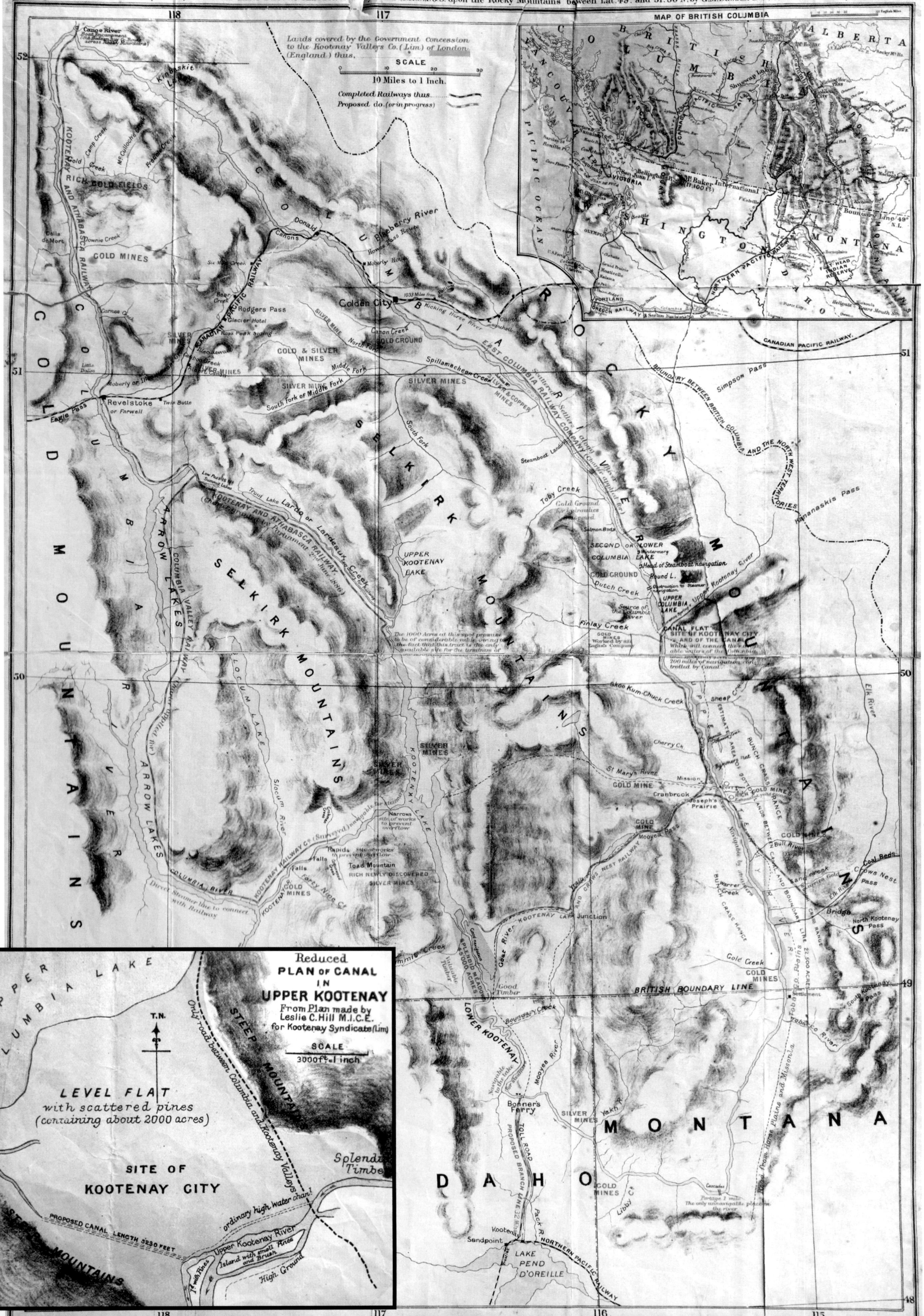

Lands covered by the Government Concession to the Kootenay Valleys Co. (Lim.) of London (England.) thus.

SCALE

10 Miles to 1 Inch.

Completed Railways thus..........
Proposed do. (or in progress)..........

MAP OF BRITISH COLUMBIA

Reduced
PLAN of CANAL
IN
UPPER KOOTENAY
From Plan made by
Leslie C. Hill M.I.C.E.
for Kootenay Syndicate (Lim)

SCALE
3000ft. 1 inch

LEVEL FLAT
with scattered pines
(containing about 2000 acres)

SITE OF
KOOTENAY CITY

UPPER COLUMBIA LAKE

T.N.

PROPOSED CANAL LENGTH 5250 FEET

Splendid Timber

High Ground

THE TRANSCONTINENTAL CONNECTION

The idea of connecting British Columbia to the east with a railway was not a new one. In 1860 Alfred Waddington had begun promoting the construction of a wagon road to the Cariboo gold mines (see page 90) and expended considerable sums trying to build it, but by 1868 his objectives had changed—to a transcontinental railway. His map showing the route of his wagon road became one showing a route for the railway. Waddington travelled to Britain to attempt to raise interest there and also sent letters to Ottawa, all to no avail, for it was an idea before its time.

At this time, a debate raged as to whether British Columbia—after 1866 the United Colony of British Columbia—should join the United States, go it alone, or join Canada. And it was the promise of a transcontinental railway that tipped the decision—Confederation with Canada. Article 11 of the 1871 Terms of Union stipulated that the Dominion government begin work within two years on a railway to connect the seaboard of British Columbia with the railway system of Canada. And it was to be complete within ten.

The Canadian Pacific Survey was authorized the same year, and Sandford Fleming, surveyor of the Intercolonial Railway (and later inventor of worldwide standard time), appointed engineer-in-chief. As early as 1863 Fleming had written a treatise titled *Observations and Practical Suggestions on the Subject of a Railway through British North America,* a detailed outline for a transcontinental railway, and had submitted it to the govern-ment of the Province of Canada. He had advocated building first a road, then a telegraph line, and then a railway, the track of which was to be laid directly on the road.

Within a few months, Fleming had dispatched twenty-one survey parties across the country; at the survey's peak some eight hundred men were involved. In British Columbia Walter Moberly, by then one of the province's most experienced surveyors, was placed in charge, and he organized four separate survey parties.

The following year Fleming determined to see the West for himself and organized a transcontinental expedition. Fleming traversed the Yellowhead Pass and ended up in New Westminster, and in so doing seems to have convinced himself that this would be the best route for

Map 300 (*below*).

This 1868 map by Alfred Waddington shows the route of his *Projected Railway,* ending at the head of *Bute Inlet,* where, not coincidentally, he had pre-empted land to establish a settlement called Waddington Harbour. Neither his route nor his town ever came to anything. The Waddington route at first follows the one he had proposed a few years earlier for a wagon road to the Cariboo and then follows the Fraser River around the "Big Bend" and crosses the Rockies via the *Yellow Head Pass.* Other possible routes across the Rockies have also been inked in, including the one across the *Kicking Horse Pass* that would be followed by the railway in 1884.

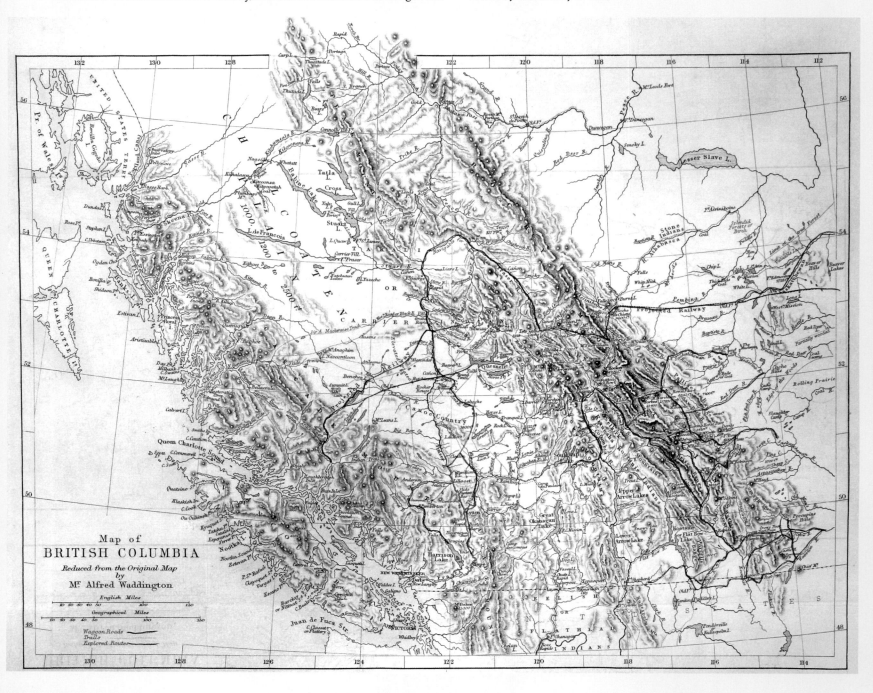

MAP 301 (*above*).
The surveys for the Canadian Pacific, directed by Sandford Fleming, for a railway route in British Columbia are shown on this 1876 map. Veteran surveyor Walter Moberly was in charge of the British Columbia surveys. A myriad of possibilities are shown, including Fleming's 1871 route through and beyond the Yellowhead Pass. Waddington's route to *Bute Inlet* is one of the pathways shown, and *Waddington Harbour* is marked. There is a written notation (off the map to the left) that "Waddington Harbour is no harbour at all." Fleming's preferred route through the Yellowhead Pass is shown as a thicker solid red line, but beyond *Fort George* this route splits into three possibilities (shown with dashed lines), none of which were all that practical: to Bute Inlet, to *Dean Channel* (just north of today's Bella Coola), and *Gardner Ch[annel]* (today the site of the Kemano powerhouse; see page 322). The route ultimately selected for the transcontinental railway is shown except for two sections. The first omitted is that across the Selkirk Range that would prove so troublesome to the Canadian Pacific; here is the written notation: *impracticable.* The second is that through Kicking Horse Canyon and Kicking Horse Pass; the route instead cuts across to the North Saskatchewan River farther north.

MAP 302 (*below*).
Published in 1876, this American map shows a single route for the transcontinental railway, entering British Columbia somewhere between the Kicking Horse and Yellowhead Passes and terminating at New Westminster.

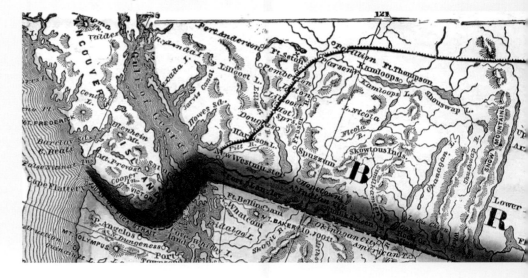

Map 303 (*above*).

The results of the Canadian Pacific surveys on an 1878 map. The routes surveyed and considered are shown in red, though some routes shown on Map 301 (*previous page*) have now been removed. Note how the route to Bute Inlet now connects with a line down the east coast of Vancouver Island to Victoria, the preferred route of Victoria businessmen. Also of interest is an alternate route from Kamloops to Hope, using the valleys of the Coldwater and Coquihalla Rivers.

the railway. This was indeed the routing Fleming recommended, and it remained the accepted route for the proposed railway for nearly a decade.

A private Canadian Pacific Railway was incorporated in 1873 but was essentially a front for American investors and American control. That same year Prime Minister John A. Macdonald was found to have received improper campaign contributions in return for railway concessions—the so-called Pacific Scandal—and resigned. His successor, Alexander Mackenzie, was lukewarm to the Pacific railway idea, fearing its potentially enormous costs. The Pacific railway was put on hold. This infuriated many in British Columbia, for the Terms of Union had guaranteed a start to the railway within two years. In 1874 a delegation was preparing to go to London to see the British government when the colonial secretary, forewarned, agreed to personally arbitrate the dispute. Mackenzie reluctantly agreed, and the result was the Carnarvon Terms: in return for an extension of the date of completion of the Pacific railway to 1890, the federal government would build the Esquimalt & Nanaimo Railway on Vancouver Island (see page 164).

The federal government attempted in 1874 to entice private interests to build the transcontinental line. The Canadian Pacific Railway

Map 304 (*left*).

A cartoon by J.W. Bengough that appeared in the *Canadian Illustrated News* on 9 September 1876 displayed a map of the *Line of Projected Canadian Pacific Railway* in *British Columbia,* and the Yellowhead route is shown. "Uncle Aleck"—Alexander Mackenzie—talking to "Miss B. Columbia," says, "Don't frown so my dear, you'll have your railway by and by." To which Miss B. Columbia replies, "I want it now. You promised I should have it, and if I don't, I'll complain to Maw." It would be nine more years before Miss B. got her railway.

The Railway Belt

Under the Terms of Union under which British Columbia joined Confederation, the province was obliged to reserve and transfer lands for 20 miles (32 km) on either side of the proposed route. From this, the Dominion government was to grant lands to the Canadian Pacific and sell the rest to finance the vast subsidies awarded to the railway. The company could select the land it was granted, but none was chosen from the mountainous areas of British Columbia.

It was not until 1881 that the final route was decided; up until that time the intended route was the one Sandford Fleming recommended—through the Yellowhead Pass and down the North Thompson River. A provincial Settlement Act, ratified in April 1884, transferred the land along the final railway route to the federal government. The same act transferred 1.4 million ha of land in the Peace River to the Dominion government as compensation for alienated or non-arable land in the Railway Belt.

MAP 305 (*right*).
This is an assessment of the merits of parts of a railway belt encompassing a line through the Yellowhead Pass, which in May 1881 the Canadian Pacific decided against using in favour of Kicking Horse Pass. The belt has been drawn on an 1871 base map (the Trutch map; see MAP 913, *pages 358–59*) by Joseph Pope, John A. Macdonald's private secretary after September 1882. The dark green–coloured areas are those with agricultural potential; the light green, mountainous except for some areas of timber value. Note also that the land grant zone on Vancouver Island is roughly marked (see page 165).

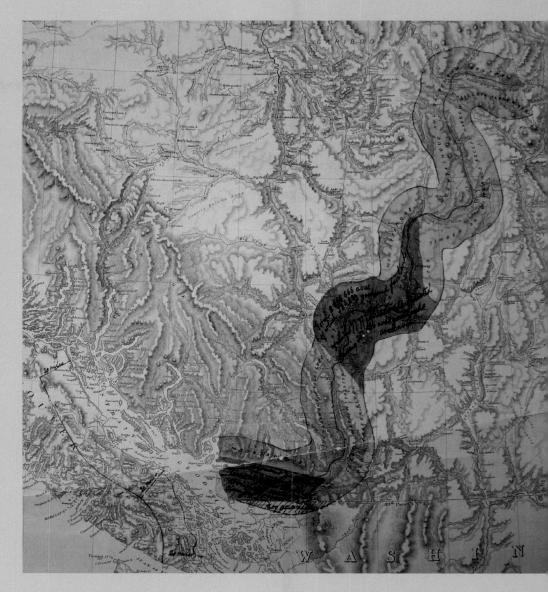

MAP 306 (*below*).
The Railway Belt as defined in April 1884, recognizing the change of route of the railway through Kicking Horse Pass instead of through the Yellowhead Pass, as originally planned. Transferred under the Terms of Union between Canada and British Columbia in 1871, the dual jurisdiction of the Railway Belt lands was to cause confusion for many years, with governments suing each other from time to time, especially over matters of mineral rights. Land in the Railway Belt not granted was returned to the provincial government in 1930. The area from which the railway selected its land grant is shown in orange.

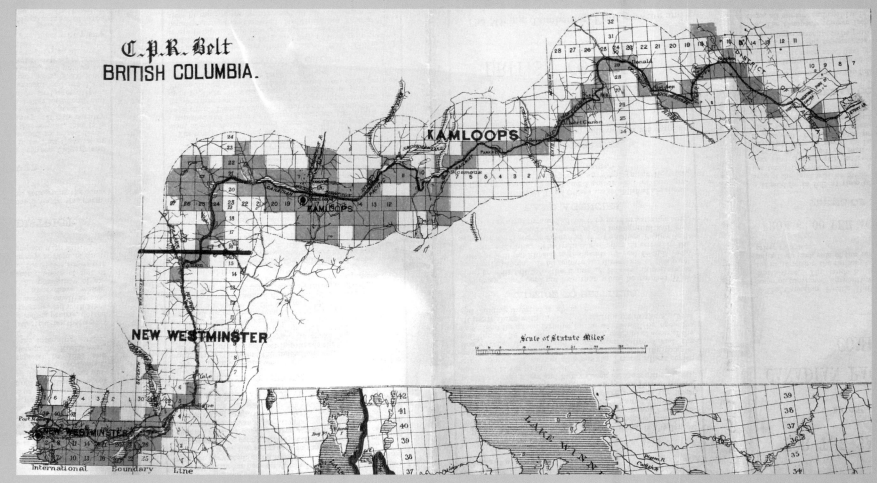

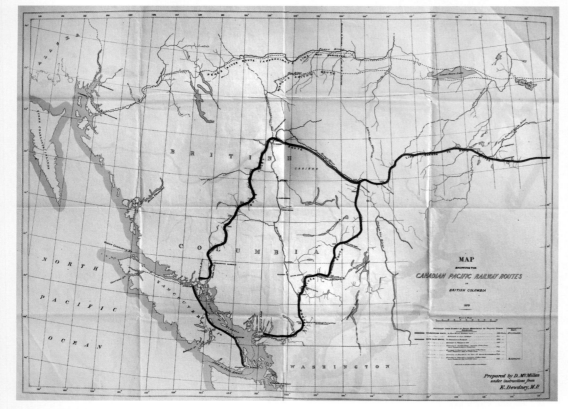

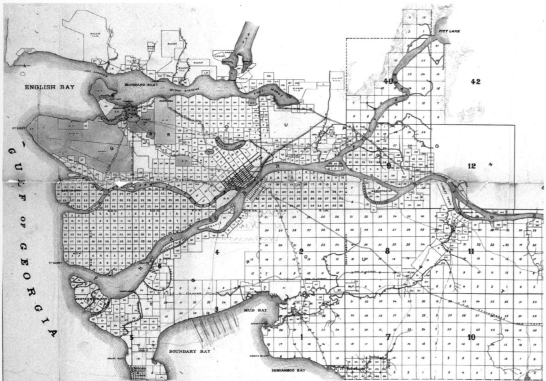

Map 307 (*left, top*).
By 1879, the date of this map, the government had already decided on the red line route shown here, which followed the North Thompson and Fraser to tidewater at Port Moody. The other main contender had been the blue line route following the Fraser to *Ft. George* and reaching tidewater at Bute Inlet, allowing for a ferry service to connect with a line down Vancouver Island. The latter route was the preferred choice of Victoria residents. The government officially chose the route to the Lower Mainland in July 1878, which allowed them to award the Fraser Canyon–to–Kamloops Lake construction contract to Andrew Onderdonk in 1879. The decision to use Kicking Horse Pass instead of the Yellowhead to cross the Rockies was to come later (see page 122).

Map 308 (*left, centre*).
The line of the (government's) Canadian Pacific Railway is depicted in red, terminating at *Port Moody* on this 1876 map of the Lower Mainland. Also marked, in pencil, is a branch to New Westminster. Note also the dock drawn in pencil in *Boundary Bay*, a possibility first shown by the Royal Engineers' Colonel Moody (see Map 194, *page 70*) but one that would have required an inordinate amount of dredging to build and to maintain. Governmentally reserved land is shown in red; much of this would form the land granted to the railway in 1885 to extend the line to English Bay.

Left, bottom.
Once railway construction began in the Fraser Valley in 1880, the head of navigation, Yale, became a busy place. Here the sternwheeler *R.P. Rithet* is docked at Yale about 1882.

Act offered cash and land subsidies, but there were no takers. Planning went on, and a start was made on the line north of Lake Superior, but there was much political interference, and cost overruns led to Fleming's forced resignation in February 1880. The route to the Lower Mainland was chosen in 1878, but it was not until John A. Macdonald returned to office in September 1878 that real progress on the line west of Winnipeg began to be made.

Firstly, the government awarded contracts for two sections of the line—a section west of Winnipeg and a section from the head of navigation on the Fraser to Savona's Ferry, on Kamloops Lake. The British Columbia contract was awarded to an experienced American railway engineer, Andrew Onderdonk, and construction began in May 1880.

Secondly, new offers were made to any private builder: a cash subsidy of $20 million and 30 million acres (12 million ha) of land. A group of entrepreneurs led by George Stephen, president of the Bank of Montreal, offered to build the line for $25 million and a land grant of 25 million acres (10 million ha). This the government accepted and also agreed to hand over the sections of line already built or under contract, totalling some 1,100 km. A contract was signed on 21 October 1880. The bill confirming the contract received royal assent on 15 February 1881, and the following day the Canadian Pacific Railway was incorporated.

The City of Emory

When railways are proposed, those who seek to profit from the increase in land value are never far behind. Such was the case for Emory, a townsite laid out on the west bank of the Fraser several kilometres south of Yale in 1879. Andrew Onderdonk had been awarded the government contract to build part of the transcontinental line from Yale to Savona, on Kamloops Lake, and so prosperity seemed assured. Onderdonk did lay rails to Emory and did unload men and materials for the railway construction there. Emory soon sported two hotels, nine saloons, a brewery, and a newspaper. Yet it was all a fleeting vision. Within a short time it was apparent that Yale would benefit most from the railway, and even that prosperity would be temporary, for the (private) Canadian Pacific Railway was chartered in 1881 to build past Emory all the way to tidewater. The railway would run through Emory but would not terminate there.

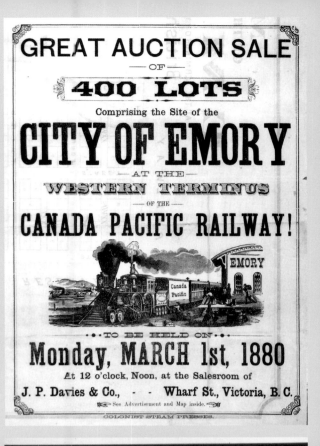

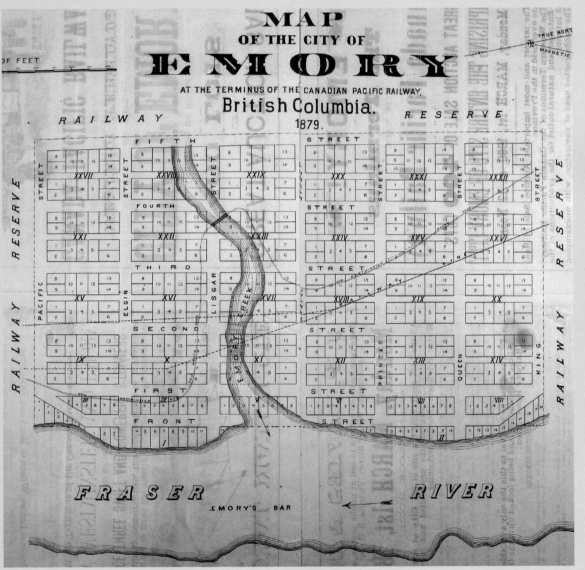

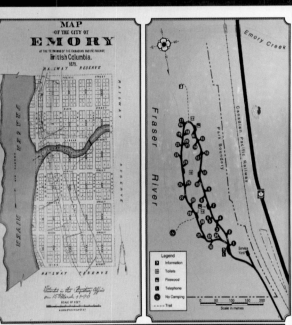

MAP 309 (*left, top*).
The townsite of the City of Emory. Yale is 5 km north (to the right). The map is from promotional material for an auction of lots.

MAP 310 (*above*).
Today the townsite of Emory is a provincial campground. A sign at the entrance contains an old map of Emory similar to MAP 309, this time a copy registered at the Land Title office on 15 March 1880, two weeks after the advertised auction of lots.

Above, top and left.
Panels from the townsite auction promotional material. Note on the locomotive the obviously pasted-on name Canada Pacific rather than Canadian Pacific.

Murder on the Canadian Pacific

By the end of 1882 Andrew Onderdonk had imported about six thousand Chinese workers for his railway-building contracts between Port Moody and Savona. The first arrivals came from working on the Northern Pacific in 1880 and the second from the Southern Pacific in California the following year. Then, in 1882, he brought in more Chinese men on ships directly from China. His reason was economic. He had been forced to accept contracts at lower prices than he had originally calculated as necessary, and thus he had to keep his costs down. The Chinese would work—and work very well—for just a dollar a day, half the normal American rate. And in any case, other railway workers were hard to find, because of the many lines then being built in the United States.

The Chinese workers were divided into gangs of thirty with a cook and a bookkeeper under a EuroCanadian or American foreman.

Anti-Chinese sentiment was rampant among the Euro-Canadian population, and the Chinese workers, though good workers and usually compliant, would resort to mass action or violence to achieve redress if they felt they had been wronged. In 1883 one Chinese worker had his head blown off when a work gang was not given sufficient warning of an impending blast; the foreman was lucky to escape with his life. He dived into the river, hotly pursued by workers armed with tools and rocks.

One of the worst incidents of violence took place as the track laying approached Lytton. At Camp 37 on 8 May 1883 an American foreman, one Mr. Gray, fired two workers he believed were lazy and then refused to pay them for the work they had completed that morning. The rest of the gang then attacked him and three others, injuring one of them.

That night twenty workers, all of them thought to be Americans who had worked together on the Northern Pacific, armed themselves with clubs and stole into the Chinese camp, surrounding it. As the inhabitants tried to escape in the dark, they were clubbed and beaten. The attackers then set fire to the camp and departed. Nine Chinese workers were left on the ground for dead, all with serious head injuries from which some, including a worker named Ah Fook, would not recover. One, Yee Fook, died on the spot—a severe case, the *Daily Colonist* in Victoria gruesomely reported the next week, where "the brain appeared to be oozing from the skull."

Unusually, some of the accused murderers of Ah Fook and Yee Fook, including Gray, were brought to trial, but a jury, lacking, of course, Chinese jurors, refused to convict them, a not unexpected outcome for this time. The map reproduced below appears to have been drawn for the trial and shows the geographical details of the case.

MAP 311 (*right, top*).
This 1882 map has had the *Railway Now Under Contract* overprinted in red. Note that the line of the railway has been drawn rather roughly, missing *Lytton* altogether; the line as built crossed the river on the Cisco Bridge about 8 km south of Lytton. This was a steel cantilever structure of very advanced design for its time; Onderdonk's engineers had convinced him that Cisco (now Siska) was the best crossing place, as the river would have to be crossed to continue north along the Thompson Valley. See also MAP 911, *page 354*.

MAP 312 (*right*).
This is the Yee Fook murder map, which seems to have been drawn up to use at the trial of his accused murderers, probably in June 1883. In the bottom right corner is the *Chinese Camp* and *Camp 37*, the EuroCanadian camp; between the two is the *Hole where the body of Yee Fook was found*. The red line across the middle of the map is the line of the *Railway Track* being constructed. At left is *Lytton* and the *Thompson River*, with the *Fraser River* at bottom. This is an exceptionally rare map now in a private collection; it was unusual for EuroCanadians to be brought to trial for the abuse of Chinese workers, and this is the only map illustrating such a case known to exist. The facts of the case were difficult to unearth because there seems to be no official court record of the trial, possibly because no one was convicted of the murder. The main details of the case were found in contemporary newspaper reports, which have to be read cautiously owing to their possible anti-Chinese bias.

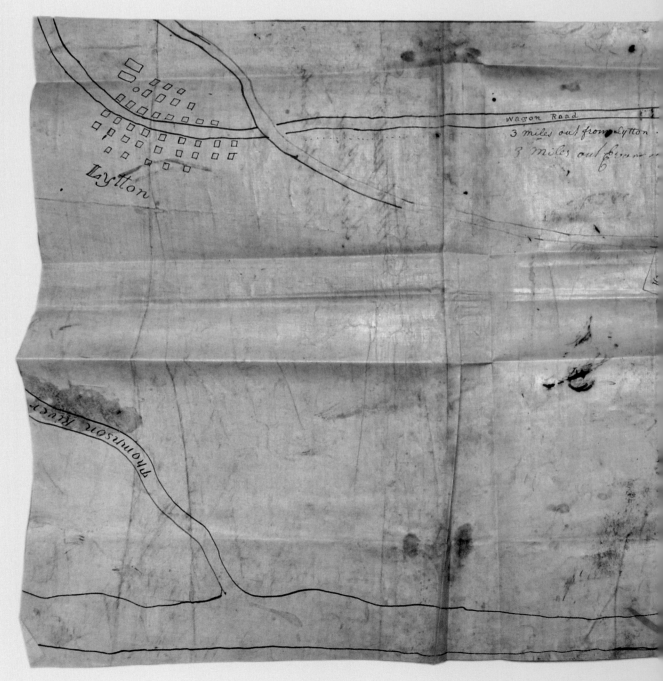

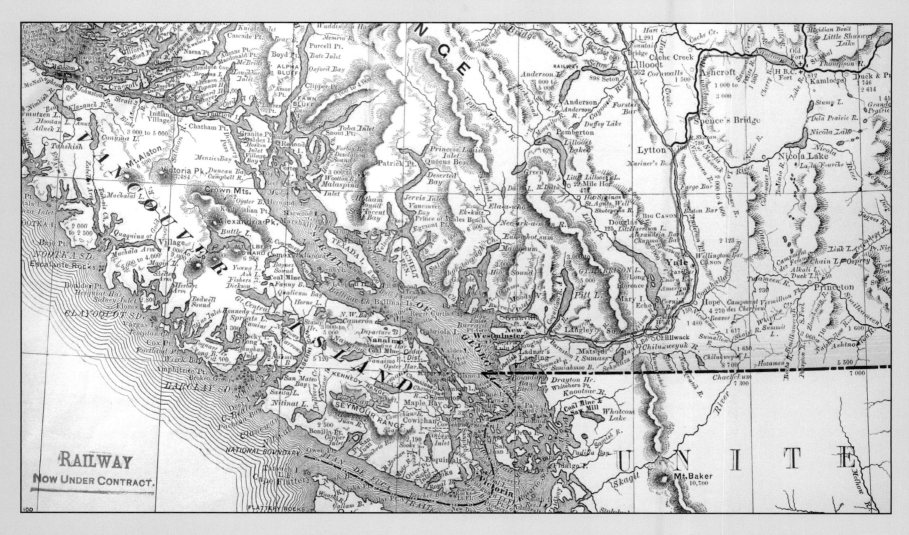

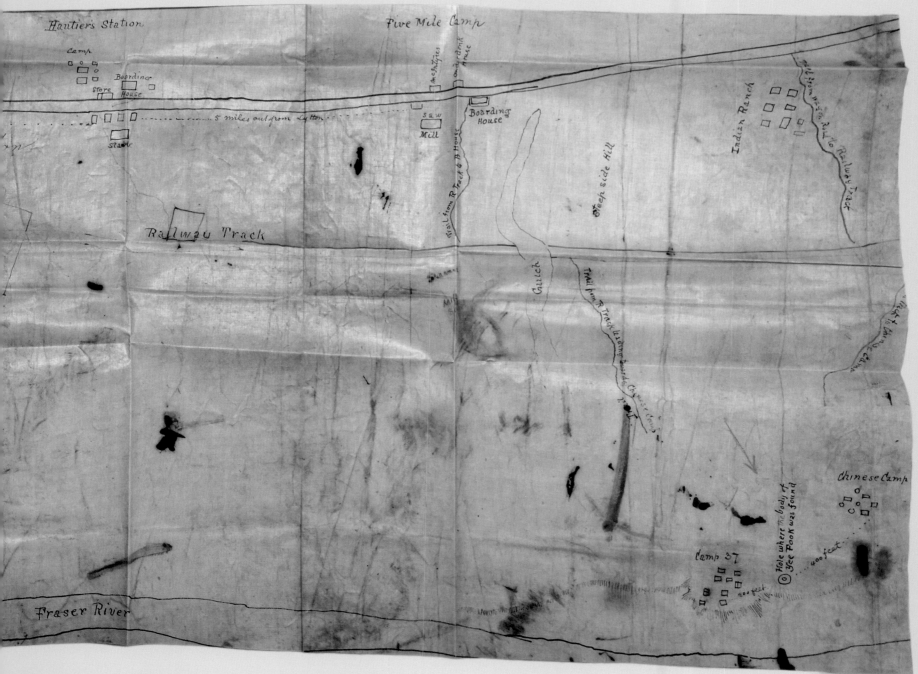

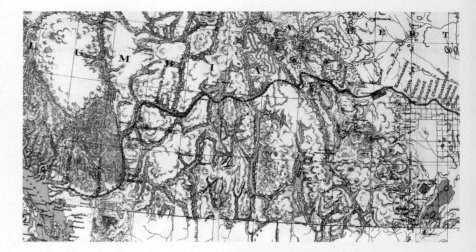

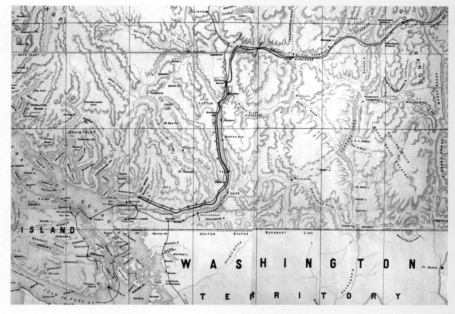

One of the first decisions made was to reroute the line on a more southerly path and use Kicking Horse Pass to cross the Rockies rather than the Yellowhead. Noted botanist John Macoun met with the directors and convinced them, contrary to popular thought at the time, that the southern route would encompass land suited for agriculture. With this decision most of the Fleming surveys were rendered useless—at least for the Canadian Pacific. An American railroad surveyor, Major Albert Bowman Rogers, was hired in February 1881 to locate paths through the Rockies and the Selkirk Range, which now stood, without a discovered pass, in the path of the railway.

Construction began in May 1881, but the pace was slow, and in October the company recruited an experienced American railroad builder, William Cornelius Van Horne. He began work in January 1882, and the pace of construction picked up. Van Horne hired Thomas Shaughnessy as his purchasing agent, charged with ensuring that all the materials required were in the right place at the right time. Shaughnessy's name would later be synonymous with a railway-developed upscale area of Vancouver. George Stephen, meanwhile, was spending much of his time trying to raise the vast sums of money required for the enterprise.

Rogers located the line through Kicking Horse Pass in the summer of 1881 and explored the Selkirks, where he thought he saw a way though, before running out of supplies. The following year he followed the Beaver River and its tributary Bear Creek west from the Columbia, and on 24 July 1882 reached a pass in the mountains that led to the Illecillewaet River—Rogers Pass. The railway's path was defined.

Onderdonk, meanwhile, completed a contract with the Canadian Pacific to build the line from Yale to Port Moody, driving the last spike on 22 January 1884. He finished an iron-and-steel cantilever bridge across the Fraser at Cisco (now Siska) that year and satisfied his government contract by completing the track laying from Yale to Savona's Ferry. In 1885 Onderdonk began another Canadian Pacific contract, building east from Kamloops Lake to meet the oncoming track from the other direction.

MAP 313 (*above, top*).
On this 1882 map the final path of the transcontinental line in British Columbia has been defined. The route through the Selkirk Range—using Rogers Pass—was the final link, found that year by Major Rogers.

MAP 314 (*above*).
An 1883 government map shows the Onderdonk-built, government-contracted portion of the line as cross-hatched—though it only went as far as Savona's Ferry at the western end of Kamloops Lake—and the Canadian Pacific–contracted portion east as a dashed line. The green line is the existing telegraph to the Cariboo and north (see page 98) while the solid thin black line is the railway telegraph. Note that the mapmaker has erroneously placed Moodyville adjacent to Port Moody; it was in fact in what is today North Vancouver (see page 94).

MAP 315 (*left*).
Land speculation is all about anticipating what might happen in the future. In 1882 a group of investors purchased the so-called Bricklayers' Claim (MAP 250, *page 93*)—the land that is now the West End of Vancouver—and laid out a new city they called Liverpool—in anticipation that the Canadian Pacific would not stop at Port Moody but continue to English Bay. When the railway's surveyor, Lauchlan Hamilton, laid out its townsite on the adjacent land grant (MAP 341, *page 133*), the road pattern did not mesh with that of Liverpool's. One of the investors could not be located in time to agree to a change, and so to this day only every other street connects across Burrard Street, one of the major north–south roads in Vancouver's downtown peninsula. The map has the names of the investors written at different locations, presumably representing their internal division of the land. The road pattern of the West End today is almost identical to the one originally laid out here.

Below. The reality of Liverpool was quite different from what the ordered rectangular pattern on the map suggested, the roads then being no more than surveyed lines through dense forest. This photo, thought to have been taken in the 1860s, shows a cabin in the wilderness among the detritus of cleared trees approximately at the location of Georgia and Denman Streets.

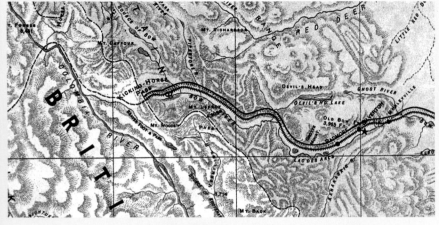

MAP 316 (*left*).
The railway (tracked line) and the telegraph (solid line) halt at the Continental Divide on this map showing the situation at the end of 1883. The track is poised to enter British Columbia—and the most difficult part of the construction work—the following season.

MAP 317 (*left, below*).
Kicking Horse Pass, the route across the Rocky Mountains used by the transcontinental line, was discovered in 1858 by geologist James Hector, a member of the scientific Palliser Expedition of that year, and was named after a horse that fell into the river and, after being pulled out, kicked Hector in the chest, knocking him temporarily unconscious. This map, signed by Hector himself, notes the *Kicking Horse R.* and the *Height of Land 5420'* (1,652 m). Note also the *Columbia R.?*—Hector was not certain this river was the Columbia.

MAP 318 (*below*).
In order to cope with the steep grades of the "Big Hill"—at 4.4 per cent double the accepted maximum—the railway installed three safety switches with lengths of track onto which a runaway train could be diverted. This plan is of the first safety switch, *No. 1*. The photo below shows one of the many helper locomotives simmering on one of the safety switches as it waits for its next assignment. The other photo shows this safety switch as it was in 1890.

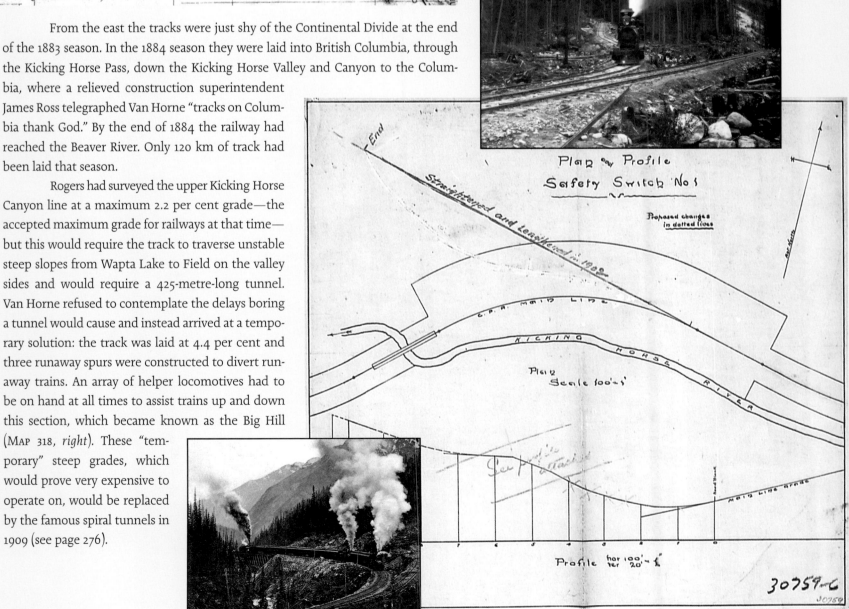

From the east the tracks were just shy of the Continental Divide at the end of the 1883 season. In the 1884 season they were laid into British Columbia, through the Kicking Horse Pass, down the Kicking Horse Valley and Canyon to the Columbia, where a relieved construction superintendent James Ross telegraphed Van Horne "tracks on Columbia thank God." By the end of 1884 the railway had reached the Beaver River. Only 120 km of track had been laid that season.

Rogers had surveyed the upper Kicking Horse Canyon line at a maximum 2.2 per cent grade—the accepted maximum grade for railways at that time—but this would require the track to traverse unstable steep slopes from Wapta Lake to Field on the valley sides and would require a 425-metre-long tunnel. Van Horne refused to contemplate the delays boring a tunnel would cause and instead arrived at a temporary solution: the track was laid at 4.4 per cent and three runaway spurs were constructed to divert runaway trains. An array of helper locomotives had to be on hand at all times to assist trains up and down this section, which became known as the Big Hill (MAP 318, *right*). These "temporary" steep grades, which would prove very expensive to operate on, would be replaced by the famous spiral tunnels in 1909 (see page 276).

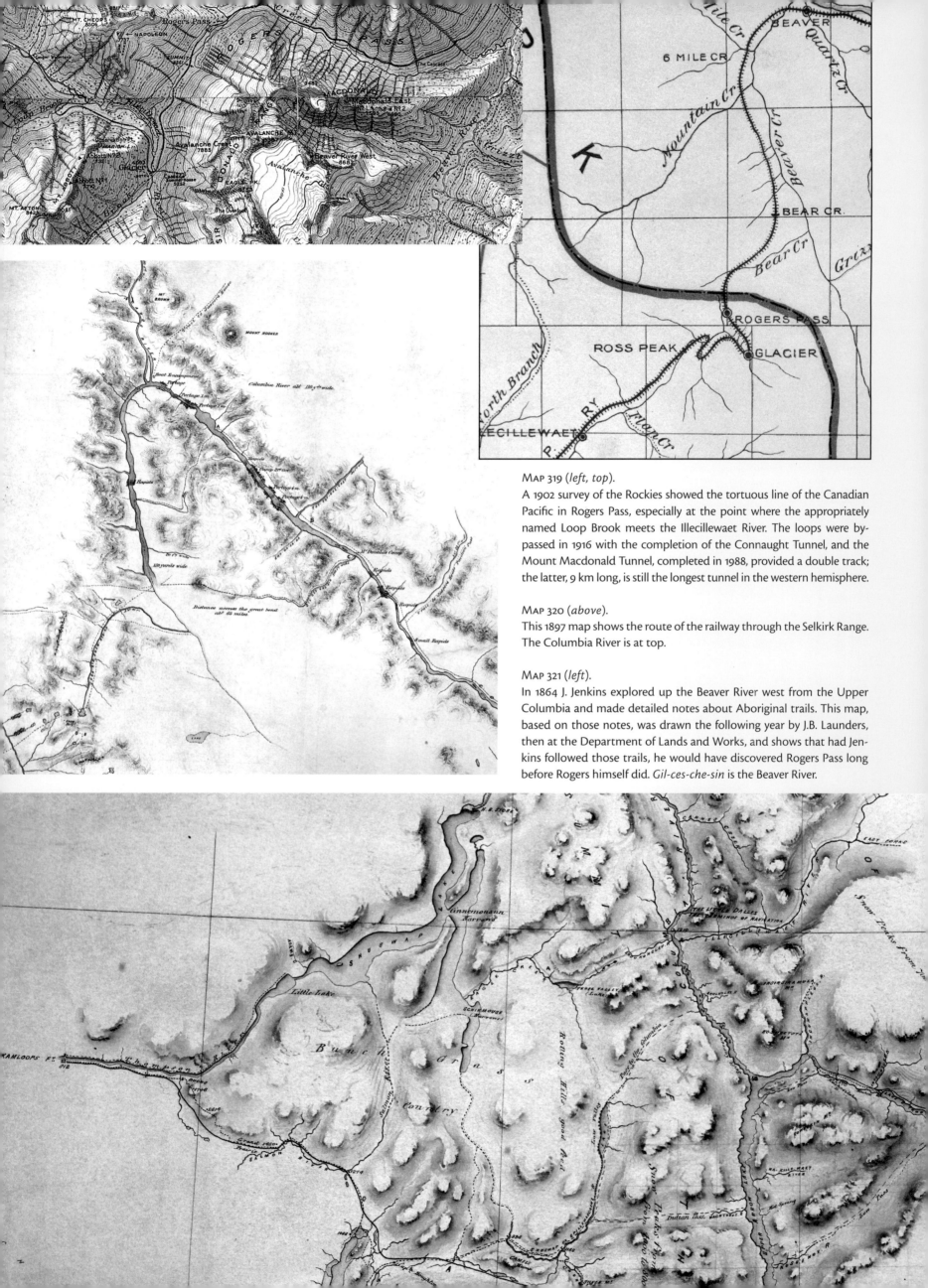

Map 319 (*left, top*).
A 1902 survey of the Rockies showed the tortuous line of the Canadian Pacific in Rogers Pass, especially at the point where the appropriately named Loop Brook meets the Illecillewaet River. The loops were by-passed in 1916 with the completion of the Connaught Tunnel, and the Mount Macdonald Tunnel, completed in 1988, provided a double track; the latter, 9 km long, is still the longest tunnel in the western hemisphere.

Map 320 (*above*).
This 1897 map shows the route of the railway through the Selkirk Range. The Columbia River is at top.

Map 321 (*left*).
In 1864 J. Jenkins explored up the Beaver River west from the Upper Columbia and made detailed notes about Aboriginal trails. This map, based on those notes, was drawn the following year by J.B. Launders, then at the Department of Lands and Works, and shows that had Jenkins followed those trails, he would have discovered Rogers Pass long before Rogers himself did. *Gil-ces-che-sin* is the Beaver River.

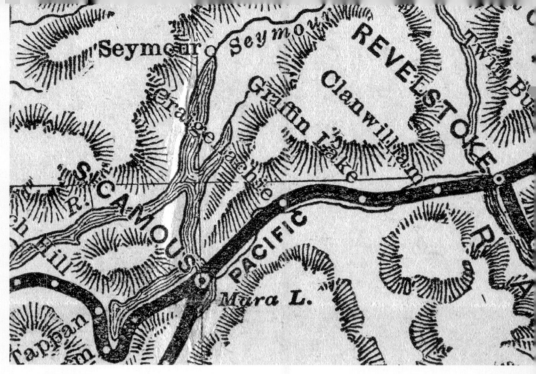

The railway was fast running out of money by this time, as the engineering works required were proving enormously expensive. George Stephen was spending most of his time selling stock to raise money, but the stock was declining owing to investor worries as to whether the line might ever be completed. Then, in the spring of 1885, came Louis Riel's North-West Rebellion. The Canadian Pacific was used to rush troops to the scene, and the entire public perception of the railway's worth changed, enabling Macdonald to gain passage of a bill to provide further assistance to the railway.

In 1885 the Canadian Pacific crews built up the Beaver River to Rogers Pass, which Major Rogers had only finished surveying the year before. The line was still difficult, requiring a number of hairpin bends, but the pass was reached on 17 August. The difficult gorge of the Illecillewaet River was next, and this required yet more loops, totalling a dizzying 2,500 degrees.

Crossing the Columbia once more at Revelstoke, recently renamed from Farwell (see next page), the crews built to Three Valley Gap and over Eagle Pass, discovered in 1865 by Walter Moberly. Finally, on the western slope of the pass they approached the point where Onderdonk's crews had halted because they had run out of rails. Here, at a place Van Horne named Craigellachie, the last spike was driven by company director and original 1880 syndicate member Donald Smith on 7 November 1885. Behind Smith and next to Van Horne in the famous photograph of the event stood Sandford Fleming, since 1884 also a director of the Canadian Pacific and certainly the man with the longest involvement in the building of the transcontinental link.

The Pacific railway was a reality, although innumerable improvements would yet be required to allow commercial operation to begin—track ballasting, repairs, bridge strengthenings—and, as the railway would soon find out—many snowsheds would need to be built in the mountain sections. For now, it was too late in the season. A demonstration shipment was staged—forty drums of oil from the naval base at Halifax to the one at Esquimalt. They had been held in Québec pending completion of the line and were now shipped, taking seven days to reach Port Moody and two more to Esquimalt. It was not until 4 July the following year that the first passenger train from the east arrived at Port Moody.

Right, centre and bottom.
Arguably Canada's most famous photograph, the scene at Craigellachie at 9:22 am on 7 November 1885 as the last spike of the Canadian Pacific Railway is driven by Donald Smith; and the scene today: a granite monument, placed by the railway in 1927, marks the spot. Van Horne had long ago decided on the name Craigellachie, and so it was perhaps the only place named *before* its location was fixed.

Map 322 (*left*).
This is the map recording surveyor Walter Moberly's September 1865 discovery of Eagle Pass, two decades later used by the Canadian Pacific. He found the pass after he observed the path taken by flying eagles towards the mountains. *Eagle Creek* flows into *Shuswap Lake* at *Schikmouse Narrows* (Sicamous). *Three Valley (Lake)* is at Three Valley Gap. This valley now also carries the Trans-Canada Highway. Surveys were being carried out in the 1860s as a result of the gold strike at Wild Horse Creek (see page 104).

Map 323 (*above*).
The Canadian Pacific always had a place for *Craigellachie* on its maps despite the fact that there has never been a settlement there worthy of a station. Here in the Eagle Valley, a few kilometres east of *Sicamous*, the last spike was driven on 7 November 1885.

Farwell

Arthur Stanhope Farwell, a wide-ranging surveyor who had previously been employed on the Canadian Pacific surveys, knew a good thing when he saw it. He figured he knew exactly where the Canadian Pacific railway line would have to be located as it crossed the Columbia River for the second time, and in October 1883 he applied for a Crown grant of land at precisely this spot. The railway registered its route a month later and, sure enough, the line ran right through Farwell's land.

Farwell had already partly laid out his townsite (Map 326, *below, right*) by January 1885, when he received title to his grant. Some settlement had already occurred here—the location had been called both Columbia City and Second Crossing. When the Canadian Pacific arrived, the company wanted land free of charge from Farwell in return for establishing a station there, but Farwell, convinced that the railway had no other option, refused. A long and expensive lawsuit followed, with the company maintaining that Farwell should not have been granted the land in the railway's path in the first place. Farwell at first won the court battle, but the decision was over-

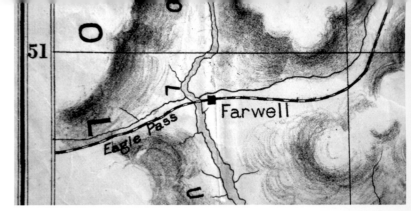

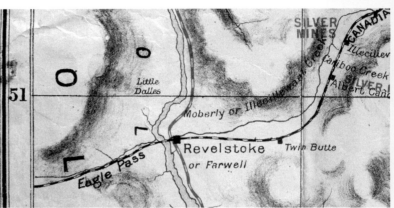

Map 324 (*above, top*) and Map 325 (*above*).
Two editions of the same map, the first from 1885 and the other from 1886, respectively, document Farwell's conversion to Revelstoke.

turned, though without an order removing his name from the provincially registered land grant title. This situation caused great confusion over the titles of those who had purchased lots from Farwell and delayed the incorporation of the city until 1899.

In any case, for all practical purposes, Farwell lost out. In 1885 the company simply built its station and laid out a townsite two kilometres away. This became the heart of the business section of its town, named Revelstoke in honour of Edward Baring, Lord Revelstoke, head of the Baring Brothers bank that had recently bailed out the railway just as it was running out of money to complete the line.

Map 326 (*right, centre*).
Arthur Farwell's *Town of Farwell*, drawn in 1885, correctly showed the path of the railway—but no station; this the Canadian Pacific could and would locate elsewhere. *Above* is a photo of Farwell in 1885.

Map 327 (*right*).
Revelstoke, about 1898. This map shows how Farwell's townsite—at left, sans station—was incorporated into the city; the railway station is far away at top right, while the Canadian Pacific's own townsite is at right.

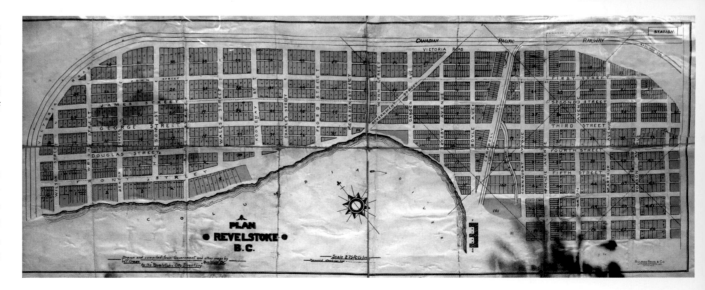

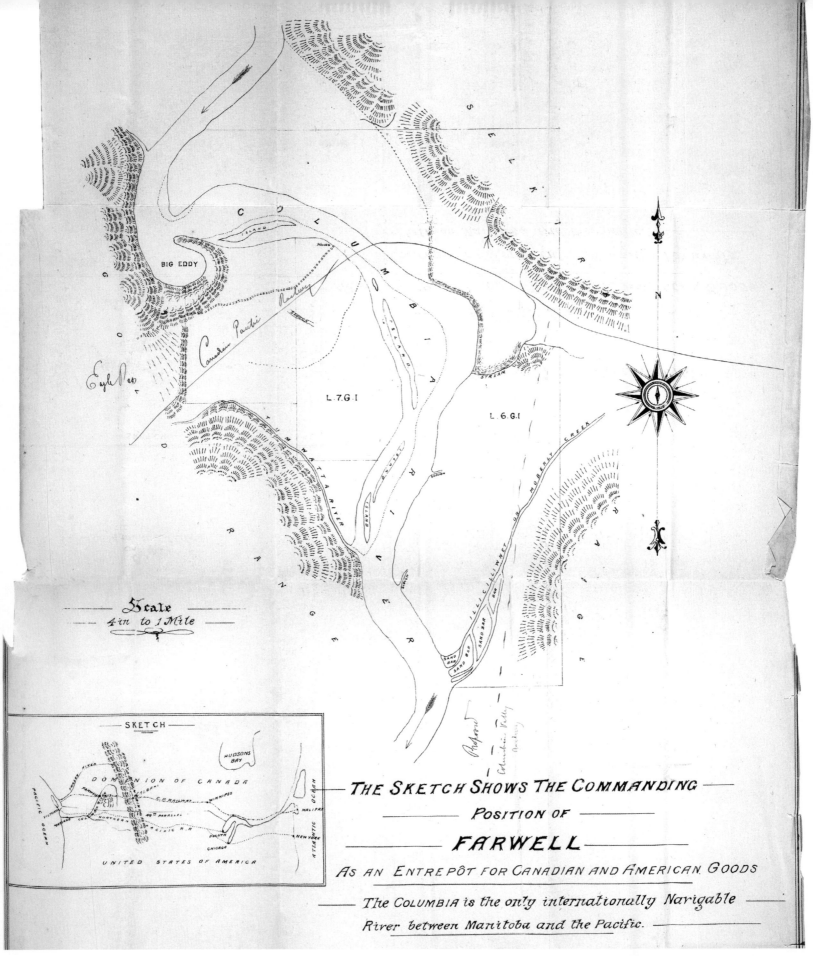

Scale
4 in to 1 Mile

THE SKETCH SHOWS THE COMMANDING —
— POSITION OF —
FARWELL
AS AN ENTREPÔT FOR CANADIAN AND AMERICAN GOODS
— The COLUMBIA is the only internationally Narigable
River between Manitoba and the Pacific. —

MAP 328 (*above*).
This map, showing the "commanding position" of Farwell, was published by Arthur Farwell in 1885 with a booklet promoting his townsite. The line of the *Canadian Pacific Railway* has been added with pen and ink after the map was printed, as has a dashed line to the south marked *Proposed Columbia Valley Railway*. Farwell was located exactly where the railway would cross the *Columbia River* on its way to *Eagle Pass*. The new railway, the booklet emphasized, "makes it certain that Farwell . . . will be the chief city of the interior of British Columbia."

Right.
Despite the Canadian Pacific's dislike of Arthur Farwell, the name of his city survives to this day as an official company railway station name 1,500 m west of Revelstoke station.

Kamloops

The first settlement in Kamloops independent of the fur trade forts (see page 44) was on a narrow bench above the Thompson River immediately opposite the North Thompson. When it became known that the Canadian Pacific would build its line on the south bank of the South Thompson (instead of the Yellowhead route, which would have utilized the North Thompson, as the Canadian Northern did a few years later) local businessman J.A. Mara realized that this original settlement would have its growth restricted by the bluffs above it. In 1884 he put together a syndicate to buy property to the east, which was not so restricted.

The syndicate hired local surveyor Robert Lee to lay out a new townsite (MAP 329, *right, top*). This plan covers most of what is today downtown Kamloops. Unlike some townsite promoters, Mara and his associates sensibly persuaded the Canadian Pacific to place its station and yards in its townsite by offering to give the railway about a third of the other lots. MAP 331, *far right, top*, shows the lots given to the Canadian Pacific coloured green.

New businesses naturally tended to locate around the railway station, and some of those in the old town moved east; one, the Arlington Hotel, was physically moved to a site near the station. Serious fires in the old townsite further contributed to its decline.

A few years later the townsite syndicate was still trying to sell its lots, this time emphasizing the nearby mining camps (MAP 330, *right, bottom*).

MAP 329 (*right, top*).
The 1885 survey of Kamloops by Robert Henry Lee. The lots crossed out on the waterfront represent an amalgamation of a site for the Shuswap Milling Company, shown as such on two other maps reproduced here. The red-coloured area is that reserved for the Canadian Pacific station and yards. Lee, a surveyor who drew both this map and MAP 332, *far right*, moved to Kamloops in 1884, was mayor from 1894 to 1896, and served as city engineer from 1898 to 1928.

Right, centre. An early-twentieth-century view of Kamloops, looking east. The road in the foreground is the wagon road to Nicola.

MAP 330 (*right*).
Kamloops and the nearby mining claims, from an 1897 booklet published by the townsite syndicate, which also included the ad. Note that on the map south is at the top.

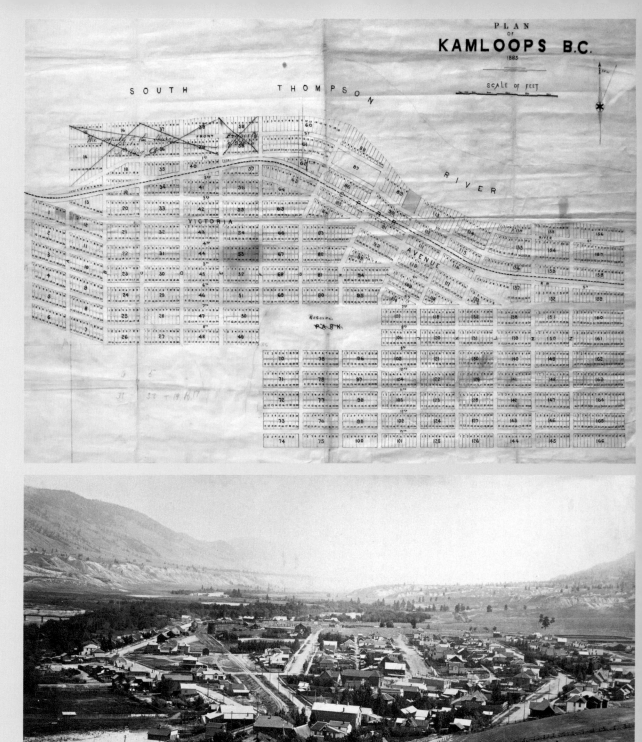

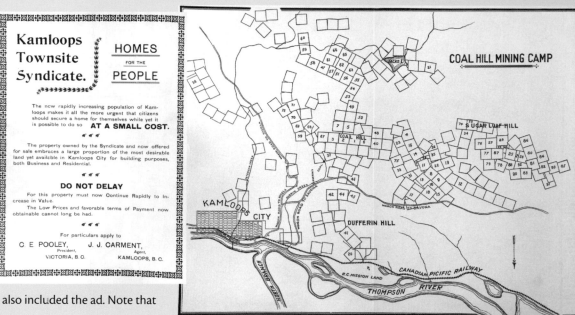

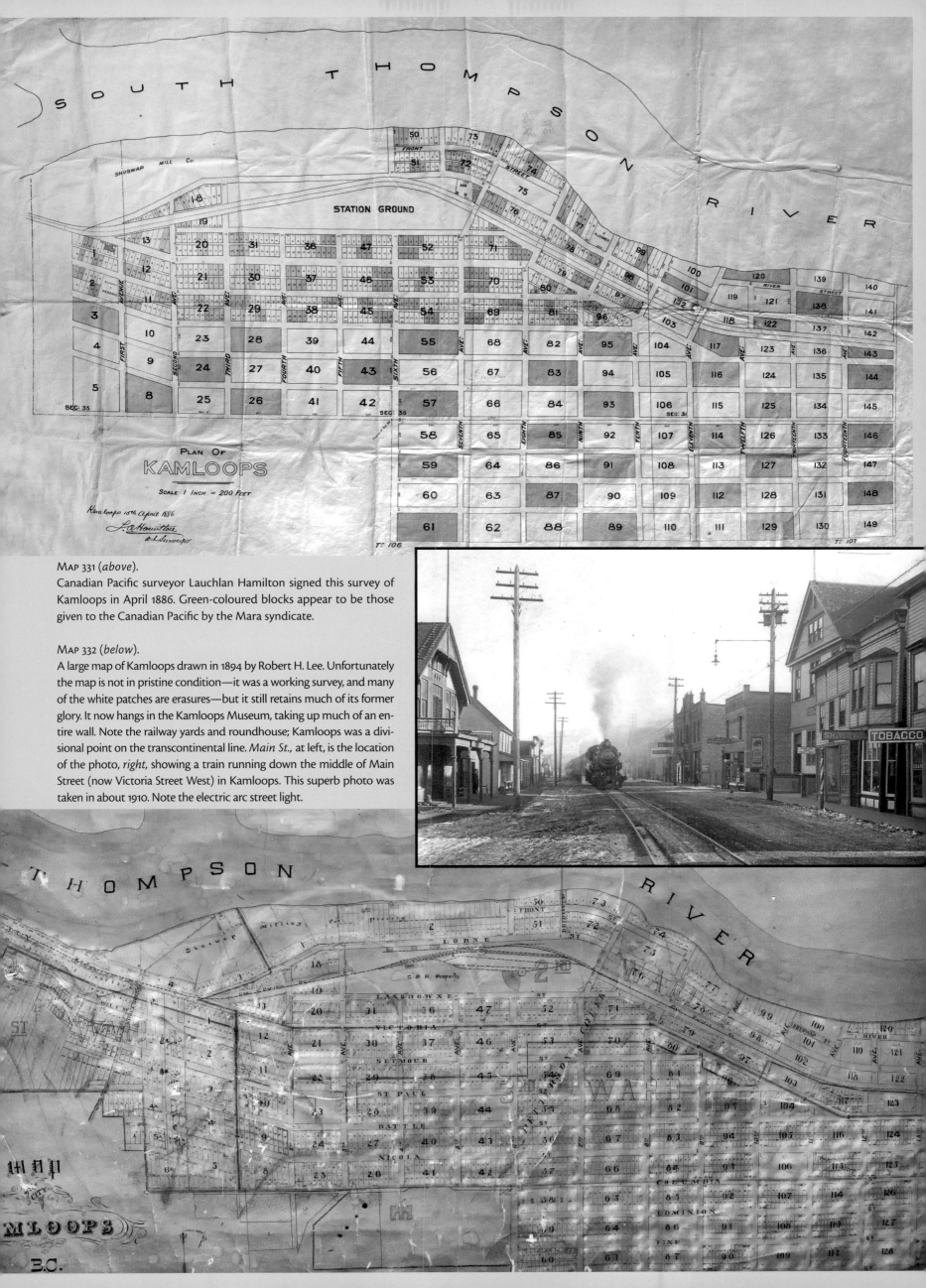

PLAN OF
KAMLOOPS
SCALE 1 INCH = 200 FEET

Map 331 (*above*).
Canadian Pacific surveyor Lauchlan Hamilton signed this survey of Kamloops in April 1886. Green-coloured blocks appear to be those given to the Canadian Pacific by the Mara syndicate.

Map 332 (*below*).
A large map of Kamloops drawn in 1894 by Robert H. Lee. Unfortunately the map is not in pristine condition—it was a working survey, and many of the white patches are erasures—but it still retains much of its former glory. It now hangs in the Kamloops Museum, taking up much of an entire wall. Note the railway yards and roundhouse; Kamloops was a divisional point on the transcontinental line. *Main St.*, at left, is the location of the photo, *right*, showing a train running down the middle of Main Street (now Victoria Street West) in Kamloops. This superb photo was taken in about 1910. Note the electric arc street light.

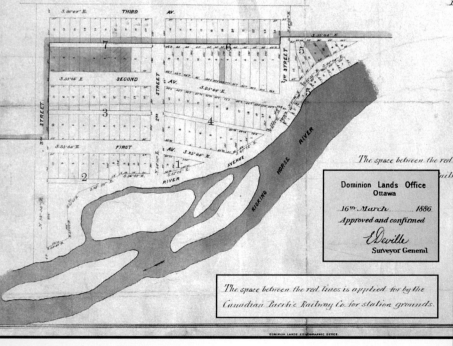

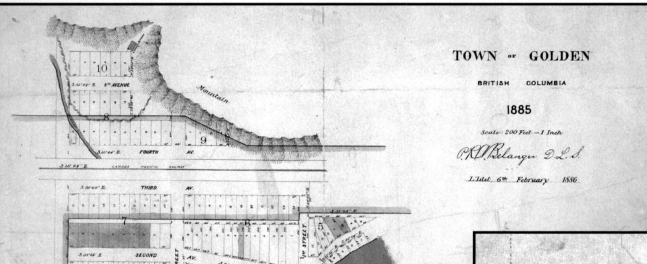

TOWN OF GOLDEN

BRITISH COLUMBIA

1885

Scale: 200 Feet = 1 Inch

P.R.W.Belanger D.L.S.

L'Islet 6ᵗʰ February 1886

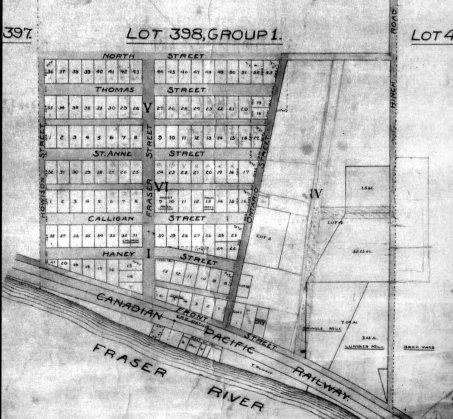

MAP 333 (*above*).
Surveyor Edward Mohun prepared this elegant map of British Columbia for the Department of Lands and Works in 1884, and it shows the transcontinental's state of affairs as of the end of that year. Cross-hatched lines depict finished track. The railway is poised at the Continental Divide to enter the province from the east, and the Onderdonk-built section between *Savona's Ferry* and *Port Moody* is complete.

MAP 334 (*above*).
The Canadian Pacific line spawned many new townsites in its path. This is the newly laid-out town of Golden, at the point where the *Kicking Horse River* meets the Columbia. The map was prepared by the Dominion Lands Department, which was responsible for Railway Belt lands, and is signed as approved by newly appointed surveyor general Edward Deville and dated 16 March 1886.

MAP 335 (*right*).
Hoping to cash in on the coming railway, brick maker Thomas Haney laid out the Haney Townsite on his previously pre-empted land in 1882.

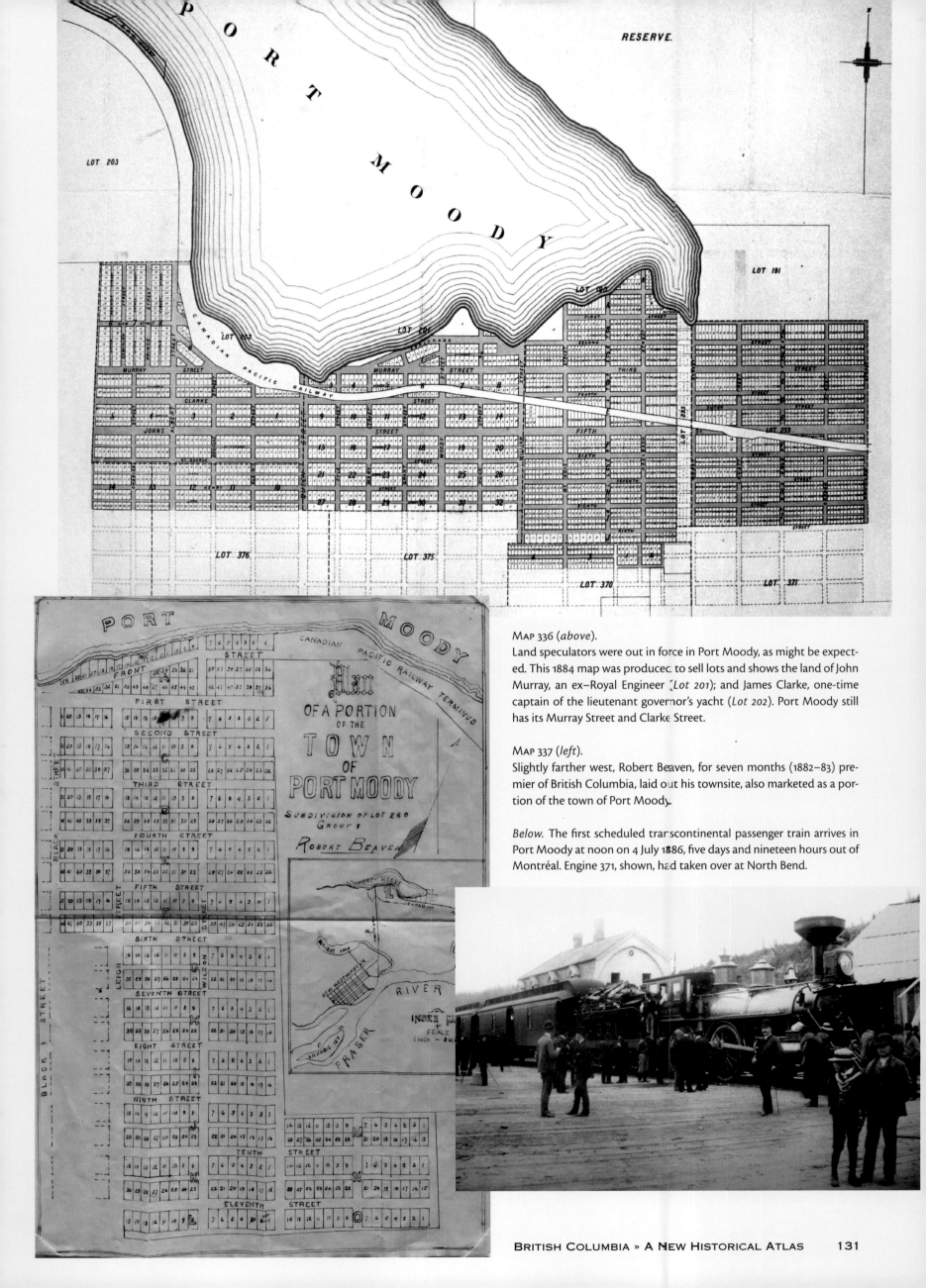

Map 336 (*above*).
Land speculators were out in force in Port Moody, as might be expected. This 1884 map was produced to sell lots and shows the land of John Murray, an ex–Royal Engineer (*Lot 201*); and James Clarke, one-time captain of the lieutenant governor's yacht (*Lot 202*). Port Moody still has its Murray Street and Clarke Street.

Map 337 (*left*).
Slightly farther west, Robert Beaven, for seven months (1882–83) premier of British Columbia, laid out his townsite, also marketed as a portion of the town of Port Moody.

Below. The first scheduled transcontinental passenger train arrives in Port Moody at noon on 4 July 1886, five days and nineteen hours out of Montréal. Engine 371, shown, had taken over at North Bend.

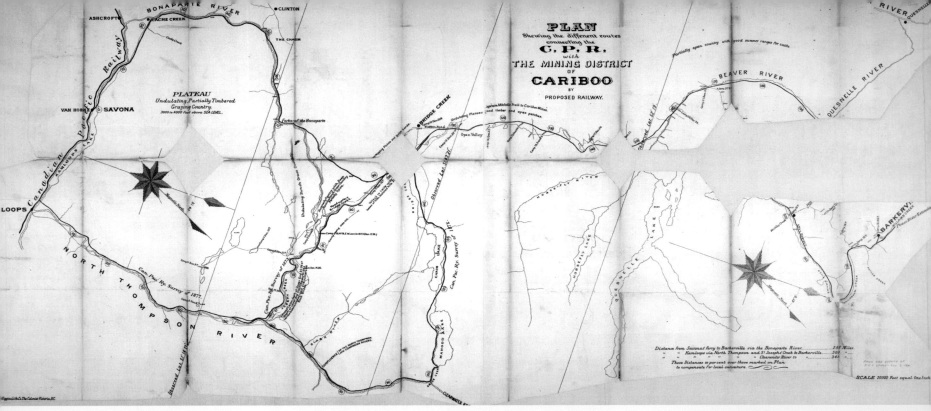

MAP 338 (*above*).
Few know that the Canadian Pacific was planning an extension from its main line to the gold mines of the Cariboo even as the first transcontinental train drew into Vancouver. But this map, drawn in 1887, shows that it was. Unfortunately the map is in poor condition and is missing parts where it has been folded, yet it has been restored as much as possible by its repository, Library and Archives Canada. Note that north is oriented to the right. The line was never built. In 1894 the CPR came to an agreement with the Cariboo Railway Company, chartered to build from Ashcroft to Barkerville, but this plan also came to nothing. The map shows two potential routes: one up the *North Thompson River*—which would have utilized one of the original Canadian Pacific surveys from the Sandford Fleming days—with two further routes then cutting across

to the line of the Cariboo Road (shown as *Wagon Road*); and a route leaving the main line at *Savona* and using the north bank of the Thompson to reach *Cache Creek* before turning north. Both routes merge near *Bridge Creek* and end up at *Barkerville*. The North Thompson route was used by the Canadian Northern (Pacific) Railway, completed in 1915 (see page 272). The map also shows Savona as *Van Horne*. The residents of Savona's Ferry, as it was known before the coming of the railway, were so grateful to the Canadian Pacific that when they heard that the general manager, William Cornelius Van Horne, was to visit their village, they renamed it in his honour and put up a sign at the station. The unprepossessing Van Horne, however, was not amused. "Tear it down," he barked at a railway employee. And so Savona it remained.

The first order of business the following spring was to begin the ballasting work often overlooked in the race to connect the line. In Rogers Pass, work was begun on some thirty-one snowsheds, which would have an aggregate distance of 8 km and take two years to complete. Here and in Kicking Horse Pass, cuts had to be widened and graded to try to control landslides. The first transcontinental passenger train left Montréal on Monday, 28 June, connected with a train from Toronto, and arrived in Port Moody on Sunday, 4 July 1886. Service eastward from Port Moody began two days later. However, the second *Atlantic Express*, which left Port Moody for Montréal on 8 July, was caught in a forest fire in the Selkirk Range and almost totally destroyed. The event underlined the vast amount of work that still needed to be done on the line to make it a both safe and reliable continental connection.

THE TERMINAL CITY

Even before the transcontinental line's completion, the Canadian Pacific had come to the conclusion that the line should end not in the restrictive and prone-to-silting Port Moody but farther along Burrard Inlet near Granville or even on English Bay, given concerns about the po-

tential difficulties large ships might have entering the harbour through First Narrows.

Van Horne began negotiations with British Columbia premier William Smithe, and in February 1885 agreements were concluded; some 6,245 acres (2,527 ha) were granted to the railway (MAP 348, *page 135*) in return for the westward extension and the building of a major hotel and an opera house. Another 213 acres (86 ha) were privately donated to the company. The province hoped thereby to create a new major city and make a lot of money from the sale of land it still owned in the vicinity (MAPS 347, 349, and 350, *page 135*). And the railway, of course, also hoped to cash in from the sale of the land it was given.

Construction of the extension to the new townsite Van Horne had resolved to name Vancouver was delayed when Port Moody landowners filed suit to prevent the line from crossing their land, but, with legal objections overruled, the track was completed into Vancouver in May 1887, and the first scheduled transcontinental train, wildly decorated to celebrate not only the event but also the Golden Jubilee of Queen Victoria, pulled into the new city on 23 May.

MAP 339 (*below*).
The 1887 extension of the transcontinental line from Port Moody to Vancouver is shown on this 1888 map.

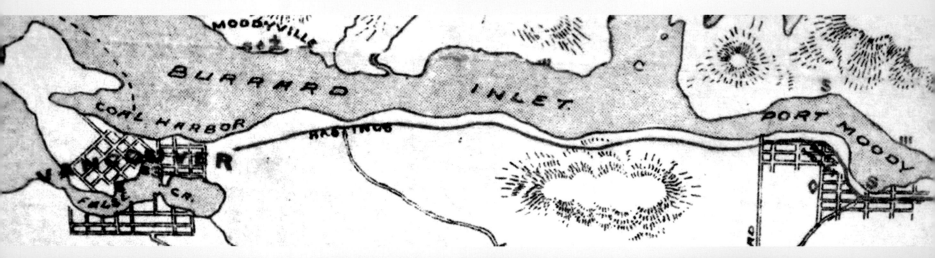

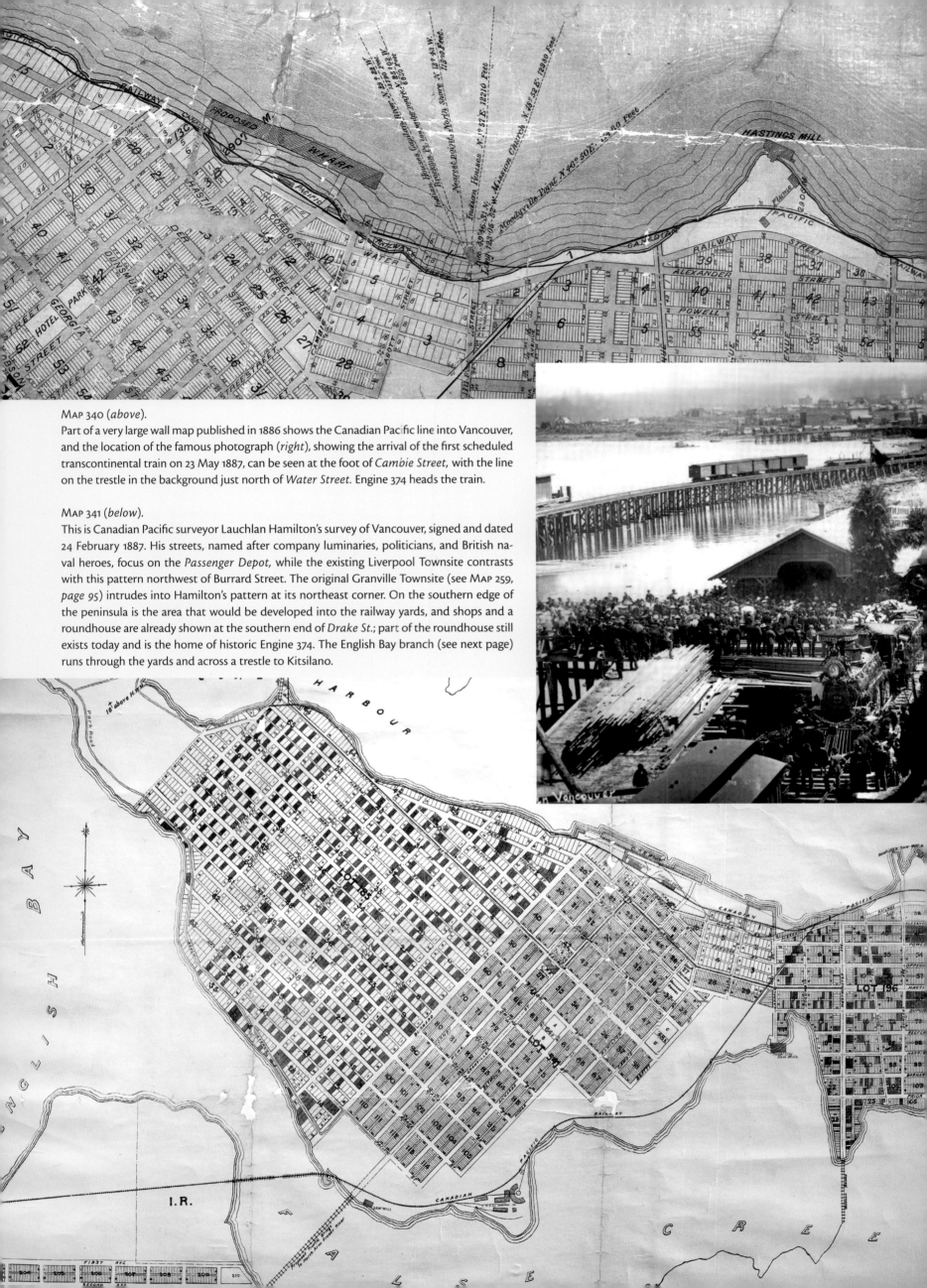

MAP 340 (*above*).
Part of a very large wall map published in 1886 shows the Canadian Pacific line into Vancouver, and the location of the famous photograph (*right*), showing the arrival of the first scheduled transcontinental train on 23 May 1887, can be seen at the foot of *Cambie Street,* with the line on the trestle in the background just north of *Water Street.* Engine 374 heads the train.

MAP 341 (*below*).
This is Canadian Pacific surveyor Lauchlan Hamilton's survey of Vancouver, signed and dated 24 February 1887. His streets, named after company luminaries, politicians, and British naval heroes, focus on the *Passenger Depot,* while the existing Liverpool Townsite contrasts with this pattern northwest of Burrard Street. The original Granville Townsite (see MAP 259, *page 95*) intrudes into Hamilton's pattern at its northeast corner. On the southern edge of the peninsula is the area that would be developed into the railway yards, and shops and a roundhouse are already shown at the southern end of *Drake St.;* part of the roundhouse still exists today and is the home of historic Engine 374. The English Bay branch (see next page) runs through the yards and across a trestle to Kitsilano.

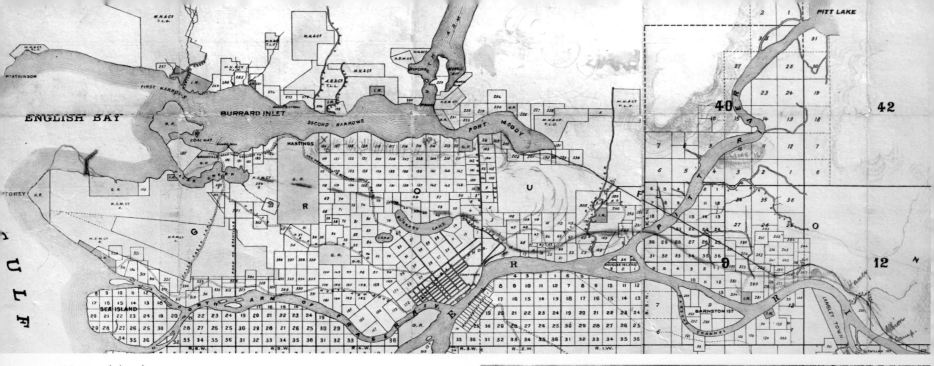

Map 342 (above).

In the contract between the Canadian Pacific and the provincial government signed on 23 February 1885, both Coal Harbour and English Bay were specified as terminals for the transcontinental line. This map, an 1876 printed base with ink and pencil additions dating to 1885, shows not only the line to Port Moody (in red ink) but also a projected line (in purple crayon) connecting the main line to False Creek, with a large breakwater at Jericho Beach on English Bay suggesting this might be the ultimate intended destination. The crayon addition to English Bay is said to be the work of Van Horne himself, though this is uncertain. The purple line route is very similar to the one later used by the Great Northern (see page 209). Compare also Map 491, page 204, which shows slightly later suggestions for branch line routes in the Lower Mainland, one of which reaches English Bay via an extension of the line to New Westminster.

Map 343 (right).

Detail of a large 1886 wall map shows the English Bay branch route, through *Railway Yard Shops & Roundhouse* located on Kits Point to a terminus *2909½ Miles from Montreal*, not coincidentally the western edge of the railway's land grant. Track was laid along this route, part of which later became a streetcar line.

Map 344 (right).

The boundaries of the new city of Vancouver, incorporated on 6 April 1886, included the peninsula to the west, then a *Government Reserve* and soon to be Stanley Park. Even this 1886 map notes that it is a *Proposed Public Park*; the new city council that year had requested the Dominion government lease the land to the city for use as a park. By the following year Map 351, page 136, shows the reserve as a *Public Park*; it was dedicated in 1888 and officially opened and named by Lord Stanley, then governor general of Canada, the next year.

Map 345 (below).

This interesting map was contained in a letter from Van Horne to company surveyor Lauchlan Hamilton in January 1885. Van Horne tells Hamilton that the railway is asking the Dominion government for the strip of land marked on the map in green—what is today a sizeable

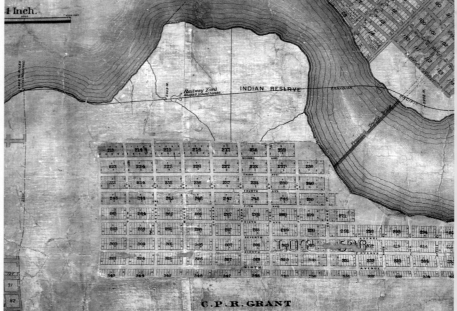

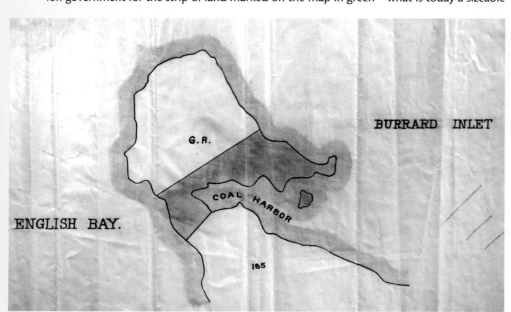

chunk of Stanley Park. Van Horne does not tell Hamilton what the company requires the land for, but it seems likely that it was simply for possible future dock extension. Van Horne asks Hamilton to draw a better map because the minister of militia needs to assess the topography of the peninsula to determine, as Van Horne writes, "how much of the ground should be retained by the Government for defensive purposes and how much they can spare us." The federal government, of course, determined that it needed to retain all of the land, yet it was not for defensive purposes, for the land was leased to the city three years later for use as a park. Gun batteries were set up in the park during both World War I and World War II, however (see Map 822, page 315).

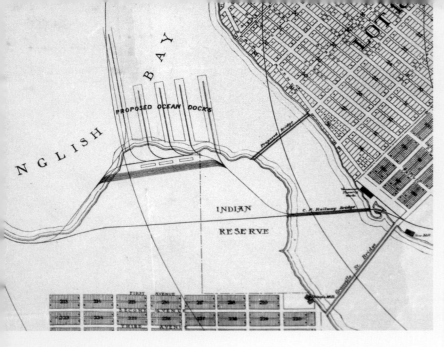

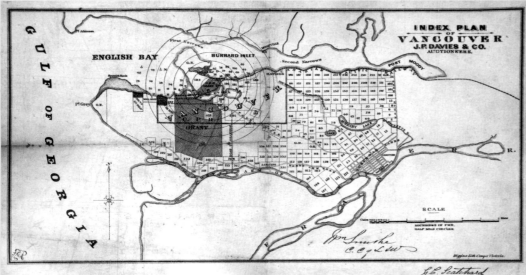

Map 346 (*above*).
There were a number of proposals for docks on English Bay, both by the Canadian Pacific, as here in 1891, and other railway companies (see Map 609, *page 246*). The idea was to allow large ships to avoid having to pass through the treacherous First Narrows or having to await a change of tide.

Map 347 (*above*).
The provincial government was willing to give the Canadian Pacific a large land grant partly because it hoped to thereby sell a lot of other land it owned in the vicinity. This was the auction map for a land sale of lots at Hastings, near the site of today's Pacific National Exhibition. *Victoria Avenue* is today's Renfrew Street.

Map 348 (*left*), **Map 349** (*below, centre*), and **Map 350** (*below, right*). These three maps and the advertisement (*below, left*) are part of a sheet published in advance of an auction of provincial government lands in 1887. The locations of the two blocks involved are shown in red on the Index Plan (**Map 348**). Note that one is within the city boundary and the other outside; both are adjacent to the Jericho *Naval Res^{ve}.*, which was originally put aside as a source of spars for sailing ships. The map also shows the *C.P.R. Grant* shaded grey. **Map 349** is of an area behind Locarno Beach; *Queen Avenue* is Northwest Marine Drive, and *Salisbury Street* is Belmont Avenue. **Map 350** shows lots already sold shown in red. *Campbell Street* is Alma Street, then the western boundary of the city; *Victoria Street* is Point Grey Road, and *Richards Street*, on the eastern boundary of this map, is today Balaclava Street. Then as now, waterfront property was prime—and all the lots along the beach have been sold

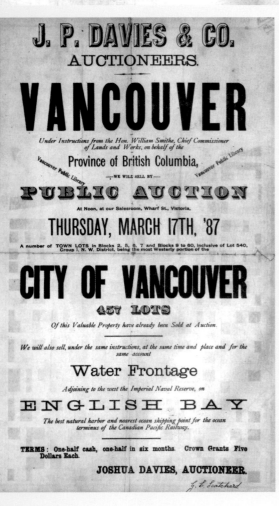

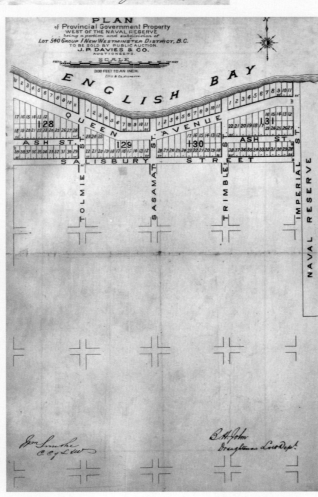

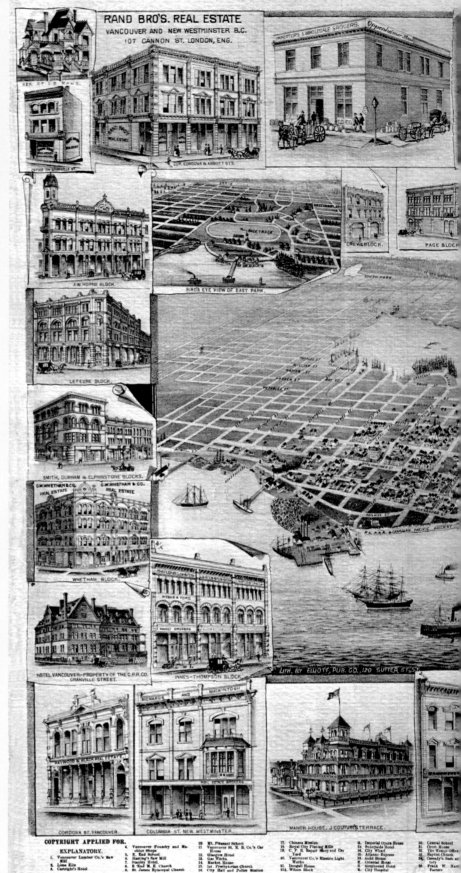

Map 351 (above).

Published by real estate brokers Rand Brothers in 1887—and depicting projected features as well as those existing—this map of Vancouver clearly shows the differing patterns of lots at the CPR townsite, the ex-Liverpool West End, the original Granville Townsite, and the peripheral lots subdivided by private investors. The *C.P.R. Grant* is shown, as are both the Coal Harbour and English Bay rail lines and the Drake Street shops and roundhouse. The *Traffic Bridge to North Arm Fraser River* is the first Granville Bridge, opened in 1889. A bridge also crosses False Creek on *Westminster Avenue* (now Main Street), already defining the east end of the creek that would later be filled in (see page 274). At the extreme right edge is a *Government Reserve (Townsite of Hastings)*.

Before the railway arrived, Granville had perhaps 300 inhabitants; indeed all Burrard Inlet had only about 900. By the end of 1886 Vancouver had a population of 2,000, which by the first official census in 1891 had reached 13,709. A water line was laid across the First Narrows to bring fresh water from the Capilano River, and garbage disposal was centralized—garbage was loaded on a scow and dumped in the middle of Burrard Inlet. And, of course, a fire department had been created in 1886 following the city's initial brush with devastation that year. One of the first acts of the new city council had been to ask the federal government for a lease on the government reserve that became Stanley Park in 1888, though already established as a park the year before. And the first tracks of an extensive streetcar system were laid in 1889 (see page 140).

Vancouver, created as the terminus of the transcontinental railway, was well on its way to commercial pre-eminence in the West. The promotional bird's-eye map shown here (Map 353, *right*) testifies to the promise of the new city as early as 1890. The economy took a downturn after 1892 and experienced slower growth until 1897, when the city revived as the principal supply point for the Klondike Gold Rush (see page 236).

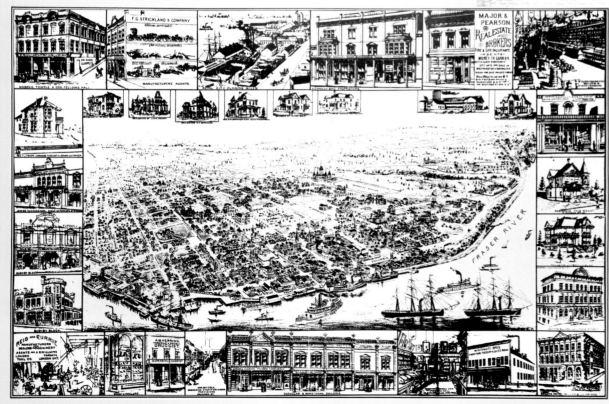

Map 352 (left).

This map, also published in 1890, is the direct competitive equivalent to the Vancouver map above. It is not certain which map followed which, but it seems likely this one was a competitive response to the other. Unfortunately the original seems to have been lost, and only this less-than-optimal copy remains. *Below.* Vancouver burned to the ground a few weeks before the first transcontinental arrived in Port Moody but was quickly rebuilt. New Westminster had its great fire in 1898; this is a contemporary print of the event.

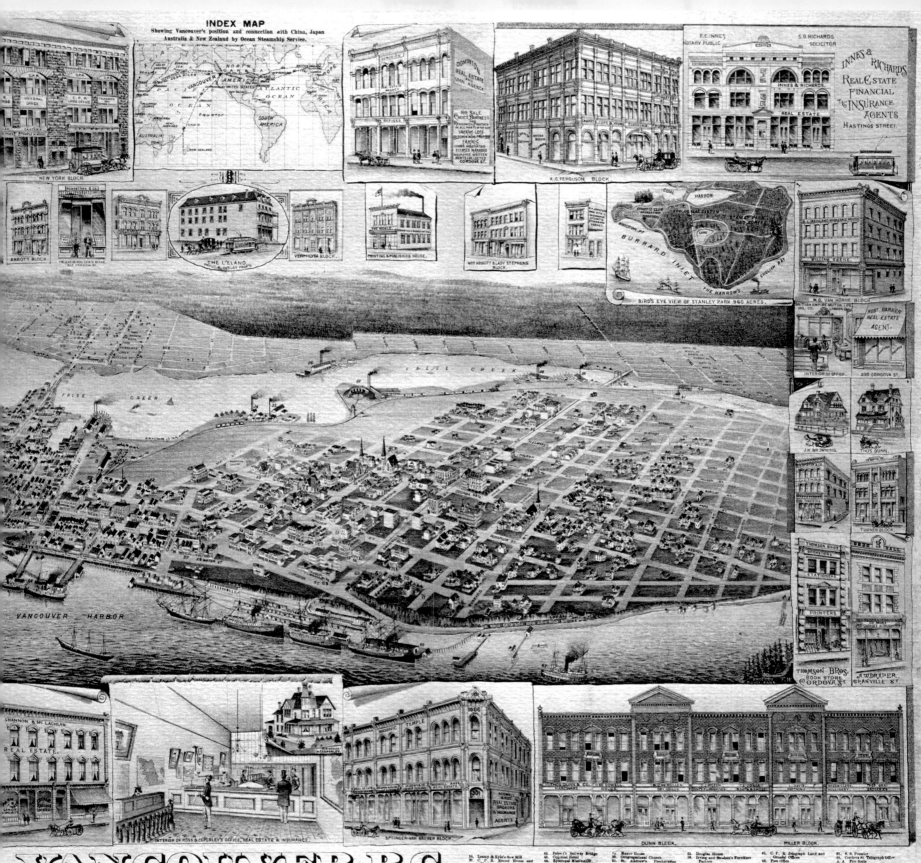

VANCOUVER, B.C.
1890.
Holiday Supplement to the "Vancouver Daily and Weekly World".

Map 353 (above).

Progress was so rapid in Vancouver following the arrival of the Canadian Pacific Railway that by 1890 the *Vancouver World* published this superb promotional bird's-eye map, complete with detailed views of advertisers' buildings around its perimeter. False Creek penetrated far to the east, crossed by the *Westminster Av. (Main Street) Bridge*. Prominent at centre is the railway's depot and wharves. This view shows well how the Canadian Pacific deliberately established its terminus to the west of the existing Granville Townsite in order to create and control its own land development. In addition to its own land grant the company was donated about one block in every four in the West End (about 135 acres, or 55 ha) but only about one lot in six to the west of Granville (about 40 acres, or 16 ha) owing to the multiplicity of landowners. The land grant included 39 lots totalling about 8 acres (3 ha) in the Granville townsite itself. The land grant south of the station amounted to about 480 acres (194 ha) and was the primary focus of the company's development efforts; the grant south of False Creek, then largely forested—and most is shown here in that state—amounted to 5,795 acres (2,345 ha).

Map 354 (overleaf).

Thought by many people to be the finest map of Vancouver ever produced, this magnificent bird's-eye was published by the *Vancouver World* newspaper in 1898, celebrating the city's return to prosperity following the discovery of gold in the Klondike. There are in fact not many changes from the 1890 bird's-eye because the period between 1892 and 1897 was one of economic recession, but one new feature is the first Cambie Bridge, a simple wooden trestle. Like Map 353 this view shows the original eastward extent of False Creek very well, partitioned off by the *Westminster Ave.* Bridge (now Main Street). The harbour is full of ships to the point of unbelievability—but this was a promotional map, of course. At *right* is an ad for this map that appeared two years later; only a "few hundred copies" of the map remain for sale.

UPPER FALSE CREEK FLATS

BURRARD INLET

CONTINUATION OF ALEXANDER & POWELL STREETS, SHOWING THE VANCOUVER CITY FOUNDRY & MACHINE WORKS & SUGAR REFINERY.

Streetcars for Growth

It was clear to most people that Vancouver was going to rapidly grow into a major metropolis. Facilitating that growth in the era before cars and buses was the streetcar, which had the ability to direct and funnel growth like no other factor. Developers were known to pay the streetcar company to extend a line to their property to ensure its success.

The Vancouver Electric Railway & Light Company began laying track in 1889, and streetcars began running on 27 June 1890. The company soon extended its lines out of the downtown peninsula to Fairview, lured by the offer of a gift of sixty-eight lots from the Canadian Pacific. The Fairview Belt Line was just the first of a number of lines to precede development rather than service the existing population.

Despite the initial promise, however, the company fell into bankruptcy with the onset of the 1892–97 recession. Service on the Fairview line was even suspended by the trustee. Another company took over but also went bankrupt, and it was not until 1897, when a British company, the British Columbia Electric Railway Company (BCER), took over that the business stabilized. BCER also took over an interurban to New Westminster (see below and page 270), New Westminster local lines, and the Victoria system. Ridership increased until 1914 and then began a slow decline, the victim of the automobile and the bus; despite the introduction of sleek new models in 1939—the PCC streetcars—the tide could not be turned, and streetcars disappeared from Vancouver in 1955.

Victoria had managed to begin streetcar service four months before Vancouver, in February 1890. The National Electric Tramway & Lighting company began operating over an 8-km-long system and by October service was extended to Esquimalt. BCER took over in 1897 following the failure of its predecessor company after one of the streetcars plunged off the Point Ellice Bridge on 26 May 1896. Victoria's streetcars lasted until 1948.

Not to be left behind, work began in New Westminster late in 1890, within the city by one company and between New Westminster and Vancouver by another; the companies amalgamated as the Westminster & Vancouver Tramway Company the following year. Service between New Westminster and Vancouver commenced on 8 October 1891. But by 1893 the company was bankrupt. Its assets were sold in 1895 and added to the new BCER system two years later. The BCER system was further expanded in 1907 by the addition of 8 km of lines in North Vancouver.

Outside the southwestern part of the province, the only other British Columbia city to acquire a street railway system was Nelson. Here another British company, the British Electric Traction Company, received a thirty-five-year franchise to build and operate streetcars, which were to prove a considerable benefit on the steep streets of that city. Streetcars lasted in Nelson until 1949, another victim of buses.

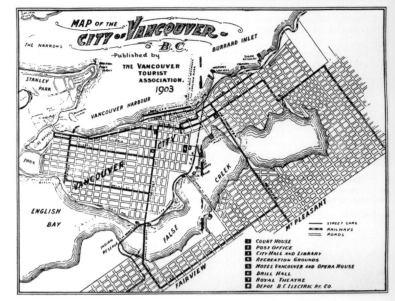

Map 355 (*above*).
Vancouver's streetcar system on a 1903 tourist map.

Map 356 (*below*).
The Westminster & Vancouver Tramway on a map dated 1899 but depicting about 1892. *Inset* is a photo of the line near today's Central Park, taken in 1892.

Map 357 (*below*).
The red lines are the streetcar routes in the Kitsilano area in 1911. The photo, *inset, left,* is a classic in its illustration of the way the streetcar drove urban growth. In this boom period of Vancouver's history (see page 236) the forest would soon be replaced by houses and businesses. The photo was taken by James Quiney, a local real estate agent, in 1909, the year the line opened. His sign is prominent at right. The scale of development expected was such to warrant a double track despite the obvious initial lack of patrons on this route. The photo is of 4th Avenue at Waterloo Street.

Above.
A view of Hastings Street as depicted on a postcard about 1914. Note some of the automobiles have signs on the front of them. These are jitneys, which stopped at the streetcar stops to pick up passengers; they were a thorn in the side of BCER until they were banned in 1918.

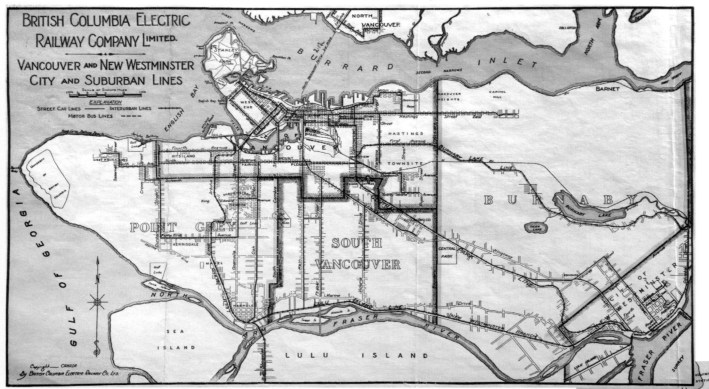

Map 358 (*left*).
The streetcar system in and around Vancouver and New Westminster in 1923. Note the municipalities of *South Vancouver*, incorporated in 1891; *Burnaby*, incorporated in 1892; and *Point Grey*, incorporated in 1908 out of the western part of South Vancouver. Streetcar lines now reach south to the Fraser and east to Capitol Hill; the Burnaby Lake Line connects to New Westminster. At the peak there were some 300 km of streetcar tracks.

Map 359 (*right*).
This map of Esquimalt in 1893 shows the streetcar line to the settlement there and can be related to the photo, *inset*; the streetcar is just where the line cuts across a small bay. St. Paul's Garrison Church behind the streetcar is the larger black building on the map. The church was later moved nearer to Victoria.

Map 360 (*below*).
A simplified map of Victoria's streetcar routes in 1923, which now reach Oak Bay. The line with the red dots is the Saanich interurban (see Map 685, page 271).

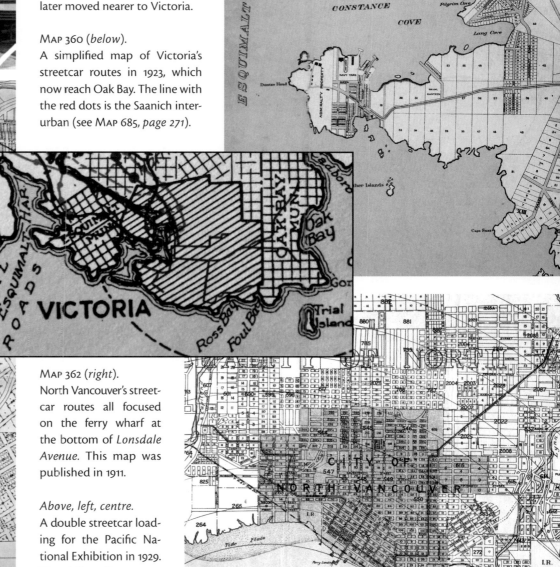

Map 361.
Nelson's streetcar system is shown by the black lines on this 1912 map. The track runs north to Fairview.

Map 362 (*right*).
North Vancouver's streetcar routes all focused on the ferry wharf at the bottom of *Lonsdale Avenue*. This map was published in 1911.

Above, left, centre.
A double streetcar loading for the Pacific National Exhibition in 1929. The ability of streetcars to double up when traffic warranted was one of their strengths.

Lower Mainland Municipalities

The Royal Engineers subdivided large areas of the Lower Mainland so that the land could be sold (see page 70). Once pioneers began settling on the land, they needed roads—or trails at least—to get their produce to the Fraser and then to market and to bring in supplies. But their taxes went to the provincial government, not to local road building.

In 1872 the province passed the Municipality Act, which allowed a local government to be created if thirty resident men petitioned for it. This offered the opportunity to divert local taxes to the building of local roads and so led to the creation of a number of municipalities in the ensuing years.

The agricultural communities of Langley and Chilliwhack became the first municipalities outside of New Westminster (1860) and Victoria (1862) when they were created a year later. The Langley petitioners made a mistake in defining the boundaries of their municipality as ten miles (16 km) square from the Fraser River, leaving out a 4-km-wide strip of land along the international boundary (seen on Map 364, *below*). This area remained unincorporated until 1894, when the mistake was rectified.

Maple Ridge became a municipality in 1874 and included the area that is now Pitt Meadows and part of Port Coquitlam. In 1892 residents of Pitt Meadows, seeking to lower their taxes, petitioned to revert to unorganized territory, and the area remained as such until incorporation as the separate municipality of Pitt Meadows in 1914.

In 1879 the settlers of Richmond, Delta, and Surrey petitioned and were granted municipal status. Those of Surrey also made a mistake, failing to define their eastern boundary coincident with the western boundary of Langley, creating yet another strip of no man's land. This required the surrender of the original letters patent and the issuance, in July 1882, of new ones. Clearly it was difficult for the untrained to define boundaries in a forest.

All these municipalities predated Vancouver, incorporated in 1886. At Westminster Junction, where the branch line of the Canadian Pacific left the main line for New Westminster, landowners petitioned in 1891 to create Coquitlam. Both 1891 and 1892 were heady years for the incorporationists: a battle ensued between landowners south and east of Vancouver to form their own municipalities. South Vancouver was granted municipal status in April 1891; Burnaby was created the following year, but not before an abortive attempt to create a municipality covering all the area to Point Grey had failed (Map 367, *right, bottom*). North Vancouver, covering the entire North Shore from Deep Cove to Horseshoe Bay, was also created in 1891.

Farther up the valley, Huntingdon was created the same year, while Sumas, Nicomen, Matsqui, and Mission saw life in 1892; only the last still exists independently. (Abbotsford, a village, had to wait until 1924.)

Later, in 1907, North Vancouver City separated from North Vancouver District (see page 242), and West Vancouver in 1912. Point Grey split from South Vancouver in 1907. Port Coquitlam separated from Coquitlam in 1913 (page 249), and Port Moody became a separate city in 1913.

Map 363 (*above*).
A map of land subdivisions in Delta about 1900. Note the holdings of W.M. Ladner and T.E. Ladner. William registered a townsite on his property that became Ladner. The more grandiose Lorne Estate, never an actual estate but an attempt to consolidate land, ended up as the Vancouver city dump on Burns Bog. *Inset* is a 1907 ad for farmland in Delta at $150 to $200 per acre ($370–$500/ha).

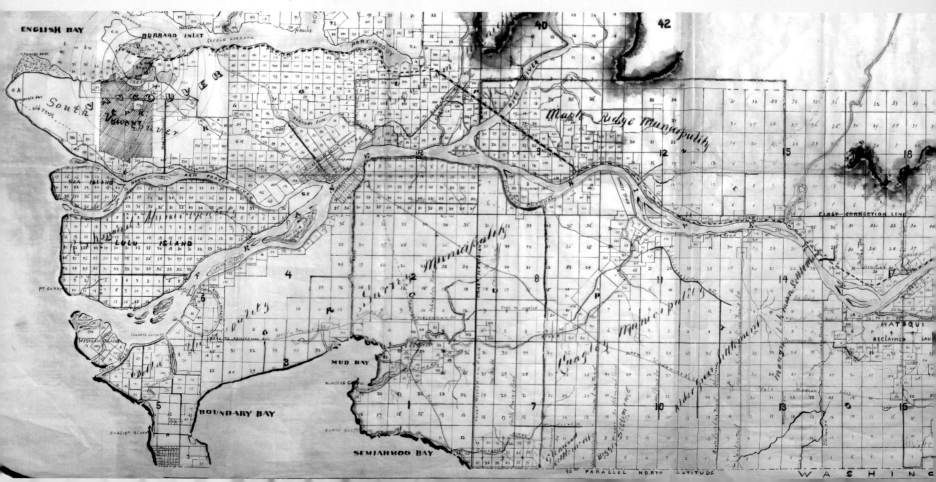

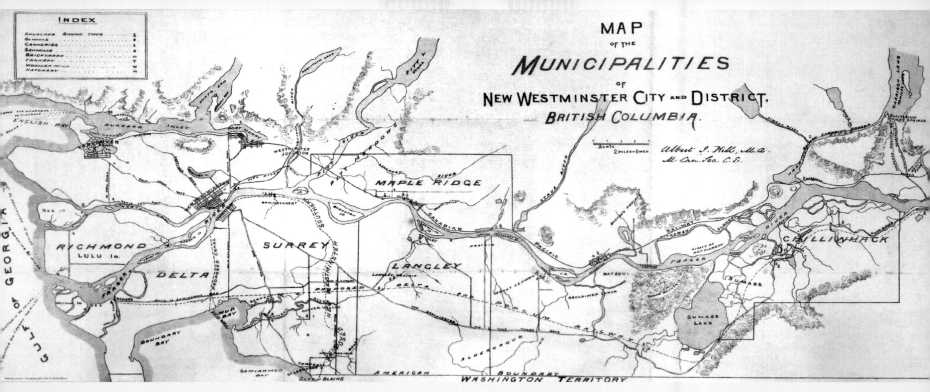

MAP
OF THE
MUNICIPALITIES
OF
NEW WESTMINSTER CITY AND DISTRICT,
BRITISH COLUMBIA.

Map 364 (below, left).
The boundaries of municipalities created up to and including 1891 are shown on this map. Langley's boundaries do not include the *Glenwood Settlement* and *Biggar(?) Settlement* close to the international boundary. *Maple Ridge Municipality* extends over today's Pitt Meadows and into Port Coquitlam.

Map 365 (above).
Drawn in 1888, this map shows the municipalities of the Lower Mainland existing at that date. Note *Westminster Junction*, which would become Coquitlam three years later. Langley is shown extending to the forty-ninth parallel, though the 4-km-wide strip close to the boundary was not officially added to Langley until 1894.

Map 366 (right, centre).
South Vancouver, incorporated in 1891, excluding the City of Vancouver and *Hastings Townsite.* South Vancouver's name was chosen as promoters thought it would make real estate easier to sell. Point Grey split from South Vancouver in 1907; Hastings joined the City of Vancouver in 1911, and both Point Grey and South Vancouver joined with the city in 1929 to finalize today's boundaries.

Map 367 (right, bottom).
This was the map drawn in 1891 as an attempt by some landowners to persuade the province that Burnaby should incorporate all of the area west to Point Grey as well as the area we know as Burnaby today. The area south of the City of Vancouver instead became South Vancouver Municipality.

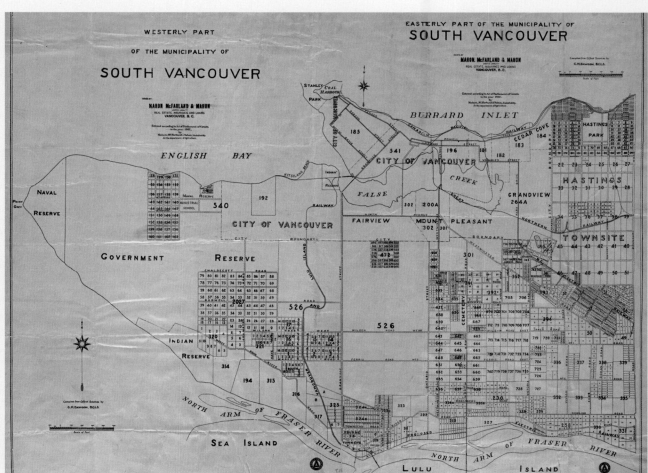

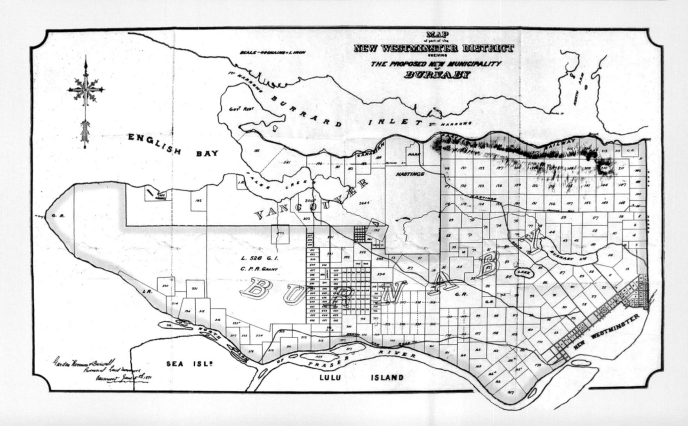

OPENING THE OKANAGAN

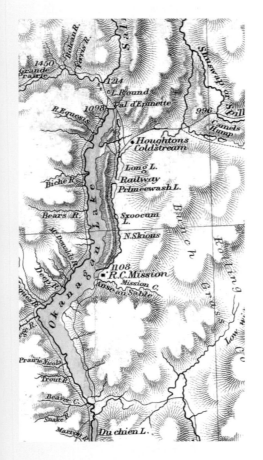

The potential of the Okanagan Valley for agriculture had been apparent since the Hudson's Bay Company used it as the main route connecting New Caledonia with the Columbia.

In 1864 Captain Charles Frederick Houghton claimed land along the banks of a stream he called Coldstream Creek under a special military land grant scheme that made land available to veterans of the Crimean War. Houghton never fought in that war, being on a troop ship en route to the Crimea when the war ended, but nonetheless he qualified. Houghton soon found out that the land he had claimed was part of a land reserve Governor James Douglas had set aside for a road, but Houghton was able to use his connections to obtain title to the land anyway.

Houghton's cousins, the brothers Charles Albert and Forbes George Vernon, had travelled to British Columbia with him, having also been attracted by the promise of free land. Their interest was initially diverted to silver mines at Cherry Creek (shown on MAP 370, *below*), but by 1865 they returned to claim land grants and buy further land at the north end of Okanagan Lake at a place that became Okanagan Landing. Six years later the brothers exchanged their land for Houghton's Coldstream property and began extensive improvement and expansion, not least of which was to arrange a water lease and begin irrigating the land. They also began buying cattle driven north from Oregon and began supplying gold-mining areas in the Cariboo and elsewhere. In 1883 Charles sold his share in the ranch to his brother Forbes, who became the sole owner of more than 53 km².

Some men who drove cattle from Oregon, such as Cornelius O'Keefe, Thomas Wood, and Thomas Greenhow, in 1867 decided instead to remain in the North Okanagan Valley, and all three assembled large land holdings for their ranches.

When it became clear that the Canadian Pacific Railway was going to route its transcontinental line along the shores of Shuswap Lake, Vernon joined a syndicate set up to build an extension south from the main

MAP 368 (*above*).
This detail of one version of the Trutch map of 1871 (MAP 913, *pages 358–59*) shows *Houghtons Coldstream* near the north end of *Okanagan Lake*. *Long L.* is today's Kalamalka Lake. Note also the *R.C. Mission* at *Anse au Sable*, today's Kelowna.

MAP 369 (*below, left*).
The 640-acre survey blocks on this 1878 survey map are interrupted by the previously granted irregular land parcel of the Vernon brothers' Coldstream Ranch along the *Cold Stm.* The site of the future city of Vernon is midway between *Swan L.* and Kalamalka Lake, at bottom.

MAP 370 (*below, right*).
In a prospectus published for potential investors in 1888, this map showed the proposed route of the Shuswap & Okanagan Railway from the *Canadian Pacific Railway* at *Sicamous* to *Vernon*, omitting the line's continuance to *Okanagan Lake*. *Enderby*, its name recently changed from Belvedere, is shown, as is *Lansdowne*, which would actually be bypassed by the rail line. Note *Cherry Creek* and *Rich Silver Mines*, where the Vernon brothers first tried their luck, and the *Mission* near the future Kelowna.

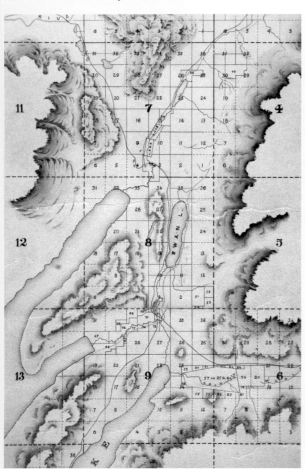

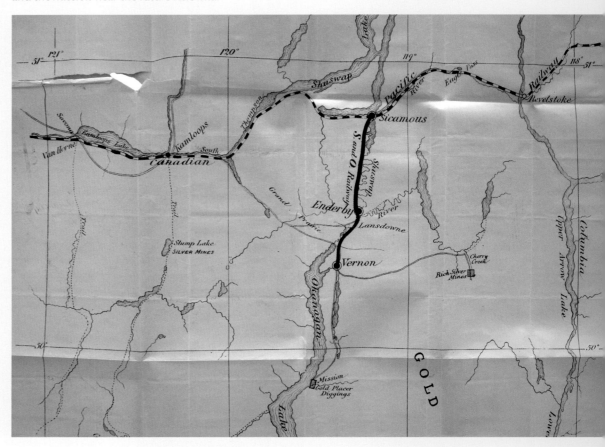

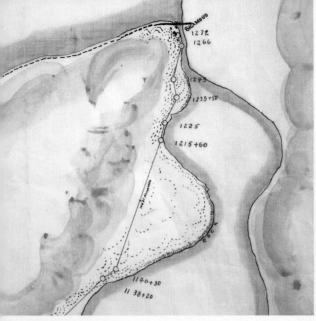

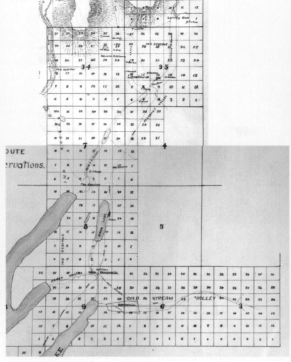

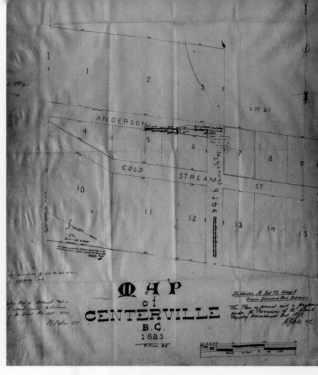

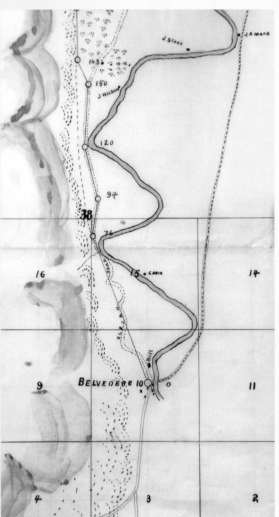

Map 371 (*left, top*) and Map 372 (*left*).
Details of a survey carried out in 1884 to determine the best route to Okanagan Lake down the valley of Mara Lake and the Shuswap River from the Canadian Pacific main line at Sicamous. Map 372 features *Belvedere*, now Enderby. This location, home to a large flour mill, was originally a steamboat landing place, the head of navigation southward. It had many short-lived names: Spallumcheen, Steamboat Landing (naturally enough), Fortune's Landing, Lambly's Landing, and Belvedere or Belvidere. It had this last name when in 1887 the residents decided on Enderby, a name derived from poetry.

Map 373 (*below, left*) is an official survey on which the name *Belvidere* has been crossed out and the new name of *Enderby* pencilled in. Also shown is a *Line of Proposed Canal to Okanagan Lake*, a proposed Spallumcheen and Okanagan Canal surveyed in 1882 by the federal government but shelved owing to high cost; it was ultimately made unnecessary by the railway.

Map 376 (*below, right*).
The northernmost end of a survey of the Shuswap & Okanagan Railway drawn by Frank Latimer, a local surveyor. South is at the top of this map. The railway was completed in 1892, joining the *Main Line Can. Pac. Railway* at this point. Canadian Pacific built one of its fine hotels here, initially called the *Lake View Hotel* (and marked as such on the map) and later the Sicamous Hotel (shown on page 291). The hotel sits on *Sicamous Narrows*, where the Shuswap River enters Shuswap Lake, at bottom. Further sections of this long survey map are shown overleaf.

Map 374 (*above, centre*).
Another survey map from about 1885 showing the North Okanagan area. *Centreville* has now appeared, at the head of an arm of Okanagan Lake in *Priest['s] Valley*, the first name of the settlement here, as a branch of the Okanagan Mission had been built in 1862. The Coldstream Ranch is marked with the name of its owner, *F.G. Vernon*. On the northern segment of this map is *Lansdowne*, a community bypassed by the railway in 1892. The settlement virtually disappeared when owners physically moved their buildings nearer to the line.

Map 375 (*above*).
Centerville in 1885. Two years later the town was renamed after Forbes George Vernon, who was twice elected to the provincial legislature and was British Columbia Commissioner of Lands and Works; his name appears in this capacity on several of the maps in this book (for example, Map 378, overleaf).

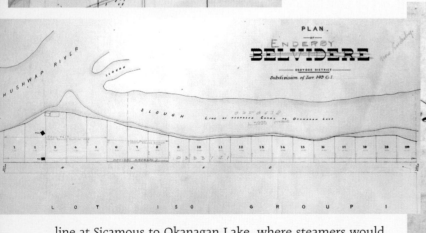

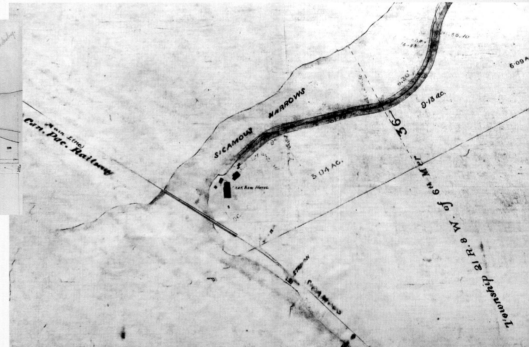

line at Sicamous to Okanagan Lake, where steamers would be able to connect, thus opening up the entire valley (see page 154). "With 50 miles of railway, 150 miles of country is opened up," boasted a prospectus. The first steamer on the lake, the SS *Sicamous*, was launched in 1886.

Another group of investors, led by a Scottish land promoter, George Grant Mackay, and also including Forbes Vernon,

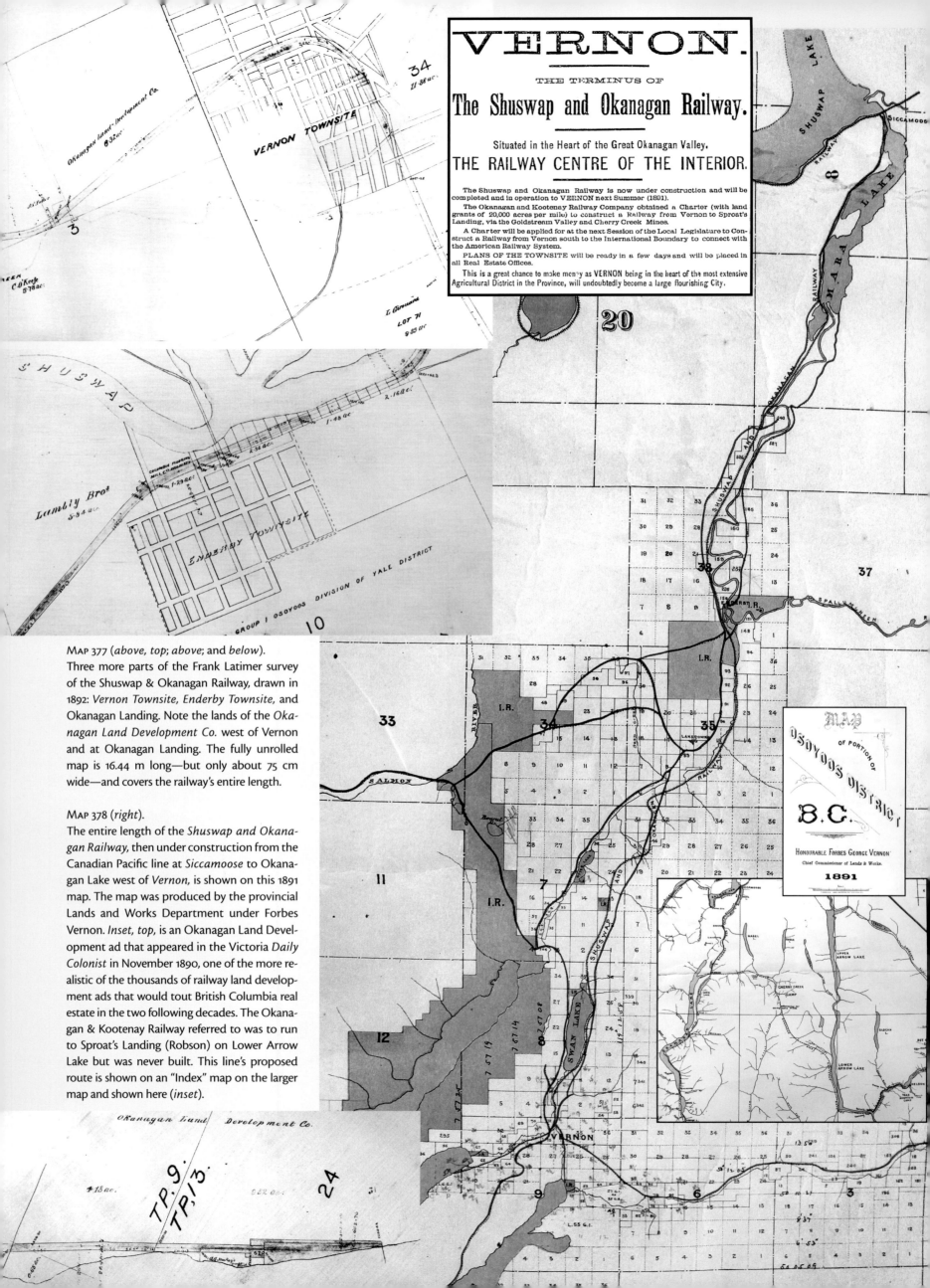

MAP 377 (*above, top; above;* and *below*).
Three more parts of the Frank Latimer survey of the Shuswap & Okanagan Railway, drawn in 1892: *Vernon Townsite, Enderby Townsite,* and Okanagan Landing. Note the lands of the *Okanagan Land Development Co.* west of Vernon and at Okanagan Landing. The fully unrolled map is 16.44 m long—but only about 75 cm wide—and covers the railway's entire length.

MAP 378 (*right*).
The entire length of the *Shuswap and Okanagan Railway*, then under construction from the Canadian Pacific line at *Siccamoose* to Okanagan Lake west of *Vernon*, is shown on this 1891 map. The map was produced by the provincial Lands and Works Department under Forbes Vernon. *Inset, top,* is an Okanagan Land Development ad that appeared in the Victoria *Daily Colonist* in November 1890, one of the more realistic of the thousands of railway land development ads that would tout British Columbia real estate in the two following decades. The Okanagan & Kootenay Railway referred to was to run to Sproat's Landing (Robson) on Lower Arrow Lake but was never built. This line's proposed route is shown on an "Index" map on the larger map and shown here (*inset*).

set up the Okanagan Land and Development Company to develop and sell land to the settlers that the railway was expected to bring. A townsite named Centreville had been laid out in 1885 (Map 375, *page 145*), and most of it was also purchased by the company. A waterworks system was constructed and a hotel built, and the company began advertising Vernon as "The Railway Centre of the Interior" (ad, *left*). The Shuswap & Okanagan Railway began construction in 1890 and was completed to Okanagan Lake in May 1892. The railway was officially taken over by the Canadian Pacific in 1903.

The Earl of Aberdeen (governor general of Canada 1893–98) had purchased land on the east side of Okanagan Lake (land he named the Guisachan Ranch, now within the City of Kelowna) in 1890, and the following year also purchased Forbes Vernon's ranch, renaming it the Coldstream Ranch. Lord Aberdeen was intent on subdividing some of his land for sale as orchards. Some 25,000 apple, pear, and cherry trees were planted on forty hectares of his land in 1892, together with hops, which were to provide a cash crop until the trees began production. The first fruit matured in 1896, and the first crop of apples was purchased by the Canadian Pacific Railway for use in its dining cars and hotels. This was the beginning of commercial fruit growing in the North Okanagan.

Aberdeen's enterprise was also used as proof that orchards were viable, and the fact that a member of the British aristocracy was involved enhanced sales potential, since most of the purchasers of the land Aberdeen subdivided were of recent British origin. In Vernon a jam factory was built, and milk cows were imported to begin a dairy.

Vernon had begun an irrigation system, and this was considerably expanded by Aberdeen, but it was not until 1905 that work began on what would become the province's longest irrigation canal, the Grey Canal, an aqueduct leading water some 30 km from Lavington to Okanagan Lake via the water-starved benchlands surrounding Vernon. Another canal, the South Canal, was built along the Coldstream Valley and a network of distributor ditches dug within the Coldstream Ranch. The Grey Canal was opened as far as the Coldstream Ranch in October 1906 by Earl Grey, then governor general of

Map 379 (*above*).
This 1892 map of Vernon shows the line of the Shuswap & Okanagan Railway sweeping through town in a broad arc, and the original settlement of Centreville. To the west is the land of Luc Girouard, who registered a pre-emption in 1867 and planted one of the finest early orchards; to the east and south is the land of Price Ellison, an Englishman who had arrived in 1876 and amassed considerable land holdings. Later, as a provincial government minister, he was responsible for the preservation of Strathcona Park, on Vancouver Island, the first provincial park (see page 311).

Map 380 (*right*).
This 1907 map shows the line of the *Grey Canal*, flowing from the east (right side) to *Okanagan Lake*, at left, by a circuitous route encircling *Vernon* and *Swan Lake* that allowed its water to flow downhill to irrigate the benchlands. The map was produced by the Land and Agricultural Company, a Belgian-owned enterprise that purchased and developed a considerable amount of the land in this area. The orange-coloured area is land that had been sold by 1914. Note the *Belgian Orchard*. The same company developed much of the orchard land near Kelowna (see Map 393, *page 151*).

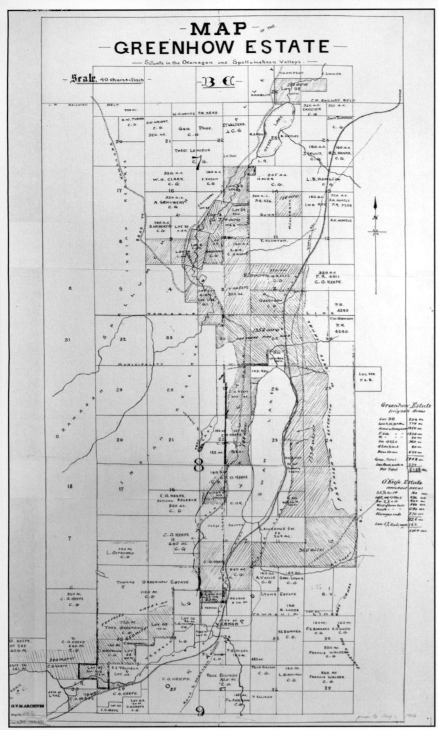

MAP
GREENHOW ESTATE
— Situate in the Okanagan and Spallumcheen Valleys —

Scale, 40 chains=1inch.

B. C.

RU **Coming to Buy A Home**

B4 **It is Too Late?**

? ?

Send for list of my Properties.

Yours truly,

H. P. LEE,
REAL ESTATE,
Vernon, B.C.

Map 381 (*left*).
The intermingled ranchlands of Thomas Greenhow and Cornelius O'Keefe have been assessed for irrigability on this marketing map dating from about 1900. Water was to be supplied by the Grey Canal, shown as *Surveyed Line of Grey Canal* and *Proposed Extension of Grey Canal* along the east side of *Swan Lake.* Many of the large landowners of the North Okanagan are represented here. Of note are the lands of *B.X.* and *F.S. Barnard* of the Barnard family. Francis Jones Barnard was the founder of the BX Express mail delivery service, renowned for its pony express and then stagecoach service from Yale to Barkerville during the Cariboo Gold Rush (see page 85). Barnard needed a supply of fresh horses for his stagecoaches, and to this end he purchased land in the North Okanagan and created the BX Ranch, which stretched east beyond land shown on this map, some 25,500 ha in all. Some of the BX Ranch was converted to orchards soon after Lord Aberdeen planted the first 80 ha at his Coldstream Ranch.

Map 382 (*above*).
Much early orchard marketing was directed at British immigrants, but the Prairies were also an important source of buyers. This 1910 ad was targeted to buyers from the Prairies, depicted as the man here, shown with bags stuffed with money. Fecund Okanagan orchards are contrasted with cold, bleak, and bare land to the east.

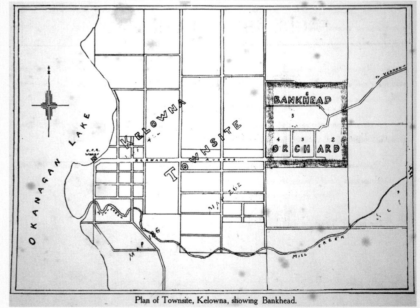

Plan of Townsite, Kelowna, showing Bankhead.

Map 383 (*below*).
Armstrong was a small community named Aberdeen before the building of the Shuswap & Okanagan Railway. It was renamed in honour of W.C. Heaton-Armstrong, head of the London bank that had sold the bonds to build the line. This map is from a local board of trade booklet published about 1910.

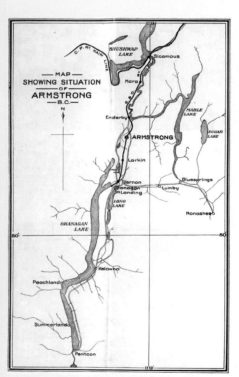

MAP
SHOWING SITUATION
OF
ARMSTRONG
B.C.

Map 384 (*right*).
The *Bankhead Orchard*, immediately adjacent to the *Kelowna Townsite*, was planted about 1900 by Thomas W. Stirling, later a principal of the Kelowna Land and Orchard Company and who was responsible for creating the Kelowna Shipper's Union, the first co-operative fruit association in Kelowna. This map comes from a booklet from about 1910 promoting Kelowna as "The Orchard City."

Canada and an owner of nearby land. The canal was extended beginning in 1909 and, following an agreement with the Land and Agricultural Company, completed round Swan Lake and into Okanagan Lake by 1914. The Grey Canal lasted until 1970, when it was replaced with buried pipelines. Today portions of the canal's route have been converted into trails. Much of the Grey Canal's route is shown on MAP 380, *previous page.*

Lord Aberdeen's orchards in the Kelowna area were not at first as successful as those on his Coldstream lands. His Guisachan Ranch fruit trees did not thrive owing to poor planting, and in 1896, five years after they had been planted, had to be uprooted. Orchards in the Kelowna area were perfectly viable, however, and this setback did not stop the sale of land for this use. George Grant Mackay was again busy, and, indeed, it was he who sold Aberdeen his first land.

As in the North Okanagan, ranchlands gave way to orchard land. The Kelowna Land and Orchard Company bought 2,650 ha (shown on MAP 386, *right,* and on MAP 388, *overleaf*) from early settler Bernard Lequime in

Above. The Pandosy Mission, today preserved as a historical site.

Early Kelowna

Father Charles Pandosy, an Oblate missionary who had arrived in the Oregon Territory from France in 1847, travelled north and set up Okanagan Mission on the banks of Mission Creek in 1859; the mission is shown on many early maps. Pandosy and his fellow missionaries were pioneer agriculturalists, growing considerable amounts of wheat, barley, and potatoes, all without irrigation, thus demonstrating the fertility of the region.

MAP 385 (*above*).
Father Pandosy's *Mission,* shown on an 1861 map.

Trader Eli Lequime arrived at the mission in 1861 and was persuaded to stay by fellow Frenchman Father Pandosy. Lequime pre-empted land, built a store, and operated a sawmill, a cattle and horse ranch, a blacksmith shop, and a post office. Other Catholic, French-speaking settlers arrived over the ensuing years, creating a distinct enclave. Lequime and his son Bernard purchased other pre-emptions from time to time, and by 1879 they had acquired 320 acres (130 ha) each and had set up a substantial farming operation.

In 1890 land salesman George Grant Mackay negotiated the sale of the land that became Lord Aberdeen's Guisachan Ranch. Anticipating where the planned railway would run, Mackay surveyed a townsite on the Mission Trail, naming it Benvoulin. He managed to sell a number of lots there, but the proposed railway did not arrive and the settlement stagnated.

In 1892 Bernard Lequime, seeing that the future lay with lake transportation, had a townsite surveyed on his land at the mouth of Mill Creek (MAP 387). The Lequime land was purchased in 1904 by the Kelowna Land and Orchard Company.

MAP 386 (*right, centre*).
This 1890 map shows the early land holdings in the Kelowna area, generally dating from the 1880s. *Eli Lequime*'s land is shown, as is that of his son, *Bernard. Mc Dougall* is the land that Lord Aberdeen purchased for his Guisachan Ranch; it previously belonged to ex–Hudson's Bay Company employee John McDougall.

MAP 387 (*right*).
This is a townsite plan of Kelowna, presumably dating from 1892, the year the townsite was registered. There are spelling and other mistakes: at left, *Mission Creek* is really Mill Creek, and *Pendozi* is clearly a misspelled Pandosy. Note also *Eli Av.* and *Bernard Av.* The townsite was laid out on the land pre-empted by August Gillard (and shown as his on MAP 386) but absorbed by the Lequimes during the 1880s.

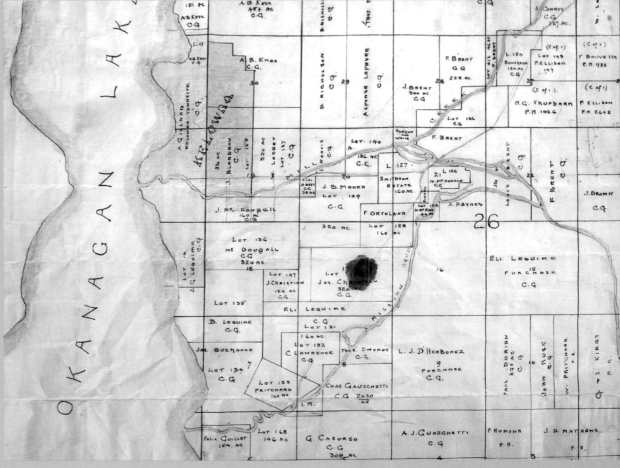

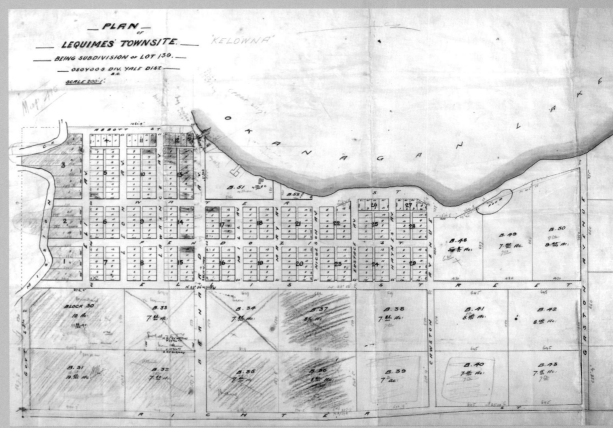

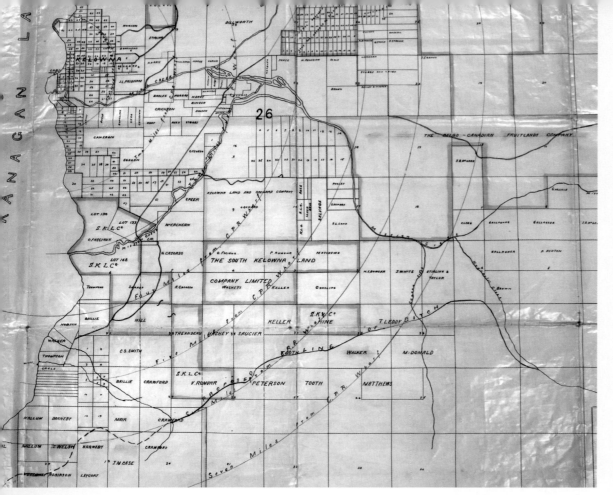

The Central Okanagan Land and Orchard Company published this little map about 1907, illustrating the position of its land relative to the critical lake steamer route and railway. *Inset* is its booklet cover.

MAP 388 (*above*).

This 1900 map shows the holdings of the *Kelowna Land and Orchard Company* (in green outline), which bought up the Lequime family lands, and its offshoot the *South Kelowna Land Company* (red outline). Also shown are the southern part of the large holdings of the *Belgo-Canadian Fruitlands Company* (yellow outline; on the later map the Land and Agricultural Company of Canada) and part of the land of the Central Okanagan Land and Orchard Company (purple outline). These were all companies that purchased ranchland, brought water to it, and then resold it as orchard land. Irrigation plans for the South Kelowna company's land is indicated by the *Proposed Line of Ditch* from *Hydraulic Creek*, a tributary of *Mission Creek*. The circles on the map show the distance from the CPR wharf in Kelowna, the point from which fruit could be shipped up the lake to the railhead at Okanagan Landing (see photo accompanying MAP 400, *page 155*). Until 1923, when a railway link finally arrived, this would be the principal method by which fruit would travel to markets.

1904. Three years later an affiliated company, the South Kelowna Land Company, was formed by some of the directors of the first but with some new investors that added much-needed capital for the expensive irrigation works required. In 1907 the Belgian-controlled Land and Agricultural Company of Canada began developing 2,400 ha east of the Kelowna Townsite (MAP 393, *far right, bottom*), constructing a ditch 22 km long from the North Fork of Mission Creek (now Belgo Creek). In 1907 the Central Okanagan Land and Orchard Company subdivided just over 600 ha and built a ditch some 8 km long from Mill Creek (now Kelowna Creek).

All the land companies spent vast sums installing irrigation systems: aqueducts and ditches, which typically needed to be concrete lined; pipes; siphons; and flumes, the latter on trestles when needed to maintain the flow of water. The cost was prohibitive, and when market conditions changed, disaster loomed. By 1912 thousands of acres were being offered for sale. American producers tapped Prairie markets. Then came World War I, which led to the collapse of most of the companies and the failure of some lenders, notably the Dominion Trust Company, and in 1918 the provincial government passed legislation creating irrigation districts that co-operatively owned and managed water distribution systems. In Kelowna this has led to the unique situation today of its having five water suppliers, each with a different source for its water.

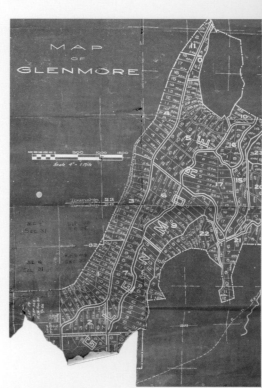

Above.

An ad by the Kelowna Fruitlands Commission, a joint effort of the South Kelowna Land Company and the Belgo-Canadian Fruit Lands Company. It appeared in 1912.

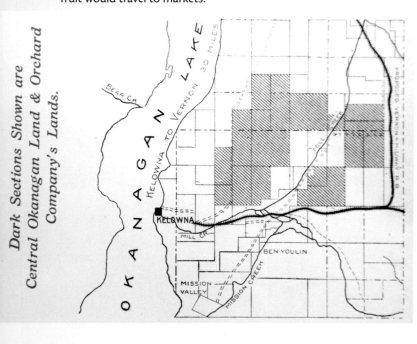

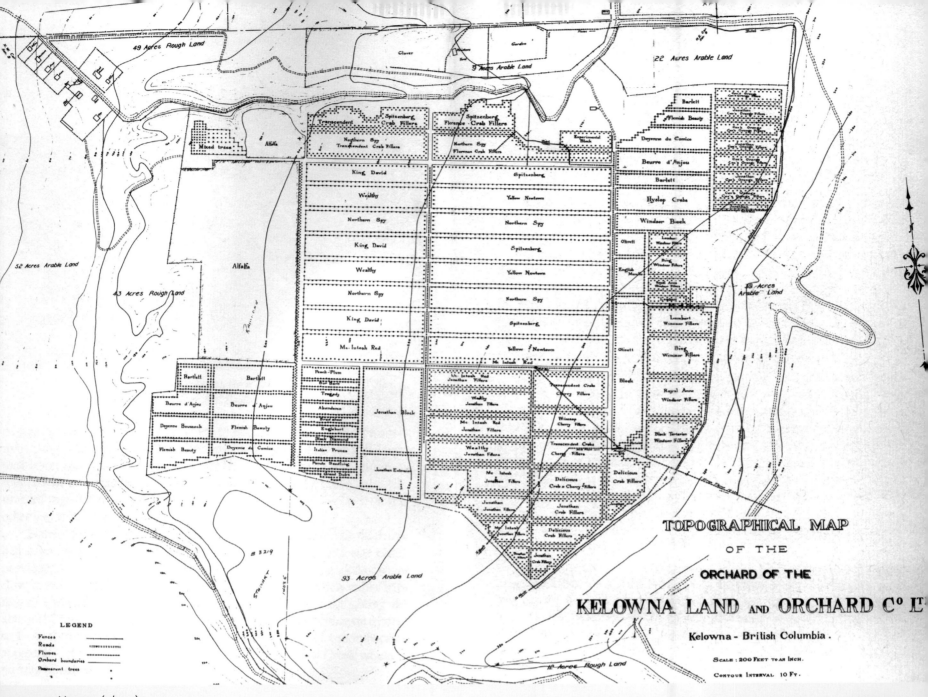

TOPOGRAPHICAL MAP
OF THE
ORCHARD OF THE
KELOWNA LAND AND ORCHARD Cº Lᵗ

Kelowna - British Columbia.

SCALE : 200 FEET TO AN INCH.

CONTOUR INTERVAL 10 FT.

LEGEND

Fences
Roads
Flumes
Orchard boundaries
Permanent trees

Map 392 (above).

This was the Keloka Orchard, a model farm established by the Kelowna Land and Orchard Company on the lower bench and shown here in 1912. It was used for several purposes: to assist in sales by showing what was possible; to test types and varieties of fruit for suitability; and to provide trees that the company could use to establish orchards for land buyers, mainly from Britain, who had not yet arrived. The map is detailed, right down to the pattern of trees; although the interiors of the blocks have not been filled in, in most cases it would be possible to count the totals of each fruit. Varieties of apples, pears, plums, prunes, peaches, and cherries are being grown. Water is flowing to the orchard from the right in the *Drop Flume* and distributed by a main ditch down the middle pathway. Contours are shown, as the shape of the land is critical to water distribution. The farm office and houses are at top left. As a sales tool, the orchard may have been valuable, but it was somewhat deceptive, for it had been established on some of the best land—land with clay that would retain more water than most. *Right.* A 1912 Canadian Pacific brochure showing Okanagan apples.

Map 390 (far left).

Published in 1907 in a promotional booklet, this map shows the location of the lands of the Central Okanagan Land and Orchard Company near *Kelowna*.

Map 391 (left).

Drawn about 1922, this is the subdivision map of orchard lots in *Glenmore*, incorporated that year but since 1960 part of Kelowna; it still contains orchards as well as housing. This map covers the western part of the Central Okanagan Land and Orchard Company's land shown in **Map 390**, but the subdivision curves around the higher land such as at right, centre.

Map 393 (right).

A map shows the Land and Agricultural Company of Canada (Belgo-Canadian Fruit Lands) land about 1923. Now the *Canadian National Railway* is *under construction* south from Vernon, which will belatedly tie *Kelowna* into the national rail network. Note the *Irrigation Intake* to the right, and the little inset map showing the *dam* site. *Inset* is the cover of the promotional booklet that contained this map.

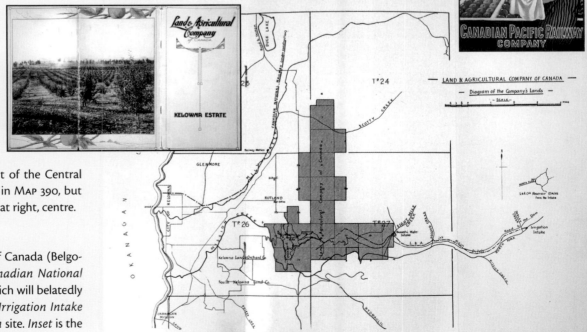

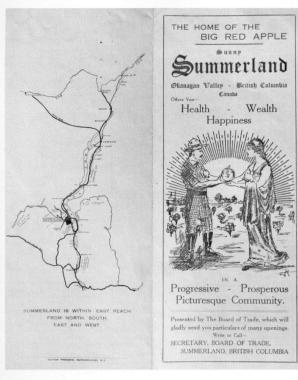

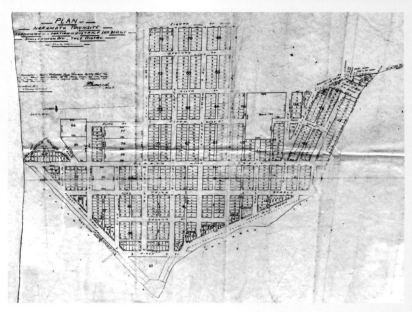

Map 394 (*left*). Health, wealth, and happiness feature in this promotional brochure and map from Summerland, published about 1914. Note the *Dominion Experimental Station,* an agricultural test facility opened that year. Summerland was laid out in 1902 by John M. Robinson, a land salesman who was also responsible for Naramata (1907) and Peachland (1897).

Map 395 (*above*).
The 1907 Naramata Townsite, one of three townsites laid out by John Moore Robinson.

Map 396 (*below*).
A nicely illustrated 1910 sales map of Kaleden Irrigated Fruit Lands laid out in 1909 by developer James Ritchie, who incorporated a "Garden of Eden" marketing theme into his development's name. Ritchie laid out Kaleden on Skaha Lake, shown here as *Lower Okanagan Lake.* Free water is offered until 1913, "when the system will become co-operative."

In the southern Okanagan it was Irishman Tom Ellis who, in 1865, pre-empted land at Penticton and began building a vast empire of cattle lands, holding in the end some 12,500 hectares—125 km². For many years Ellis brought in all his supplies by pack train over the Dewdney Trail but in 1890 partnered with Captain Thomas Shorts to build a steamboat, the *Penticton,* to carry freight to and from Okanagan Landing. Of course, this neatly coincided with the building of the Shuswap & Okanagan Railway and its connections to the Canadian Pacific main line.

Ellis realized that his land at the southern end of the lake was the ideal place for a townsite, and in 1892 he joined with a small group of investors to form the Penticton Townsite Company, which surveyed a townsite around the government wharf (Map 397, *right, top*).

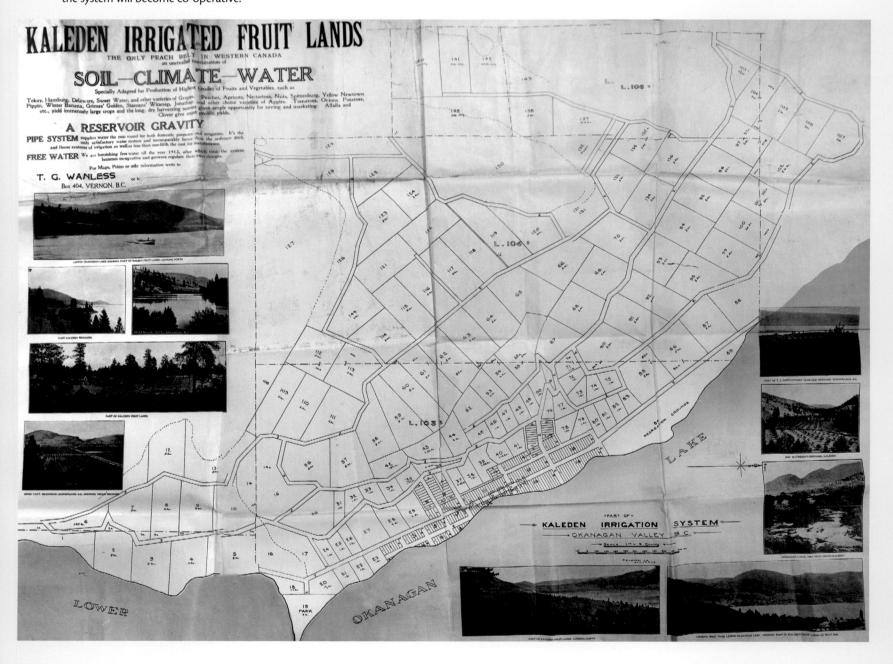

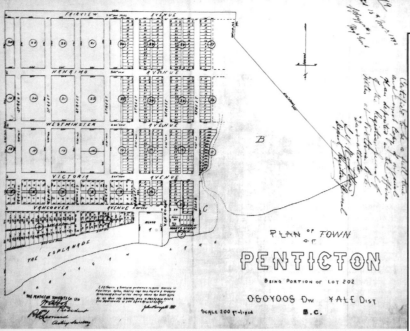

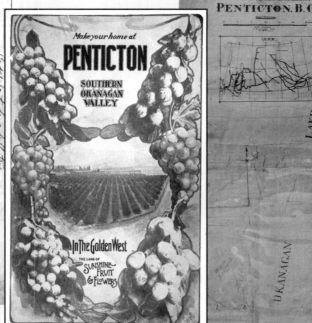

MAP 397 (*above*).
The original townsite plan of Penticton, at the southernmost end of Okanagan Lake, drawn for Ellis and the Penticton Townsite Company in 1892. North is at bottom. The wharf (not shown) was at the end of *Van Horne Street.*

MAP 399 (*below*).
The expanded Penticton Townsite laid out by the Southern Okanagan Land Company in 1905. Strangely for an obviously sales-related map, the government wharf—the point of connection with the critically important lake steamers—is still not shown. *Inset.* Front Street (*Smith Street* on the map) in 1910, showing wooden sidewalks being installed.

Above, centre.
The cover of a Southern Okanagan Land Company sales brochure showing an orchard and advertising Penticton as the "Golden West."

MAP 398 (*above*).
Red shows land still for sale on this Latimer-drawn land sales map of the Southern Okanagan Land Company, dated 1915. Although covering only the Penticton area, the map gives a good idea of the considerable extent of Tom Ellis's original land holdings.

Ellis purchased land at Fairview later in the decade, in an attempt to replicate his Penticton success, but this venture failed when the gold mining there wound down (see page 199).

Ellis sold his land in 1904 to Lytton and Walter T. Shatford, storekeepers who saw big money in converting ranchlands to irrigated fruit lands, and they created the Southern Okanagan Land Company. The pair engaged surveyor Frank Latimer

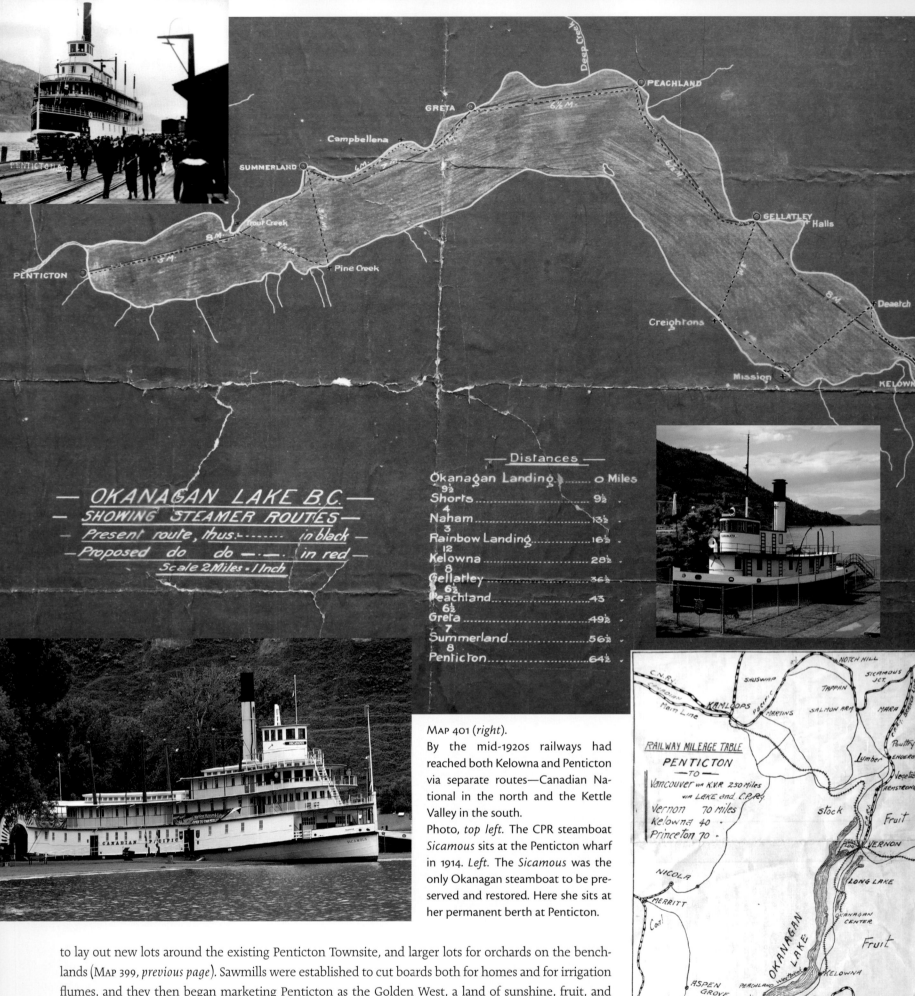

— OKANAGAN LAKE B.C.—
— SHOWING STEAMER ROUTES —
— *Present route, thus* ------- *in black* —
— *Proposed do do* —·— *in red* —
Scale 2 Miles = 1 Inch

— Distances —

Okanagan Landing	0 Miles
Shorts	9½
Naham	13½
Rainbow Landing	16½
Kelowna	28½
Gellatley	36½
Peachland	43
Greta	49½
Summerland	56½
Penticton	64½

Map 401 (*right*).
By the mid-1920s railways had reached both Kelowna and Penticton via separate routes—Canadian National in the north and the Kettle Valley in the south.
Photo, *top left*. The CPR steamboat *Sicamous* sits at the Penticton wharf in 1914. *Left*. The *Sicamous* was the only Okanagan steamboat to be preserved and restored. Here she sits at her permanent berth at Penticton.

to lay out new lots around the existing Penticton Townsite, and larger lots for orchards on the benchlands (**Map 399**, *previous page*). Sawmills were established to cut boards both for homes and for irrigation flumes, and they then began marketing Penticton as the Golden West, a land of sunshine, fruit, and flowers (brochure, *previous page*).

Penticton did not incorporate as a city until 1948, but its future was assured by the coming of the Kettle Valley Railway in 1912 (see page 216). In 1918 the company sold large tracts of its southernmost lands to the provincial government for the latter's South Okanagan Lands Project for returning soldiers, and the process of converting ranchlands to irrigated farmland continued (see page 288).

Steam navigation on Okanagan Lake began in 1886 with the launch of a ten-metre-long boat, the *Mary Victoria Greenhow*, by Captain Thomas Dolman Shorts, who had for the previous three years *rowed* up and down the 111-km-long lake! His short-lived venture was succeeded by the *Penticton*, which Shorts jointly owned with Tom Ellis. The *Penticton* began service in 1890 and was the first *real* steamboat on the lake, the first of a number of ever-more-luxurious sternwheelers, to which were added steam tugs

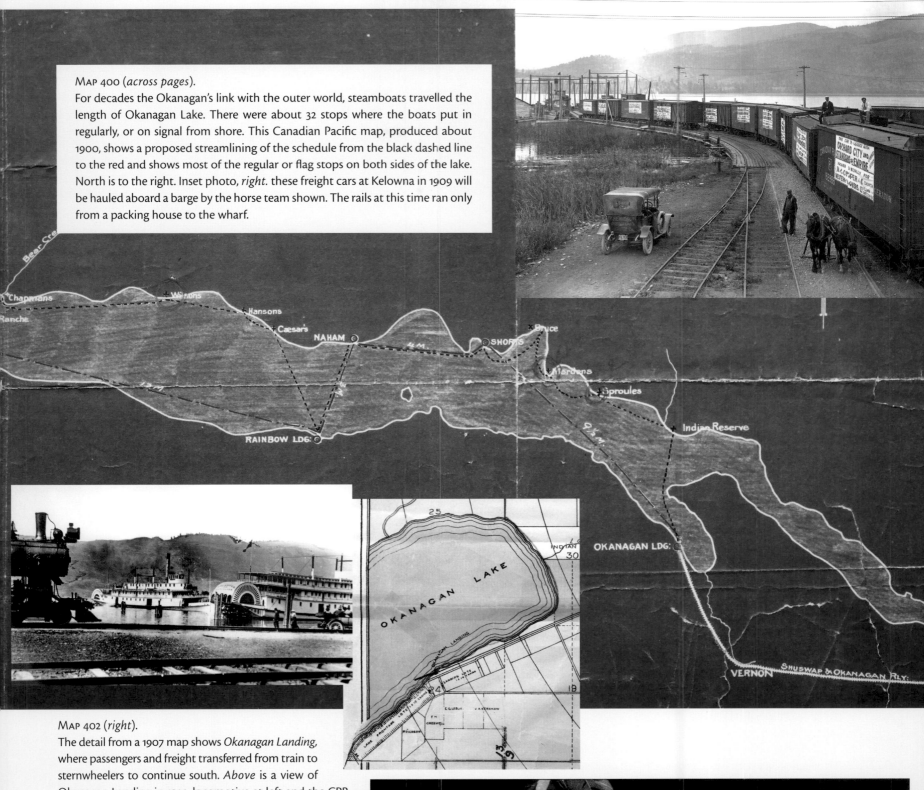

Map 400 (*across pages*).
For decades the Okanagan's link with the outer world, steamboats travelled the length of Okanagan Lake. There were about 32 stops where the boats put in regularly, or on signal from shore. This Canadian Pacific map, produced about 1900, shows a proposed streamlining of the schedule from the black dashed line to the red and shows most of the regular or flag stops on both sides of the lake. North is to the right. Inset photo, *right*. these freight cars at Kelowna in 1909 will be hauled aboard a barge by the horse team shown. The rails at this time ran only from a packing house to the wharf.

Map 402 (*right*).
The detail from a 1907 map shows *Okanagan Landing*, where passengers and freight transferred from train to sternwheelers to continue south. *Above* is a view of Okanagan Landing in 1914, locomotive at left and the CPR steamboats *Okanagan* (at centre) and *Sicamous* (at right) behind. *Above left, centre*, is the *Naramata*, a steam tug launched in 1914. It ran until 1967 and is now preserved at Penticton. Its steel hull was sometimes used for icebreaking.

Map 403 (*right*).
This apple crate label from the 1920s features a map of British Columbia and a view of Okanagan Lake. British Columbia Growers was a co-operative packing house.

and barges. Canadian Pacific, operating the Shuswap & Okanagan initially on a lease, launched the *Aberdeen*, its first steamboat, in 1892. The big steamboats lasted until 1935, when railways—to Penticton in 1912 and to Kelowna in 1925—and roads finally led to their demise. Tugs lasted until 1972. Nevertheless, the steamboats had performed a vital role in opening the Okanagan Valley to settlement.

Between 1904 and 1914 some 16,200 ha— 162 km²—of Okanagan land was irrigated, and by this time the apple crop alone was 750,000 boxes a year.

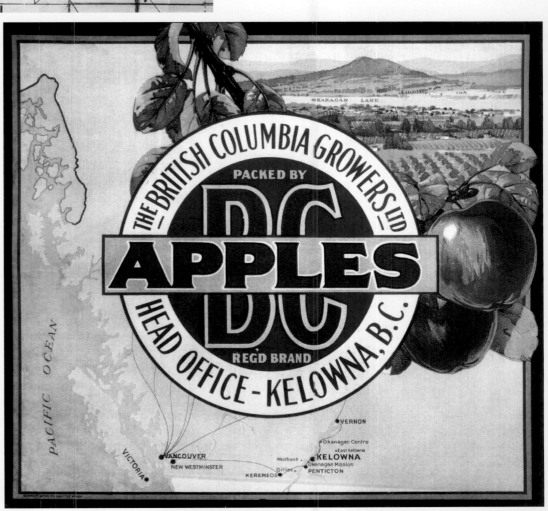

VICTORIAN VICTORIA

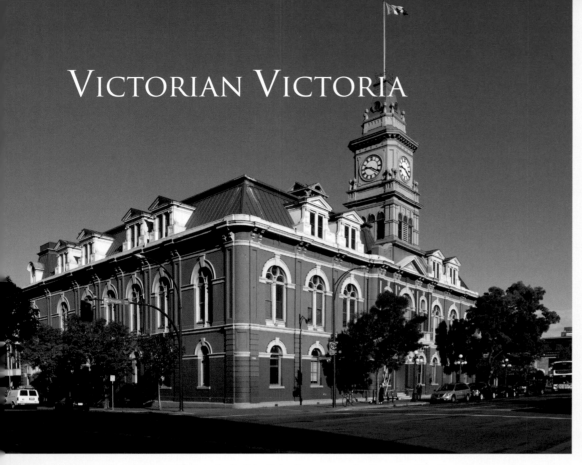

Victoria's transition from a Hudson's Bay Company trading post to city began when a townsite was laid out behind the fort in 1851–52 and the name "Victoria" adopted. But development really only took off with the arrival of the gold rush in 1858. Then lots no one wanted before suddenly were sold for a minor fortune.

Victoria was, of course, made the capital of the Colony of Vancouver Island in 1849 and also became the capital of the mainland Colony of British Columbia after its creation in 1858 (see page 62). After the two were joined in 1866, Victoria lost its capital status, but only for two years. The first legislative assembly had been convened by Governor James Douglas inside the fort in 1856. In 1859 legislative buildings had been constructed and were promptly dubbed *the birdcages* for their distinctive style (MAP 405, *right*).

Victoria was incorporated as a city in 1862 and had no rivals until the growth of first New Westminster

Above. Victoria's magnificent city hall, built in three stages, in 1877–78, 1880–81, and 1890–91. Built of red brick with a tin mansard roof, its architecture style is known as Second Empire.

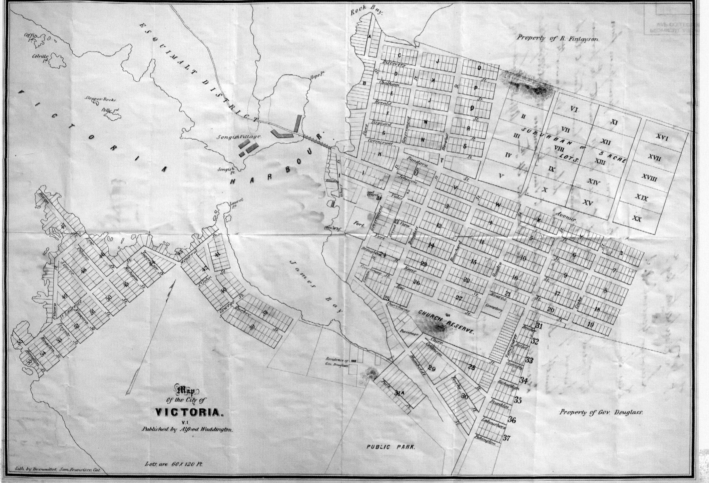

MAP 404 (*left*).
This is road builder and later railway planner (see page 114) Alfred Waddington's map of Victoria, published in 1859. The townsite of Victoria had been laid out seven years earlier behind the *Fort.* Other lots have been subdivided by the date of this map. Note the *Residence of Gov. Douglass* (*sic*) and *Property of Gov. Douglass,* and also the *Property of R. Finlayson,* who became chief factor at Fort Victoria on Douglas's resignation.

Below.
A view of Victoria in 1860. The fort is at centre, and the Lekwammen (Songhees) village on the opposite shore in the foreground.

and then Vancouver, following much rancorous debate as to the proper terminus for the transcontinental railway; the choice of Vancouver ensured that Victoria would eventually be commercially eclipsed by its mainland nemesis. Victoria's destiny was to be of a more genteel nature, a government city instead. In 1858 Governor Douglas set aside 75 ha for a park. In 1882 the province transferred ownership to the city, and the formal Beacon Hill civic park was laid out on most of the land beginning in 1889. The exact provenance of the design is confused and may be an original plan that was altered (see MAP 410, *overleaf*).

MAP 405 (*right*).
Part of a nicely coloured map of Victoria drawn by surveyor Hermann Otto Tiedemann in 1863. The multi-talented Tiedemann was a surveyor, engineer, architect (he designed the first legislative buildings, the birdcages), artist (he drew the view of Victoria opposite), and explorer of Alfred Waddington's road from Bute Inlet (see MAP 214, *page 79*). Tiedemann worked in the office of the surveyor general under ex–Hudson's Bay Company and now colonial surveyor general Joseph Despard Pemberton. This map, however, was found in the Hudson's Bay Company Archives. Note that the fort is no longer shown on this map; it was torn down in stages between 1859 and 1864 and the land sold for development. Tiedemann's birdcages, shown in the photo *below*, are the *Governm.t. Buildings* on the map, next to *Governor Douglas 12 Acres*. Across the harbour the unnamed black buildings are the Lekwammen (Songhees) village visible in the view.

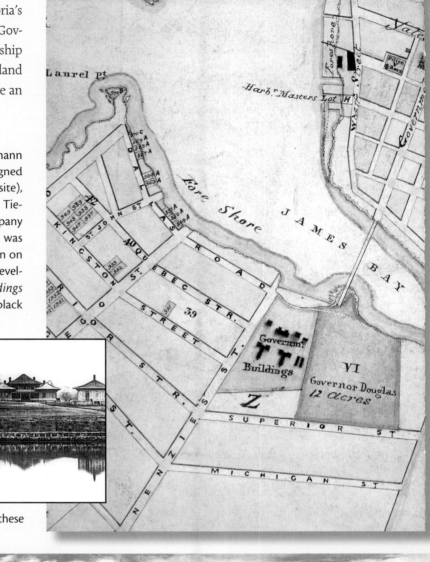

MAP 406 (*below*).
A bird's-eye map of Victoria published in 1878. The prominent mountain in the distance is Mount Baker, positioned with some artistic licence. James Bay cuts deeply into the city. The racetrack at right is part of the reserve that would become Beacon Hill Park. The legislative buildings, the birdcages, are visible in front of the James Bay bridge, and the Lekwammen village is across the harbour. The map was drawn by Eli Glover, who produced many of these bird's-eyes, mainly of American cities. Compare this map with MAP 407, *overleaf*.

BIRD'S-EYE VIEW OF
VICTORIA,
Vancouver Island, B. C. 1878.

Map 407 (*right*).
Clearly influenced by Eli Glover's 1878 original (Map 406, *previous page*), though not drawn by him, is this 1889 Victoria bird's-eye. The viewpoint is similar, but the whole of the Oak Bay shoreline can now be seen; Mount Baker again looms on the horizon. Beacon Hill Park has now been laid out, and the Esquimalt & Nanaimo Railway (*E.&N.R.R.*) cuts through the Songhees Indian Reserve and enters the city via a bridge across the Inner Harbour (see page 166). This map, currently in the collection of the Library of Congress, is stamped *General Staff Map Collection.* There are a number of American military maps from this period that depict details of Canadian cities, and this map appears to be one of them, held in case the United States went to war with Britain, an event that appeared likely after the Civil War in 1865 and threatened again over the Alaska boundary at the beginning of the twentieth century.

VICTORIA, B.C.
1889.

Map 409 (*below*).
An 1890 map now shows Beacon Hill *Public Park* as laid out, incorporating the *Race Track* (the large oval at its southern end).

Map 410 (*right, top*) and
Map 411 (*right, bottom*).
This is the 1888 design for Beacon Hill Park. This large, beautifully coloured plan may be the one drawn by Henry Cresswell, who won a competition arranged by the City of Victoria, or it may be an altered plan by John Blair, the Scottish landscaper who was given the job of implementing the park's design. Blair apparently added buildings, a number of carriage roads, including one up Beacon Hill, and another lake on the east side of the park (though that was never created). The racetrack has cleverly been incorporated into the design. The detail shows a flower garden and fountain (*F*) and a conservatory (*C*), that classically Victorian dalliance.

Map 408 (*above*).
James Bay was well silted up by 1886, the date of this map. Subdivided land covers the area where the fort once was. The birdcages overlook James Bay, and a detailed layout of the *Indian Village* is noted. At its northern end are *Indian Graves*. The photo, taken by Richard Maynard, who with his wife, Hannah, was responsible for many of the classic historical photographs of British Columbia in this period, shows the Lekwammen (Songhees) village with numerous canoes on the beachfront, contrasting with the now urban area of Victoria across the harbour.

PLAN
OF
BEACON HILL PARK
VICTORIA, B.C.

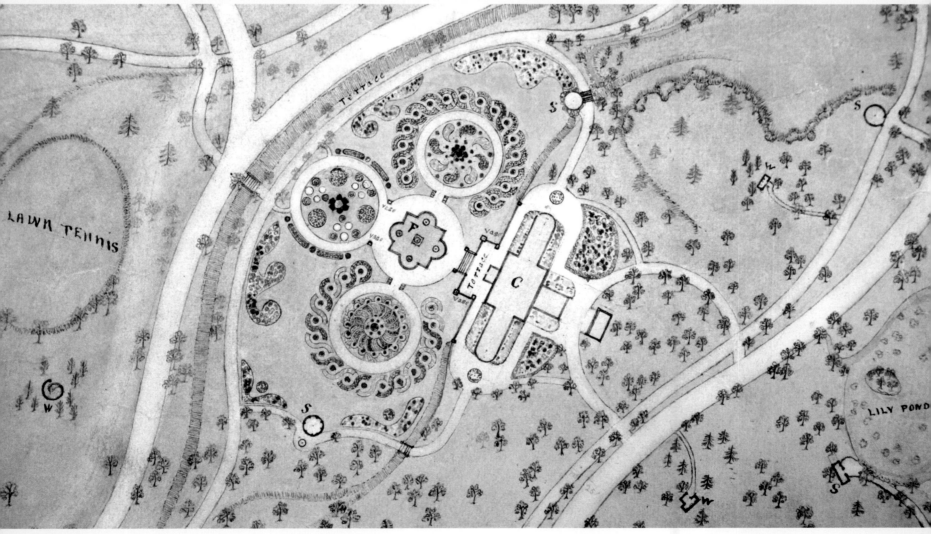

LAWN TENNIS

LILY POND

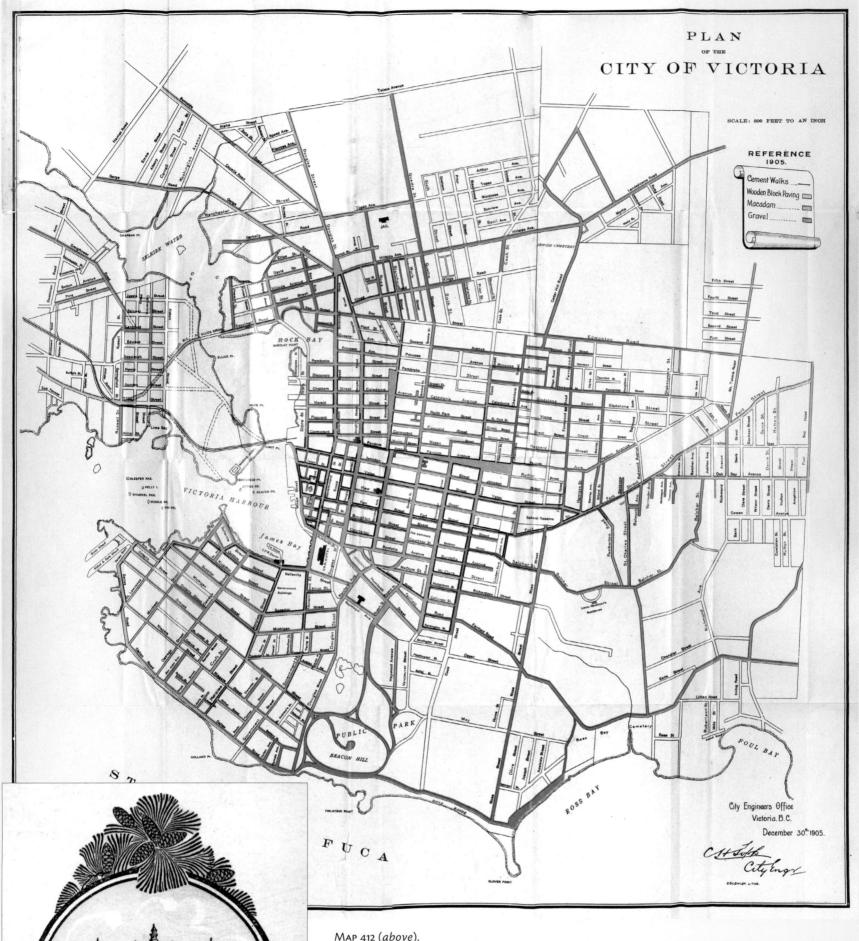

PLAN
OF THE
CITY OF VICTORIA

SCALE: 800 FEET TO AN INCH

REFERENCE
1905.

Cement Walks
Wooden Block Paving
Macadam
Gravel

City Engineers Office
Victoria, B.C.
December 30ᵗʰ 1905.

City Engr

THE NEW TOURIST HOTEL to be erected on JAMES BAY EMBANKMENT. VICTORIA, B.C.

MAP 412 (*above*).

Drawn in the Victoria city engineer's office and signed by the city engineer is this 1905 map documenting the state of the roads in that year. Some of the downtown roads are made of wooden block paving (shown in yellow). The majority are "macadamized." This was a process invented by John Macadam in Britain in 1815, where layers of successively smaller stones were laid on top of each other, which yielded a tolerable surface, at least for the traffic of the day. This should not be confused with "tarmacadamizing," which created the hard road surface we take for granted today—"tarmac." The green-coloured roads are gravelled, while those left white were just dirt. The demand for better roads would soon make itself heard, since a new vehicle had just appeared on the roads—the automobile—and they tended in the early days to belong to wealthier individuals who would insist on better surfaces for their new toys. On this map the east end of *James Bay* has now been filled in and a *C.P.R. Hotel* built overlooking the harbour—the *New Tourist Hotel* shown on the illustration, *left*, from the back cover of the tourist booklet shown *opposite*. The original trestle across James Bay was replaced with a stone causeway in 1900, and the already silting-up area behind it was infilled shortly after. The Canadian Pacific's Empress Hotel, intended as a terminal hotel for the company's steamships, was initially completed in 1908, though it has been extensively expanded and altered since.

Map 413 (*right*), with cover (*below, left*) and illustration (*below, right*). The Tourist Association of Victoria published this booklet promoting the city as a tourist destination in 1905. Victoria has always been known for its "Britishness," and it was no different in 1905; here the best promotion was Victoria as "An Outpost of Empire." It was perhaps, then as now, designed to appeal to Americans, for the connections—all rail and steamship—from the United States are emphasized. The Northern Pacific had been completed to Tacoma in 1883 and the Great Northern to Seattle in 1893, making travel to Victoria much more feasible for eastern Americans than previously, when a steamship trip north from San Francisco, at the end of the Union Pacific and Central Pacific's transcontinental line, would have been necessary. Canadian Pacific connections from the mainland are also shown, as is the *E & N R*[y] north as far as *Wellington Coal Mines*, which one might think a strange thing to note on a map stated to be for tourists. It was all to demonstrate how developed Victoria and its hinterland were; it was the same thought process that created countless bird's-eye maps with harbours full of ships and cities full of smokestacks. *Copper Mines*, *Iron Mines*, and also *Hunting* and *Fishing* are noted on the map. The photo, also in this booklet, is clearly proud to show off the same automobile three times!

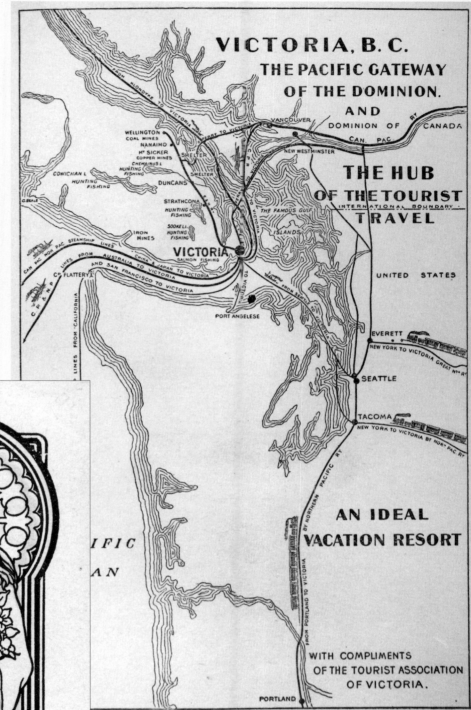

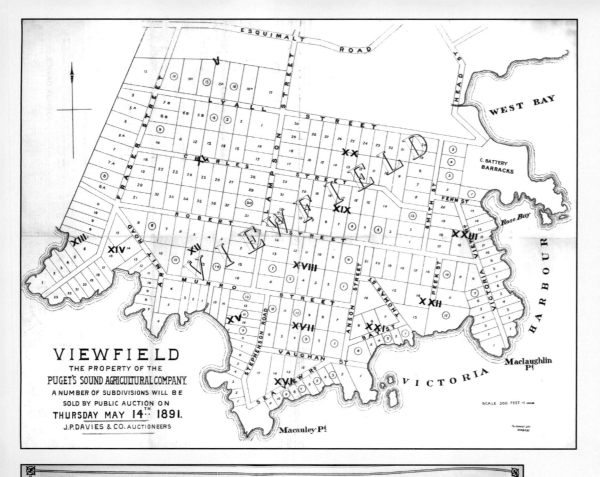

VIEWFIELD
THE PROPERTY OF THE
PUGET'S SOUND AGRICULTURAL COMPANY.
A NUMBER OF SUBDIVISIONS WILL BE
SOLD BY PUBLIC AUCTION ON
THURSDAY MAY 14TH 1891.
J.P. DAVIES & CO. AUCTIONEERS

Map 414 (*left*).
On the west side of the harbour entrance was the Hudson's Bay Company's 243-ha Viewfield Farm, created by its subsidiary, the Puget's Sound Agricultural Company, in 1850. In 1891 the land was subdivided and auctioned. The street plan, although not exactly as it was originally laid out, is still recognizable today.

Map 415 (*left, bottom*).
Victoria, like Vancouver in the same period (see page 236), expected to profit from increased maritime traffic due to the opening of the Panama Canal in 1913. The canal was essentially a "sure thing" after 1907, as construction was underway. Nowhere was the boom more obvious than in the real estate market. The newspapers of those years are full of ads for various real estate schemes. Uplands was the premier upmarket subdivision in greater Victoria and was designed by famous landscape architect John Olmsted in 1911. A total of 188 ha, with nearly 5 km of water frontage, was divided into 375 lots and offered for sale at prices ranging from $3,000 to $55,000. No house was to cost less than $5,000. The Royal Victoria Yacht Club moved to Flamborough Head in Cadboro Bay in the Uplands in 1913. This ad appeared in April 1912.

Map 416 (*below*).
A 1910 ad for lots at *Shoal Bay* (McNeill Bay in Oak Bay): waterfront at $1,100 to $1,300 and other lots for $375 to $450. *Shoal Bay Road* is today Beach Drive.

Victoria
COMMERCIALLY

MAP 417 (above).

This unusual bird's-eye map of Victoria was published about 1912 and features a vast array of docks at Ogden Point, together with (Authorized) City Industrial Sites right behind them. Had this plan been followed, industry would have come within a couple of blocks of the legislative buildings and rimmed a significant part

of the Inner Harbour. The docks, for *Foreign Service*, were to take advantage of all the shipping expected to flood the West Coast after the Panama Canal's completion, but like many other schemes of this period (note especially similar dock projects proposed for Vancouver; see page 245) they foundered with the start of World War I. A 762-m-long breakwater was completed in 1916, and two piers and a cargo warehouse were completed two years later. Ogden Point is now used principally for cruise ships. There are a number of other *City Industrial Sites* marked, and the railways serving Victoria by this time are prominently displayed.

MAP 418 (right).

Lots could be purchased for only $150 in the Gorge area in May 1909. Note that "handy to the [street] cars" is a selling point. *City Park* is Gorge Park. The street pattern is essentially identical today.

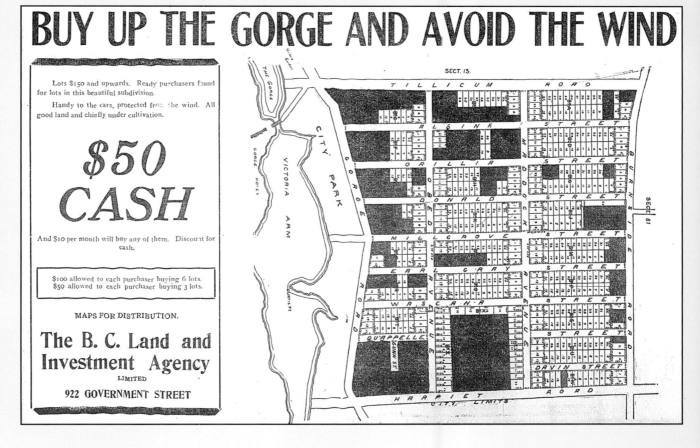

BUY UP THE GORGE AND AVOID THE WIND

Lots $150 and upwards. Ready purchasers found for lots in this beautiful subdivision.

Handy to the cars, protected from the wind. All good land and chiefly under cultivation.

$50 CASH

And $10 per month will buy any of them. Discount for cash.

$100 allowed to each purchaser buying 6 lots.
$50 allowed to each purchaser buying 3 lots.

MAPS FOR DISTRIBUTION.

The B. C. Land and Investment Agency
LIMITED
922 GOVERNMENT STREET

THE ISLAND RAILWAY

The compromise reached between British Columbia and the federal government in 1874 after the transcontinental line was delayed produced the Carnarvon Terms (see page 116), which ensured that a federally subsidized railway would be built on Vancouver Island. This was the Esquimalt & Nanaimo (E&N), the first of several railways on the island.

Initially it was intended that the transcontinental railway would end at Esquimalt, and the federal government made an official statement to this effect in 1875. The provincial government authorized the grant of public land twenty miles on either side of the line, but the bill to allow construction to begin did not pass, and the grant was rescinded.

In 1882 the provincial government made a deal with an American company called the Vancouver Land and Railway Company, led by ex–Central Pacific engineer Lewis Clement, to build the island line in return for a massive land grant, but the company was unable to provide the necessary financial security and was unable to proceed. John A. Macdonald, then both member of parliament for Victoria and prime minister, managed to persuade coal baron James Dunsmuir (see page 224) that his interests would be well served by connecting his coal mines to the Victoria market using a railway. Dunsmuir called on friends who had been part of the building of the first American transcontinental and persuaded them to take a half share in the line. The group included three of the so-called Big Four of the Central Pacific—Leland Stanford, Charles Crocker, and Collis Huntington.

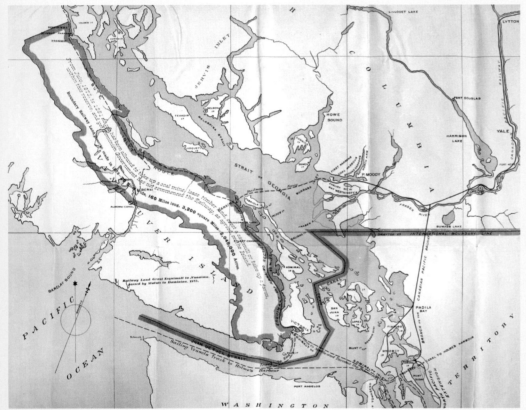

MAP 419 (*left*).
This map appeared in a broadsheet published in 1881 by Amor de Cosmos, one-time premier of British Columbia (1872–74), to promote *Victoria* as the terminus of the transcontinental railway. His idea was to connect a line running to *English Bay* with *Nanaimo* by a steam ferry, and hence by the *Esquimalt & Nanaimo* Railway to Victoria. The map also notes a complaint about the extent of the governmental reservation resulting from the first attempt to encourage railway building in 1875 by granting land. He also worried about a potential extension of the American Northern Pacific Railroad (which would be completed to Tacoma in 1883) to a location opposite the Strait of Juan de Fuca (to which he gave the name *Holmes Harbour*), where it could rob Port Moody of its ocean-going business. Indeed, the Canadian Pacific was soon extended to English Bay, and a steamer connection with the island was established (though from Coal Harbour). Amor de Cosmos had been a major supporter of the Esquimalt terminus for the transcontinental railway, advancing the case with both the federal government and the British government.

Above, left.
The imposing Esquimalt & Nanaimo Railway logo from a 1914 sales booklet.

MAP 420 (*right*).
The central part of the Esquimalt & Nanaimo Railway, shown on a map from about 1902. *Crofton* was created in 1902 to house smelter workers from the Lenora, Mt. Sicker Copper Mining Company, owned by Henry Croft. He also built the *Lenora Railroad* (the Lenora, Mt. Sicker Railway) to haul copper ore from Lenora Mine on Mount Sicker, which came into production in 1900. An aerial tramway was first constructed to bring ore to the E&N but this proved inadequate and was replaced by a railway. This line took ore first to a siding on the E&N, from where ore was taken to Ladysmith, and then shipped to a smelter on Texada Island. In 1902 a smelter was built in Crofton and ore was shipped there. Note the switchbacks on the line between the E&N siding and Crofton. The company was in receivership by 1904, and although attempts were made to revive it, it was closed again in 1907. In Horseshoe Bay near *Chemainus Station* is a sawmill; several sawmills had operated there since 1862.

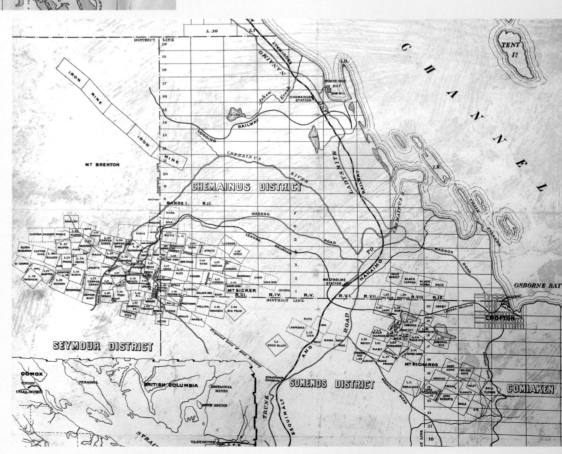

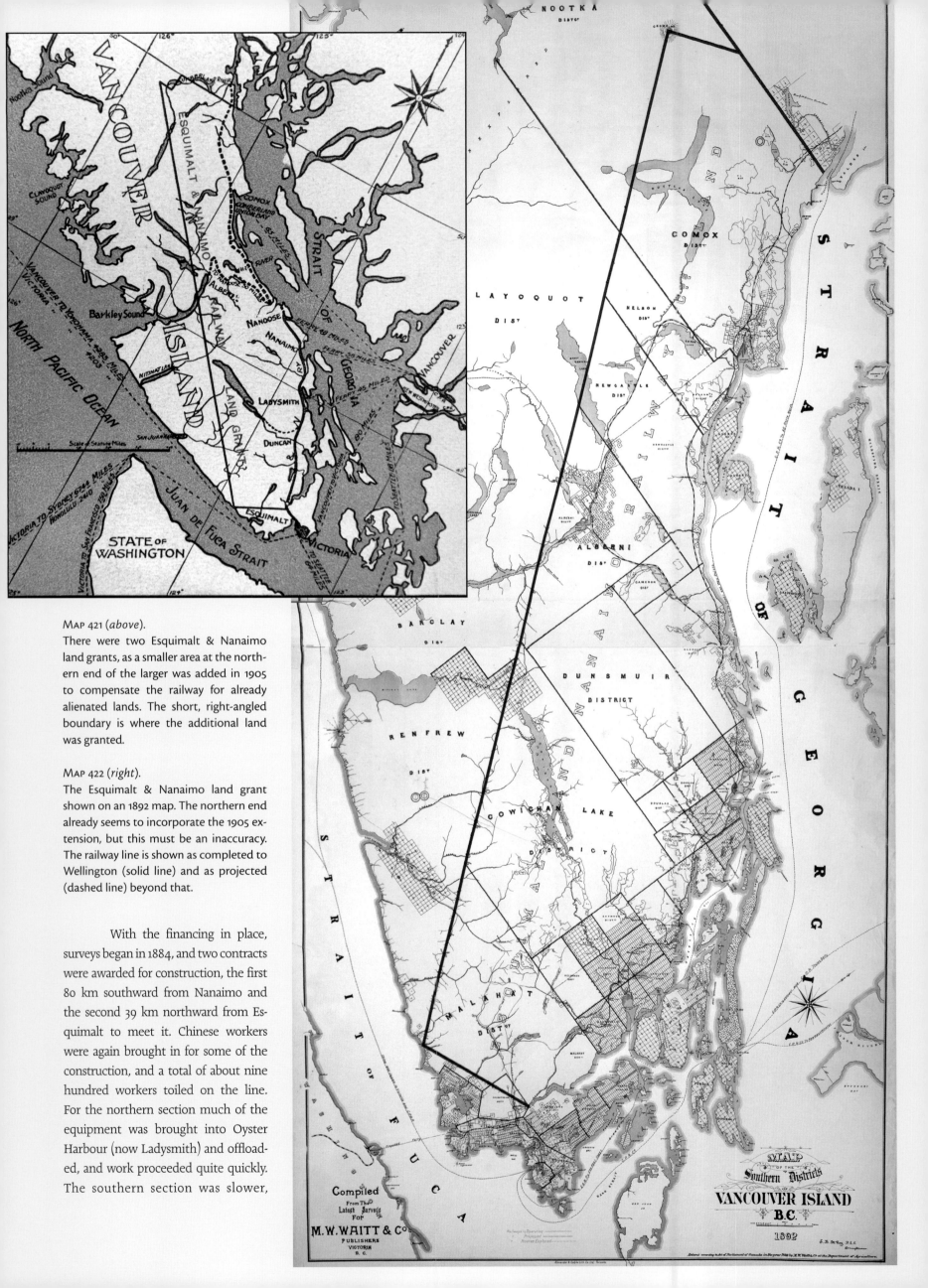

MAP 421 (*above*).
There were two Esquimalt & Nanaimo land grants, as a smaller area at the northern end of the larger was added in 1905 to compensate the railway for already alienated lands. The short, right-angled boundary is where the additional land was granted.

MAP 422 (*right*).
The Esquimalt & Nanaimo land grant shown on an 1892 map. The northern end already seems to incorporate the 1905 extension, but this must be an inaccuracy. The railway line is shown as completed to Wellington (solid line) and as projected (dashed line) beyond that.

With the financing in place, surveys began in 1884, and two contracts were awarded for construction, the first 80 km southward from Nanaimo and the second 39 km northward from Esquimalt to meet it. Chinese workers were again brought in for some of the construction, and a total of about nine hundred workers toiled on the line. For the northern section much of the equipment was brought into Oyster Harbour (now Ladysmith) and offloaded, and work proceeded quite quickly. The southern section was slower,

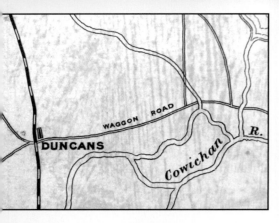

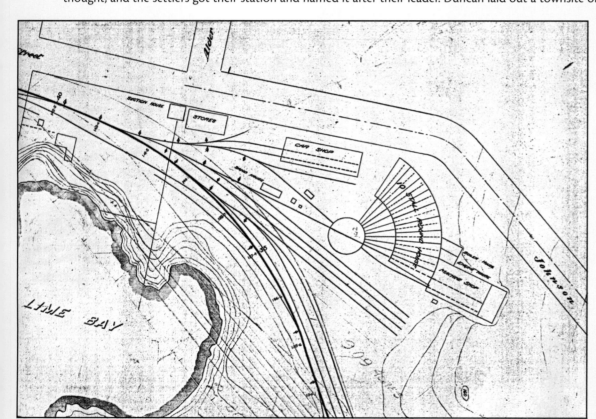

Map 423 (*above*) and Map 424 (*right*).
The inaugural train from Nanaimo to Esquimalt carrying Prime Minister John A. Macdonald to drive the last spike on the railway on 13 August 1886 was met near the crossing of the Cowichan River by a huge gathering of settlers, reputed to be two-thousand strong. Led by William Chalmers Duncan, the settlers wanted a station near their land and determined to convince Dunsmuir that one should be built. And so Duncan gathered virtually the entire population from the area and erected a ceremonial arch, as was the custom; the official train was met by a cheering throng and came to an unplanned halt. As Duncan had planned, this gave Dunsmuir the idea that the area was far more populated than he had thought, and the settlers got their station and named it after their leader. Duncan laid out a townsite on his land the next year and called it *Alder-lea* after his farm (Map 424). The name, however, never caught on; first it was Duncan's Crossing, then Duncans, and later just plain Duncan. His plan shows *Duncan's Station* and the all-important railway *Siding,* which allowed freight to be assembled or unloaded without obstructing the main line.

Map 425 (*left*, in two parts).
Two details of a plan of the Esquimalt & Nanaimo Railway in Victoria, the first showing the shops and *10 Stall Round House* at Russell's (*Lime Bay*) and the second showing the *swing bridge* across the Inner Harbour and the station in downtown Victoria. After about 1905 the railway line continued to freight yards shown on Map 429, *page 168.* The modern photo, *below,* shows today's end of track as the line begins its curve to go north on Store Street; the *General Offices* building shown on the plan is the red brick structure in the photo.

involving as it did the climb across the Malahat and the construction of a number of long trestles nearer to Esquimalt. The last spike was driven on 13 August 1886 by John A. Macdonald at Cliffside, on the eastern side of Shawnigan Lake, close to the meeting point of the two contracts.

Dunsmuir, pleased with the fast progress of his railway, decided to extend the line to within reach of Victorians, and in thirty-five days the line was built to Russell's, at Lime Bay, where the track arrived on 24 September. The line was also extended north to Dunsmuir's coal mines at Wellington by June the following year.

The citizens of Victoria did not like the inconvenient location of Russell's, however, and city council petitioned Dunsmuir to extend the line a further 1,200 m right into the business district of Victoria. Russell's became the location of the shops and roundhouse, while a new depot was built on Store Street, across the Inner Harbour. The extension may have been short, but it necessitated a 250-m-long rock cut and construction of a swing bridge (replaced by the Johnson Street Bridge in 1924). The line was completed into downtown Victoria, and the first train pulled into the new station on 29 March 1888.

In 1905 the Canadian Pacific purchased the assets and land grant (except for coal and fire clay rights) of the Esquimalt & Nanaimo Railway and leased the company; thus the railway retained its name. In 1909 the track was extended north to Parksville and by the end of 1911 had been extended west to Port Alberni over a route that featured grades to 2.2 per cent, the maximum allowed. Earlier that year a branch from Hayward Junction, just north of Duncan, to Lake Cowichan had been completed. Finally, by the summer of 1914 the line was laid as far north as Courtenay. The intention had been to continue north to Campbell River and Hardy Bay, now Port Hardy (Map 430, *overleaf*), the latter chartered in 1912, but World War I intervened, and the track laying stopped just 13 km north of Courtenay, never to proceed farther.

Map 426 (*right*).
The Esquimalt & Nanaimo line into Victoria is well shown on this 1893 map. A circle denotes the roundhouse at Russell's at *Lime Bay* (which was partially filled in the 1950s). Another bridge across the Inner Harbour now carries the Esquimalt streetcar west, crossing under the railway line near Russell's.

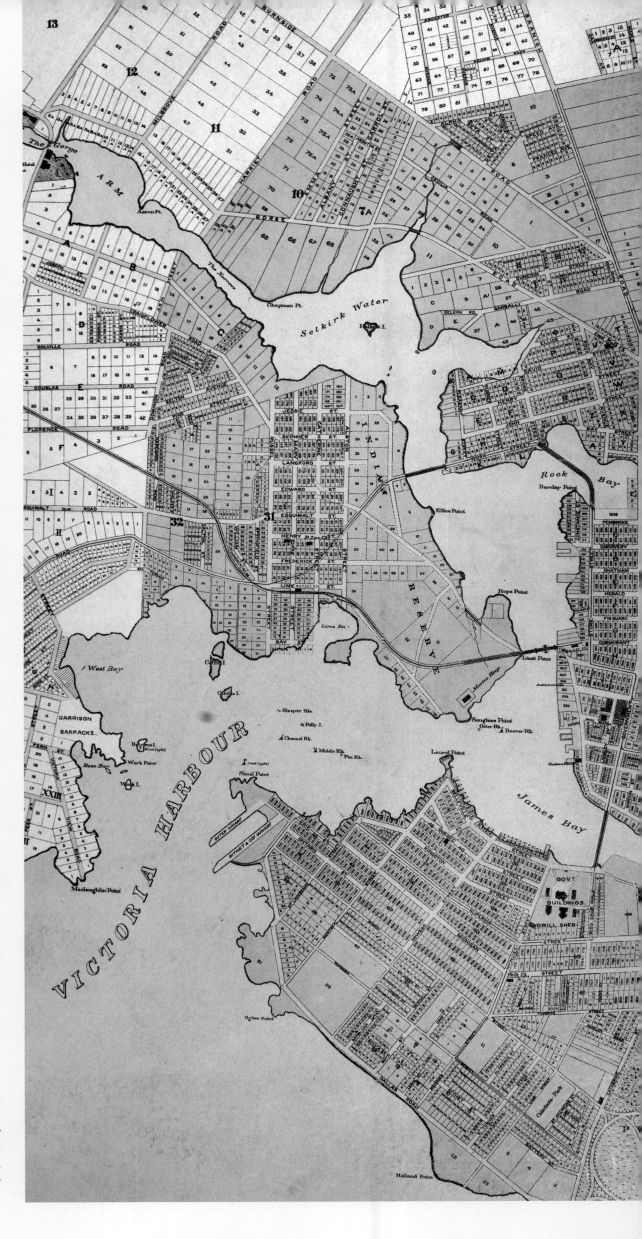

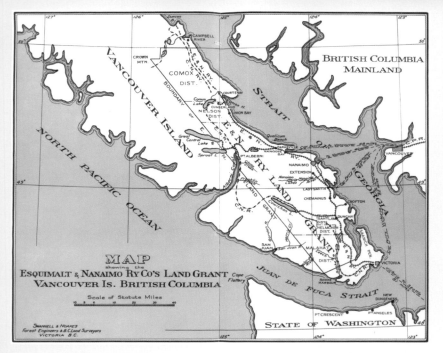

MAP 427 (above).
The Esquimalt & Nanaimo land grant shown on a 1922 map, together with the railway and other existing and still-proposed lines on Vancouver Island. The constructed *E & N Ry.* lines (cross-hatched) to *Courtenay* and *Port Alberni* and on to *Great Central Lake* are shown.

MAP 428 (right).
The Esquimalt & Nanaimo's extension to *Port Alberni* is shown on this 1912 map, which was published to promote a townsite on Kleecoot Arm of nearby *Sproat Lake*. The townsite was called *Kleecoot Park*. The map also shows a proposed railway extension to Great Central Lake; this was built soon after this map was produced.

MAP 429 (right).
This 1908 map shows the Esquimalt & Nanaimo Railway extended into Victoria to reach *E & N Freight Yards* by running along *Store Street*. The Songhees Indian Reserve is now shown unmarked and empty; within a few years it would become the site of Canadian Northern rail yards (see MAP 698, *page 275*).

MAP 430 (below).
This wonderful promotional map appeared in July 1912, placed by promoters of lands at *Hardy Bay*—now Port Hardy—at the northern end of Vancouver Island. Hardy Bay appears as the centre of all the action. The inset text at *left,* from the same ad, states definitively that the "C.P.R. Will Build To Hardy Bay." Its *E.& N. Ry.* extension runs to Hardy Bay, as does the *C.N.P. Ry.* (Canadian Northern Pacific). Existing landowners are listed and include the lieutenant governor of British Columbia and Francis Rattenbury, architect of both the legislative buildings and the Empress Hotel in Victoria. The Hardy Bay Development Company, however, turned out to be fraudulent. Note the railway to Bute Inlet and the *Grand Trunk Pacific Ry.* following a route down Howe Sound much the same as the later route of the Pacific Great Eastern (see page 278).

C. P. R. Will Build To Hardy Bay

By Taking Over Lease of E. N. R. R. Transcontinental Route to Orient is Shortened

Legislation will also be provided authorizing the lease of the Esquimalt & Nanaimo Railway to the Canadian Pacific, which now owns its capital stock, after which that railway will be operated as the Island Division of the Canadian Pacific. The company agrees forthwith to extend its line from a point on the Alberni extension, near Parksville, to Comox, a distance between 40 and 50 miles.

As has been already announced, the plans of the Canadian Pacific contemplate the extension of its line to Hardy Bay. Presumably the line shown upon the map will be followed.

Jeremiah H. Kugler, Sole Agent,
Hardy Bay Development Co.

When Railroads with Assets of over $600,000,000 Seek the same Terminal Seaport—THERE MUST BE A REASON—in this case

HARDY BAY

IS THE REASON

THE RICHEST RESERVOIR OF RICHES YET UNTAPPED

HARDY BAY

This townsite will soon be placed upon the market for public sale. In the meantime all those who seriously want to make money should read this and future announcements.

Every few days finds some fresh townsite or subdivision launched. They are taken as a matter of course. But! Apart from the multitude of minute guns stands out the occasional "boom" of the mighty artillery. In other words, out of the numerous investments offered to the public only once in a long while does the really big thing present itself.

THE "BIGGEST GUN" OF A HEALTHY BOOM IS

HARDY BAY

SYMBOLIZES A NEW ERA IN THE HISTORY OF VANCOUVER ISLAND

Public sentiment on Vancouver Island and the Mainland has now grown to such a pitch that the long deferred Seymour Narrows Bridge now looks like a sure thing. The time has come when it must be done and done mighty quick, too. When you see gigantic corporations like the Canadian Pacific Railway, Canadian Northern, Great Northern, the Northern Pacific, Chicago, Milwaukee and St. Paul and the Union Pacific and other transcontinental lines all seeking a competitive point to Prince Rupert, the Grand Trunk Pacific seaport, you can bank on it that big things are just about to happen.

For many years the government has been promising Vancouver Island a terminal point for the transcontinental lines. Keep your eyes open. Conditions are due to change in the "twinkling of an eye."

Those mighty railroads have just got to secure a more economical terminal point where they can compete with the G. T. P. That point is on Vancouver Island, just as sure as you

That being the case, how far do you think the combined efforts of six transcontinental lines, backed by the government, will go towards making this bridge a reality?

Hardy Bay is a **veritable gold mine** to these railroads and when they go after anything they get it—every time. Watch Hardy Bay—and see.

Hardy Bay affords ample anchorage for all large and small vessels and is pronounced the safest port on Vancouver Island.

It affords a shorter route to the Orient, which country, teeming with millions for trade, is awaiting the shortest and most economical route.

Millions of dollars will be saved by using Hardy Bay. The loading and discharging of cargoes at this point should revolutionize the shipping of the Pacific Coast. Why?

Because: A vessel leaving Hardy Bay will reach open sea one hour after departing and will strike out upon its course direct.

Whereas a vessel bound for Victoria or Vancouver has to slow up to eight knots an hour

The Victoria & Sidney Railway

Map 431 (below) and Map 432 (inset).
The *Victoria and Sidney R.R.* on an 1895 map of the Saanich Peninsula. The *E. & N. R. R.* (Esquimalt & Nanaimo Railway) is also shown. After 1903 the railway connected at Sidney with a steam ferry to Port Guichon (Ladner). The line ended in Victoria at today's Topaz Avenue and was extended farther south to a depot near city hall in 1900 (inset, 1910 edition of same map) under the charter of the Victoria Terminal Railway and Ferry Company.
Right. A train about to leave Victoria, probably about 1910.

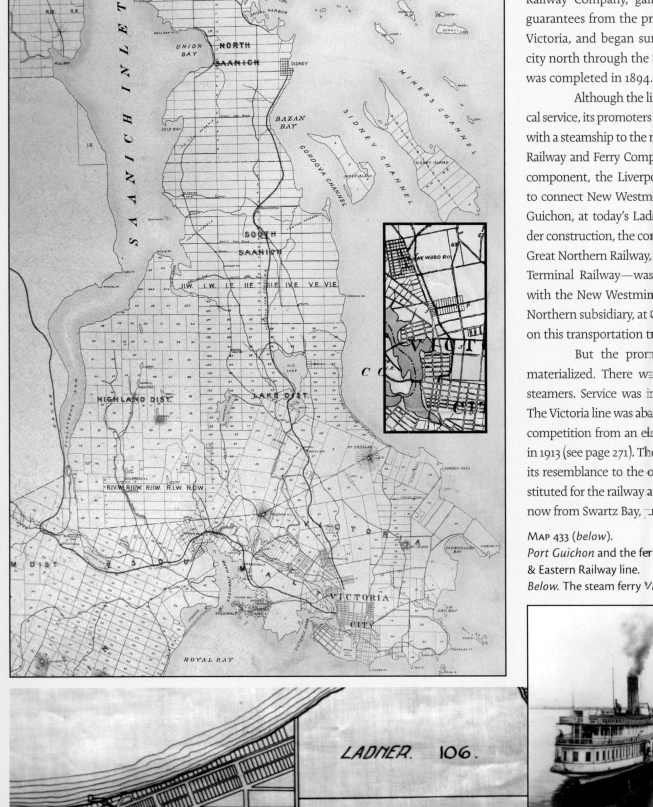

In 1892 a group of investors formed the Victoria & Sidney Railway Company, gained tax concessions and mortgage guarantees from the provincial government and the City of Victoria, and began surveying and grading a line from the city north through the Saanich Peninsula to Sidney. The line was completed in 1894.

Although the line was at first intended to provide a local service, its promoters hoped that eventually it could connect with a steamship to the mainland. In 1900 the Victoria Terminal Railway and Ferry Company was created; this had a mainland component, the Liverpool & Canoe Pass Railway, which was to connect New Westminster with steamers to Sidney at Port Guichon, at today's Ladner. While the mainland line was under construction, the company was taken over by the American Great Northern Railway, and the line—now called the Victoria Terminal Railway—was modified to connect Port Guichon with the New Westminster Southern Railway, another Great Northern subsidiary, at Cloverdale (Map 500, *page 208*). Service on this transportation triumvirate commenced in 1903.

But the promise of the connection never really materialized. There was competition from Canadian Pacific steamers. Service was inadequate and profits were marginal. The Victoria line was abandoned in 1919 following a few years of competition from an electric interurban that began operating in 1913 (see page 271). The Victoria & Sidney service is notable for its resemblance to the one in place today: cars have been substituted for the railway and B.C. Ferries for the steamer service, now from Swartz Bay, just to the north of Sidney.

Map 433 (below).
Port Guichon and the ferry terminal with the *Vancouver Victoria & Eastern Railway* line.
Below. The steam ferry *Victorian* at Port Guichon in 1903.

A Mineral Bonanza

The mountains of the West Kootenay region had long been known to contain mineral wealth. One of the earliest discoveries is attributed to Archibald McDonald of the Hudson's Bay Company (see Map 126, page 46) in 1844. This was the lead and copper deposit that became the rich and long-lasting Bluebell Mine, at today's Riondel.

The pioneering Bluebell (Map 453, page 180) had a complex beginning that is the stuff of mining legend. The claim was staked in 1882 by Robert Sproule, an American prospector. When Sproule left near the end of the season, his claim was jumped by another prospector, Tom Hammill, who was financed by wealthy Columbia River steamboat captain George J. Ainsworth. The latter created Ainsworth, the first town in the Kootenays, on his land grant across the lake the following year. A court awarded the Bluebell claim back to Sproule after lengthy proceedings in which Sproule was represented by William Baillie-Grohman (see page 112), who acquired part of the claim in payment for his services, and then proceeded to sell it to Hammill! Then Hammill was shot, and Sproule was held responsible—though on flimsy evidence—and was hanged for the murder. Such were the wild beginnings of West Kootenay mining.

Nelson and the Silver King

In 1886 a group of American prospectors from Colville, Washington, led by brothers Osner and Winslow Hall, found gold late in the season on what became Toad Mountain, just south of today's Nelson. But they were unable to register their claim and had to return the next year, by which time the news had leaked out and they were followed by dozens of others hoping to piggyback on their discovery. The Hall group, calling themselves the Kootenay Bonanza Company, established themselves at their claims—known as Kootenay Bonanza, American Flag, and Silver King—in 1887, and soon others staked claims all around them.

A group of cabins grew up nearby on the shores of Kootenay Lake. Henry Anderson, newly appointed mining commissioner who arrived in the spring of 1888, gave the settlement the name "Salisbury," but the gold commissioner, Gilbert Malcolm Sproat, named it "Stanley." The community was known by both names until a post office was applied for; "Stanley" was used on the application, but there was already a Stanley elsewhere, and so it was renamed "Nelson" after the sitting lieutenant governor of the province, Hugh Nelson.

Map 434 (below, left).
Cabins are shown at the steamboat *Landing now used* on this 1888 survey. It was drawn by Gilbert Malcolm Sproat, assistant commissioner of lands and works from Revelstoke, who was appointed gold commissioner, and shows the area that became the City of Nelson—the rectangle with *Proposed position of Town Site*. At bottom is *East End of Toad Mt Camp*.

Map 435 (below).
A component map on *Perry's Mining Map* (Map 436, right) shows *Nelson* and the *Silver King* mining area on *Toad Mountain* in 1893, before a tramline, which commenced operations in 1896, was built to carry ore to the smelter at Nelson.

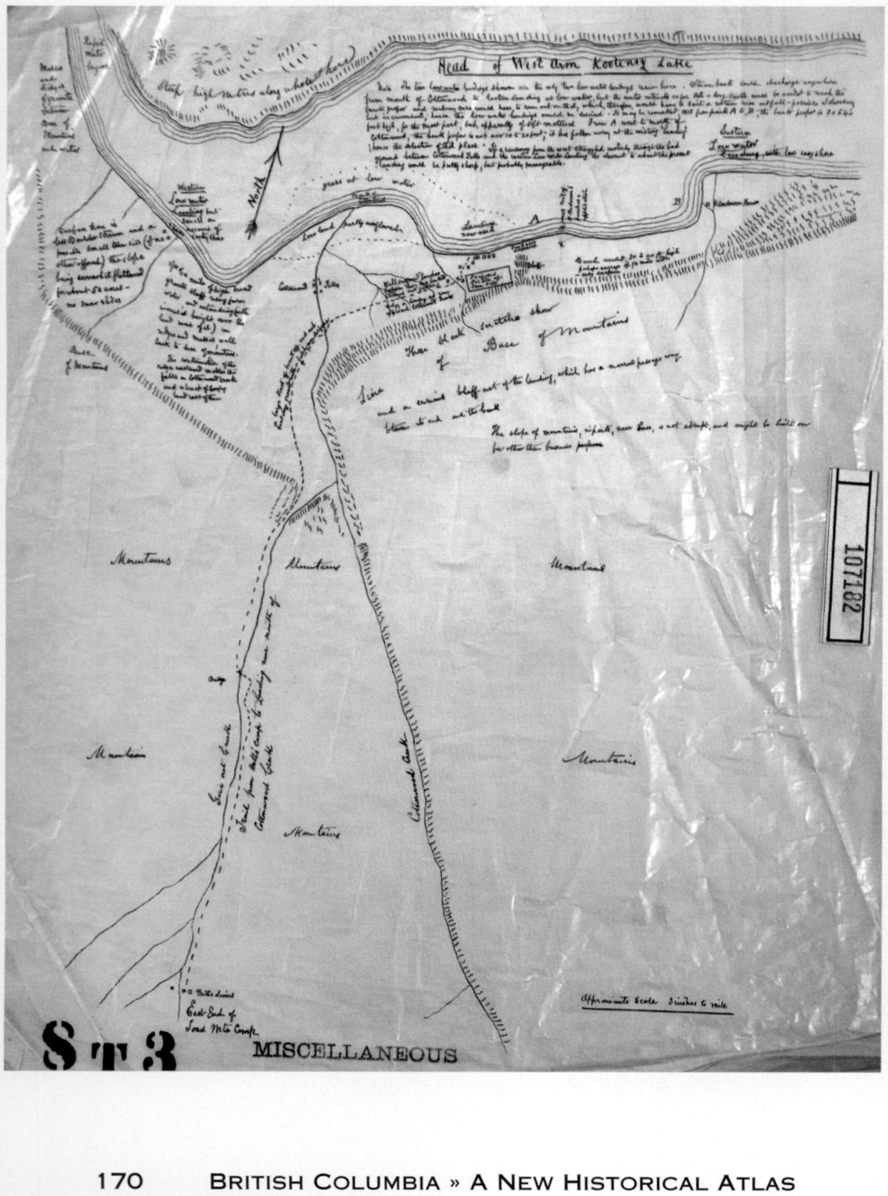

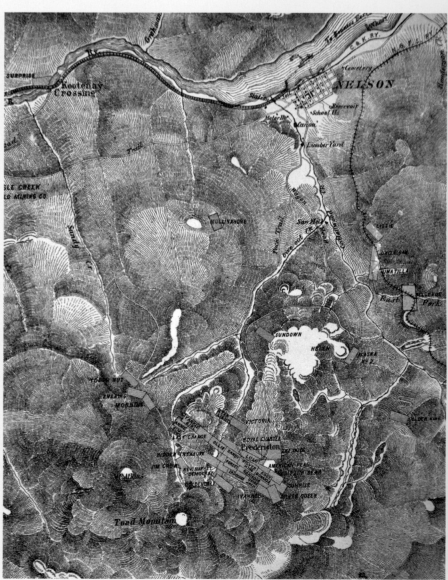

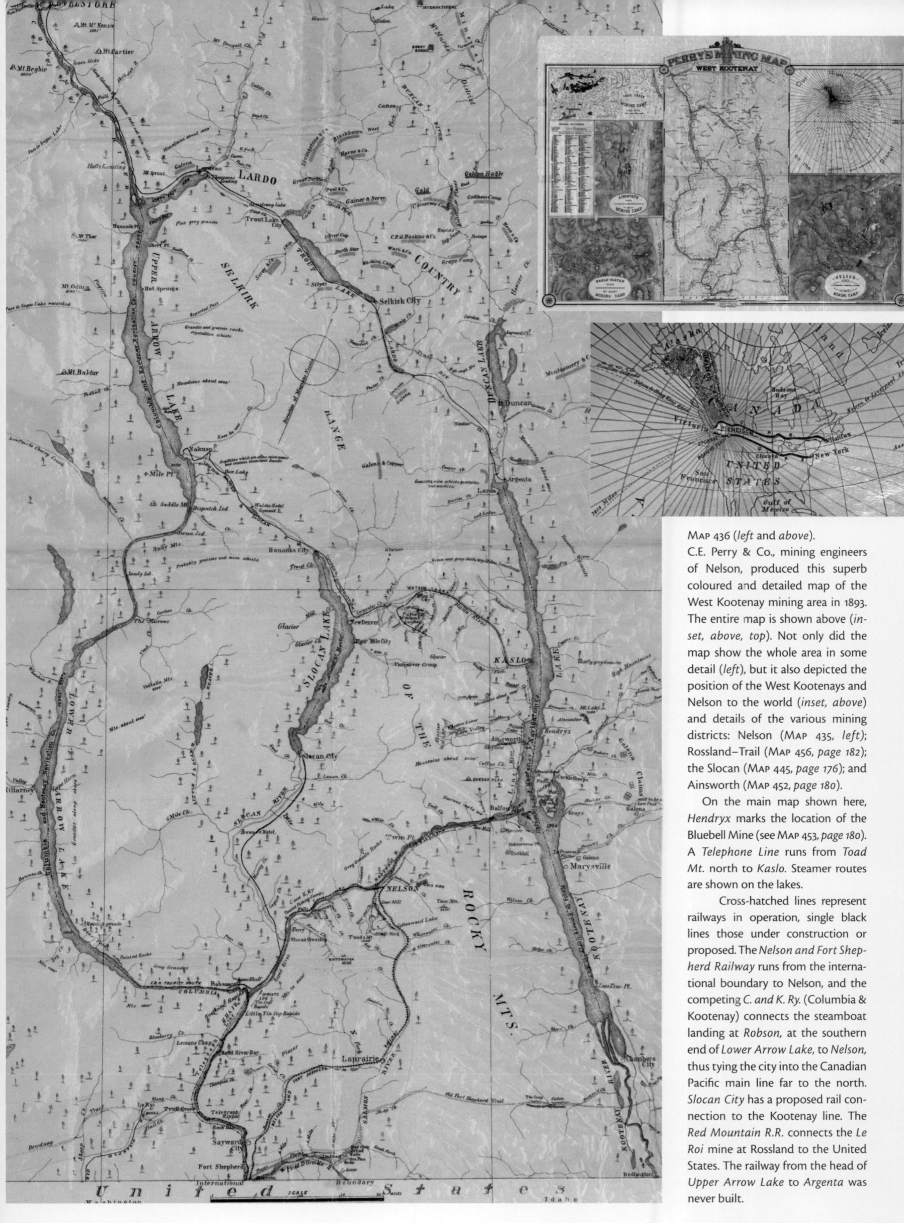

Map 436 (*left* and *above*).

C.E. Perry & Co., mining engineers of Nelson, produced this superb coloured and detailed map of the West Kootenay mining area in 1893. The entire map is shown above (*inset, above, top*). Not only did the map show the whole area in some detail (*left*), but it also depicted the position of the West Kootenays and Nelson to the world (*inset, above*) and details of the various mining districts: Nelson (Map 435, *left*); Rossland–Trail (Map 456, *page 182*); the Slocan (Map 445, *page 176*); and Ainsworth (Map 452, *page 180*).

On the main map shown here, *Hendryx* marks the location of the Bluebell Mine (see Map 453, *page 180*). A *Telephone Line* runs from *Toad Mt.* north to *Kaslo*. Steamer routes are shown on the lakes.

Cross-hatched lines represent railways in operation, single black lines those under construction or proposed. The *Nelson and Fort Shepherd Railway* runs from the international boundary to Nelson, and the competing *C. and K. Ry.* (Columbia & Kootenay) connects the steamboat landing at *Robson*, at the southern end of *Lower Arrow Lake*, to *Nelson*, thus tying the city into the Canadian Pacific main line far to the north. *Slocan City* has a proposed rail connection to the Kootenay line. The *Red Mountain R.R.* connects the *Le Roi* mine at Rossland to the United States. The railway from the head of *Upper Arrow Lake* to *Argenta* was never built.

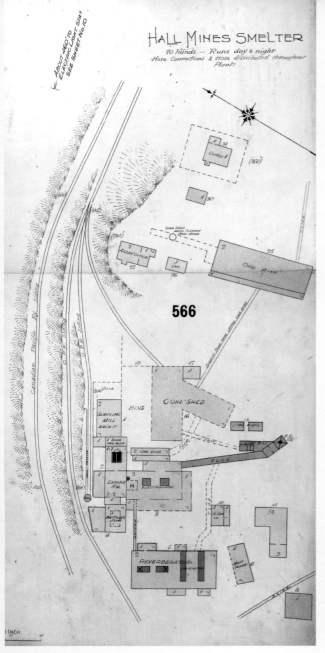

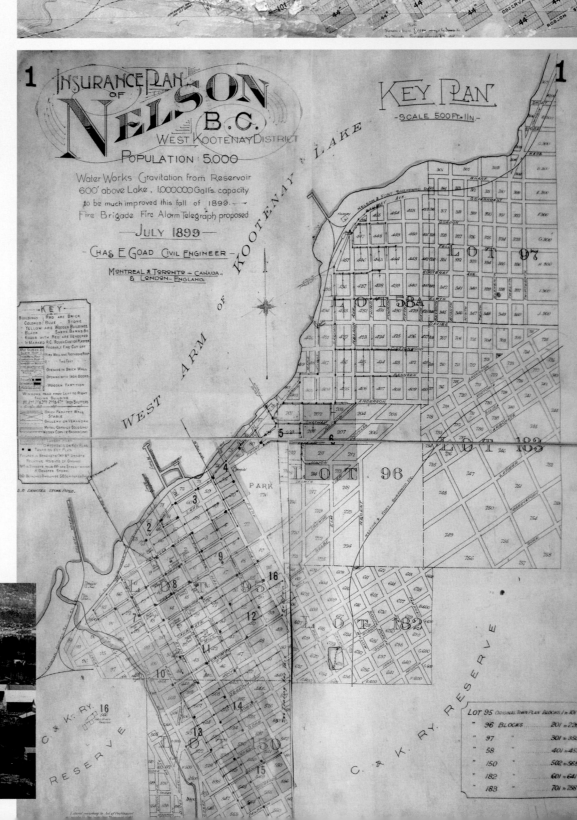

MAP 437 (*above*).
This detail of a fire insurance plan of Nelson produced in 1899 shows the Hall Smelter and an aerial turntable for the tramway that brought ore from down the mountainside from the Silver King. The smelter opened in 1896–97 and closed in 1907.

MAP 438 (*above, right*).
A survey of Nelson drawn about 1893. The Columbia & Kootenay Railway approached from the west and ended at a long wharf on Kootenay Lake. A *line located* proceeds along the waterfront; this would eventually connect with the rival Nelson & Fort Sheppard Railway.

MAP 439 (*right*).
This index map of Nelson from an 1899 fire insurance atlas shows the *Hall Mines Smelter* on the Columbia & Kootenay land grant west of the city. The railway was initiated by a Canadian Pacific director and immediately leased to that company.

Below.
Baker Street, Nelson, in 1890. Today Nelson has more historic buildings than any other British Columbia city.

It was not sufficient to find and mine mineral ores. A means of getting them to the outside world was also required. A trail had been hacked through the forest to the lake, and in the spring of 1888 a veteran American packer, Richard Fry, purchased a rusty Lake Pend Oreille steamboat, the *Idaho,* and, incredibly, had it hauled 64 km overland to Kootenay Lake. In September, 22½ tons of Silver King ore containing gold, silver, and copper were packed down the trail and loaded onto the *Idaho* and shipped first to Sandpoint, Idaho, and then by rail to a smelter at Anaconda, Montana.

In 1891 the Canadian Pacific, annoyed that ores were being sent to smelters in the United States, completed its Columbia & Kootenay Railway (MAP 481, *page 198*) connecting Nelson with its Arrow Lake steamboats at Robson. Then American railway entrepreneur Daniel C. Corbin built the Nelson & Fort Sheppard Railway in 1894 to connect with his lines and smelter across the international boundary. Initially it stopped at Five Mile Point east of Nelson because the Columbia & Kootenay had been given exclusive access to the city in its charter. The two companies' lines were later connected along the waterfront (see maps on this page).

There remained the problem of getting the ore from the mine to the lake. The mine was 1,460 m above and 7 km from Nelson, a very steep grade. A wagon road was completed in 1891 after a first contractor had found it impossible to finish at the contracted price. A wire tramway was constructed in 1895, with buckets holding 45 kg each. Even this proved problematic, for the weight was too great for some of the spans, and the tramway had to be divided into two sections. Finally, in 1896 a smelter was built to the west of Nelson to process the ore, and the following year it was expanded with a refinery and a furnace that was the largest in the entire Northwest.

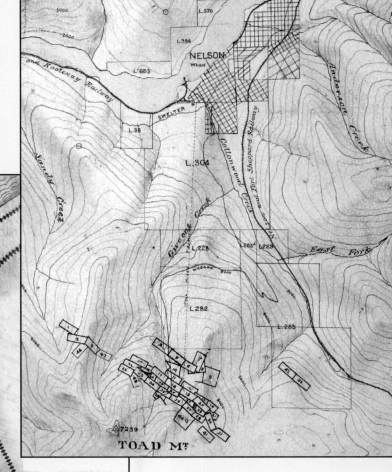

MAP 440 (*above*).
This 1896 map shows the relationship of the two railways serving Nelson before they were connected. The circuitous route of the *Nelson and Fort Sheppard Railway* was due to the necessity of descending the steep grades into the city. A rival townsite, Fairview, is depicted but not named northeast along the lakeshore. The mineral claims on Toad Mountain are shown.

MAP 441 (*left, centre*).
Railways, the *Wagon Road,* the *Smelter,* and the *Wire Tramway* from *Silver King Mine* are shown on this map published in 1896.

MAP 442 (*left, bottom*).
This map of the claims on Toad Mountain was published in 1897. The original claims are shown—the Hall group's *Silver King, Kootenai Bonanza,* and *American Flag.*

Above, left, centre. The mid-station of the Silver King Tramway in 1896, after the line had been split into two sections. Nelson can be seen far below.
Left. The mining settlement at the Silver King on Toad Mountain in 1896.

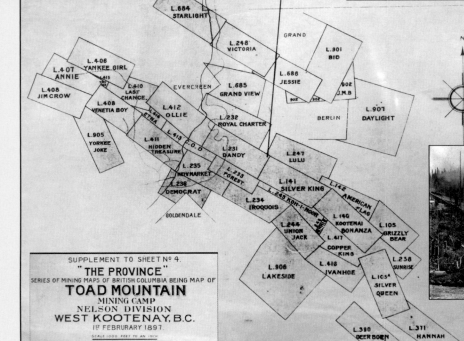

The Silvery Slocan

What proved to be one of the most profitable mineral claims of all was staked on Payne Mountain, in the region north of Nelson that became popularly known as the "Silvery Slocan." In 1891 Eli Carpenter, a circus high-wire artist out to find his fortune, and John L. Seaton, a miner, found silver on Payne Mountain, a short distance east of today's New Denver, on Slocan Lake. Both tried to exclude the other from the find, and, returning separately, Seaton staked the Noble 5 claim and some twenty-six others before Carpenter arrived. Both sold their interests to others, and in the end neither benefited much

from their discovery. But their find set off the predictable stampede of other fortune seekers.

Many came to Kane's Landing, a townsite on Kootenay Lake pre-empted by George Kane in 1889. The little settlement, which was renamed Kaslo in 1892, suddenly boasted a largely transient population of 600. It seems to have taken a few months before prospectors generally realized that Slocan Lake was nearer to the centre of mining activity, and they began choosing that access route. In the spring of 1892 businessmen from Nelson financed the construction of a rough trail from the Columbia & Kootenay Railway line up the Slocan valley to the southern end of Slocan Lake, where Slocan City arose. From there boats could transfer prospectors

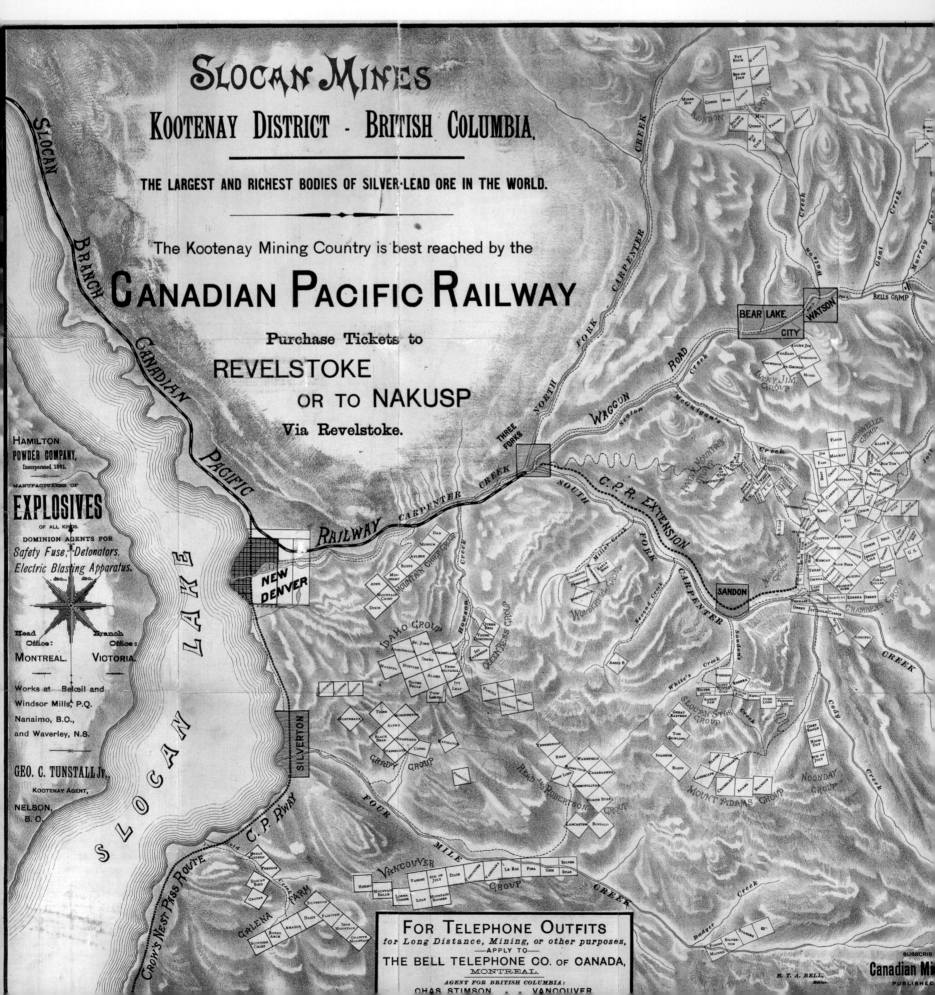

to the mouth of Carpenter Creek, where another small settlement sprang to life. At first called Eldorado, its name was soon changed to New Denver. Yet another mining town was established at Three Forks, where the trail from Kaslo met the one from New Denver. This community assumed more importance in April 1893 when the Canadian Pacific Railway announced that it would build a line from Nakusp, on Upper Arrow Lake, that would terminate at Three Forks.

The railway was a direct and immediate response to a threat from the other direction; the Kaslo & Slocan Railway, a narrow-gauge line chartered in 1892, was rumoured to have been acquired by James J. Hill of the Great Northern Railway, who was intent on siphoning off the Slocan's

Map 443 (below).

It is not hard to guess that this 1895 map of the Slocan mining region was published for the Canadian Pacific Railway. Although the Nakusp & Slocan is marked as the *Slocan Branch Canadian Pacific Railway* and the *C.P.R. Extension* to *Sandon* and Cody is shown (dashed line), the railway's competitor—and the reason for the CPR extension—is not shown at all, either on the main map or the smaller-scale inset location map. The only connection between the Sandon mines and Kaslo is the *New Denver–Kaslo Waggon Road*; the pioneering Kaslo & Slocan, then well under construction, is nowhere to be seen. The projected *Crow's Nest Pass Route C.P. R'way* along the shores of *Slocan Lake* was never built; a steamer service connected with Slocan City, at the southern end of Slocan Lake, from where a branch line was completed in 1897, connecting with Slocan Junction, on the Kootenay River and the main Crow's Nest line (the line is shown on Map 481, *page 198*). The map is liberally sprinkled with ads, for everything from telephones to explosives.

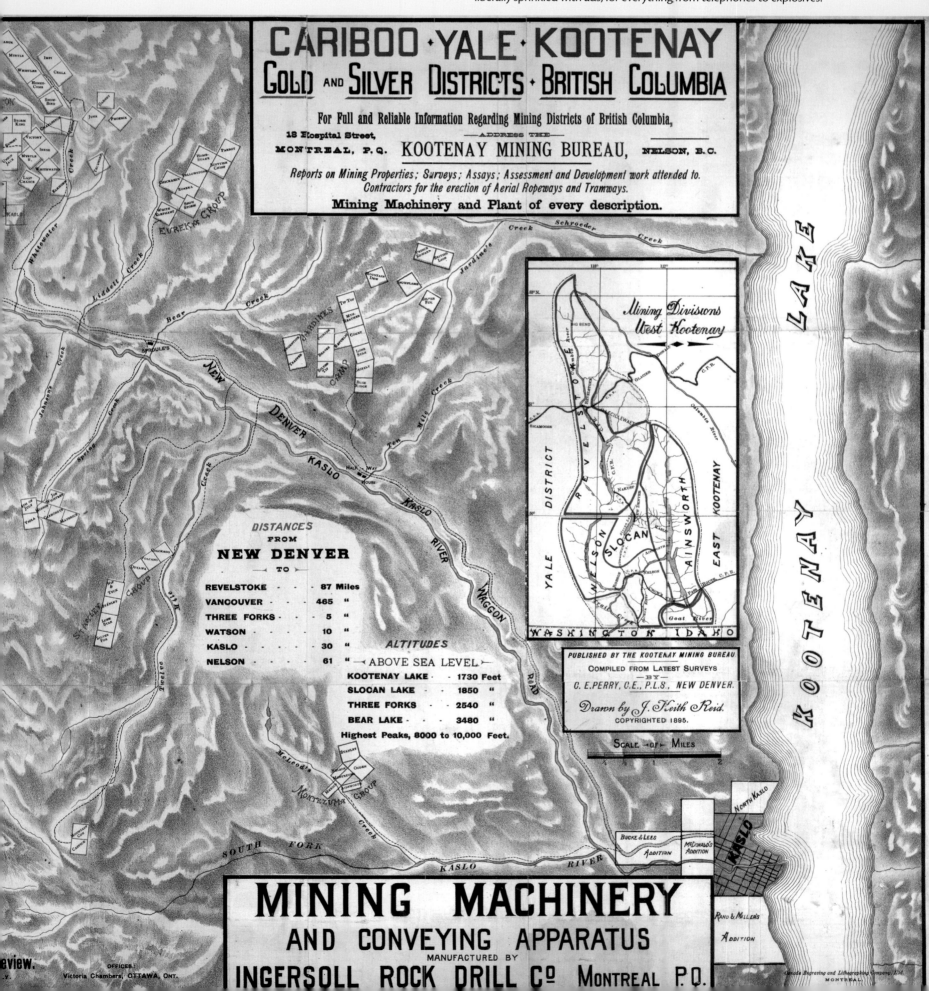

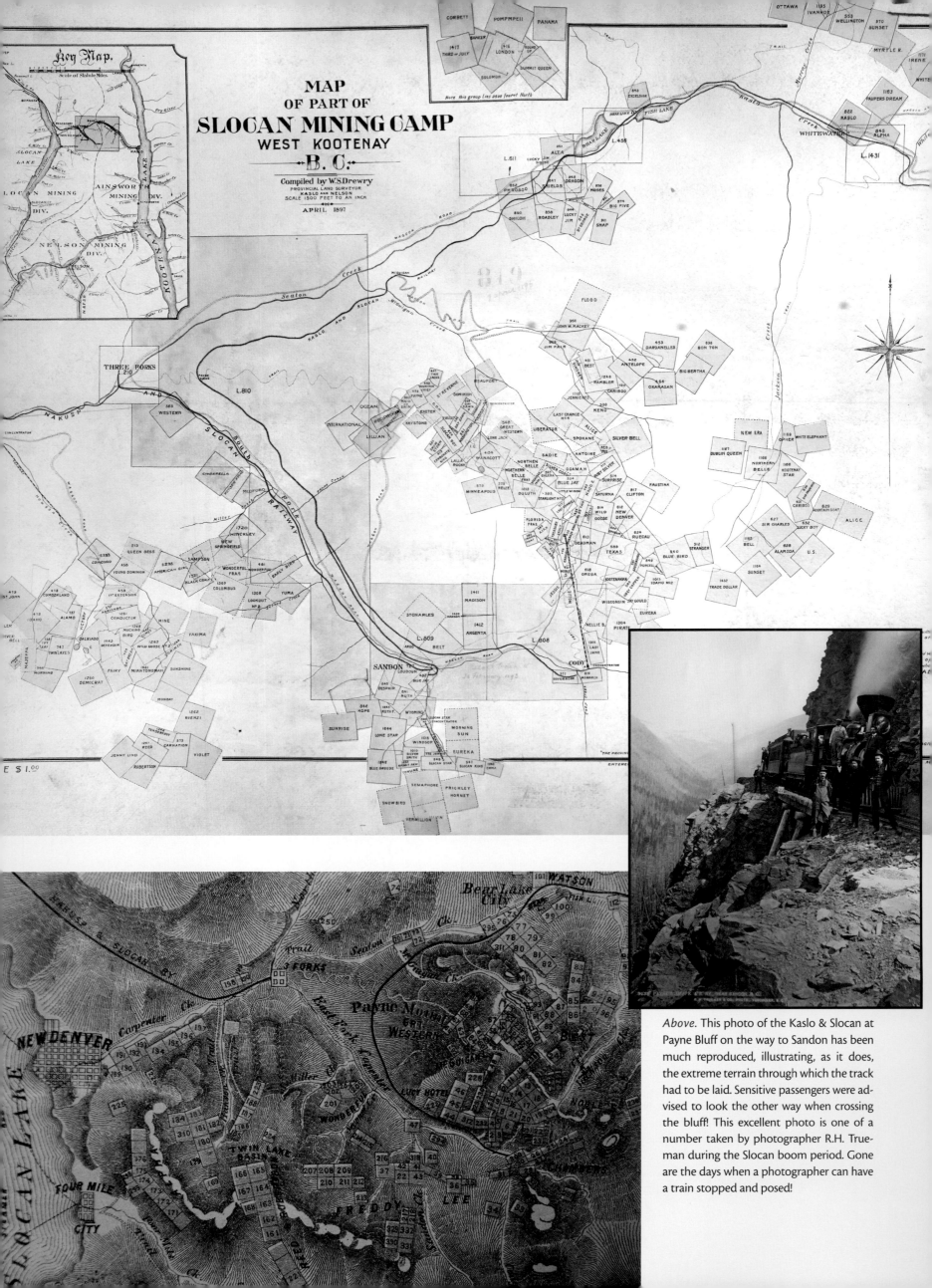

MAP
OF PART OF
SLOCAN MINING CAMP
WEST KOOTENAY
B. C.

Compiled by W. S. Drewry
PROVINCIAL LAND SURVEYOR
KASLO AND NELSON
SCALE 1500 FEET TO AN INCH
APRIL 1897

Above. This photo of the Kaslo & Slocan at Payne Bluff on the way to Sandon has been much reproduced, illustrating, as it does, the extreme terrain through which the track had to be laid. Sensitive passengers were advised to look the other way when crossing the bluff! This excellent photo is one of a number taken by photographer R.H. Trueman during the Slocan boom period. Gone are the days when a photographer can have a train stopped and posed!

riches southward to smelters in the United States using, of course, his rail system to transport them.

The Nakusp & Slocan tracks reached Three Forks in October 1894, but not before the wooden town had been all but wiped out by windstorms and fire—yet quickly rebuilt. A year later the Kaslo & Slocan Railway line reached beyond Three Forks right into the mining area via the valley of the south fork of Carpenter Creek, to Sandon in October and Cody in November.

The Canadian Pacific, not to be outdone, had also begun an extension to Sandon, and by December 1895 the two companies were at war with one another. Men of the Kaslo & Slocan descended upon the newly built facilities of their rival in Sandon, derailing freight cars, ripping up track, and pulling down the depot by corralling it with a steel cable attached to a locomotive!

Canadian Pacific admitted defeat for a while but then began planning an extension up Seaton Creek to tap into the Kaslo & Slocan's business there. But after a few more years nature decided the battle. Landslides took out a long section of the latter's line, and in 1910, after a forest fire had also wiped out its stations, the Kaslo & Slocan halted operations. In a final twist of the knife, in 1912 the Canadian Pacific took over its former rival, building a connection between the two, standardizing the gauge, and operating a line from Kaslo to Nakusp, which lasted until 1955, when torrential rains washed out many sections of track.

It was this storm that destroyed much of Sandon, and today it is one of the most fascinating of the Silvery Slocan's many ghost towns.

Map 444 (*left, top*) and Map 445 (*left, bottom*).
Two versions of a map of the so-called Silvery Slocan mining area. Map 445 is one of several maps on the 1893 *Perry's Mining Map* (Map 436, *page 171*), while Map 444 is one of a series of mining maps published by the *Province* newspaper in 1897. Map 445 shows the Canadian Pacific's *Nakusp & Slocan Ry.* terminating at *3 Forks*; in fact it was only under construction at this time. The other rail line, the narrow-gauge Kaslo & Slocan, also then under construction, loops around *Payne Mount[ain]* and into the valley of the south fork of Carpenter Creek, here misspelled as *East Fork Capenter*. Numerous mining claims, many with exotic names, cover the mountainsides. The Nakusp & Slocan would soon decide to extend its line to match that of its rival. At the mouth of Carpenter Creek is *New Denver*, and just to its south *Four Mile City*, later renamed Silverton. Both names hint at the fact that many of the American miners came from earlier mineral finds in Colorado. Map 444, depicting the situation four years after Map 445, shows both railways completed to *Sandon* and the *Kaslo & Slocan Railway* continuing to *Cody*. Several tramways were constructed to bring ore down to the railway line—and three are shown on this map—one to Cody, one to the eastern edge of New Denver, and one bringing *Slocan Star* ore to the *Slocan Star Concentrator* just south of Sandon; at both other locations a *Concentrator* is noted. *Three Forks* is now served by a branch line or, more realistically, a siding.

Map 446 (*above, top*).
Detail of *Kaslo* from the 1893 Perry map, with steamboat routes and wharves shown and even illustrated with a little drawing of a steamboat. The railway line is the Kaslo & Slocan. Note the *Thereapeutic Springs*.

Above, left. Another superb photo by R.H. Trueman, taken in 1896. The on-photo caption says that this train of the Kaslo & Slocan is in Carpenter Creek Canyon, but the photo appears to have been taken in the Kaslo River Canyon, 25 km west of Kaslo.

Above. The sternwheeler *Kokanee* on Kootenay Lake in 1896, the year it was built at Nelson. Here bags of ore are stacked on a railway transfer barge. More photos and maps of steamboats on the Kootenay lakes are shown on pages 212–14.

The townsite for Sandon had been awarded in 1892 as a Crown grant to miner John Harris, who had located the Loudoun claim (on Map 444, *far left, top*) and purchased the rich galena Reco Mine nearby. Harris laid out a plan for the town (Map 447, *overleaf*) and named it Sandon after an old miner, John Sandon, who had recently drowned in Carpenter Creek. The decision by both railway companies to build lines to Sandon sealed the fate of rival Three Forks. Sandon began a period of phenomenal growth and was incorporated as a city in 1897. By this time it had a population of about 5,000 and serviced another 2,000 miners living nearby. It possessed the most advanced electrical system of any city in North America, with electricity generated from flumed water. The town had a full array of services any city would be proud of, with seventeen hotels, three banks, two newspapers, and, of course, two railways. It also had a fire brigade, a service that would be required but that was quite overwhelmed when a major fire broke out in May 1900, scorching most of the town. It was quickly rebuilt, and what turned out to be a fateful decision was made: to build over Carpenter Creek, which henceforth would run in a flume under the main street. Always problematic and often needing repairs, it was the creek, bursting from its channel, that all but destroyed Sandon in 1955.

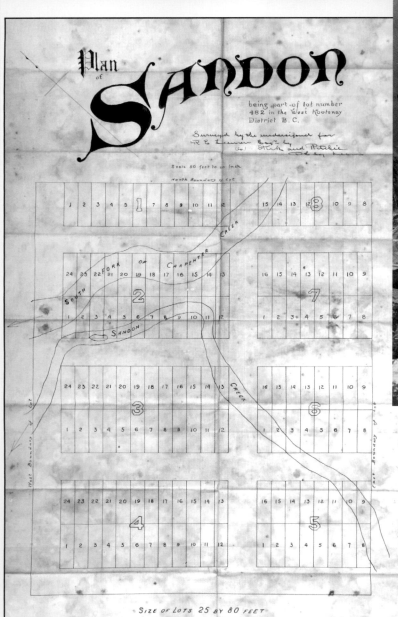

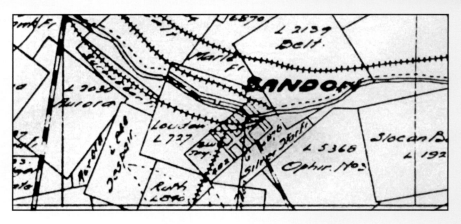

Map 447 (*above*) and Map 448 (*above, right*).
The subdivision of Sandon, laid out for John Harris in 1892, shows how *South Fork of Carpenter Creek* and *Sandon Creek*, which flows into the former, were ignored; lots are drawn right over them. The reason is that there was very little land suitable for building in the narrow valley bottom. Map 448 shows the position of this plan within the claims and rail lines on a later map, drawn about 1905.

Above, top.
A view of Sandon in 1896 looking down the valley. The Nakusp & Slocan Railway line is in the foreground, and the Kaslo & Slocan line runs along the slightly higher ground to the right. Carpenter Creek runs through the centre of the settlement.

Left.
Another superb photo by R.H. Trueman. This is Reco Avenue (parallel to and one street north of the creek) in Sandon in 1898. The photographer really was king: everyone on the street has been posed for this photo, commissioned by a hotel owner seen standing on the hotel steps to the left.

Below.
All that was left of Sandon by 2010; this is the city hall, built after the 1900 fire. Carpenter Creek is in the foreground.

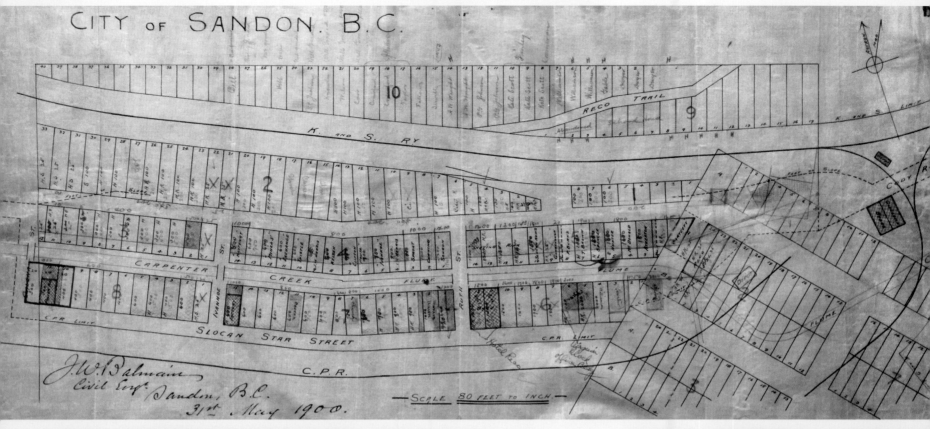

CITY of SANDON. B.C.

J.W. Balmain
Civil Eng. Sandon, B.C.
31st May 1900.

—SCALE 80 FEET TO INCH—

MAP 449 (*above, top*).
Sandon as it was after the 4 May 1900 fire that destroyed most of the original town. Harris's initial subdivision is shown in the angled lots at the right side. Notice how *Carpenter Creek flume* goes right down the middle of the street. It was a disaster waiting to happen, but a creative way to deal with the lack of available building land. This plan was drawn up on 31 May 1900, 27 days after the fire, as a blueprint for the rebuilt town. The

Kaslo & Slocan line (*K. and S. Ry.*) is at top and the Canadian Pacific's Nakusp & Slocan (*C.P.R.*) at bottom (compare with the photo, *left, top*). The photos (*above, left* and *above, centre*) were taken in the 1920s and show how Carpenter Creek emerged from under the street at the lower end of the town. The other photo (*above, right*) shows the state of the flume and boardwalk today—an old mass of wood piled high after decades of neglect.

MAP 450 (*below, left*).
An 1894 plan of Three Forks clearly shows the three forks that it was named for. The Nakusp & Slocan Railway right-of-way is at bottom. The names of many streets reflect their environment: *Silver Street, Ore*

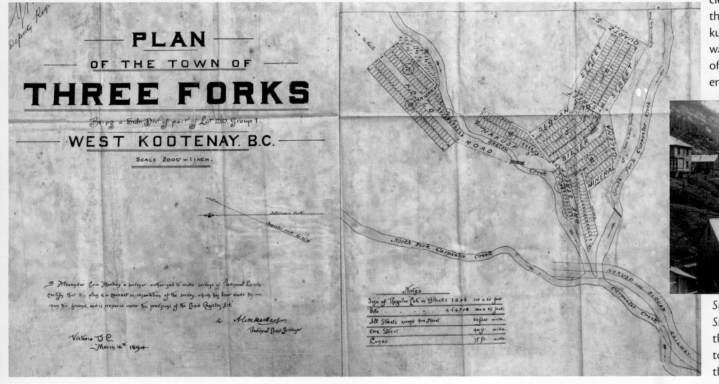

Street, Galena Street, Mineral St., and *Quartz St.* North is to the left. *Inset* is a photo of the town taken in 1905, well after the town's heyday.

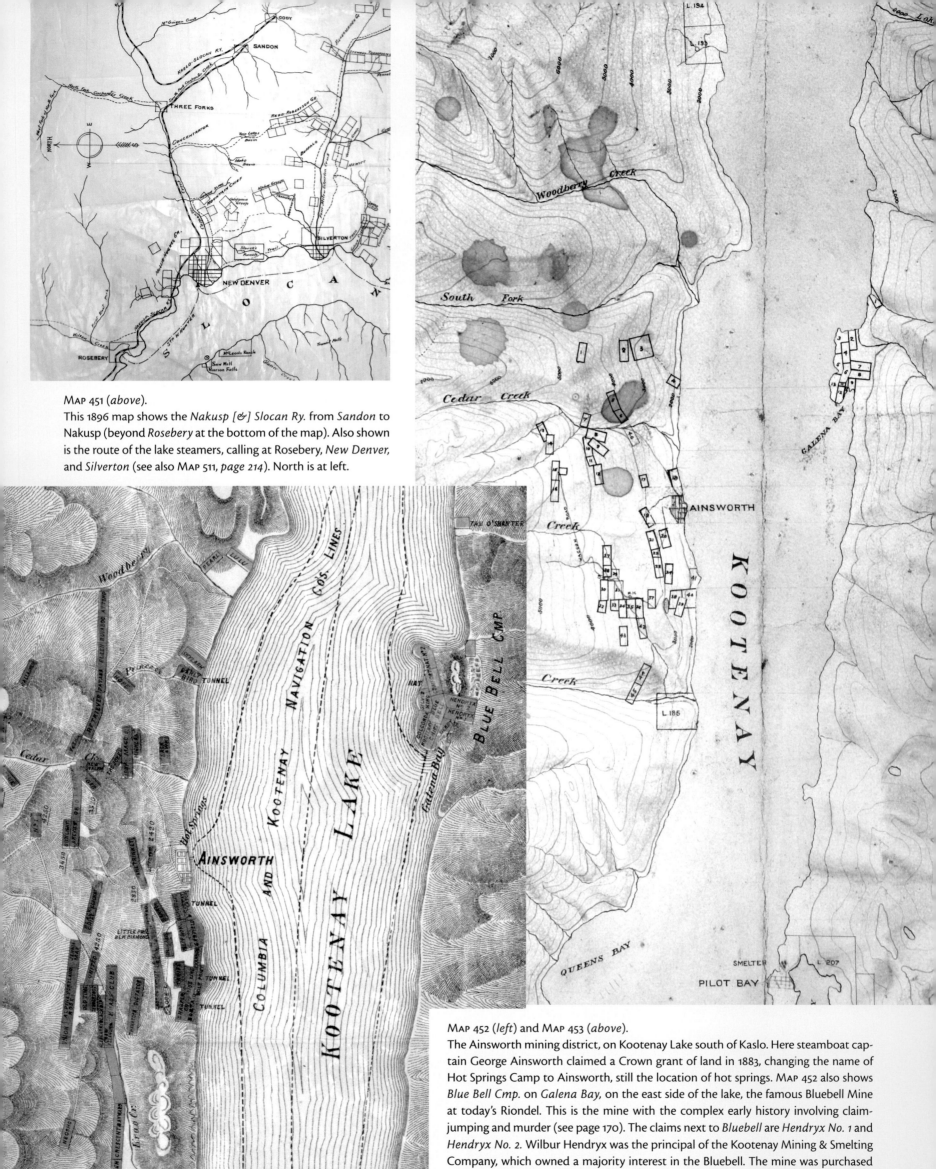

MAP 451 (*above*).

This 1896 map shows the *Nakusp [&] Slocan Ry.* from *Sandon* to Nakusp (beyond *Rosebery* at the bottom of the map). Also shown is the route of the lake steamers, calling at Rosebery, *New Denver*, and *Silverton* (see also MAP 511, *page 214*). North is at left.

MAP 452 (*left*) and MAP 453 (*above*).

The Ainsworth mining district, on Kootenay Lake south of Kaslo. Here steamboat captain George Ainsworth claimed a Crown grant of land in 1883, changing the name of Hot Springs Camp to Ainsworth, still the location of hot springs. MAP 452 also shows *Blue Bell Cmp.* on *Galena Bay*, on the east side of the lake, the famous Bluebell Mine at today's Riondel. This is the mine with the complex early history involving claim-jumping and murder (see page 170). The claims next to *Bluebell* are *Hendryx No. 1* and *Hendryx No. 2*. Wilbur Hendryx was the principal of the Kootenay Mining & Smelting Company, which owned a majority interest in the Bluebell. The mine was purchased in 1905 by the Canadian Metal Company, whose principal was Edouard Riondel, from whom the modern name of the community comes. Note the steamboat routes.

Map 454 (*above*).

Believed to date from the mid-1920s, this map of all mining claims in the northwest Slocan area displays the amazing density of mining activity in the region as claimants jostled each other in the scramble for pay dirt. *Sandon* and *Cody* are at the centre of this jumble. The *C.P. Ry.* now continues east beyond *Three Forks*. Canadian Pacific had established a through line from Nakusp to Kaslo in 1912, when it leased the Kaslo & Slocan and abandoned the difficult Payne's Bluff section the following year. The line lasted until 1955.

Map 455 (*right*).

Although the hub of Kootenay mining activity in the late nineteenth and early twentieth century was the Slocan, prospectors continued to search in less lucrative areas. Here, nearer to the Canadian Pacific main line (shown in red) are mining claims of the Big Bend area, the Upper Columbia and its tributaries. The map was published in 1897 in *Golden* as a guide to the existing mines of the region, with the numbers referring to lists at either side of the map. The latter, by use of a colour code, indicate the minerals mined, the majority being gold mines (quartz), with a lesser number mining copper and galena (lead ore).

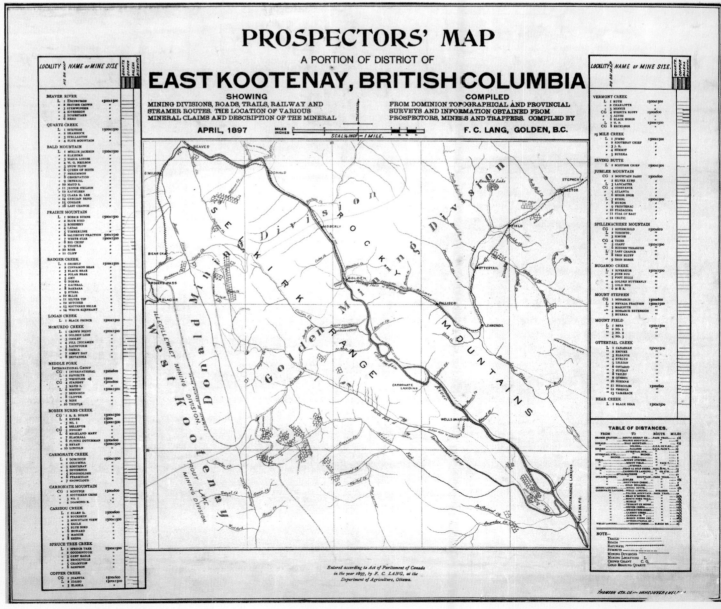

ROSSLAND AND TRAIL

An American fortune seeker, Eugene Sayre Topping, had made his way to Nelson in 1888, attracted by the potential of the Silver King. While there he wounded himself by discharging a gun into his wrist. Unable to continue prospecting, Topping acquired Canadian citizenship and got himself a job—as deputy mining recorder. In the summer of 1890 two prospectors came to his office to record finds they had made on Red Mountain, near the Dewdney Trail to the south. The pair, Joe Morris and Joe Bourgeois, staked five claims—named Centre Star, War Eagle, Virginia, Idaho, and LeWise—but the law did not allow them to register more than two claims each, and so they offered one to Topping if he would pay the recording fees. Topping paid $12.50 for ownership of the LeWise claim—and changed its name to Le Roi. It would prove to be one of the richest mines British Columbia ever produced.

Above. Old mine buildings located on the grounds of the Rossland Museum close to Le Roi.

Above. A share certificate for the optimistically named Big Buck Gold Mining Company of Rossland, issued on 29 December 1896. Now in the Rossland Museum, the certificate is illustrated with idealized scenes. Mining companies typically issued such important-looking share certificates.

When Topping examined his claim, he realized that it was very promising. He also saw that the only way to move ore out was down the steep valley of Trail Creek, which led to the Columbia River. And he saw a business opportunity at the location where the ore would be loaded onto steamboats, so he and a friend, Frank Hanna, pre-empted 342 acres (139 ha) there for a townsite. In 1890 Hanna built a log cabin for his family, calling it Trail House.

By the spring of 1891 Topping had interested investors to provide the capital to mine his claim, and ten tons of ore were extracted from Le Roi, packed down to the Columbia, and shipped to a smelter in Butte, Montana. The investors, pleased, created the Le Roi Gold Mining Company and offered to buy Topping out. He accepted $30,000—for a mine that would sell seven years later for $3 million.

The richness of Topping's mine on Red Mountain attracted others, and claims were staked all around it. One of the early arrivals was Ross Thompson, who in 1892 pre-empted land for a townsite. The settlement that grew up there was soon called Rossland.

After the sale of Le Roi, Topping turned his attention to his townsite, and when in 1895 F. Augustus Heinze, the Butte smelter owner and mining millionaire, showed interest in building a smelter there, Topping offered him a 16-ha site on a bench above the river at no cost and a one-third interest in the townsite. Not unexpectedly, Topping and Hanna's real estate business suddenly took off.

Heinze had had the Trail Creek valley assessed with a view to building a railway connecting mines and smelter and soon began constructing a 23-km-long narrow-gauge line, called the Trail Creek Tramway. During construction Heinze changed the name to the Columbia & Western Railway and applied for land grants that would come with a bigger enterprise—building all the way to Penticton, where a connection could be made with the steamboats on Lake Okanagan and the Canadian Pacific main line. The railway and the land grant awarded (MAP 481, *page 198*) was sold in 1898 to Canadian Pacific; the Trail Creek line was converted to standard gauge in 1898–99 and was soon after incorporated into Canadian Pacific's southern main line and the Kettle Valley Railway (see page 216).

However, Heinze also had a competitor. The Rossland mines were becoming so profitable—by 1896 the output was valued at about $1 million—that the American railroad entrepreneur Daniel C. Corbin had decided to connect them to his Spokane Falls & Northern line to Rossland. He had built this line to connect with his Nelson & Fort Sheppard line, and a relatively short spur—his Red Mountain Railway—was required to reach Rossland. This would also tie Rossland's mines into the American rail network (MAP 465, *pages 186–87*). Corbin proved to be a nuisance to Heinze, suing him at one point for incursions onto his Nelson & Fort Sheppard land grant (although this didn't stop Heinze). But, coupled with the steep terrain, finding a pathway for the Columbia & Western was difficult, and this can be seen on maps that show the rail route.

MAP 456 (*below* and *below, right*).
Two parts of the Rossland–Trail section of the 1893 Perry mining map (MAP 436, *page 171*). The left part shows *Trail* at the point where Trail Creek meets the much larger Columbia; Topping's townsite is not shown other than by name. Only a trail comes down the valley. Note the *Telegraph Line*. The part at right shows the claims that were in place on the slopes of Red Mountain by 1893 and a proposed line—far too straight as it turned out—for the Red Mountain Railway of Daniel Corbin. The first claims are shown—*Le Roi, Center Star, Idaho, Virginia,* and *War Eagle*—but have been surrounded by many others.

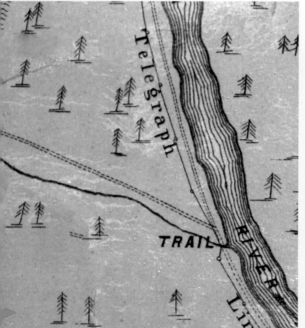

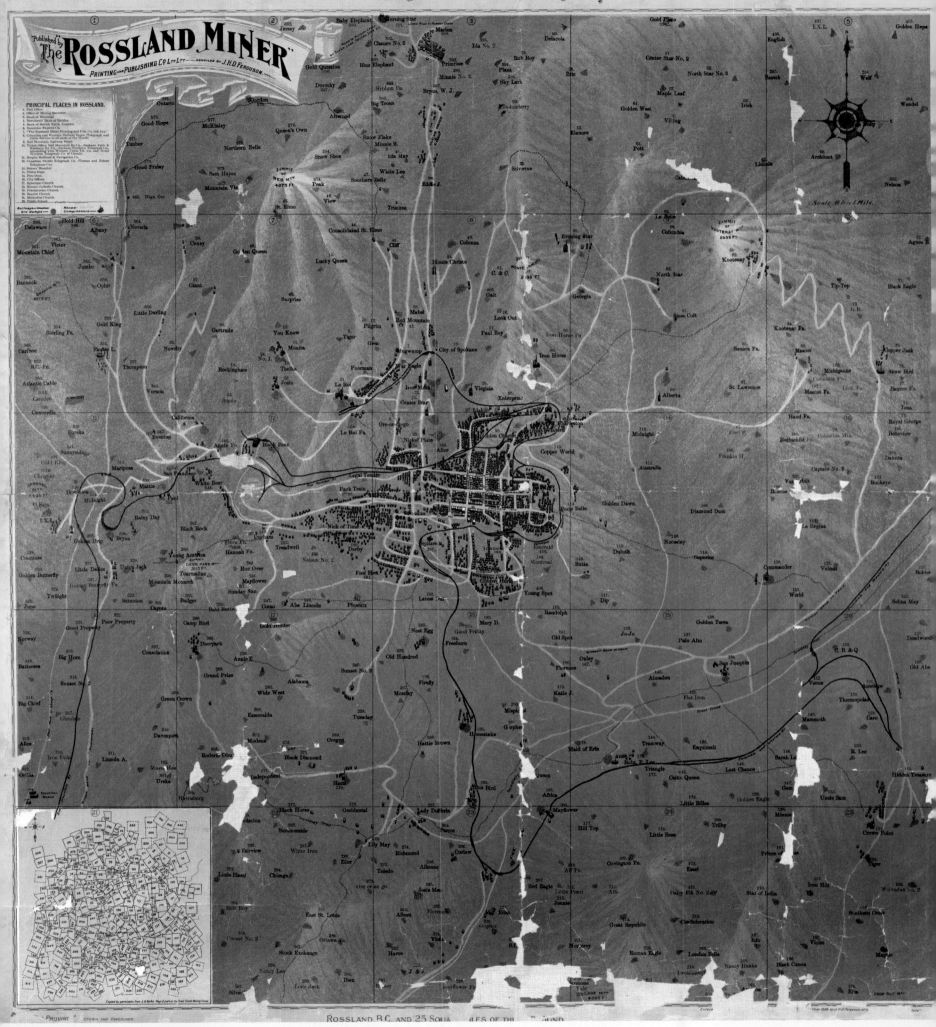

Map 457 (above).

This very unusual map was published in 1898 to show the locations of the hundreds of mines that blanketed the region. At bottom left is a far more mundane map of the same area showing just the mining claims, with the city limits and the two railways added. On the main map, the key locates ore dumps (in red). The circuitous routes taken by both rail lines are well illustrated here. The *Columbia* *& Western Ry.*, at right, which connected Rossland with the smelter at Trail, has two switchbacks shown, and the *Red Mountain Ry.* route involves an almost 360-degree loop. An attempt has been made to depict the individual buildings of the town. The key to the mines, listed by numbers that pertain both to the main map and the smaller one, is at left (but has here been repositioned from right).

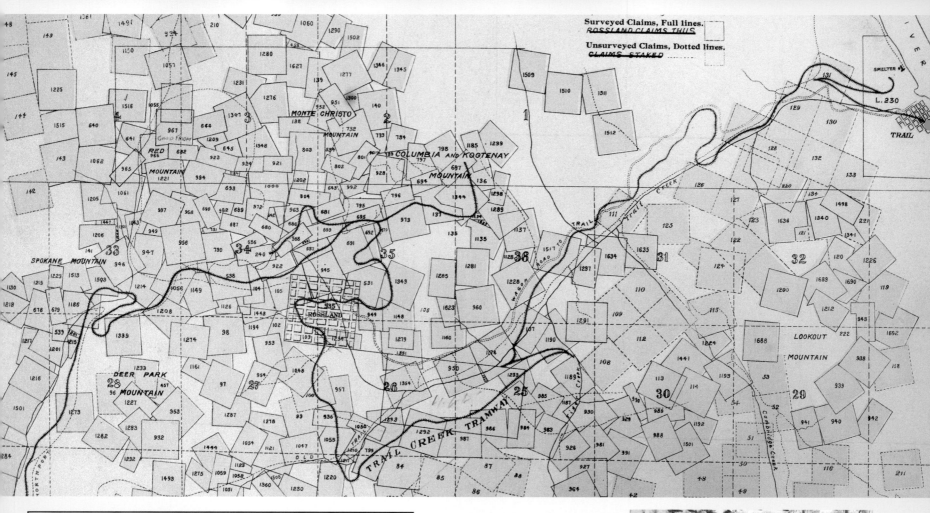

Surveyed Claims, Full lines.
ROSSLAND CLAIMS THIS
Unsurveyed Claims, Dotted lines.
CLAIMS STAKED

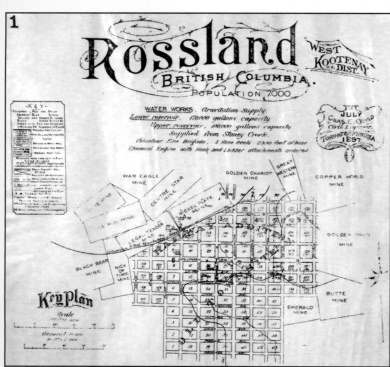

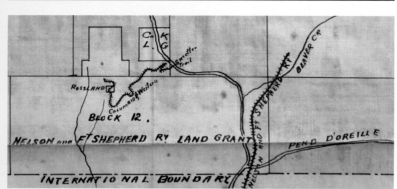

MAP 459 (*above, centre*).
An expanding Rossland in 1897 showing the close juxtaposition of the mining claims. The map is the index to a fire insurance atlas.

MAP 460 (*above*).
Surveyor Arthur Farwell made this map in 1896 showing railway land grants in the vicinity of Trail. The *Columbia & Western* is shown from *Rossland* to *Trail,* and the *C.K.L.G.* (Columbia and Kootenay land grant) and *Nelson and Ft. Shepherd Ry. Land Grant* and rail line are also shown.

MAP 458 (*above*).
This 1897 map of the mineral claims in the Rossland–Trail area shows the hairpin bend and switchbacked route of the Columbia & Western Railway from *Rossland* to the *Smelter* at *Trail;* note that here it still has its original name of *Trail Creek Tramway.* The nearly as difficult route of the Red Mountain Railway is also shown, at left.

Right, centre and *right, bottom.* The two railway contestants for Rossland. At *right, centre* is a Red Mountain Railway (then Great Northern) passenger train crossing a trestle at Little Sheep Creek below Rossland about 1920. At *right, bottom* is a Columbia & Western ore train on the so-called tail track at the bottom of the switchback below Rossland about 1910. Trains were required to halt here long enough to cool their brakes, and the photographer has taken advantage of this to take his shot. The train is short but was about all a single locomotive could manage at the time. At least the heavy loads went downhill and the empties up.

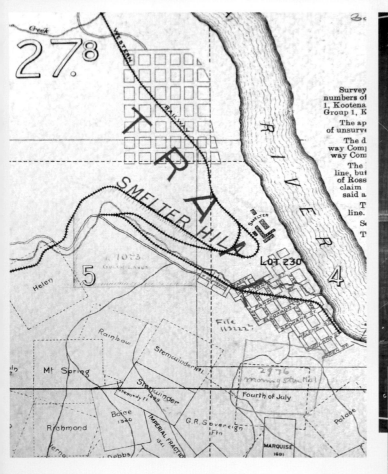

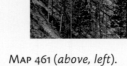

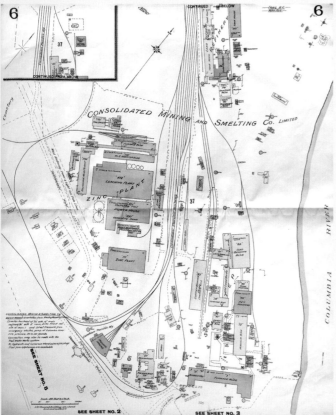

Map 461 (*above, left*).
The *smelter* and townsite of *Trail* plus the Columbia & Western rail lines, including a switchback, are shown on this 1897 map.

Map 462 (*above, right*).
Eugene Topping's plan for his townsite—Trail—surveyed in 1891 by Arthur Farwell (see page 126).

Map 463 (*left*).
This 1913 revision of an 1897 fire insurance plan shows the Trail smelter, purchased by Canadian Pacific from Augustus Heinze along with the Columbia & Western in 1898. The Consolidated Mining and Smelting Company (Cominco after 1966), once the railway's subsidiary company, is today part of Teck Resources.

Map 464 (*centre, right*).
A later bird's-eye map of the smelter site, also created for fire insurance purposes. This isometric view from the west dates from 1943.

Above, centre.
The Trail smelter on completion in 1897.

Below and *right, inset.*
A study in differences in environmental regulation, perhaps: these are photographs of the Trail smelter about 1910 and a hundred years later, in 2010.

MAP 465.

This is a very strange map because it was published by the Northern Pacific Railway in 1896 to show how its American system could serve the West Kootenay mining districts; the map even shows the *Nelson & Ft. Shepherd* as a Northern Pacific line. Yet the Northern Pacific never entered Canada. The line from Spokane to the international boundary, the Spokane Falls & Northern, was, like the Nelson & Fort Sheppard, a Daniel Corbin line. It seems that Corbin must have been intending to sell his lines to the Northern Pacific at the time, but they came under the control of James J. Hill's Great Northern two years later. The way the *Red Mtn Ry.* entered Rossland "by the back door," connecting with the American lines at *Northport,* is well illustrated. Also shown on this map is the *C. & W. Ry.* (Columbia & Western) from Rossland to Trail and the *C. & K. Ry.* (Columbia & Kootenay) from Robson to Nelson. As competing railways they were shown with thinner lines, a concession indeed, for railway maps often didn't bother to show their competitors' lines at all.

The inset map shows the American lines of the Northern Pacific tying in to the cross-border route. The line shown to *Sumas,* previously a Seattle, Lake Shore & Eastern Railroad line (see MAP 494, *page 205*), was controlled by the Northern Pacific from soon after its completion in 1891.

The technique used to depict the topography is a drawn representation of the mountain terrain shaded as with the light from the sun setting in the west. It gives an excellent view of this mountainous region and is a technique still in use today. However, whereas modern mapmakers have satellite photos and computers to aid them, in 1896 the mapmaker would have had to laboriously construct every feature, no doubt using imagination for the many parts for which information was not available in detail.

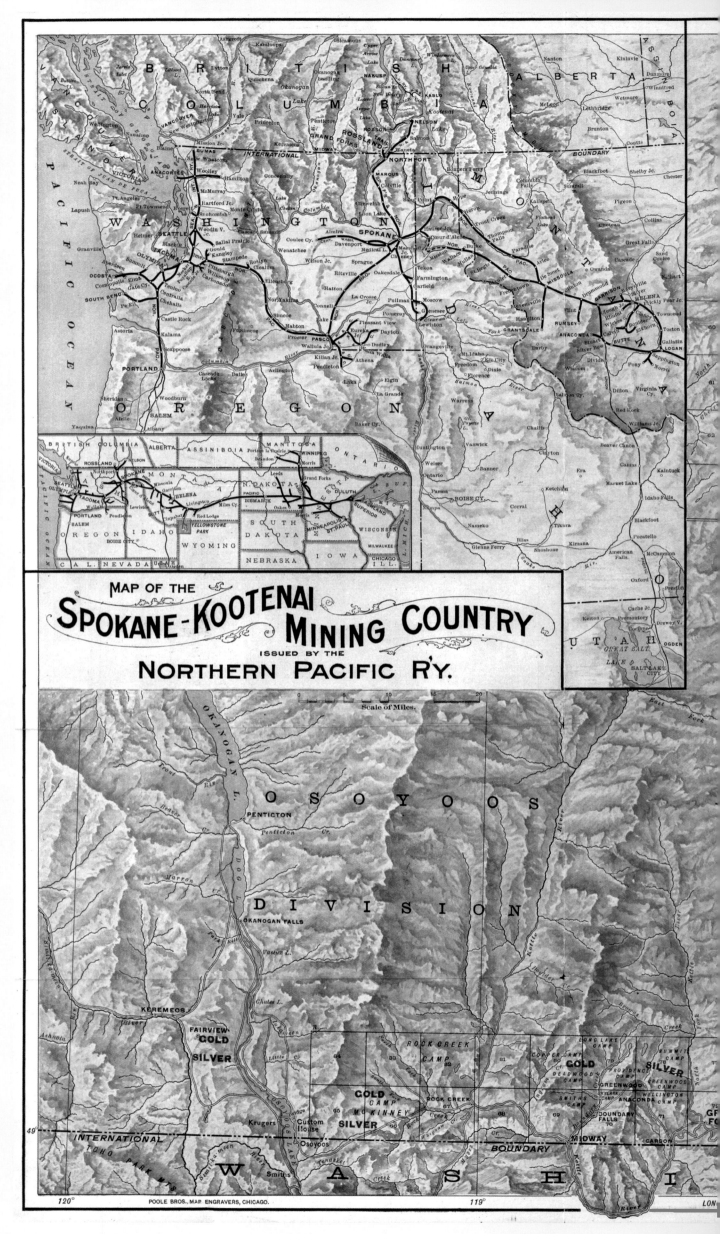

Map 466 (*above, top*).
The complexity of building a city, dozens of mines and processing plants, and two railways, both with passenger as well as freight facilities, is well illustrated by this map of Rossland and its mines, published in 1908 by the federal Department of Mines. The steep terrain and the position of waste dumps are depicted using contours. There are several switchbacks on the rail lines. In the northern part of Rossland, just east of the *Nickel Plate* Mine, are the *G.N.R. Station*, on the Red Mountain line, and the *C.P.R. Station*, on the Columbia & Western. The map is on canvas, so as to be durable. Unfortunately it lay folded in a drawer for many years, and this has accentuated the fold marks. The three photos on this page all are of points on this map:

Above, left. Black Bear Mine, with *Le Roi* behind it on the left. On the map this is the unnamed complex immediately north of the rail "Y." The photo was taken in 1897.

Above. In a 1910 photo a Canadian Pacific ore train leaves for Trail on the line immediately south of *Centre Star* Mine, which can be seen in the background.

Left. The view from Le Roi Mine, in the foreground, looking southwest over the city of Rossland in 1900.

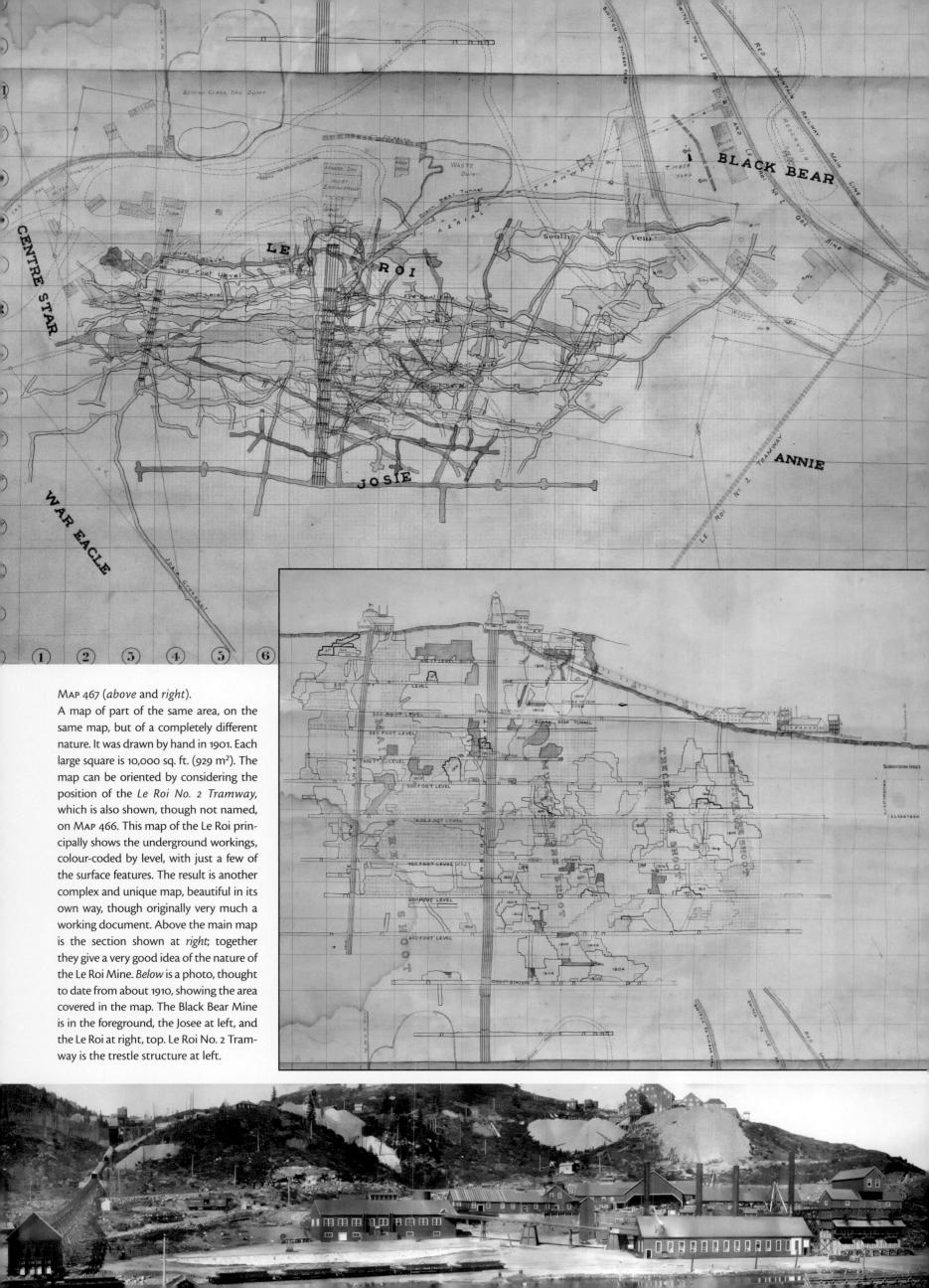

Map 467 (*above* and *right*).

A map of part of the same area, on the same map, but of a completely different nature. It was drawn by hand in 1901. Each large square is 10,000 sq. ft. (929 m²). The map can be oriented by considering the position of the *Le Roi No. 2 Tramway*, which is also shown, though not named, on **Map 466**. This map of the Le Roi principally shows the underground workings, colour-coded by level, with just a few of the surface features. The result is another complex and unique map, beautiful in its own way, though originally very much a working document. Above the main map is the section shown at *right*; together they give a very good idea of the nature of the Le Roi Mine. *Below* is a photo, thought to date from about 1910, showing the area covered in the map. The Black Bear Mine is in the foreground, the Josee at left, and the Le Roi at right, top. Le Roi No. 2 Tramway is the trestle structure at left.

KIMBERLEY AND THE SULLIVAN

The mine that became the largest lead and zinc producer in the world was discovered in 1892, but it was a later technological development that increased production dramatically. The Sullivan Mine in Kimberley was purchased by the Canadian Pacific's Consolidated Mining and Smelting Company in 1909, but it was not until 1917 that an efficient method was found to separate the lead from the zinc. From that time until closure in 2001, the mine produced an astonishing 149 million tons of ore. Led by the Sullivan, British Columbia was producing 10 per cent of the world's lead by the 1920s.

Significant finds of silver and lead were made in 1892 by a number of prospectors including Joe Bourgeois, one of the finders of the *Le Roi* mine in Rossland (see page 182); here he staked the North Star claim. The Sullivan was found by a group that included Pat Sullivan, and the mine was named for him after he was killed in a mining accident in Idaho.

From 1895 ore was hauled down to the Columbia in wagons and shipped to American smelters, and then in 1903 to a smelter built at Marysville, farther down Mark Creek, but in 1908 the smelter became unable to deal with the complex ore found here and was closed down. The Sullivan also was closed down while the company searched for a new method of processing the ore, which it found in 1917. Ore was then shipped to the smelter at Trail.

Above.
The Sullivan Mine had an extensive system of underground railways to move both men and materials. Here an electric locomotive pulls a trainload of ore from the mine portal in 1937.

MAP 468 (*left*).
The claims of the *Sullivan M. Co.* are the top group on this 1900 claim map. The *North Star* group, now a ski area, is at bottom centre.

MAP 469 (*below*).
A map of some of the underground workings at the Sullivan Mine, shown on a map drawn in 1944. The blue areas are pillars left to support the seam roof. In 1975 a form of mechanized mining was introduced that would allow even these pillars to be removed, since they also contained valuable ore.

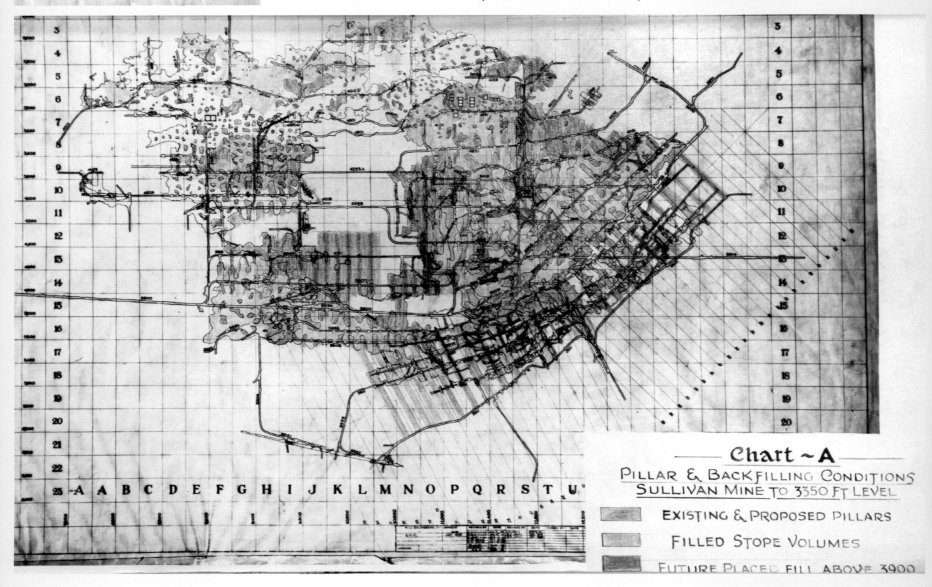

Chart ~ A
PILLAR & BACKFILLING CONDITIONS
SULLIVAN MINE TO 3350 FT LEVEL
EXISTING & PROPOSED PILLARS
FILLED STOPE VOLUMES
FUTURE PLACED FILL ABOVE 3900

SULLIVAN HILL

The
SULLIVAN MINE
KIMBERLEY BRITISH COLUMBIA

Although it is not known for sure, it is thought that Kimberley acquired its name from one Colonel William Ridpath, an American who owned some of the mining properties here in 1896 or 1897; he bestowed the name in the hope that his property would prove as rich as the diamond mines at Kimberley in South Africa. The settlement of Kimberley grew up at the base of the mountain and was first named Mark Creek Crossing, then Clark City.

FERNIE AND COAL

Michael Phillips of the Hudson's Bay Company had reported coal finds in Crowsnest Pass in 1860 and in the Elk Valley in 1873. In 1887 William Fernie, a prospector who had worked on building the Dewdney Trail and had been a government agent and mining recorder for the Kootenay District from 1876 to 1882, discovered coal seams on a tributary of Michel Creek near the Alberta boundary.

An excited Fernie, with his brother Peter, explored the vicinity further and found more coal elsewhere in the Elk Valley—enough to create extensive coalfields. And so, in 1897 Fernie and other investors established the Crow's Nest Coal and Mineral Company, which dug underground coal mines in one of the most promising areas, the valley of Coal Creek, a tributary of the Elk River. The townsite of Fernie was surveyed at the confluence of Coal Creek with the Elk River and named after its founder. Its streets were laid out parallel to the new railway.

In 1897 the Canadian Pacific had acquired the charter of the British Columbia Southern Railway, with a proposed line from Lethbridge, Alberta, to Kootenay Landing, at the southern end of Kootenay Lake (see MAP 472, *overleaf*), and signed the Crows Nest Pass agreement with the federal government. This later notorious agreement awarded the company $3.3 million in subsidies to build the line in return for a reduction in perpetuity in freight rates for grain and agricultural equipment for Prairie farmers, who had complained until that time about high rates to ship their grain to Vancouver. This so-called Crow Rate lasted until 1984.

Today the Elk Valley remains an important source of coal, but it is carried in unit trains to the coast and shipped from Roberts Bank Superport across the Pacific.

MAP 470 (*above*).
This unusual map is a sort of cutaway bird's-eye view of the *Sullivan OreBody*, shown with some surface features, on the slopes of *Sullivan Hill* at Kimberley. It gives an excellent comprehension of the mine workings. The map is thought to date from the 1930s. The open pit at the top was used to mine iron ore.

MAP 471 (*below*).
A map from about 1898 shows the proposed line of the *B.C. S. Ry.* and the coal claims in the Elk River valley but no settlements. Compare this map with the 1907 map *overleaf*.

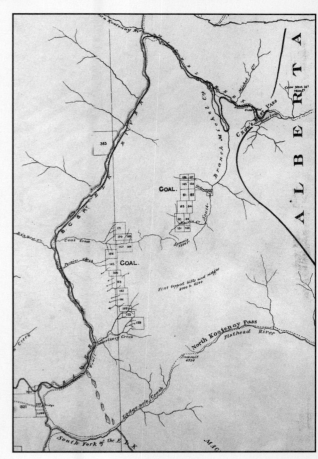

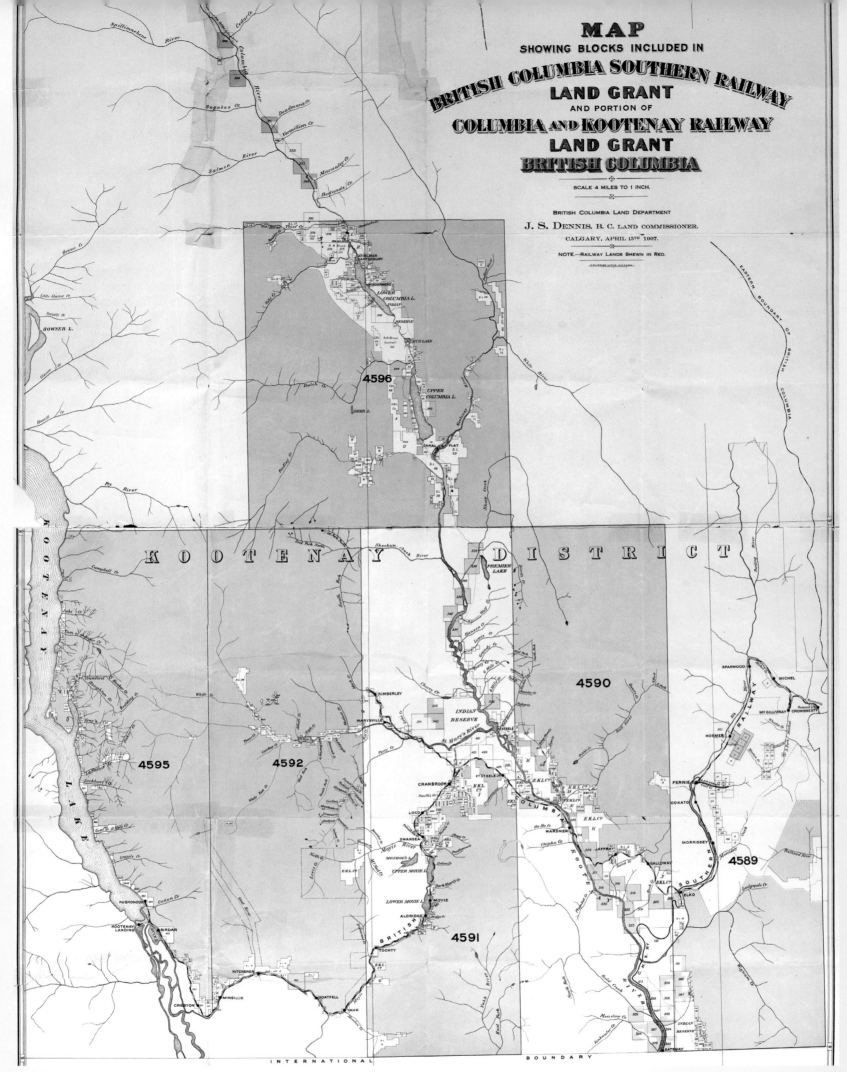

MAP
SHOWING BLOCKS INCLUDED IN
BRITISH COLUMBIA SOUTHERN RAILWAY
LAND GRANT
AND PORTION OF
COLUMBIA AND KOOTENAY RAILWAY
LAND GRANT
BRITISH COLUMBIA

SCALE 4 MILES TO 1 INCH.

BRITISH COLUMBIA LAND DEPARTMENT
J. S. DENNIS, B. C. LAND COMMISSIONER.
CALGARY, APRIL 15TH 1907.
NOTE.—RAILWAY LANDS SHEWN IN RED.

MAP 472 (*above*).

The land grant of the Canadian Pacific–controlled British Columbia Southern Railway on a 1907 map. The railway, completed in 1898, ran from the Alberta boundary at Crowsnest Pass to the southern end of Kootenay Lake, where service was continued by lake steamers. A branch line to Kimberley was added in 1899. The line allowed the Canadian Pacific to tap into the developing mineral wealth of the Kootenays and stem some of the flow southward to the United States. The land grant was awarded to the original 1888-incorporated company,

the Crow's Nest & Kootenay Lake Railway, following the provincial Railway Aid Act of 1890, but the company did not begin construction, because of a lack of investor interest following the recession of 1893. The British Columbia Southern was created in 1894 and inherited the land grant, and it was this company the Canadian Pacific acquired in 1897. Canadian Pacific thus acquired a federal subsidy of $11,000 per mile and a provincial grant of 20,000 acres per mile (5,029 ha per km). For more on Kootenay railways see page 212.

THE BOUNDARY AND PHOENIX

The search for mineral wealth was just as frenzied as the Kootenays—for a brief period—in the Boundary Country, a twenty-kilometre-wide strip of land between the Okanagan and the West Kootenay just north of the forty-ninth parallel.

In 1891 prospectors William McCormick and Richard Thompson, following reports of a large copper stain in the northern part of Boundary Creek, a tributary of the Kettle River, found and claimed a rich copper vein at what they called the Mother Lode. The find provoked a series of further claims all around the Mother Lode in the two years afterward (Map 475, *overleaf*). The Mother Lode was developed beginning in 1896 under new owners, who the following year formed the British Columbia Copper Company and hatched plans to build a smelter farther down the valley. Settlements sprang up on townsites laid out in 1899 at Deadwood, named after the ranch on whose land it was built. The same year the Canadian Pacific's Columbia & Western Railway arrived, and a spur was constructed to Deadwood and its mines. Ore could now easily be shipped to the smelter, located slightly south of Greenwood (which had been laid out as a townsite in 1895), at Anaconda. The smelter began production in 1910 and operated until 1912, when shortages of ore made it uneconomical. Today the huge smokestack towers over the area as a vivid reminder of boom days long ago.

Prospectors had also moved east in 1891, following the Mother Lode find, and staked numerous claims around the first two, the Knob Hill claim of Henry White and the Old Ironsides claim of his partner, Matthew Hotter. They are shown on Map 473, *below,* immediately south of Phoenix. The entire complex was first known as Greenwood

Above.
The remains of the copper-smelting industry of the Boundary are still very much in evidence in some locations. This is a 2010 photo of the slag heaps and smokestack of the British Columbia Copper Company at Anaconda. The stack is 36 m high and contains some quarter-million bricks.

Map 473 (below).

An overview of the mining camps and mineral claims of the Boundary, shown on a 1901 map. The darker shaded areas are townsites: *Midway, Boundary Falls, Deadwood, Greenwood, Anaconda,* and three at *Phoenix.* At Phoenix the first townsite was on the New York claim, but the city actually grew up on the Phoenix claim, just north of *Ironsides* and *Knobhill,* the original 1891 claims of Henry White and Matthew Hotter. Phoenix had incorporated as a city in 1900, the year before this map was drawn. Other townsites are at *Eholt,* the point at which the Canadian Pacific branch to Phoenix left the main line and which was a busy railway junction for a brief span of years (see photos, *page 196*), and *Summit Camp.* At lower right are *Grand Forks, Columbia,* and *Carson.* The *Granby Smelter* is located here. The *B.C. Copper Co. Smelter* is at Anaconda, just south of Greenwood, a mining community that stretched northward along the valley of Boundary Creek. The *Mother Lode* claim is shown at the end of the branch railway line to Deadwood.

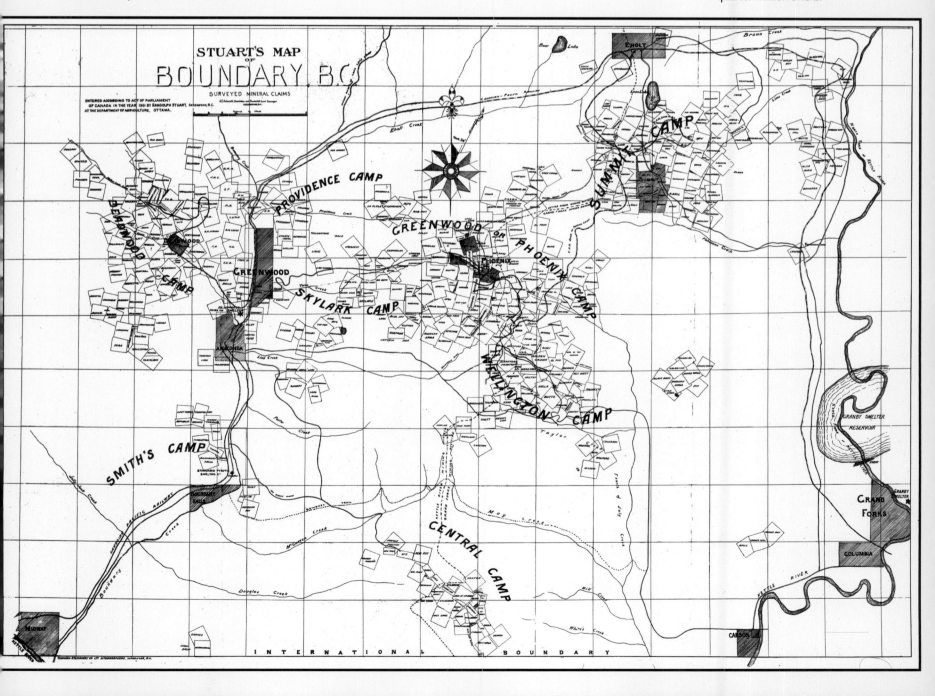

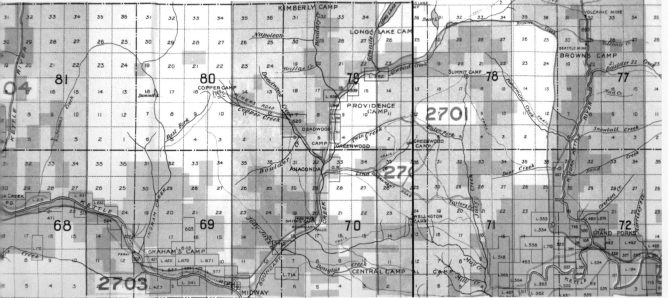

Camp and is shown on MAP 473 with both names. This is confusing because the name Greenwood was later given to a townsite developed to the west—one deliberately selected in 1895 as being a centre for many mining operations.

In 1896 a syndicate of American and Québec investors began developing the Old Ironsides and adjacent Knob Hill claims at what became—through a name change when a post office opened in 1898—Phoenix. By 1899 buildings were springing up everywhere, despite the fact that there was no official townsite. Three separate claims housed parts of what became the quite substantial town, with over a thousand inhabitants at its peak. By January 1900 Phoenix had both telephone service and electricity and, in October that year, incorporated as a city.

That year the syndicate—now the Granby Consolidated Mining, Smelting and Power Company—opened a large smelter at Grand Forks, capable of handling 700 tons of ore a day and proudly touted as the largest non-ferrous smelter in the British Empire—indeed, the second-largest in the world. It needed to be large, for the ore was of a low grade, but it had the property of being self-fluxing and thus needed only coke to be added, which improved the smelter's efficiency. The company operated the Phoenix mine, and all the ore was shipped to the Granby smelter via a Canadian Pacific Railway line that was completed to Phoenix in May 1900. The Great Northern followed in 1904 with a line, part of its Vancouver, Victoria & Eastern (see MAP 521, *page 217*) that threatened the Canadian Pacific's pre-eminence because it had lower grades.

By 1903 the company had begun open-pit mining, which

MAP 474 (*above*).
The Boundary on an 1897 map, from *Rock Creek*, site of an 1859 gold rush, to *Grand Forks*, surveyed two years before. The map shows a number of the first mining camps. The *Government Road* is the Dewdney Trail. *Greenwood Camp* is the original name for Phoenix. The dull orange–coloured areas are the Columbia & Western land grant, with the orange lines being the grant boundary; this is shown in much greater detail than on the maps showing the entire grant (such as MAP 481, *page 198*). The yellow areas are pre-empted and purchased land.

MAP 475 (*above*).
At the end of the Deadwood branch line was the Mother Lode Mine, on a rich copper vein discovered by William McCormick and Richard Thompson in 1891, touching off a frenzy of prospecting over the entire region. This map was published in 1911 by the federal Geological Survey. The photo, taken about 1910, shows a panoramic view of the Mother Lode Mine. In 1913 an unconventional mining method was tried here. A huge explosion triggered from nearly five thousand 4-metre-deep holes packed with TNT reduced half a million tons of the hillside to rubble. The idea was that it would then be easier to extract the copper ore than it was from the bedrock. In fact the reverse was true; the ore was too dispersed in the rubble. The mine closed in 1918, though it was reopened for a brief period since. The mine produced 78 million pounds (35 million kg) of copper during its lifetime, along with 174,000 ounces of gold and 700,000 ounces of silver.

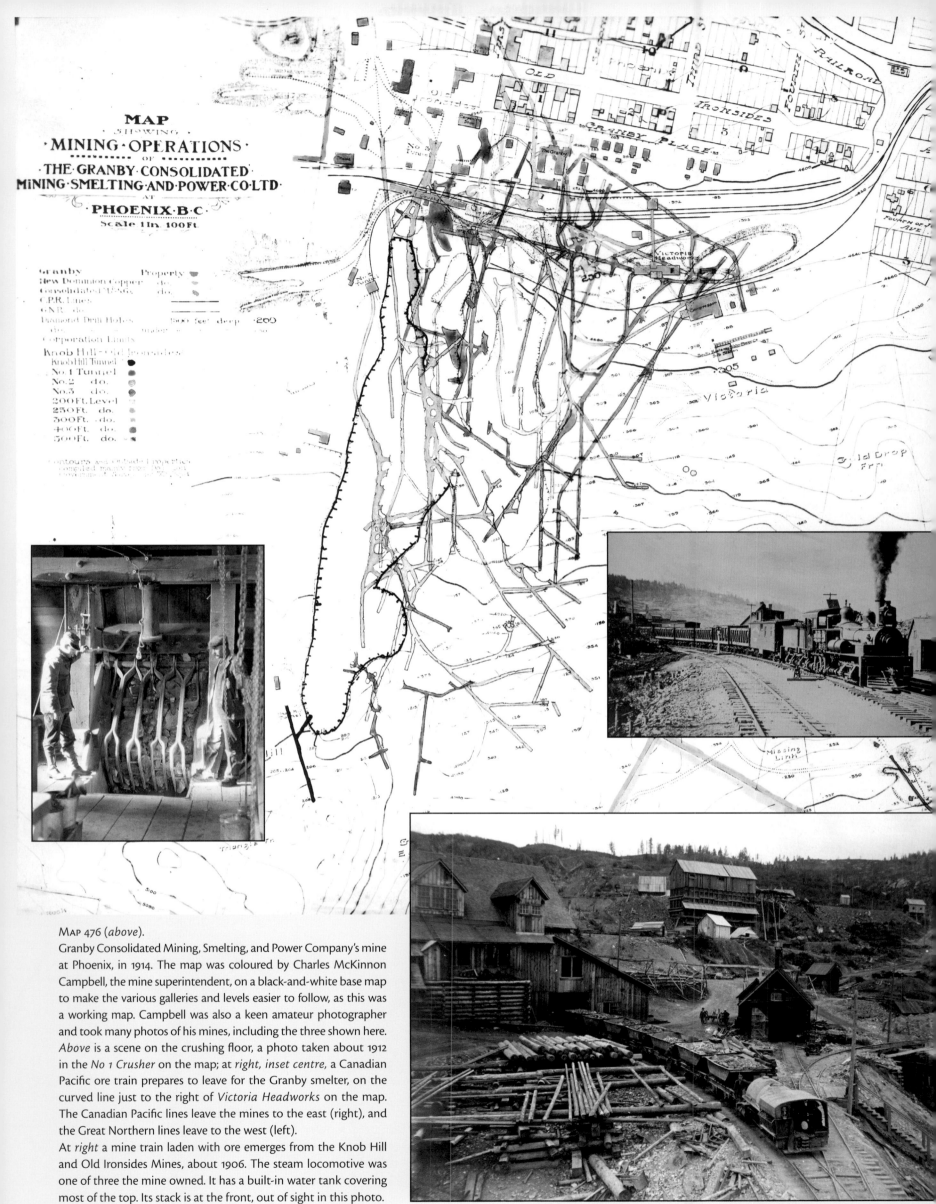

MAP
SHOWING
• MINING • OPERATIONS •
OF
• THE • GRANBY • CONSOLIDATED •
MINING • SMELTING • AND • POWER • CO • LTD
AT
• PHOENIX • B • C •
Scale 1 In. 100 Ft.

Granby Property
New Dominion Copper do.
Consolidated do. do.
C.P.R. Lines
G.N.R. do.
Diamond Drill Holes 500 feet deep 200
do. under
Corporation Limits

Knob Hill-Old Ironsides
Knob Hill Tunnel
No.1 Tunnel
No.2 do.
No.3 do.
200 Ft. Level
250 Ft. do.
300 Ft. do.
400 Ft. do.
500 Ft. do.

MAP 476 (*above*).

Granby Consolidated Mining, Smelting, and Power Company's mine at Phoenix, in 1914. The map was coloured by Charles McKinnon Campbell, the mine superintendent, on a black-and-white base map to make the various galleries and levels easier to follow, as this was a working map. Campbell was also a keen amateur photographer and took many photos of his mines, including the three shown here. *Above* is a scene on the crushing floor, a photo taken about 1912 in the *No 1 Crusher* on the map; at *right, inset centre,* a Canadian Pacific ore train prepares to leave for the Granby smelter, on the curved line just to the right of *Victoria Headworks* on the map. The Canadian Pacific lines leave the mines to the east (right), and the Great Northern lines leave to the west (left).

At *right* a mine train laden with ore emerges from the Knob Hill and Old Ironsides Mines, about 1906. The steam locomotive was one of three the mine owned. It has a built-in water tank covering most of the top. Its stack is at the front, out of sight in this photo.

MAP
OF
GREENWOOD and
WELLINGTON CAMPS
INCLUDING PORTIONS OF KETTLE RIVER AND
GRAND FORKS MINING DIVISIONS OF YALE
AND FORMING PART
of the BOUNDARY CREEK DISTRICT
Compiled and Published by Sydney M. Johnson B.A.Sc. P.L.S.
Greenwood, B.C.
SMITH & McRAE, GREENWOOD, B.C.
Selling Agents

accounted for about half the mine's production. This was one of the earliest uses of this mining method in the province. The mines closed in 1919 following a drop in production at the end of World War I, after having produced, by some estimates, over $100 million worth of copper ore. In 1959 the mine lived up to its name and opened again. This time open-pit mining totally erased all vestiges of the city of Phoenix from the landscape. This second-generation mining effort ended in 1976, and the area is now used as a ski hill.

MAP 477 (left).
A detailed map of the Phoenix area in 1898 showing the Canadian Pacific lines then *as projected*. The railway reached Phoenix in May 1900. The line at top right leads from the junction at Eholt.

MAP 478 (below).
The Great Northern arrived in Phoenix in 1904; the line is shown on this map, which dates from 1914, although it is not distinguished here from the Canadian Pacific lines. The Great Northern line entered Canada at *Laurier* and then moved north near *Grand Forks* to *Weston*. It then ran west and north, then circled *Phoenix* to the north before entering the city from the west. This circuitous route enabled the line to be built with better grades than its rival. The main line re-entered the United States before again crossing the border near *Midway*, on its way to Princeton (see MAP 523, *page 217*).

Below. This is the Canadian Pacific junction from the main line to Phoenix at Eholt, also shown on MAP 478, *below.* This photo, taken about 1905, shows an ore train coming off the Phoenix branch en route to the Granby smelter. *Below, right.* This is Eholt today, with the rubble of the depot office and station building shown in the middle of the 1905 photo. The view here looks up what was the Phoenix branch.

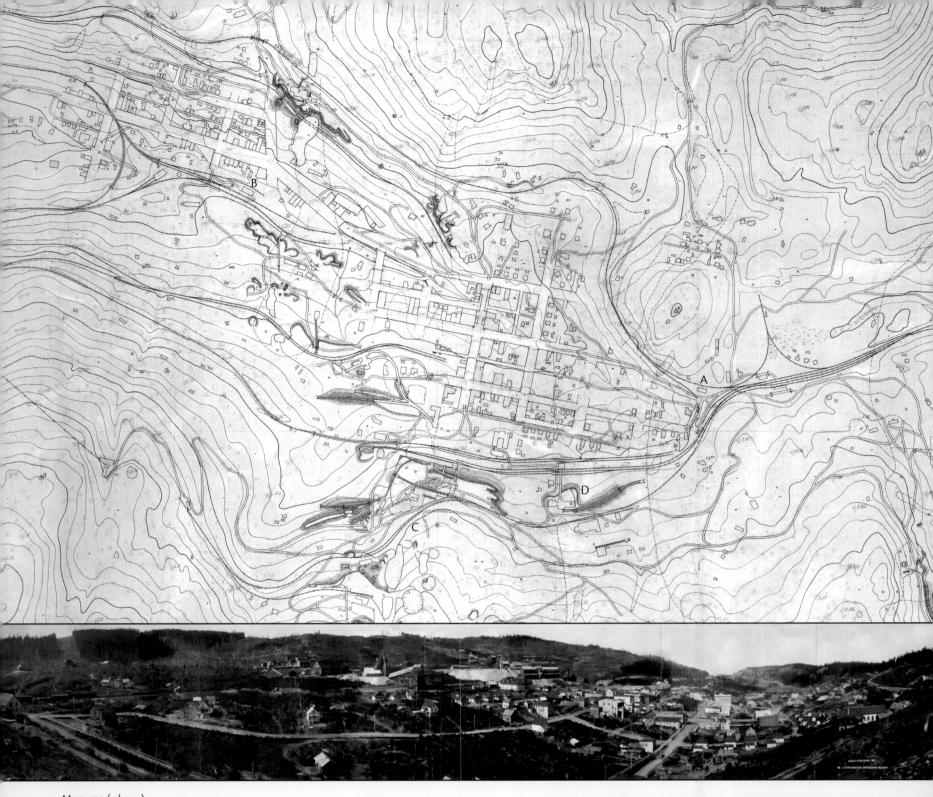

MAP 479 (*above*).

This intriguing and detailed map was found in Library and Archives Canada. It was drawn on what turned out to be cardboard, despite the fact that it was rolled up. Over time the cardboard had hardened, and conservators were required to unroll it so that it could be reproduced. It shows Phoenix in 1912. This map was a manuscript from which a printed federal Geological Survey map was made. Canadian Pacific rail lines are those entering from the east. The passenger station is at *A*. Great Northern lines enter from the west, with a station at *B*. Rail wyes were for turning trains. Old Ironsides, the first claim, is at *C*. The Victoria Shaft of the Granby Mine is at *D*. The panoramic photo depicts Phoenix in 1907. Taken from the hill above the CPR station, the view is to the southwest; the CPR station (*A* on the map) is at far left and the Victoria Shaft (*D*) at centre.

MAP 480 (*right*).

An 1896 survey of Greenwood. This was the first part of the city, at its southern end. In 1895, Robert Wood, a merchant, visited Boundary mining camps and became convinced that this location would be an ideal place to establish a service centre for the region. He purchased land and laid out a townsite. Greenwood still exists and boasts a magnificent collection of historic buildings. The photo shows Greenwood about 1900. The adjacent Anaconda smelter is in the background.

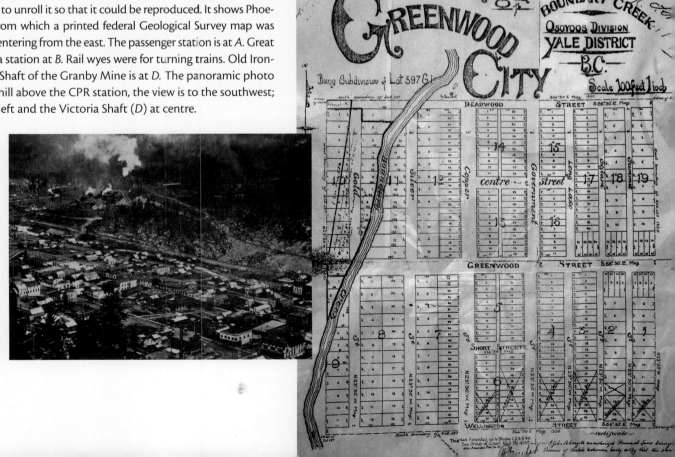

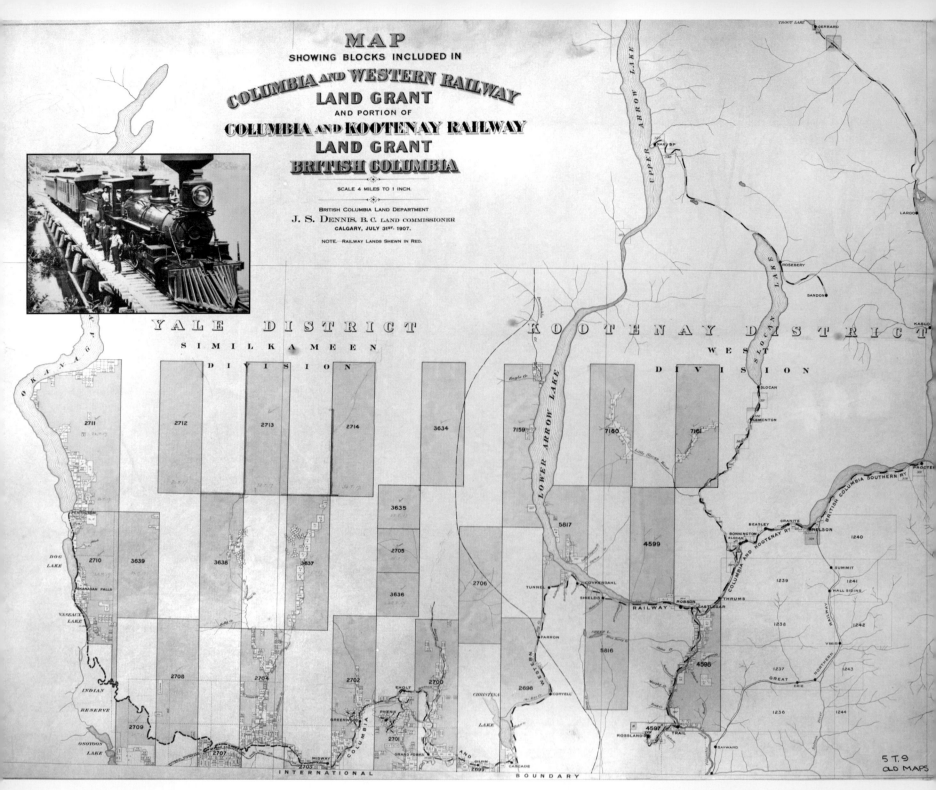

MAP

SHOWING BLOCKS INCLUDED IN

COLUMBIA AND WESTERN RAILWAY

LAND GRANT

AND PORTION OF

COLUMBIA AND KOOTENAY RAILWAY

LAND GRANT

BRITISH COLUMBIA

SCALE 4 MILES TO 1 INCH.

BRITISH COLUMBIA LAND DEPARTMENT

J. S. DENNIS, B. C. LAND COMMISSIONER
CALGARY, JULY 31ST. 1907.

NOTE.—RAILWAY LANDS SHEWN IN RED.

MAP 481 (*above*).

The Columbia & Western Railway was granted 1.35 million acres (546,000 ha), and its charter allowed it to build from Robson, at the southern end of Lower Arrow Lake, to Penticton, at the southern end of Okanagan Lake. The Columbia & Kootenay received a grant of 188,593 acres (76,323 ha). This 1907 map shows the land allocated to the railways, though not in every detail, as MAP 474, *page 194*, makes clear. The C&W line is shown completed to Grand Forks, which it reached in 1899. The

proposed route west of Midway was not built; the railway—this part the Canadian Pacific's Kettle Valley Railway—used an entirely different routing to reach Penticton (see page 216).

Inset is a Columbia & Kootenay train on the Kootenay bridge at Taghum, near Nelson, about 1898. This map shows the Slocan Valley branch, completed in 1897.

MAP 482 (*right*).

A representation of all the land grants of the railways acquired by the Canadian Pacific is shown on this summary map produced by the railway's land department in 1902.

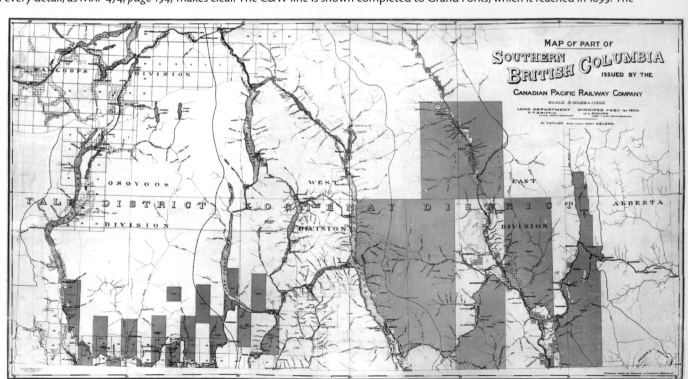

Fairview

Farther west, above the Okanagan Valley near what is today Oliver, is the short-lived gold-mining town of Fairview. Gold was discovered here in 1887, and the find was sold to a British and American syndicate in 1891. This attracted others, and before long a vast array of claims had been staked and two adjacent townsites laid out (one by surveyor Frank Latimer for local rancher Tom Ellis—see page 153). The town boasted one of the finest hotels in the province: the Fairview Hotel, otherwise known as the "Big Teepee" because of its shape. Built in 1899 in the middle of what was to be a burgeoning commercial centre, it lasted only three years before burning down—and with it went Fairview. The gold did not last, and neither did the town.

This was a familiar pattern for flash-in-the-pan mining towns. Full of the promise of easy profit they grew, prospered—and imploded.

Map 483 (right, top).
This map, thought to date from about 1905 when Fairview was already in decline, shows the various claims and the *Fairview Townsite* to their immediate south. The photo is of the townsites about 1900; the "Big Teepee" Fairview Hotel is at left—surrounded by lots of space but not many buildings—waiting for a commercial future that never arrived. The original claims at Fairview were staked at the *Stemwinder*, here grouped with the syndicate's holding, the *Fairview Amalgamated*.

Map 484 (below).
Another briefly prosperous mining town was the nearby community of *Olalla*, on *Keremeos Creek*, which flows into the Similkameen River. Here there was apparently so much copper that plans for a smelter were afoot, when miners discovered that the deposits, rich on the surface, did not go very deep. By 1905 the town was in decline. The adjacent settlement of *Keremeos Town Site* (called Keremeos Centre) simply moved south in 1906 into the Similkameen Valley (where it is today), into the path of the approaching Great Northern Railway, which arrived that year. The *C & W Ry.* shown on this map was a line west from Penticton surveyed by the Columbia & Western in case Canadian Pacific would not allow it to connect with its steamers on Okanagan Lake. The line was chartered only as "branches to mines" and was never built. The route was considered later by the Kettle Valley Railway (see page 216). This map is dated 1900.

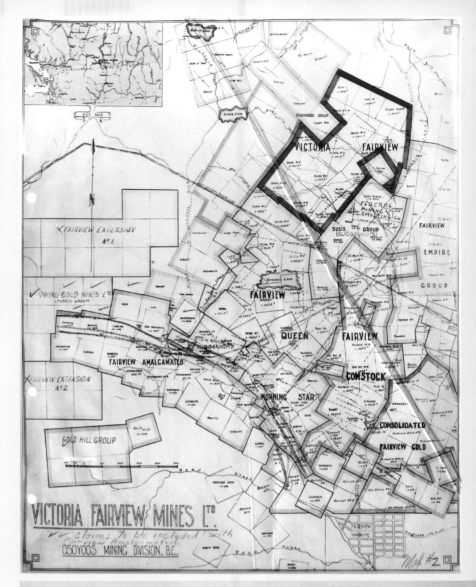

Map 485 (below).
Fairview's location above the Okanagan Valley is shown on this 1897 map.

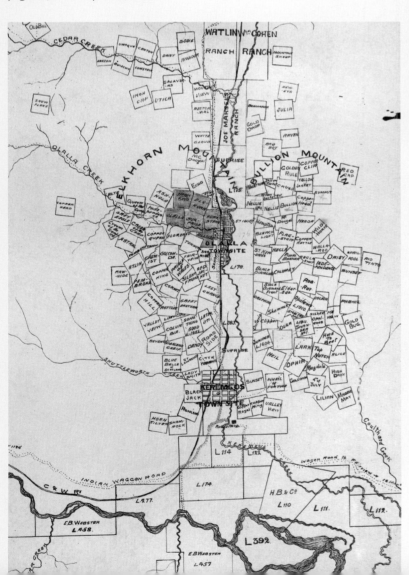

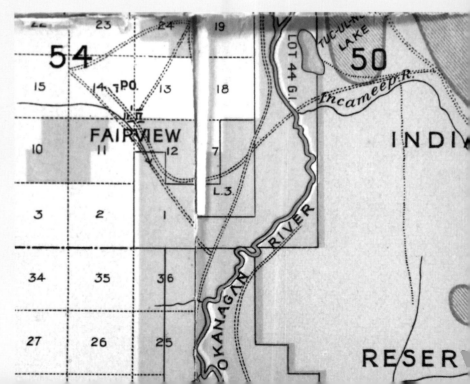

Hedley and the Nickel Plate

The vividly coloured cliffs above Hedley had been visible to travellers on the Dewdney Trail along the Similkameen River far below. Prospectors eventually investigated, and in 1897 a small group staked claims on the mountain, for here was a rich gold deposit. One was Peter Scott, who had been grubstaked by Robert Hedley, then manager of the Hall smelter in Nelson, and his name was bestowed on the little settlement that grew up here about 1899. Claims covered the mountain when a latecomer, George Cahill, staked a small wedge of steep mountainside in 1899 that had been left aside; this became the Mascot Mine, considered one of the richest fractions in British Columbia mining history.

The mines were more than a kilometre above the valley below, and getting the ore out was an issue. To solve this problem, an ingenious railway system was devised (Map 487, *right*). Electric locomotives pulled ore trucks to a bin at the edge of a steep descent, and the ore was then emptied into gravity-operated trains that descended the mountainside to a mill on the valley floor; the latter was controlled by cables and simultaneously pulled the empties back up the mountain on a parallel track. The mine was one of the longest lasting in the province, operating from 1902 to 1955, and then again as an open-pit mine between 1987 and 1996.

Above.
A "mucking machine"—really an underground miniature bulldozer attached to ore trucks—removed ore that had been reduced to rubble with explosives. This one is on display at the Hedley Mining Museum.

Map 486 (*below*).
Hedley is not yet shown on this 1900 map of the mining claims around the *Nichoe (sic, Nickle) Plate* (here coloured blue), though they are collectively labelled *Camp Hedley*. The townsite of *Similkameen* was just that—a townsite that never became a town. The settlement of Hedley grew up at the *Smelter Site* at *20 Mile Creek*, where it is crossed by the railway line—not the initially surveyed line of the *C & W Ry*. (Canadian Pacific's Columbia & Western Railway), as shown here, but the Vancouver, Victoria & Eastern of the Great Northern, which arrived in 1909, having taken two years to cover the 30 km from Keremeos to the east. The fractional Mascot claim—not yet on this map, was located between the *Copper Cleft* and *Iron Duke* claims. This map gives no clue about the difficult mountain topography necessitating the gravity tramway.

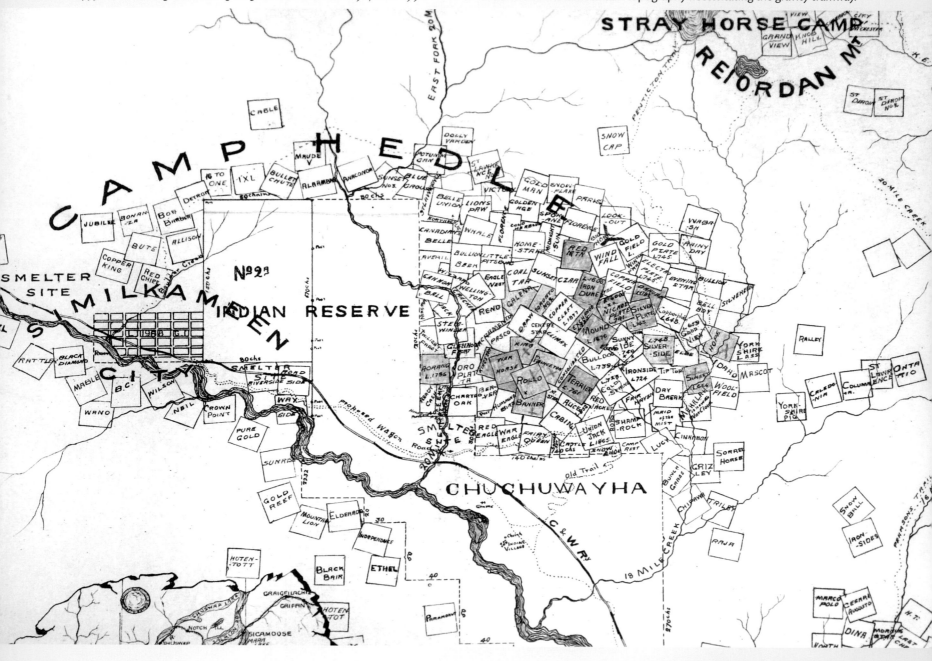

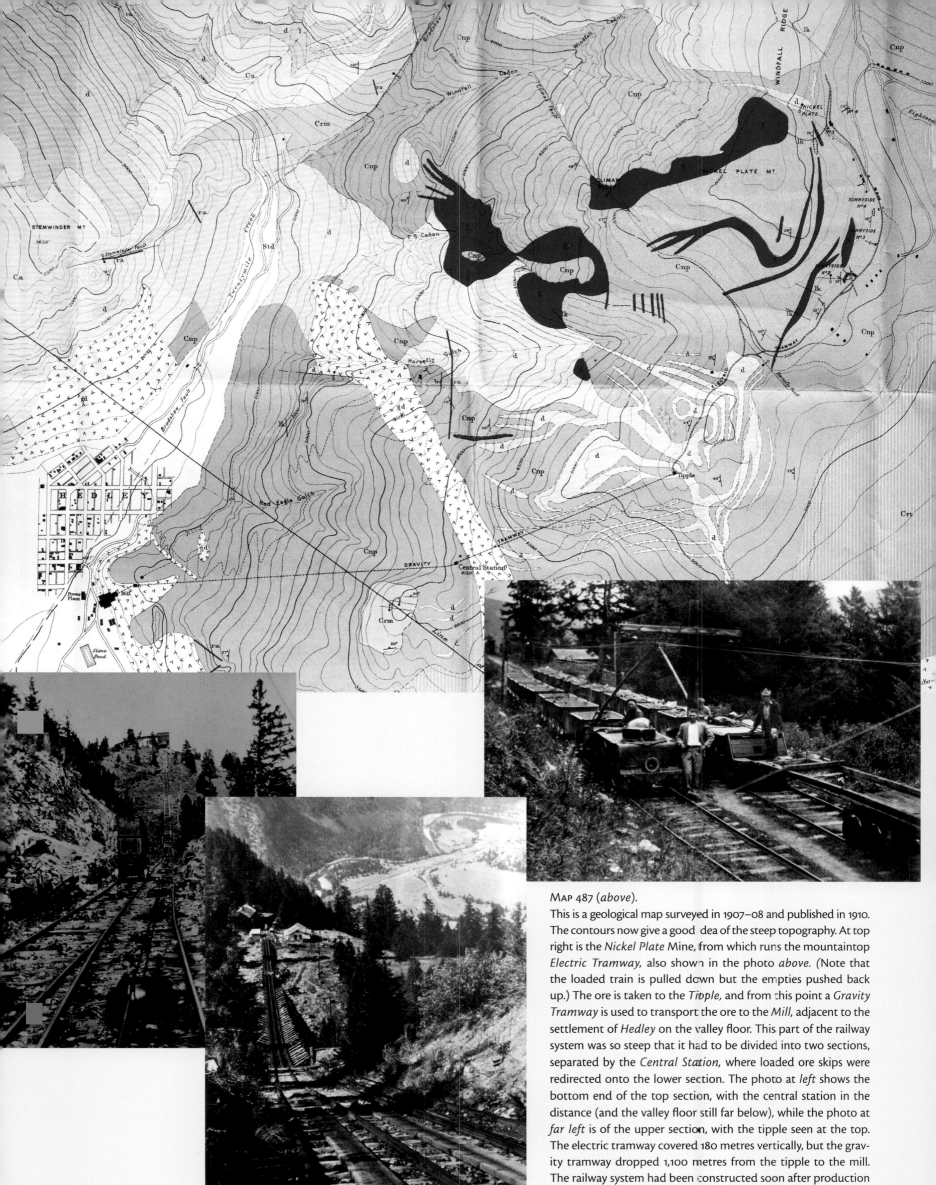

Map 487 (*above*).

This is a geological map surveyed in 1907–08 and published in 1910. The contours now give a good dea of the steep topography. At top right is the *Nickel Plate* Mine, from which runs the mountaintop *Electric Tramway*, also shown in the photo *above*. (Note that the loaded train is pulled down but the empties pushed back up.) The ore is taken to the *Tipple*, and from this point a *Gravity Tramway* is used to transport the ore to the *Mill*, adjacent to the settlement of *Hedley* on the valley floor. This part of the railway system was so steep that it had to be divided into two sections, separated by the *Central Station*, where loaded ore skips were redirected onto the lower section. The photo at *left* shows the bottom end of the top section, with the central station in the distance (and the valley floor still far below), while the photo at *far left* is of the upper section, with the tipple seen at the top. The electric tramway covered 180 metres vertically, but the gravity tramway dropped 1,100 metres from the tipple to the mill. The railway system had been constructed soon after production began, though these photos were taken sometime in the 1930s.

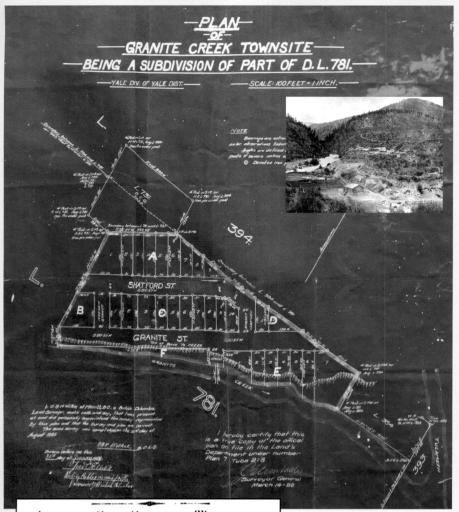

MAP 488 (*above*).
A 1921 copy of the original 1885 townsite plan for Granite City on *Granite Creek*. The photo, probably taken about 1895, looks up Granite Creek; Granite City—the area on the map—is on the bench above the creek.

ANOTHER GOLD STRIKE.—The new surface diggings recently discovered in the Semilkameen country are creating much excitement in that locality, and about twenty new claims have been located, and most of the miners are reported to be taking out from eight to ten dollars a day. The creek on which the new discoveries are made is a tributary of the Tulameen and has been named Granite creek.

Above. This is how the Victoria *Daily Colonist* announced the gold find at Granite Creek in July 1885—*another* gold strike.

A CHANCE FIND

John Chance, it seems, was the lazy member of a group of prospectors searching for gold just west of Princeton in July 1885. But every field party needs a cook, so this is the job Chance was allocated. One day, wandering about trying to shoot a grouse for dinner that night, Chance was taking a rest with his feet in a cool stream when he found, right in front of him, nuggets of gold.

Named Granite Creek, this tributary of the Tulameen River rapidly attracted gold seekers from far and wide. A settlement of tents appeared on the banks of the creek, and Chance and his associates filed a townsite plan so as to be able to sell lots (MAP 488, *above*). By the fall of 1885 some 4 km of the creek had been claimed and staked, and reports suggest the creek was yielding $50 to $75 a day per man. More substantial buildings replaced the tent city, hotels were built, and the population soared to about a thousand, many of whom had just left the construction of the Canadian Pacific; half were Chinese.

At its peak Granite City contained thirteen hotels and nine grocery stores and was reputedly the third-largest city in the province; only Victoria and New Westminster were larger. Of course, this situation did not last very long. The gold became harder and harder to find. In 1907 there was a fire that destroyed much of the city, and in 1912 the creek was irreversibly destroyed by mechanical dredges trying to extract the last ounce of gold.

BRITANNIA

Just a short distance north of Vancouver lies Britannia Beach, site of a massive copper concentrator (No. 3) completed in 1923. Copper was discovered on Britannia Mountain by a First Nations doctor and amateur prospector, A.A. Forbes, in 1888. The claims had great potential, and in 1904 the Britannia Mining and Smelting Company took over, building a mine. A year later the mine was in full production.

Britannia Beach was built as a company town around the mine to service surface operations, and another community was built higher up to service the underground operations. The latter, called Jane Camp, was obliterated in 1915 by an avalanche that resulted in the deaths of nearly sixty men, women, and children. As a result a new townsite, Mt. Sheer, was built lower on the mountainside. In 1921 much of the original community of Britannia Beach was wiped out when a company-built dam broke above it; this would happen once again in 1991.

Nonetheless, mining continued, and by the late 1920s the mine's claim to fame was that it was the largest producer of copper in the British Empire and one of the largest in the world. The extensive claims on which the mine was based are shown very well on MAP 489, *right, top*.

Low copper prices led to hard times for the mine after World War II, and in 1959 it became bankrupt. The Anaconda Mining Company purchased the mine in 1963 and operated it for another eleven years, but high operating costs forced its final closure in 1974. The British Columbia Mining Museum opened at the mine in 1975, becoming the Britannia Mine Museum after extensive renovations in 2010. The landmark No. 3 concentrator is a part of the museum.

FIRE MOUNTAIN AND TIPELLA

For each successful mineral discovery and mining venture in British Columbia there were dozens that failed, and many more that began in a blaze of glory but soon petered out, either because of lack of capital or because the deposit did not live up to expectations.

Somewhat into the latter category falls an interesting mining enterprise that began operating in 1898. The Fire Mountain Gold Mining Company was so confident of its wealth that it named its first claim the Moneyspinner. Although visible from the Douglas Road blazed at the time of the Royal Engineers (see MAP 198, *page 72*), a prominent quartz ridge on the side of Fire Mountain 25 km northwest of the northern end of Harrison Lake seems not to have been investigated until 1897, when a find of gold triggered a minor rush, resulting in many more claims being staked (MAP 490, *right, bottom*). The company laid out Tipella, a townsite on the shore of Harrison Lake, which it expected not only to service the mine but become a commercial centre for other mines in the region as well. A pack trail was cut to connect Tipella with Fire Mountain.

But the early promise quickly turned to disappointment when the gold from the mine was found to be difficult to extract because the ore was exceptionally hard. At this very time there arrived news of the gold strike in the Klondike, and many of the miners left for the richer fields north. The mine owners persevered, however, and attracted enough fresh capital to push the mine shaft 150 m into the mountainside, install a large processing mill on the lakeshore, and link it to the mine, 200 m up the mountain, with an aerial tramway. Tipella fell by the wayside, but the mine continued into the 1930s. Moneyspinner and its companions may not have produced fortunes, but they did repay some of their investors.

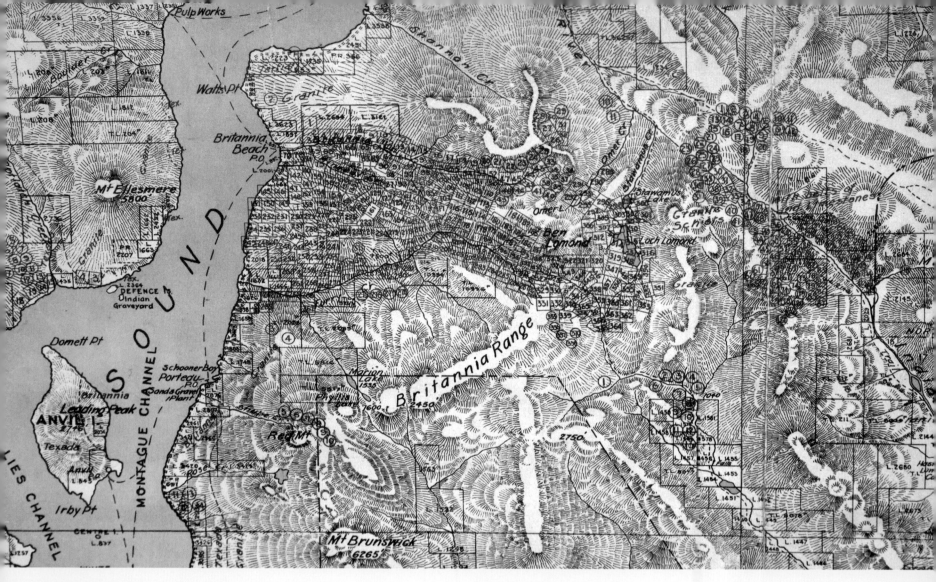

Map 489 (above).

A vast array of claims stretches miles behind *Britannia Beach* on this carefully drawn and rather beautiful 1916 map. Hachures are used to depict the topography. The dashed line shows steamer routes in Howe Sound. Until the Pacific Great Eastern Railway extension from Squamish to North Vancouver was opened in 1956, and the road two years later, Britannia could only be accessed by boat.

Map 490 (below).

This 1897 map of mineral claims on *Fire Mountain* also contains an inset map showing the location of the company's townsite, *Tipella,* at the northern end of *Harrison Lake.* The *Moneyspinner* claim is right in the middle of the cluster of other claims.

MORE RAILWAYS FOR VANCOUVER

The merchants of New Westminster had been even more horrified than those of Port Moody when they learned that the Canadian Pacific intended to extend its transcontinental line to Coal Harbour. New Westminster's city council passed a Railway Bonus Bylaw in November 1885, which offered the railway money to build a branch to their city. The railway accepted, and a branch was completed in August 1886. It is interesting to note that New Westminster thus acquired a railway before its rival Vancouver.

At Westminster Junction, where the branch left the main line, a community grew up and in July 1891 was incorporated as Coquitlam. Twenty years later the railway would decide to move its freight yards to Westminster Junction, setting off a real estate–buying frenzy (see page 249).

Canadian Pacific was extremely protective of its Vancouver monopoly and would do its best to keep rival lines out, a strategy made easier by the vast land grant it had been given. But naturally enough consideration was given to other possible routes within and around the city. MAP 491, *below,* shows some of these routes.

Canadian Pacific had had its eyes on Seattle because that city was not directly served by the first American northwest transcontinental, the Northern Pacific, which had instead terminated in Tacoma. The company would get a connection but not control. Plans were made to connect with the Seattle, Lake Shore & Eastern Railroad, which built north from MAP 491 (*below*).

This interesting map, hand-drawn in 1888 by the Canadian Pacific Railway Engineering Office, was enclosed in a letter sent by Henry Cambie, the railway's chief engineer for the Pacific Division. The proposed route of a competing line, the New Westminster Southern, is shown in brown-black ink and the CPR main line in black; the latter includes the New Westminster Branch. Unfortunately the key is missing, but the red, blue, and green dashed lines appear to be routes considered for branches and extensions. The New Westminster Branch is extended along the North Arm of the Fraser and then to English Bay partly on a route that roughly follows the one the railway later built along the Arbutus Street route and used for the Steveston interurban (see page 270). At this time Canadian Pacific was actively considering a route to join with an American line and found it three years later via Huntingdon (see next page).

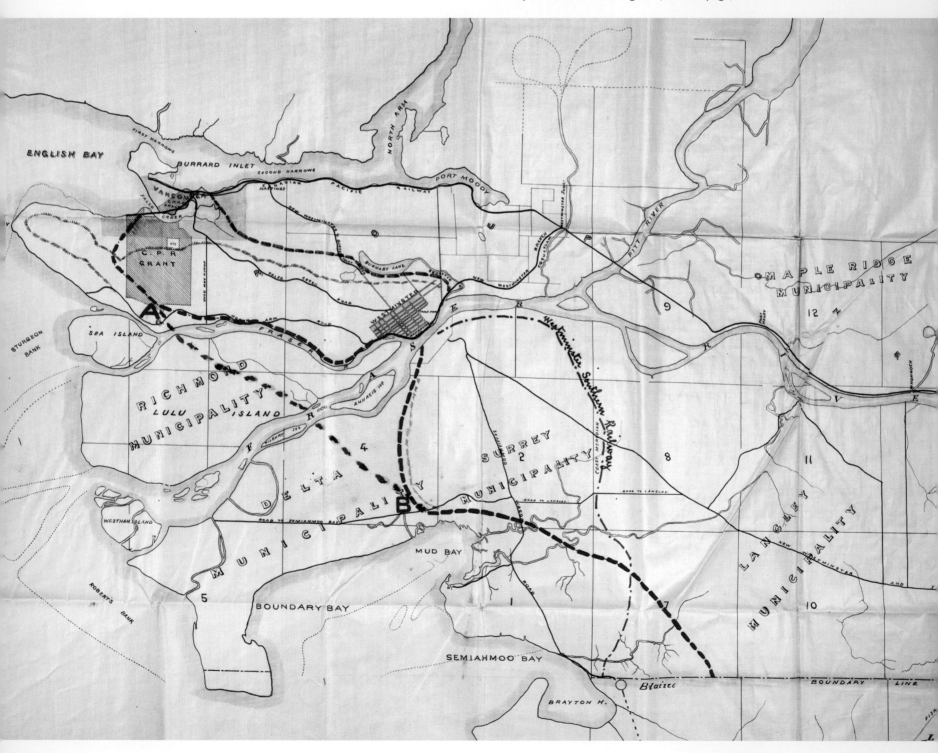

MAP 492 (*below*).
Canadian Pacific's *Mission* Junction and the Mission Bridge are shown on this detail from a 1905 map. The bridge opened on 7 February 1891.

MAP 493 (*right*).
This map, which shows the *Canadian Pacific Railway* main line and the *Esquimalt and Nanaimo Railway* completed but everything else as projected, appears to be some sort of planning map, showing how the Canadian lines might tie into the American network. The map was likely drawn about 1888. Some of the lines, such as the one through *Sumas* and the *New Westminster and Southern,* were built along the approximate route shown, while others, such as the *Blaine Branch,* were not, although the latter followed the route around the South Surrey peninsula later taken by the Great Northern.

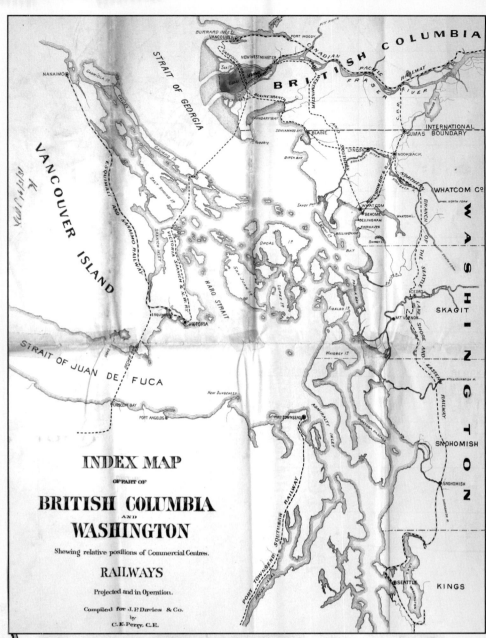

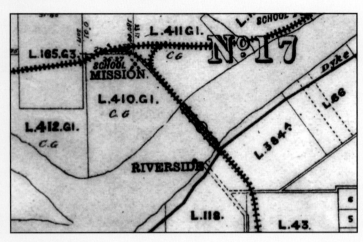

MAP 494 (*below*).
An 1888 *Seattle, Lake Shore* & Eastern Railroad map showing how its *North'n Br.* was to tie into that of the *Can. Pac.* south from Mission.

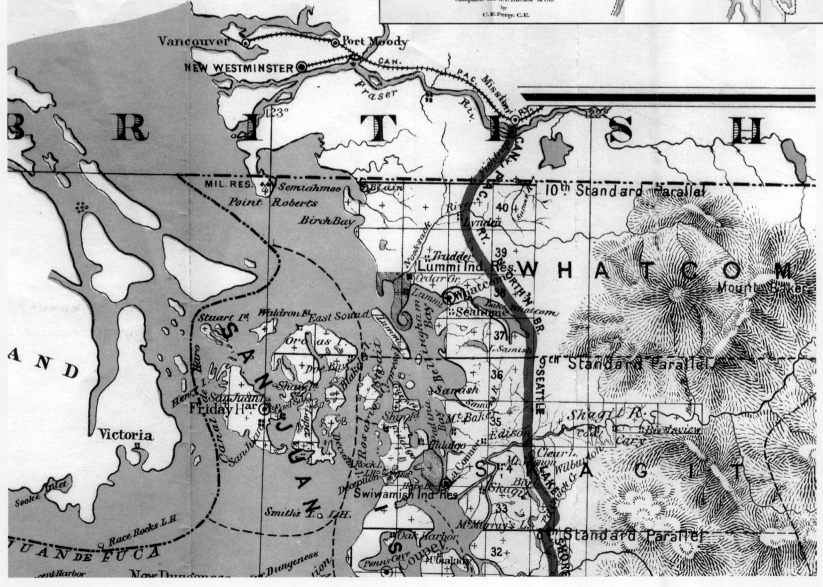

MAP 495 (*right*).

Oblate missionaries had established a St. Mary's Mission on the north shore of the Lower Fraser in 1861. The transcontinental line passed near the mission in 1885, and in 1891 it became the junction point for the first connecting line to the United States over the first bridge. The opportunity was too much for Vancouver real estate entrepreneur James Welton Horne. He purchased land around the junction (including some from James and Will Trethewey, who had registered a townsite called New Seattle in 1889) and in 1890 laid out a large townsite with over 5,000 lots, clearing 72 ha of bush and forest above the Canadian Pacific tracks. Horne organized a massive land sale on 19 May the following year, bringing potential purchasers from Vancouver by train and from New Westminster by steamboat. According to contemporary reports some $21,000 worth of lots were sold.

MAP 496 (*above, right*).
Abbotsford and *Huntingdon* have appeared on the Canadian Pacific branch south from Mission to the international boundary on this 1905 map. The Vancouver, Victoria & Eastern line is at left, although it was not completed until 1909.

MAP 497 (*left*).
This 1890 map shows the proposed (cross-hatched) line (unnamed) of the Great Northern subsidiary, the Fairhaven & Southern, completed from *Sehome*, on *Bellingham Bay*, to coal mines east of *Whatcom* and then north into British Columbia, where it met the New Westminster Southern. The Seattle, Lake Shore & Eastern's *Northern Branch* (which would become Northern Pacific line in 1898) runs north to Sumas and continues to Mission (though in Canada it was the Canadian Pacific).

MAP 498 (*below*).
The 1903 edition of the same map shows the lines as constructed. The New Westminster & Southern and the *Fairhaven & Southern* are both labelled as the latter. The line, now following the lower ground through *Port Kells*, ends at *Liverpool, Brownsville,* and *South Westminster,* all just across the river from *New Westminster.* The following year they would be connected by a new bridge. The (unnamed) line of the Vancouver, Westminster & Yukon, like the New Westminster & Southern a Great Northern–controlled line, now runs from New Westminster to Vancouver; this will be used to connect the New Westminster & Southern to Vancouver via the bridge, finally giving the Great Northern access to Vancouver. The *C.P. R.R.* branch south from *Mission* is also shown. *Inset* is a Great Northern train on the Fairhaven & Southern line about 1898.

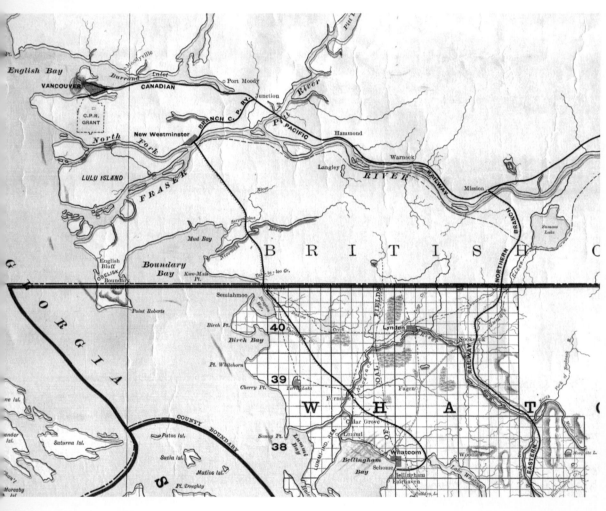

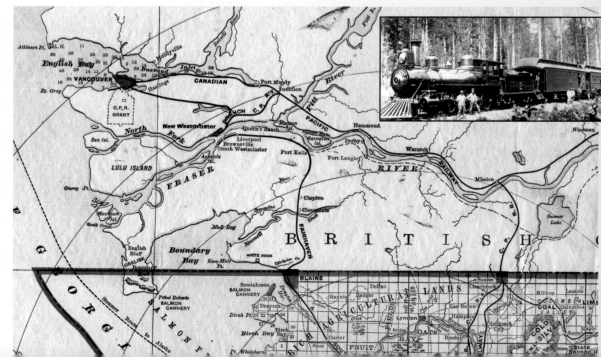

Seattle, reaching the international boundary at Sumas in April 1891. Three months later, however, Northern Pacific purchased control of the Seattle, Lake Shore & Eastern. Canadian Pacific bridged the Fraser at Mission City and built a line to the border at Huntingdon to connect. Townsites were filed that year for both Abbotsford and Huntingdon, through which the line passed, just as one had been for Mission City, at the main line junction, the year before.

Both American companies went into receivership during a general financial debacle in 1893. The name of the connecting line was changed to Seattle & International Railway, and it was sold in 1898 to a reorganized Northern Pacific—to prevent it from falling into the hands of Canadian Pacific.

The New Westminster Southern and a connecting line in the United States, the Fairhaven & Southern, were chartered by Nelson Bennett, a railway builder of some repute who had built the Stampede Tunnel through the Cascades for the Northern Pacific. The idea was to create a series of connecting lines to reach Seattle. In 1889 James J. Hill, onetime director but now competitive nemesis of the Canadian Pacific, purchased the lines from Bennett. The last spike of the two lines was driven at Blaine on 14 February 1891, and by November the same year the line was complete from New Westminster to Seattle.

Hill also was behind a later connection: the Vancouver, Westminster & Yukon Railway, from New Westminster to Vancouver, completed in 1904. This line had been chartered as a railway to the Klondike by his associate John Hendry, and Hill agreed to finance construction of the line on the understanding that it would become part of the Great Northern system. The route used the Brunette River valley, a path that had been cleared for an earlier line that had come to naught—the Burrard Inlet & Fraser Valley Railway, chartered in 1891 to build from Vancouver to Sumas.

The critical barrier was the Fraser River. In 1901 the provincial government announced the construction of a publicly owned rail and road bridge at New Westminster to open up the Fraser Valley. Hill had been ready to build his own bridge across the Fraser at Bon Accord, at the location of today's Port Mann Bridge, but could live with the notion that the government might build one for him even if it were not in exactly the spot his engineers had recommended. The double-deck New Westminster Bridge, with railway lines on the bottom and a road on the top, was opened to great fanfare on 23 July 1904.

In January 1909 an alternate route to the one New Westminster Southern took was opened. It was built ostensibly to avoid heavy grades around Hazelmere but also perhaps because Hill wanted to pre-empt a possible route north, given rumours that were abroad about other American railroads being interested in access to Vancouver. Hill also is reputed to have had ideas about establishing an international harbour in Semiahmoo Bay. The new line, legally belonging to Hill's Vancouver, Victoria & Eastern Railway and Navigation Company (VV&E), was built around the very edge of the Semiahmoo Peninsula and connected with a line north

MAP 499 (*above*).
This superb bird's-eye map was actually created to promote a subdivision in Burnaby by demonstrating its central position. Drawn in 1912, it gives a very good idea of the importance of the New Westminster Bridge, though it is vastly exaggerated with a number of non-existent "proposed" railways also converging on the bridge. *Port Mann* is farther upstream than this view would suggest. By the date of this map there was another company using the bridge—the B.C. Electric interurban to Chilliwack (see page 270)—and the Canadian Northern would be arriving soon (see page 272).
Left.
The main fixed span of the New Westminster Bridge is floated into position on 11 November 1903.

Map 500 (*left*).

Published in 1905, this detailed map of the Lower Mainland shows some railway lines that were only proposed at that date. They include the *Victoria Vancouver and Eastern Railway* (VV&E) line south of the New Westminster Bridge and the line east of *Cloverdale*; both were completed in 1909. The VV&E was used by Hill as an umbrella company for many projects, including the Coast-to-Kootenay line (see page 216). The new Great Northern line around the Semiahmoo Peninsula, which also opened in 1909, is not shown. The full route of the New Westminster Southern—now overtly here the *Great Northern Railway*—to the New Westminster Bridge is shown. The Victoria Terminal Railway serves *Port Guichon*, and there is an extension, never built, south to a location similar to today's Tsawwassen ferry terminal causeway. Hill's *Vancouver, Westminster and Yukon Railway* is correctly shown crossing *False Creek* to a terminal at Carrall and Pender Streets but is also shown running to the North Shore—en route to the Yukon! Canadian Pacific's line to *Steveston* is shown. This was built in 1901 to divert salmon cannery traffic, which was going by steamboat to New Westminster. The line would later be used by an interurban (see page 270).

of the Victoria Terminal line to the New Westminster Bridge. The new route (Map 501, *below*) was opened to traffic in March 1909. At the same time a line was built east of Cloverdale to Huntingdon.

Hill now had a clear path to Vancouver from his American lines—he not only owned the Great Northern by this time but had also acquired a controlling interest in the Northern Pacific—but his line into downtown Vancouver was difficult because of the Canadian Pacific stranglehold. The problem had been solved at first with a station at Carrall and Pender Streets (see Map 7, *page 9*) that was accessed via a trestle built straight across False Creek (Map 502, *below, bottom,* and also visible on Map 500, *left*). This somewhat unsatisfactory situation was remedied beginning in 1910, when, after a vote, the City of Vancouver agreed to allow the Great Northern to fill in the perimeters of False Creek to create a station and yards (Map 502). The railway would soon be joined by the Canadian Northern, which had much the same idea, only this time filling in the remainder of the east end of False Creek (see Map 696, *page 274*).

Map 501 (*right*).

A 1907 or 1908 planning blueprint showing, in red, alternative routes for the VV&E or

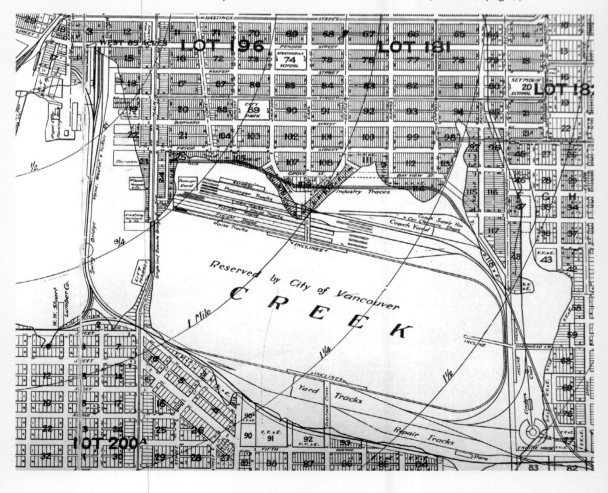

Great Northern's new route across *Mud Bay*. As amended (and built), the route runs around the bay's edge, connecting with the route north via *Victoria Terminal Railway* tracks. The original route of the railway, the New Westminster Southern, is shown at right as *Great Northern Railway*.

Map 502 (*below*).

Published in 1911, this map shows how the Great Northern reclaimed the perimeter of the eastern part of False Creek for its station and yards. The middle section marked *Reserved by City of Vancouver* is the part that Canadian Northern would soon use to locate its station and yards. A trestle and *Swing Bridge* lead the James Hill–controlled *Vanc. Westmʳ. & Yukon Ry.* to a station on *Pender Street* at *Carrall* (see Map 7, *page 9*).

Northern Gold

Gold was found along the Stikine River in 1861, prompting a minor rush. Steamboats were diverted from the Fraser River to the Stikine, providing convenient access. The following year the rush led to the creation of the Stickeen Territories (see page 84). The rush soon petered out, however, and the Stickeen (south of 60°N) became part of British Columbia.

A much larger gold rush began in 1870 with the discovery of gold on Dease Creek and Thibert Creek. Harry McDame, a prospector who had been working in the Omineca (see page 107), moved over to the Cassiar gold fields and staked a claim on a tributary of Dease Creek that became known as McDame Creek. Here, within a month, he found gold valued at $6,000. It was this creek in 1877 that yielded the largest gold nugget ever found in British Columbia. It weighed 72 ounces.

Geologist George Mercer Dawson explored the area in 1887, and his report promised continued mineral wealth. The report stimulated British explorer and entrepreneur Warburton Pike to promote a railway running from the Stikine to Dease Lake to open up the region. His Cassiar Central Railway was sufficiently promising for the provincial government to award it 700,000 acres (283,000 ha) as a land grant (Map 505, *right*).

Following the news of gold in the Klondike in 1897, other schemes, including more railways (also shown on Map 505), were proposed to give access to the remote northern area via the Stikine River. The only one that got off the ground was the White Pass & Yukon Railway, running from Skagway to Whitehorse, with 51 of its 177 km in British Columbia. The railway began construction in May 1898 and was completed in July 1900. From the beginning the railway had to contend with its workers' abandoning their jobs in favour of leaving for the Klondike.

As construction began, however, in August 1898 another gold strike was reported at Atlin, close by, causing 1,300 men, two-thirds of the workforce, to desert for the promise of greater riches close at hand. By the end of the year over two thousand miners were at work in the Atlin area. The difficult access to the new gold fields led in 1900, once the White Pass & Yukon was complete, to the building of the Atlin Southern Railway, which connected with steamboats meeting the White Pass & Yukon at Carcross. It ran across a critically located 3-km-wide neck of land, which then allowed an easy crossing of Atlin Lake to Atlin and the gold fields (Map 506, *below, right*). This little railway, always popularly known as the Taku Tram, ran from 1900 to 1951.

Map 503 (*below*).
The areas of the Stikine and Cassiar gold rushes along the *Stekin River* are shown on this 1873 map. *Discovery Claims* are noted around *Dease Creek* and *Thibert Creek*. The *Old Hudson's Bay Fort* is shown. While not obvious from this map, the Continental Divide separates the Stikine River from the southern end of Dease Lake. The map was drawn by steamboat captain William Moore.

Above.
Placer gold mining in the Atlin gold field. This is "Big Jack's" claim, photographed in 1903.

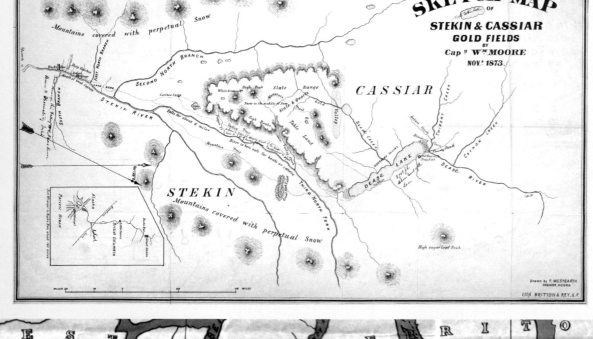
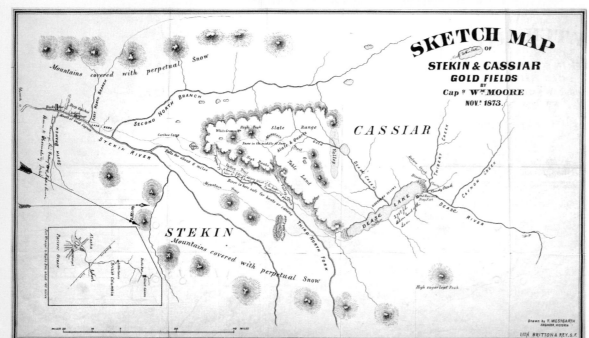

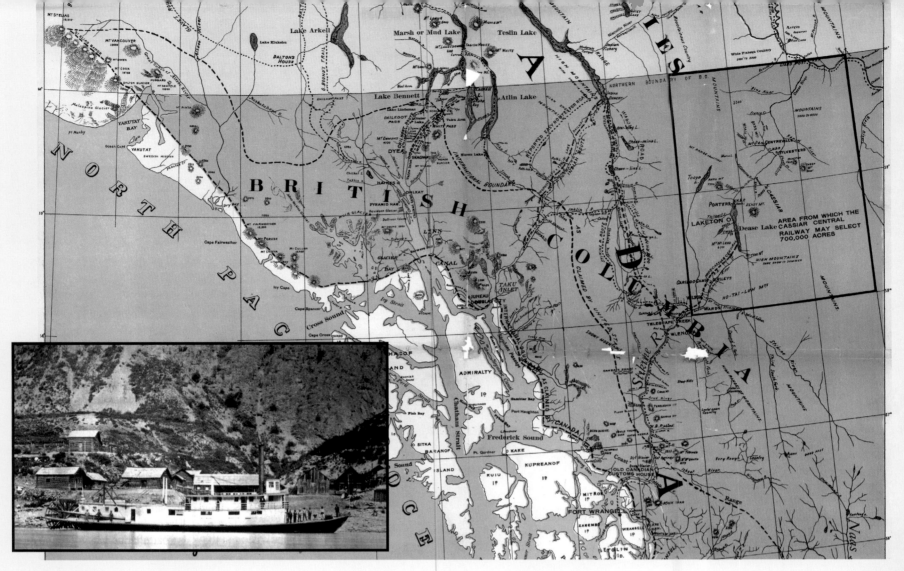

Map 505 (*above*).
One of a series of mining maps of British Columbia published in 1897, this map shows trails north and the proposed northern railways: the *White Pass Route*, the *Taku & Teslin Lake Railway*, and the *Stikine & Teslin Railway*, all following possible routes north. Also shown is the *Cassiar Central Railway* from *Telegraph Creek* to *Dease Lake*, together with a black rectangle showing the *Area from which the Cassiar Central Railway may select 700,000 acres*. Note the boundary with Alaska shown as claimed by Canada (see page 234). *Inset* is William Moore's steamboat *Gertrude* at Telegraph Creek.

Map 504 (*below, left*).
This 1899 map locates the *White Pass and Yukon Ry.*, then under construction, and, several valleys over, *Atlin Lake* and the Atlin mining district centred on *Discovery City* on *Pine Creek*.

Map 506 (*below*).
A detail of Atlin Lake—with a *New Winter Trail* right up the middle—showing the location of the Taku Tram's line crossing the narrow neck of land from *Taku*. Steamboats then crossed Atlin Lake. The mining area around Pine City (Discovery City) is also shown. *Inset*. The Atlin Southern—the Taku Tram—at Taku. A steamboat is just visible at right.

Map 507 (*right*).
The *White Pass & Yukon Route* on a 1919 map. The steamboat route from *Caribou* (Carcross) to Atlin via the Taku Tram (at *Taku City*) is shown.

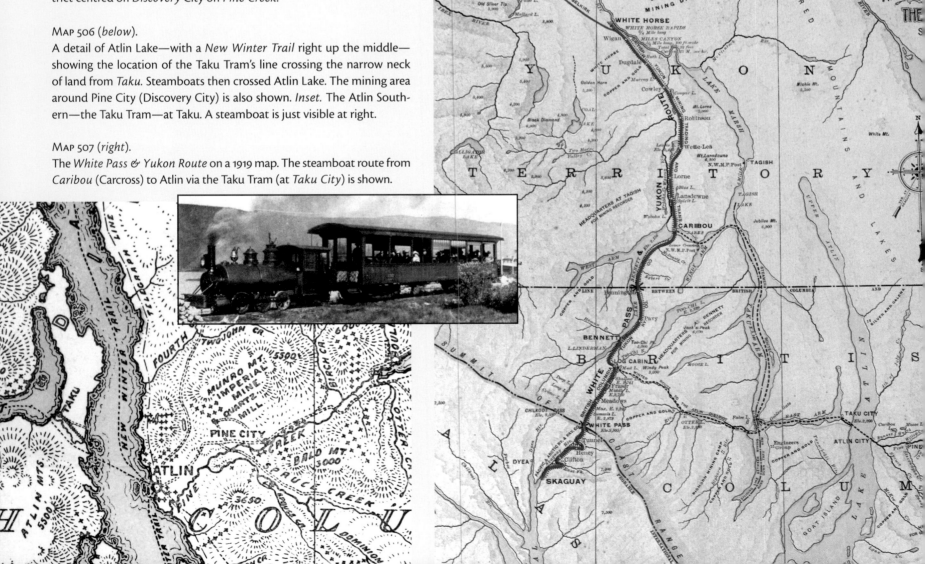

The Steamboat Connection

The Canadian Pacific's transcontinental line was, naturally enough, located where a pathway was possible. Yet, not soon after, the rich mineral deposits of the Kootenays to the south were opened up. The company needed the revenue source to make its transcontinental line profitable. In addition, it wanted to keep what it saw as intruder lines from the United States out—especially those connected to archrival James J. Hill.

The topography of the Kootenay region, with its mountain ranges and lakes and valleys aligned north to south, made east–west lines hard to build, and so the company developed a policy of building short lines south from its mainline and connecting the gaps with steamboats on the long lakes. Additionally, a complementary east–west line was contemplated much closer to the border to cut off the American railways and provide another connection point for the short lines and steamboats plying north and south (see page 216). The steamboats plied all the major lakes from Okanagan Lake (see page 154) eastward. In the Columbia Valley others ran steamboat services, overcoming navigational challenges; this route was eventually paralleled and extended by the Kootenay Central Railway.

Canadian Pacific acquired the charter of the Columbia & Kootenay Railway in 1890, and the short line from Nelson to Robson Landing, at the foot of the Arrow Lakes, was completed in 1891. This portage route over an unnavigable section of the Kootenay River was the genesis of the southern east–west route and was federally subsidized, with a land grant attached (see page 198).

In 1892 the Canadian Pacific–controlled Shuswap & Okanagan Railway from Sicamous to Okanagan Lake had been the first lake connector route to be completed (see page 144). In 1893 Canadian Pacific constructed a short line from Revelstoke to a place 27 km south called

Map 508 (*above*).
The Columbia & Kootenay Railway portaged between *Kootenay Lake* at *Nelson* and Lower *Arrow Lake* at *Robson*. This 1897 map also shows the rival *Nelson & Fort Shepherd R.R.* (see page 173).

Map 509 (*left*).
Because it is a map of railway lines only, this 1899 map demonstrates very clearly how disjointed the rail system was without the north–south connections steamboats provided.

Map 510 (*right*).
This is the railway and steamboat map of the Canadian Pacific by 1917 (the map has small revisions to 1923). The dashed lines are the steamboat routes, giving the railway a creditable network in a country otherwise very difficult to service. In addition tugboats guided barges often with entire trains on them. The only continuous rail line is that of the Kootenay Central south of Golden, today part of the vitally important route from the coalfields of the Elk Valley to Roberts Bank Superport.

Above. Steamboats carried passengers and just about everything else. Here a period truck loaded with chickens is shown inside one of the side doors on the sternwheeler SS *Moyie*, today preserved and on display at Kaslo (*below*) on Kootenay Lake. The photo *inset right* shows the *Moyie* at the same location while in operation in the early 1950s.

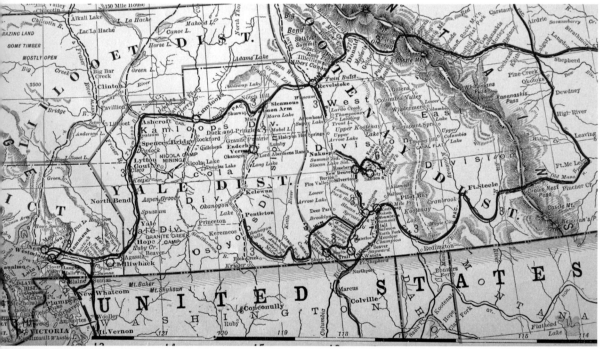

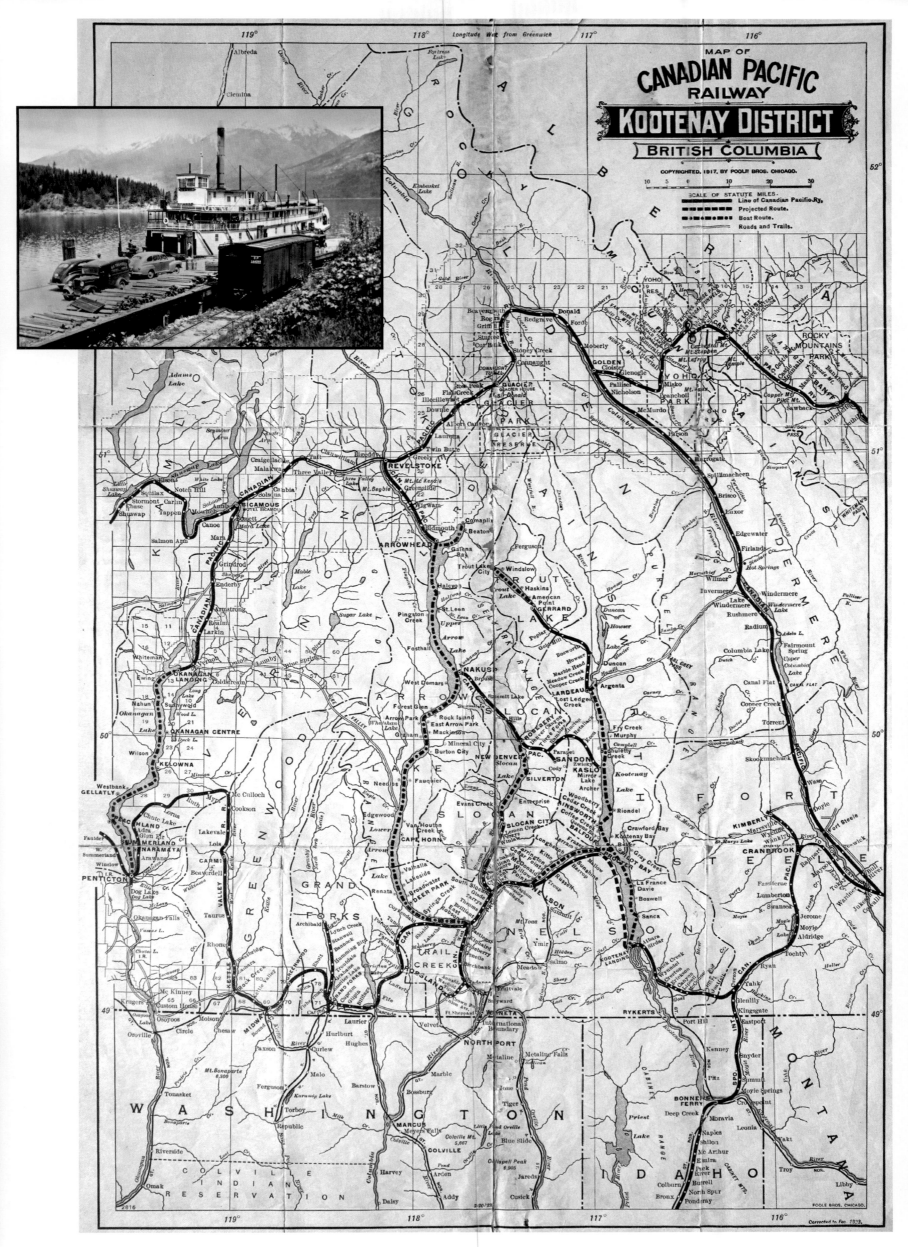

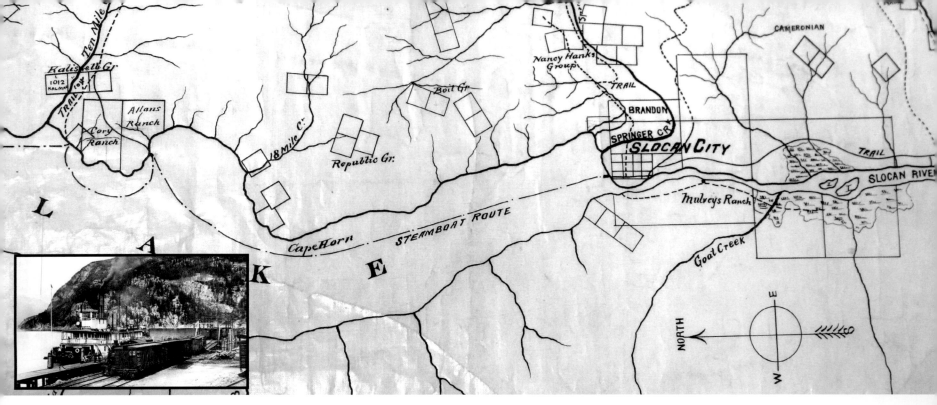

Wigwam; this was extended three years later to Arrowhead, at the northern end of Upper Arrow Lake (shown on Map 510, *previous page,* and Map 520, *page 216*). Although steamboats had been running to Revelstoke, the river presented substantial navigational difficulties in some seasons. In 1895 a line from Nakusp, on Upper Arrow Lake, to the Slocan mining area, via Rosebery, on Slocan Lake, was completed. The purpose of this line was twofold: to compete with the upstart narrow-gauge Kaslo & Slocan, completed at the same time (see page 175), which had ties to the Great Northern's emerging network; and to bypass the difficult section of the Columbia between the Upper and Lower Arrow Lakes. A line was built from Slocan City, at the southern end of Slocan Lake, to connect with the Columbia & Kootenay line 19 km west of Nelson at a place called Slocan Junction. Steamboats plied Slocan Lake (Map 511, *above*), completing the diversion.

Steamboats on the Arrow Lakes and Kootenay Lake had been provided up to this time by an independent company, the Columbia & Kootenay Steam Navigation Company (C&KSN) and several rivals. In 1896 Canadian Pacific purchased the steamboat interests of the C&KSN, thus giving it control of its own lake network. It was this that later became the company's British Columbia Lake & River Service.

At the end of the nineteenth century a silver-mining area opened up north of Kootenay Lake near Lardeau, and both Canadian Pacific and the Great Northern began constructing connecting lines. Because of the transient nature of the mining boom here, the Arrowhead & Kootenay Railway, Canadian Pacific's line, was the only one completed to the point where trains were operated (service began in 1902), and even then the route, originally intended to connect to Arrowhead via steamboats, terminated at Gerrard, on Trout Lake.

The one major north–south valley in which the Canadian Pacific did not operate steamboats was that of the Upper Columbia and Kootenay Rivers. Here the steamboat era had begun in 1886 when Frank Armstrong launched an inelegantly assembled contraption he named *Duchess.* So poorly constructed was this boat that when Major Sam Steele loaded his men and supplies later that year (see page 106) the boat promptly sank. With a succession of boats Armstrong serviced the Upper Columbia for thirty-four years, all the time fending off competitive steamboats. Other steamboats plied the Upper Kootenay as far south as Jennings, Montana, now submerged under Lake Koocanusa behind the Libby Dam. The latter were all but put out of business when the British Columbia Southern Railway was completed in 1898. Armstrong's steamboats and others on the Upper Columbia were the principal means of travel in the valley until 1913, when the Kootenay Central Railway was completed, linking Golden with Corvalli, on the B.C. Southern south of Fort Steele. The first passenger train travelled the entire route in December 1914.

Map 513 (*right*).
Many surveys were carried out for railways that never saw the light of day. This is a portion of a survey done for the Arrowhead & Kootenay Railway in 1899 to further penetrate the mining district with a branch line up the south fork of the Lardeau River, terminating at a place named *Ten Mile City,* about 12 km due east of the head of Trout Lake. The railway had been incorporated the year before to link Arrow Lake with Kootenay Lake.

Map 511 (*above*).
This 1896 map shows the *Steamboat Route* on Slocan Lake terminating at *Slocan City,* but the rail line south from there was not completed until the following year. *Inset.* The sternwheeler *Nakusp* (the first of two with that name) at Slocan City about 1896.

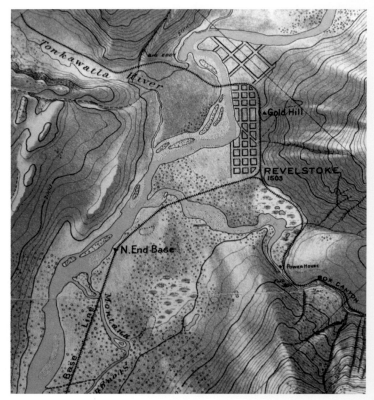

Map 512 (*above*).
This fine 1906 map of *Revelstoke* and area shows the line to Arrowhead going south. See also Map 328, *page 127.*

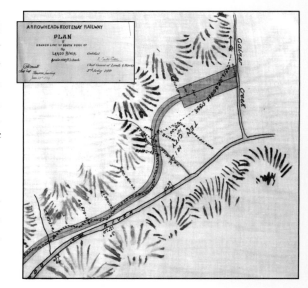

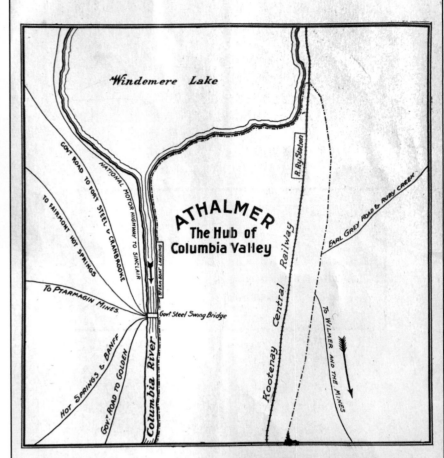

All Roads Lead to Athalmer

ATHALMER
The Hub of Columbia Valley

Athalmer the head of Navigation on the Columbia River.
Athalmer the Principal Railway Station on the Kootenay Central.

Invermere, at the north end of Lake Windermere, began life in 1890 as Copper City. Nine years later it was transformed into Canterbury by a real estate developer, the Canterbury Townsite Company. Acquired by the Columbia Valley Fruit Lands Company, the town changed its name once more, to Invermere. Adjacent to the north was Athalmer, which grew up around the steamboat landing (MAP 514, *above, left*) and became part of Invermere in 1982. The map is of a townsite laid out about 1906, only a few streets of which exist today. Athalmer was not really the head of navigation, as it claimed (MAP 515, *above*; note north is at bottom of map), for steamboats plied the waters from Golden to the southern end of Lake Windermere, where it connected to a horse-drawn tramway to portage freight and people to Upper Columbia Lake. Another portage tramway was used at Canal Flats to connect with the Upper Kootenay. It was this portage that William Adolph Baillie-Grohman sought to eliminate as part of his canal scheme (see MAP 518, *left, bottom*). The photo (*top, left*) shows the steamboat landing at Athalmer from the same 1912 promotional booklet as MAP 515, though it has been said to be dated about 1906. MAP 516, *far left,* shows the Upper Columbia–to–Upper Kootenay steamboat route as far south as Fort Steele. MAP 517, *left, centre,* is an 1899 map of Canterbury, Invermere's predecessor townsite. MAP 518, *left,* is an 1898 map showing Upper Columbia Lake with a *tramway* leading down to the lake—and steamboat transportation—from a silver mine. This map also shows Baillie-Grohman's *canal* and *Kootenay City* (see page 112).

215

THE COAST-TO-KOOTENAY RAILWAY

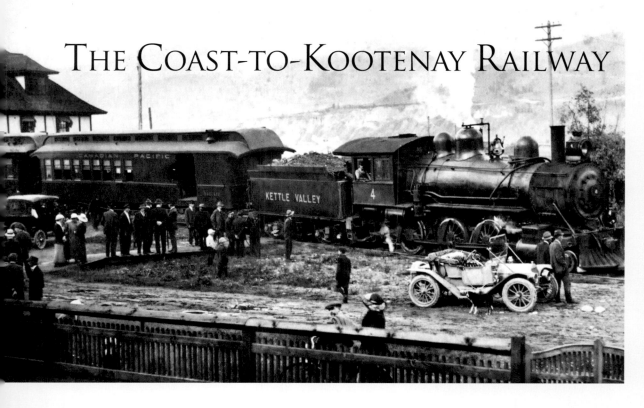

The problem began when the Canadian Pacific transcontinental line was located. It was deliberately far from the American border because of government concerns about its strategic vulnerability at a time when an American attack was still thought possible. Then when gold and silver and copper were found in vast quantities in the Kootenays, the natural lay of the land, with its north–south valleys and mountain ranges, gave the Americans far easier access to them, especially for railways. The Dewdney Trail (see page 104) was an early attempt to counteract this natural disadvantage, but what was really needed was a railway. This was the so-called Coast-to-Kootenay railway, a line to link the mining districts with Vancouver.

But the line had to be Canadian. The first proposals to build such a line came from Americans—first Daniel Corbin, who offered to continue to the coast if he was allowed to build his Nelson &

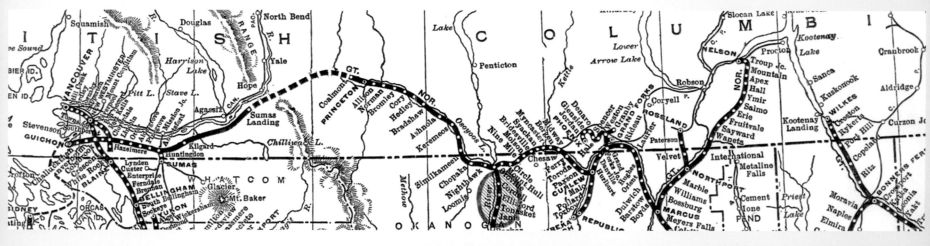

MAP 521 (*right*).

The *Vancouver, Victoria & Eastern Rly* is shown on this 1898 map as an intended line (in red) through to *Vancouver*. Here, however, it is shown joining with the Columbia & Western at *Robson*, an unlikely occurrence, since the VV&E was Hill controlled while the C&W was Canadian Pacific controlled. Note also the *Crows Nest Pass Rly* continuing the line east after passing down the west side of *Kootenay L.* This latter piece of the line, south of Procter, was in fact not completed until 1930.

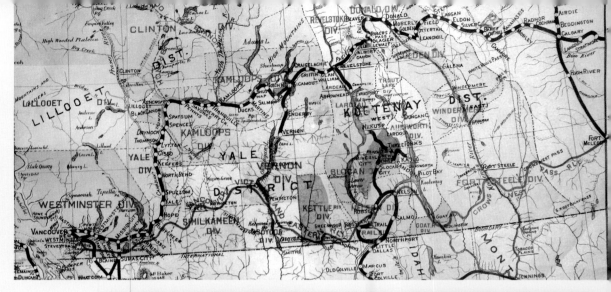

MAP 522 (*right*).

This 1912 blueprint shows the proposed KVR line between *Osprey Lake* and *Coldwater Summit* via *Five Mile Creek* (Hayes Creek) to *Princeton* and then north along *Otter Creek*. As built, the railway left Five Mile Creek and cut more directly into Princeton, albeit requiring a few hairpin bends to achieve this. The line up Otter Creek would not be built; instead, a sharing arrangement to run over the VV&E line was signed.

MAP 523 (*below*).

A year later this map, submitted by the railway to the federal Privy Council for approval, shows yet more adjustments to the KVR's route. Clearly there was an intention to miss *Princeton* altogether, and thus avoid the descent into the Similkameen Valley, and to track north again using *Summers Creek* instead of *Otter Creek*. The changes were due to the difficult nature of the terrain; chief engineer Andrew McCulloch was constantly searching for a more optimal route. Notably here the Coquihalla section is under construction in the *Coldwater River* valley south to *Coquihalla Summit*. The section south from *Merritt* had been completed in November 1911. The tortuous nature of the line west of Hydraulic Summit is clear. McCulloch had to find a way to descend 900 m in about 40 km. The result was a line with an average grade of 1.7 per cent, eighteen trestles, two spirals, and a tunnel. The route through Myra Canyon is a popular trail today, and trestles that burned in a 2003 forest fire have been rebuilt.

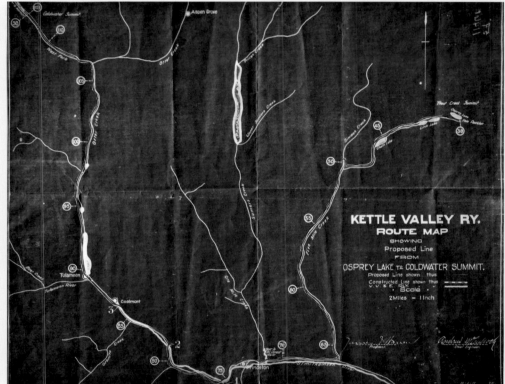

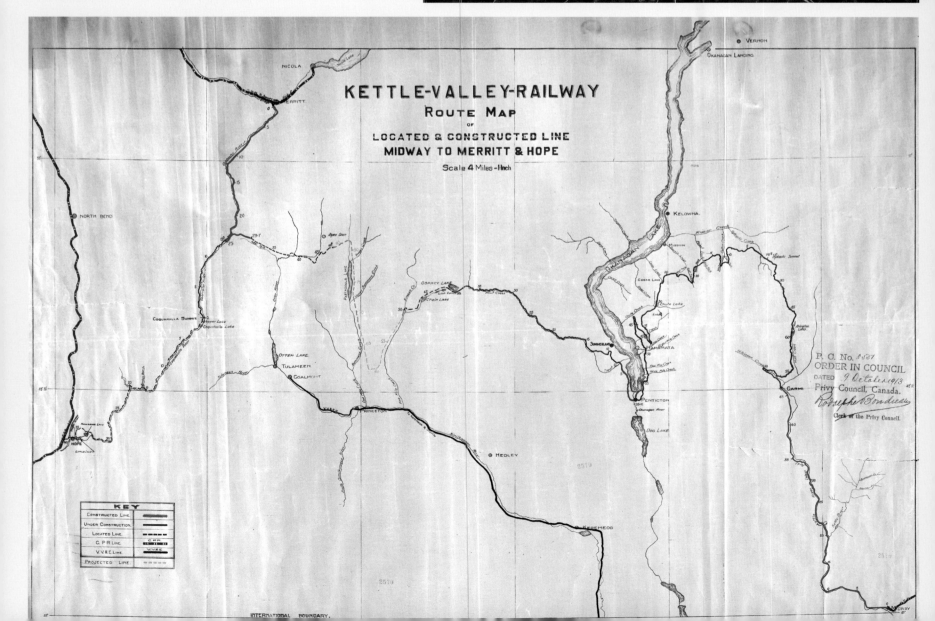

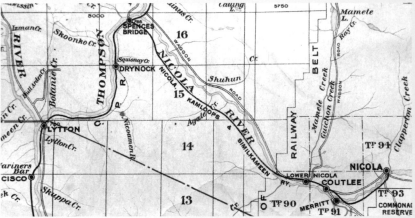

Fort Sheppard Railway (see page 173), but once that line was complete, he did not proceed farther. Then there was James J. Hill, builder of the Great Northern Railway, who had had a grudge against the Canadian Pacific ever since his suggestion, when he was a Canadian Pacific director, to route its main line through his railway hub of St. Paul, Minnesota, had been rebuffed in favour of an all-Canadian route north of Lake Superior. Hill had left the CPR board vowing to get even, and this animosity accounts for some of the less-than-rational economic decisions Hill later made.

Canadian Pacific's William Cornelius Van Horne quickly came to the decision, following the Corbin intrusion, that he needed to move quickly to head off potential further cross-border forays that seemed likely to come from Hill, who completed his transcontinental quite close to the Canadian border in 1893.

There followed the building of the Columbia & Western (see page 182); and the Crow's Nest & Kootenay Lake Railway, which became the British Columbia Southern (see page 192). By 1899 the Columbia & Western had been extended westward to Grand Forks, and the following year reached Midway.

But a year later came the announcement that Hill had acquired the charter of the Vancouver, Victoria & Eastern Railway & Navigation Company (VV&E), seemingly to build a Vancouver-to-Spokane route for his Great Northern. The VV&E had been granted to a group of Vancouver and Victoria businessmen in 1897 for the purpose of constructing a Coast-to-Kootenay railway. Concern that Hill merely meant to use the charter to funnel coal into the United States to supply the Great Northern led to demands for a subsidy to ensure completion to the coast. The British Columbia premier, James Dunsmuir, with his own coal interests on Vancouver Island (see page 224) and a customer, Canadian Pacific, that would not be amused, instead hit on the idea of a survey of the Hope-to-Princeton area, with none other than Edgar Dewdney commissioned to carry it out. All the routes were found to be very difficult and expensive; Dewdney knew that a better route was via the Nicola Valley to the CPR main line at Spences Bridge, but exploration of this route was carefully not included in his mandate.

After construction in Canada between Cascade and Grand Forks, which connected with existing line at Marcus, Washington, the VV&E entered the province at Chopaka, where the Similkameen River crosses the forty-ninth parallel, and headed north to Keremeos, which it reached in 1908. The line passed Hedley, with its Nickel Plate Mine (see page 200) and was completed into Princeton by the fall of the following year, though the first train did not run through until December 1909. At this time the VV&E intended to connect with Hope via a 13-km-long tunnel between the Tulameen and Coquihalla Valleys, but this proved to be too expensive, and instead the line was rerouted up Otter Creek and down the Coquihalla. The VV&E extended the line to Coalmont in 1911, and then, while surveying the continuing line to Otter Creek Summit, busied itself with laying track in the Fraser Valley. The VV&E completed its line between Coalmont and Otter Creek Summit in October 1914, renaming the location Brookmere.

Meanwhile, a branch that would eventually be incorporated into Canadian Pacific's Coast-to-Kootenay network, its Nicola branch, had been completed in a year, 1906, and the first train from Spences Bridge to Nicola ran through in April 1907. For the first three years of its operation the station at Merritt was just a boxcar. In the beginning the line was built to tap the supply of coal at Merritt (see page 222) following a work stoppage in the coal mines of Vancouver Island in 1905. Canadian Pacific then hastily acquired the charter of the Nicola, Kamloops & Similkameen Coal & Railway Company to build the branch.

In 1910, alarmed at the VV&E's progress, Canadian Pacific president Thomas Shaughnessy made the decision to go ahead with the railway's own line from Midway to Merritt despite poor predictions for traffic over the route. To do this the company gained effective control over the Kettle River Valley Railway (KRVR), which had been incorporated in 1901 to boost the accessibility to Grand Forks and had successfully battled Hill to gain access to the mines at Republic, Washington, famously in 1902 positioning a locomotive across a diamond crossing that had been built by VV&E crews overnight across its tracks in Grand Forks, thus preventing

Map 524 (above, top).
A rail route to *Penticton* was shown on this 1897 map, but it bore little resemblance to the one the KVR finally adopted.

Map 525 (above, centre).
A 1910 map shows the *Nicola, Kamloops & Similkameen Ry.*, a charter acquired by Canadian Pacific to access the coal mines in the Merritt area from its main line at Spences Bridge. The line north from Brookmere joined this line at Merritt.

Map 526 (above).
Grand Forks seems to have had high hopes of railway-inspired citydom in July 1913 when this ad appeared in a magazine, but, as usual for this period, its case was overstated by real estate promoters.

the VV&E from entering the town. By 1910 the KRVR had only reached a point 29 km north of Grand Forks. The CPR takeover of the KRVR, now renamed simply the Kettle Valley Railway (KVR) was accompanied by the dispatch of a talented railway engineer, Andrew McCulloch, to oversee construction of the Midway to Merritt line.

Penticton was selected as the headquarters of the railway, and a lakeshore station built to connect with steamboats (see page 154) at the end of a 2-km-long spur from the main line at South Penticton, where the main railway yard, shops, and roundhouse were built. (After 1941 the lakeshore depot was abandoned, and "South Penticton" became just "Penticton.") The KVR was built through difficult terrain that fully taxed McCulloch's engineering skills, especially in the section

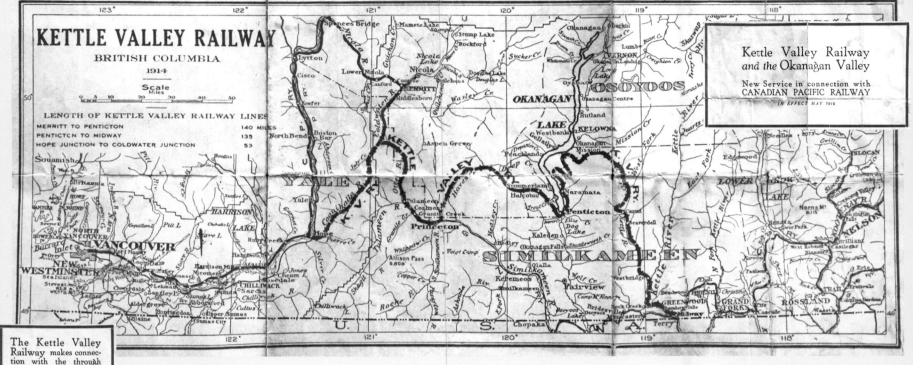

KETTLE VALLEY RAILWAY
BRITISH COLUMBIA
1914
Scale
Miles

LENGTH OF KETTLE VALLEY RAILWAY LINES
MERRITT TO PENTICTON .... 140 MILES
PENTICTON TO MIDWAY 135
HOPE JUNCTION TO COLDWATER JUNCTION ... 53

Kettle Valley Railway
and the Okanagan Valley

New Service in connection with
CANADIAN PACIFIC RAILWAY
IN EFFECT MAY 30th

The Kettle Valley
Railway makes connec-
tion with the through
Vancouver–Montreal
Main Line of the Canadian
Pacific Ry., at Spence's
Bridge, from Merritt.

The Kettle Valley
Railway makes connec-
tion at Midway with the
Canadian Pacific Crows
Nest Line, running
through Kootenay Land-
ing to Medicine Hat.

Map 527 (*above*).
The Kettle Valley Railway was not yet operating
through trains when a brochure with this map
was published in 1914, though you would never
have known it without a very careful examination
of the route, which contains some dashed lines.

Map 528 (*right*).
A 1926 freight handler's map greatly simplifies the route of the
Kettle Valley but contains all the station information. Note the
branch south of *Princeton* to *Allenby*. A branch to Copper Mountain (see page 323) had been completed in 1920 when a copper
mine opened and a smelter was built at Allenby. The smelter
closed down after only a few weeks because of poor copper prices
but in 1925 was reopened; it operated until 1954.

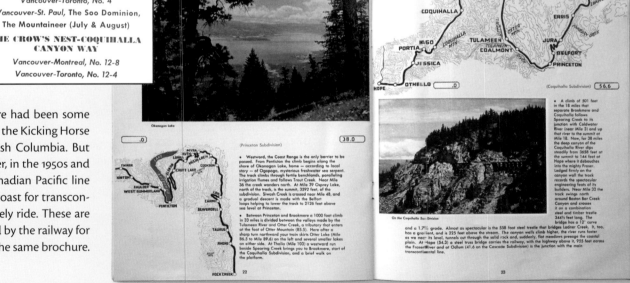

DAILY TRANSCONTINENTAL TRAINS

WESTBOUND.
THE BANFF–LAKE LOUISE WAY

Montreal-Vancouver, No. 1 & No. 7
Toronto-Vancouver, No. 3
St. Paul-Vancouver, The Soo Dominion,
The Mountaineer (July & August)

**THE CROW'S NEST-COQUIHALLA
CANYON WAY**

Montreal-Vancouver, No. 7-11
Toronto-Vancouver, No. 3-11

EASTBOUND
THE BANFF–LAKE LOUISE WAY

Vancouver-Montreal, No. 2 & No. 8
Vancouver-Toronto, No. 4
Vancouver-St. Paul, The Soo Dominion,
The Mountaineer (July & August)

**THE CROW'S NEST-COQUIHALLA
CANYON WAY**

Vancouver-Montreal, No. 12-8
Vancouver-Toronto, No. 12-4

Map 529 (*right*).
When the CPR first took over the KVR, there had been some
notion that the new line might actually usurp the Kicking Horse
Pass route as the railway's main line in British Columbia. But
the route was too circuitous for that. However, in the 1950s and
1960s, the Lethbridge-to-Hope southern Canadian Pacific line
was marketed as an alternative route to the coast for transcontinental passengers, a slower but more leisurely ride. These are
maps from a 1950s guide to the line published by the railway for
its customers. A train listing (*above*) is from the same brochure.

Map 530 (*right*).
When this map of the railways in British Columbia was published in the 1960s, the Coquihalla
line between Brodie and Hope had already disappeared; this track was pulled up during 1962
after being damaged by washouts in 1959. The
Osoyoos branch, opened as late as 1944 (see
page 288) has been scribbled out between
Okanagan Falls and Osoyoos. The rails were removed from this section in 1979. The entire line
between Midway and Penticton has also been
ignominiously erased. One of the most difficult
of the province's railways to build, the Kettle
Valley was disintegrating piece by piece.

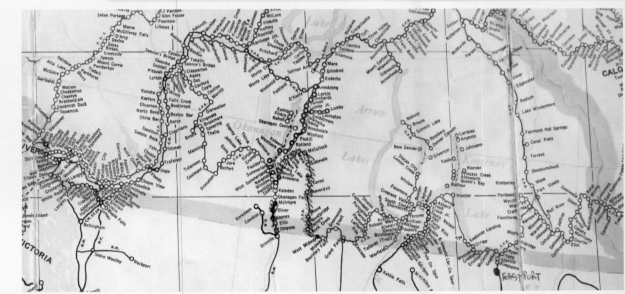

east of Penticton. Finances were buoyed in 1909 and 1912 with the election and re-election of ardent railway supporter Premier Richard McBride; in 1909 he promised a subsidy of $5,000 per mile for the line between Merritt and Penticton and in 1912 a subsidy of $12,000 per mile for the Coquihalla Pass section plus $200,000 for a bridge over the Fraser River at Hope.

The line from Merritt south to Brookmere was completed in November 1911. By 1913, with the entire route more than 50 per cent complete, economic conditions began to decline precipitously. In November 1912 and April 1913 the VV&E and the CPR came to an agreement that would allow a single line between Coalmont and Hope. The KVR was to have running rights over the VV&E's line to Brookmere, while the VV&E was to have rights over KVR tracks from Brookmere to Hope. In April 1915 the KVR was linked to the completed section of VV&E line to Brookmere, and the Midway-to-Merritt line was formally opened on 31 May 1915.

The KVR had nearly completed its line in the Coquihalla when it was stopped at the end of November 1915 by exceptionally heavy snowfall, but work, which included numerous trestles and series of tunnels, was finally completed in July 1916. The Coast-to-Kootenay connection had been made.

The KVR went on to become the second transcontinental main line for the Canadian Pacific and was marketed as an alternative route (Map 529, *previous page*). But it was a slow and relatively circuitous route, and expensive to maintain, so it was not surprising that it did not survive. The VV&E connected Vancouver and Hope via its line from Cloverdale to Abbotsford and then to Cannor, on the Canadian Northern line—from that point on running rights were negotiated—and a short connector at Hope.

Following the death of Hill in May 1916 the Great Northern quickly lost interest in the VV&E route; only a single train, carrying Hill's son Louis, made the journey from Vancouver to Spokane. As early as 1917 the Great Northern tore up the line from Hedley to Princeton after the Nickel Plate Mine temporarily closed down. In 1944 the Great Northern paid Canadian Pacific $2.5 million to get out of its track-sharing agreements through the Otter Creek and Coquihalla Valleys. The VV&E rails to Hedley survived until 1972, when a burned bridge near Keremeos was deemed not worth replacing.

The KVR closed the Coquihalla section in 1960 following washouts on the line that were too expensive to rebuild; the last passenger train from Spences Bridge to Penticton ran in 1964; and the Midway-to-Penticton section was closed in 1973. Some of the railbed is now part of the Trans Canada Trail.

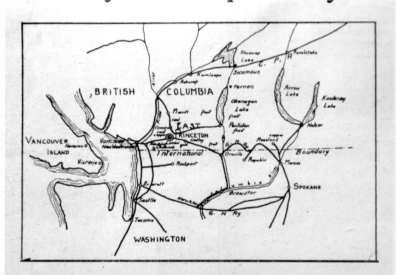

Map 531 (*right, centre*), with reduced text part of ad *above*.
Like virtually all railways, the Kettle Valley generated its share of real estate schemes. One was *East Princeton*, very close to the established centre of Princeton. This ad appeared in November 1912. Although it states that both railroads are "in now," in fact only the VV&E had reached Princeton, which it did three years earlier, and at the time it was not even certain that the KVR would run through Princeton (as shown on Map 523, *page 217*, for example). Princeton seemed to have a lot going for it, but this did not help East Princeton, which was yet another victim of a declining economic climate. Today the townsite is a largely vacant industrial area with a few blocks of residential development at its north end.

Map 532 (*right*).
East Princeton is also shown on this 1917 map, which surprisingly does not show any railways. In 1917 the VV&E track between Hedley and *Princeton* was removed, but the KVR track into town and the track shared with the VV&E to the west was in place. Two other townsites are shown on this map, and neither fared any better than East Princeton. *Princeton Heights* is today part of Princeton Airport, and *Allison Townsite*, straddling the Similkameen River, is but a few houses along the main road. In Princeton, notice the way later subdivision has ignored the orientation of the first part of the town, which dates from its early days as Vermilion Forks. The name of the settlement was changed to Princeton by Governor James Douglas in 1860 to honour the visit of the Prince of Wales to eastern Canada that year.

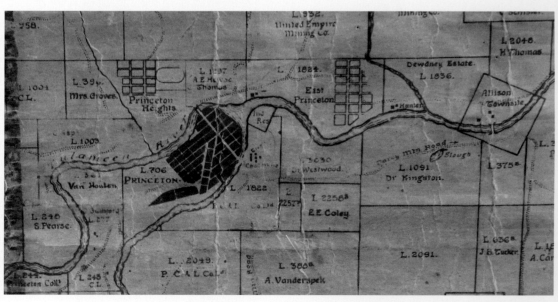

Map 533 (*right*).

Some miners from the Granite Creek gold rush settled on the shores of *Otter Lake* at a place they called Otter Flat. This more formal townsite plan was registered in 1901 with the name Tulameen. *Otter Creek* flows into the *Tulameen River* here. The valley of Otter Creek was followed by the VV&E line to Brookmere. As late as the 1930s the Great Northern operated trains to Tulameen during the winter months to access ice cut from Otter Lake.

Map 534 (*below*).

The Coalmont Coal Company's Mine No. 4, part of which is shown here, was in Blakeburn, up the hill to the south of Coalmont itself. On 13 August 1930 a build-up of gases caused an explosion that killed forty-five miners; a single miner escaped unharmed. This map was drawn by the mine surveyor as part of the investigation into the tragedy. The names of the dead men and the locations at which their bodies were found are noted. *Inset* is a photo, taken about 1938, showing the Coalmont mines. James Hill knew he could find coal for his VV&E locomotives at Granite Creek because coal, which had been reported here as early as 1858, was from 1901 being located by several mining companies. Coal was indeed found here in abundance, and by 1912, a year after the VV&E arrived, mines at Blakeburn, on Granite Creek close to where it enters the Tulameen at Coalmont, were in operation. Number 4 Mine, shown here, was opened in 1923. Mining ceased at Blakeburn in 1940. Granite Creek was the location of a spectacular gold rush in 1885 (see page 202). Coalmont's townsite plan was registered in 1911.

Above, left. The final track of the KVR is laid into Princeton on 21 April 1915.

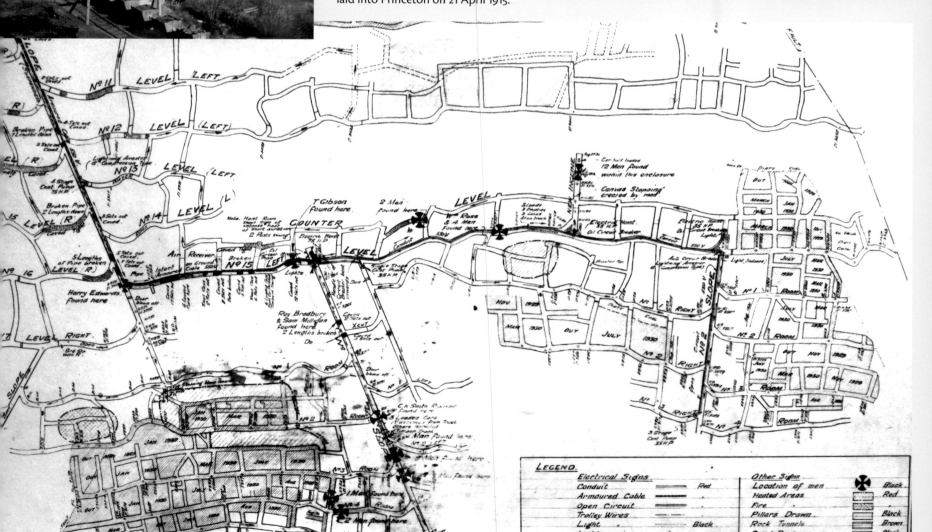

MAP OF TULAMEEN B.C.

As the Key to the Hope Pass Merritt is a Strategical Point on the Railway Map

Nicola Valley Coal

Though known about earlier, coal in the Nicola Valley appears to have first been officially noted by Sandford Fleming in 1872 during the original Canadian Pacific surveys to find a path for the first transcontinental railway. But the advantages of Nicola coal did not sufficiently offset the topographical disadvantages of the Coquihalla route.

Nevertheless, Fleming was a petitioner—along with William Hamilton Merritt and others—for a charter for the Nicola, Kamloops & Similkameen Coal & Railway Company, granted in 1891 to run a line from Kamloops to Princeton via the coalfields as well as a line from the Canadian Pacific main line at Spences Bridge. At the same time another company, the Nicola Valley Railway, was authorized to build from Spences Bridge some eighty miles up the Nicola Valley to the coalfields, a proposal apparently from individuals working with the CPR intended to counter what was seen as a competing project.

Neither project went anywhere until in 1905, when coal from Vancouver Island temporarily became unavailable because of labour problems. Canadian Pacific had let the charter for the Nicola Valley Railway lapse, but Merritt had kept his charter alive by a pretense of beginning construction. It paid off when the CPR paid Merritt many times his investment for his railway's charter. With the completion of the line from Spences Bridge in 1906 Nicola coal was now available for the CPR's locomotives and for transport to markets outside the valley. And a place variously known as The Forks, Forksdale, or Diamond Vale became Merritt.

Canadian Pacific found Nicola coal to be excellent for its locomotives and tried to acquire the entire coal output, building a spur line into the mines at Coal Gully, those of the Nicola Valley Coal & Coke Company (NVCC), which became Middlesboro Collieries in 1914. The mine was soon producing 225 tons of coal every day and was the largest mine in the region. Other coal-mining companies, shown on the maps on this page and opposite, were the Pacific Coast Colliery, next to NVCC; Inland Coal & Coke, on Coal Hill above NVCC; and Diamond Vale Collieries, across the river.

Coal shipments slowed after 1924, and the Depression of the 1930s further reduced demand. The mines closed after the final blow struck: the conversion of Canadian Pacific's (and other railways') locomotives to diesel-electric in the 1950s.

Above. British Columbia's railways are coming to Merritt and its coal, according to this cartoon, published in January 1913 in the *Nicola Valley Advocate.*

MAP 535 *(below).*
This map published about 1912 shows Merritt and the lines of the Canadian Pacific servicing the coal mines across the river. The company town of *Middlesboro,* named after the coal-mining town in northeast England, is shown as a subdivided block on the map and also in the photo, *inset.* Two companies, Nicola Valley Coal & Coke and Inland Coal & Coke, operated out of Middlesboro.

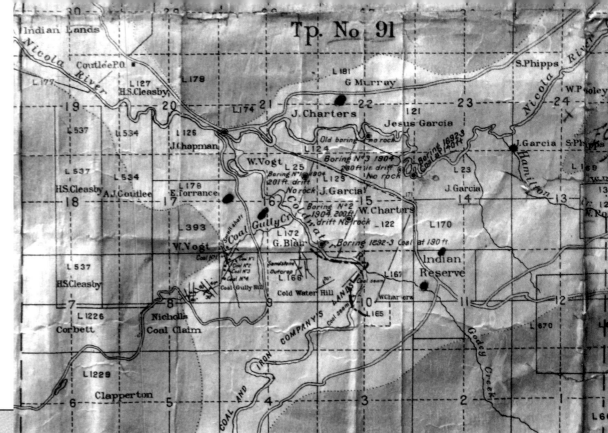

MAP 536 (*right*).
The coal resources of the Nicola Valley are plotted on this geological map, surveyed in 1904. *Coal Gully Cr.* became the prime coal-producing location, and it was here that Middlesboro was built.

MAP 537 (*below*).
This 1914 three-dimensional plan of the Nicola Coal & Coke Company's mine shows shafts and galleys honeycombing the hillside. It is a clay model.

MAP 538 (*below, bottom*).
The townsite of Merritt on a map drawn in 1907. The town was named after William Hamilton Merritt, one of the promoters of the Nicola, Kamloops & Similkameen Coal & Railway Company and one of the first to recognize the commercial potential of the coal deposits here.

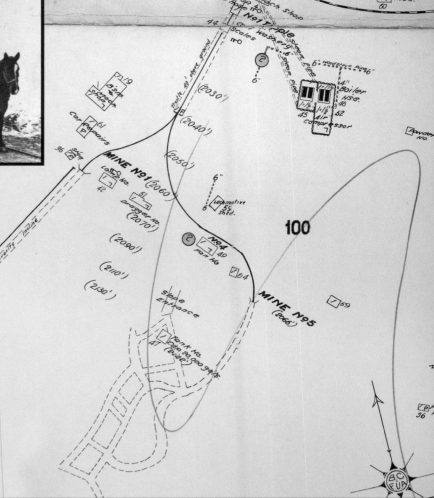

MAP 539 (*right*).
The gravity incline, trestle, and *No. 1 Tipple* used to transport and load coal into railcars on the valley floor line. The map is from a 1926 fire insurance plan.
Inset, top of map, is a 1912 view of the same area from the railway, which can be seen in the foreground.
Inset, right, about 1907, a train of coal cars exits No. 1 Mine, pulled by a horse.

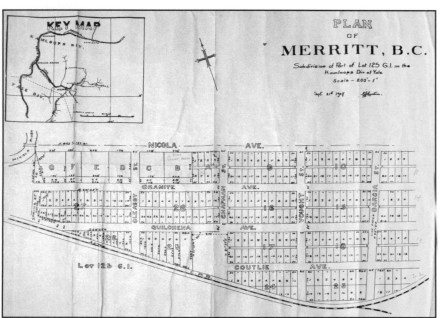

ISLAND COAL

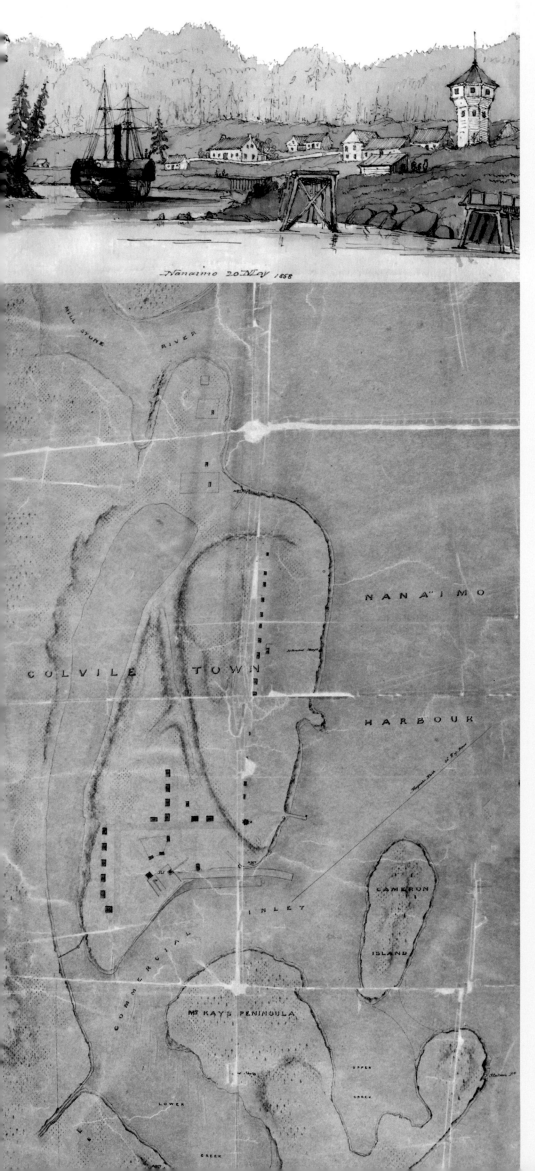

Nanaimo 20 May 1858

Early mining efforts on Vancouver Island at Nanaimo and in the Comox Valley grew in their time into major operations: coal was mined for a hundred years, a vital resource exported to a world dependent on coal-fired steam for all its power.

Coal was discovered in 1835 near today's Port Hardy. In 1849 the Hudson's Bay Company decided to try mining it and built Fort Rupert (MAP 544, *overleaf*) to protect miners brought first from Scotland and then, a year later, from England. The latter soon went on strike for improved working conditions, the first of many such work stoppages that would characterize the industry for nearly a century. This, coupled with hostility from the local Kwakwaka'wakw and poor-quality coal, ensured that the Fort Rupert mines did not last long. In 1852 coal mining began in the Nanaimo area, and the miners were transferred.

In 1851 Snuneymuxw chief Che-wech-i-kan, later known as Coal Tyee, told the Hudson's Bay Company blacksmith in Victoria of coal found on the beach at Nanaimo. After the chief returned with coal in his canoe, James Douglas visited the area and in August 1852 sent clerk Joseph William McKay to "Nanymo Bay" to formally take possession of the coal beds there for the company. Aboriginal workers aided the transferred Fort Rupert miners, and the first coal was shipped from Nanaimo on 9 September.

The following year the company built a blockhouse to serve as a refuge and trading post. The Bastion, as it soon became known, is shown on all the maps and survives to the present day. The settlement was named Colviletown after Andrew Colvile, the HBC governor at the time, though by 1858 the name used seems to have become Nanaimo, the same as the bay on which the town stood, and this was the name used when the city incorporated in 1874.

In 1854 the Hudson's Bay Company purchased the land containing the coal mines and harbour from the local Snuneymuxw people for 668 blankets and then obtained exclusive rights to

COAL
FIELD

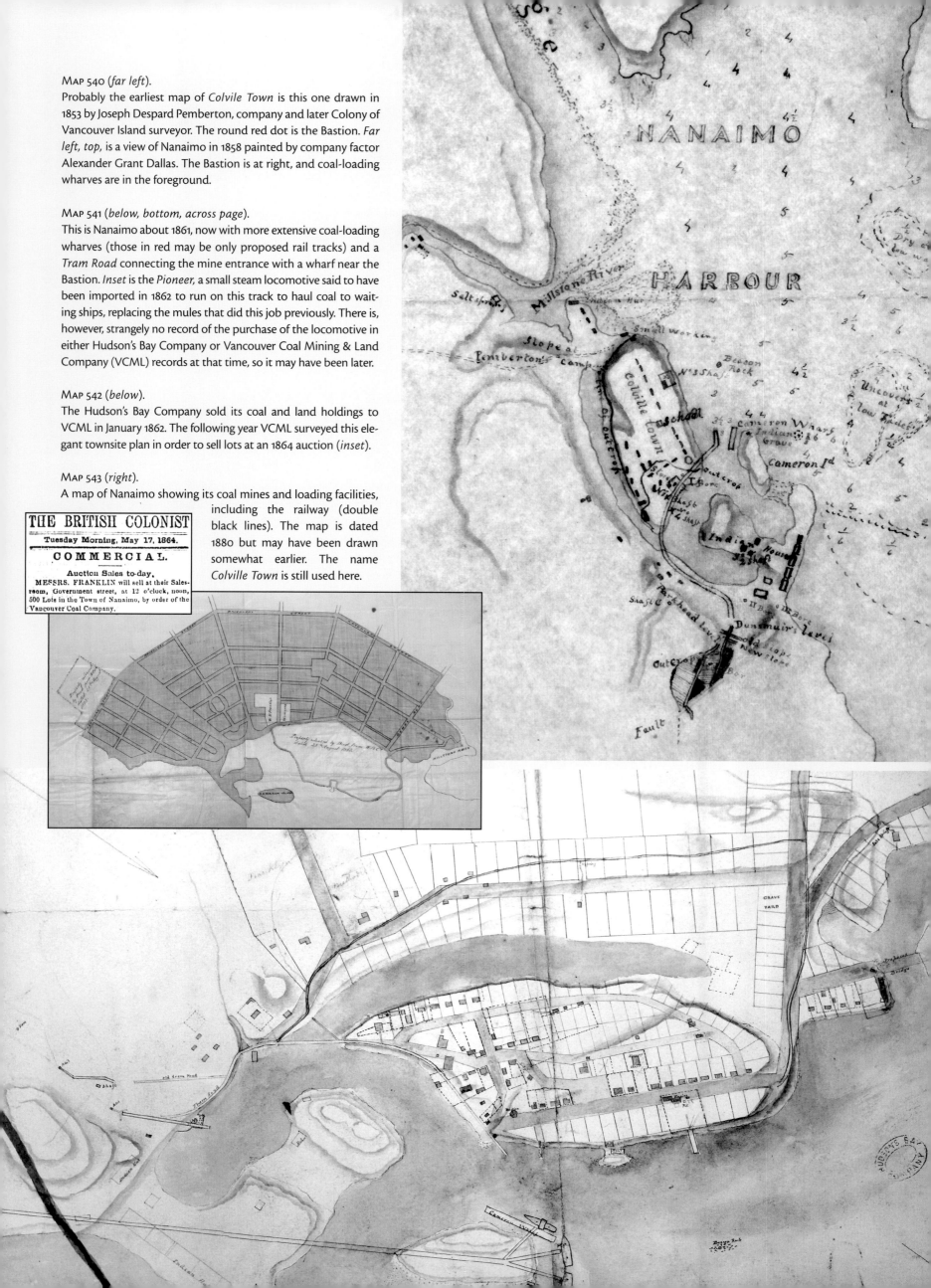

MAP 540 (*far left*).

Probably the earliest map of *Colvile Town* is this one drawn in 1853 by Joseph Despard Pemberton, company and later Colony of Vancouver Island surveyor. The round red dot is the Bastion. *Far left, top,* is a view of Nanaimo in 1858 painted by company factor Alexander Grant Dallas. The Bastion is at right, and coal-loading wharves are in the foreground.

MAP 541 (*below, bottom, across page*).

This is Nanaimo about 1861, now with more extensive coal-loading wharves (those in red may be only proposed rail tracks) and a *Tram Road* connecting the mine entrance with a wharf near the Bastion. *Inset* is the *Pioneer*, a small steam locomotive said to have been imported in 1862 to run on this track to haul coal to waiting ships, replacing the mules that did this job previously. There is, however, strangely no record of the purchase of the locomotive in either Hudson's Bay Company or Vancouver Coal Mining & Land Company (VCML) records at that time, so it may have been later.

MAP 542 (*below*).

The Hudson's Bay Company sold its coal and land holdings to VCML in January 1862. The following year VCML surveyed this elegant townsite plan in order to sell lots at an 1864 auction (*inset*).

MAP 543 (*right*).

A map of Nanaimo showing its coal mines and loading facilities, including the railway (double black lines). The map is dated 1880 but may have been drawn somewhat earlier. The name *Colville Town* is still used here.

THE BRITISH COLONIST

Tuesday Morning, May 17, 1864.

COMMERCIAL.

Auction Sales to-day.

MESSRS. FRANKLIN will sell at their Salesroom, Government street, at 12 o'clock, noon, 500 Lots in the Town of Nanaimo, by order of the Vancouver Coal Company.

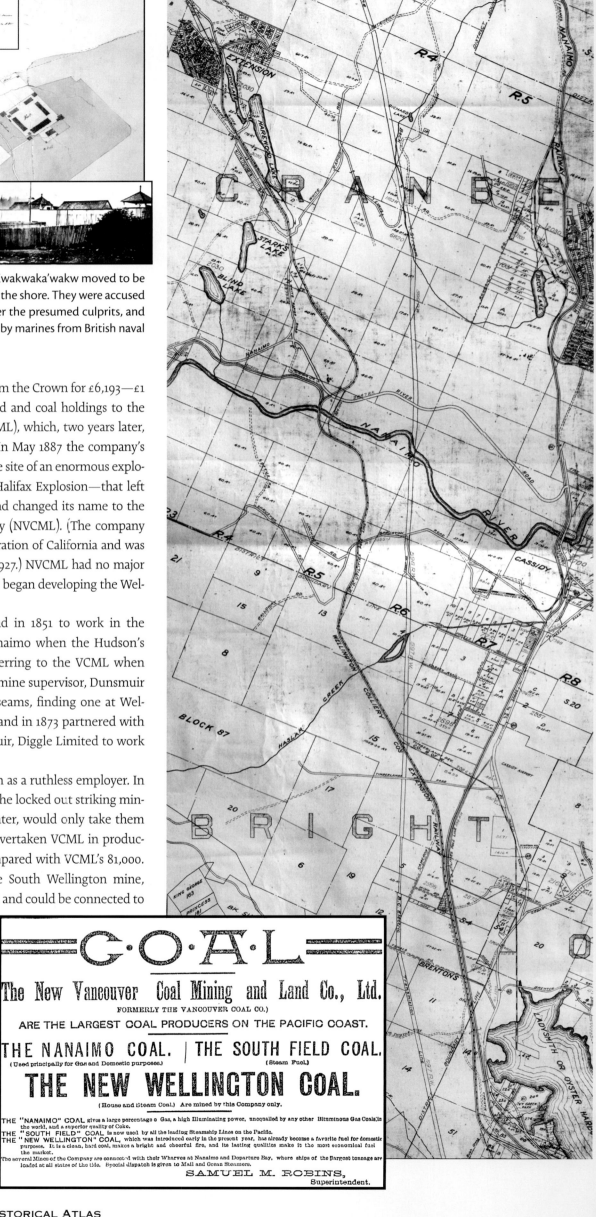

Map 544 (*above*) and photo, *inset*.
Fort Rupert, built in 1849 to protect coal miners. Kwakwaka'wakw moved to be close to the fort, and their houses are shown along the shore. They were accused of murdering three sailors but refused to surrender the presumed culprits, and as a consequence their village was twice set on fire by marines from British naval ships and bombarded by another later.

the 2,500 ha of land and the underlying coal from the Crown for £6,193—£1 per acre. In 1862, the company sold all its land and coal holdings to the Vancouver Coal Mining & Land Company (VCML), which, two years later, sold lots in the town (Map 542, *previous page*). In May 1887 the company's No. 1 Mine (shown on Map 548, *page 228*) was the site of an enormous explosion—the largest in the world until the 1917 Halifax Explosion—that left 157 miners dead. VCML reorganized in 1889 and changed its name to the New Vancouver Coal Mining & Land Company (NVCML). (The company was purchased in 1902 by Western Fuel Corporation of California and was in turn bought out by Canadian Collieries in 1927.) NVCML had no major competition until 1871, when Robert Dunsmuir began developing the Wellington mine to the north.

Dunsmuir had arrived from Scotland in 1851 to work in the mines at Fort Rupert and had moved to Nanaimo when the Hudson's Bay Company began its mining there, transferring to the VCML when the company sold its mining interests. Now a mine supervisor, Dunsmuir searched on his own account for other coal seams, finding one at Wellington that looked promising. He left VCML and in 1873 partnered with naval officers to raise capital, forming Dunsmuir, Diggle Limited to work the mine.

Dunsmuir quickly gained a reputation as a ruthless employer. In 1877 Dunsmuir added to this reputation when he locked out striking miners and, having defeated them four months later, would only take them back at lower wages. By 1878 Dunsmuir had overtaken VCML in production, that year mining 88,000 tons of coal compared with VCML's 81,000. The following year Dunsmuir purchased the South Wellington mine, which had coal in the same seam as Wellington and could be connected to

Map 545 (*right*).
The *Wellington Colliery Co's Extension Railway* is the line built by James Dunsmuir in 1899 to carry coal from his new mine at *Extension* to Oyster Harbour, here *Ladysmith or Oyster Harbour*. The town of Ladysmith is just off the bottom of this map. The main line of the Esquimalt & Nanaimo Railway bears east towards South Wellington, near *Beck's Lake*, at top right.
Inset. An ad from the Dunsmuir's competitor, the New Vancouver Coal Mining & Land Company, which appeared in the Victoria *Daily Colonist* in February 1893.

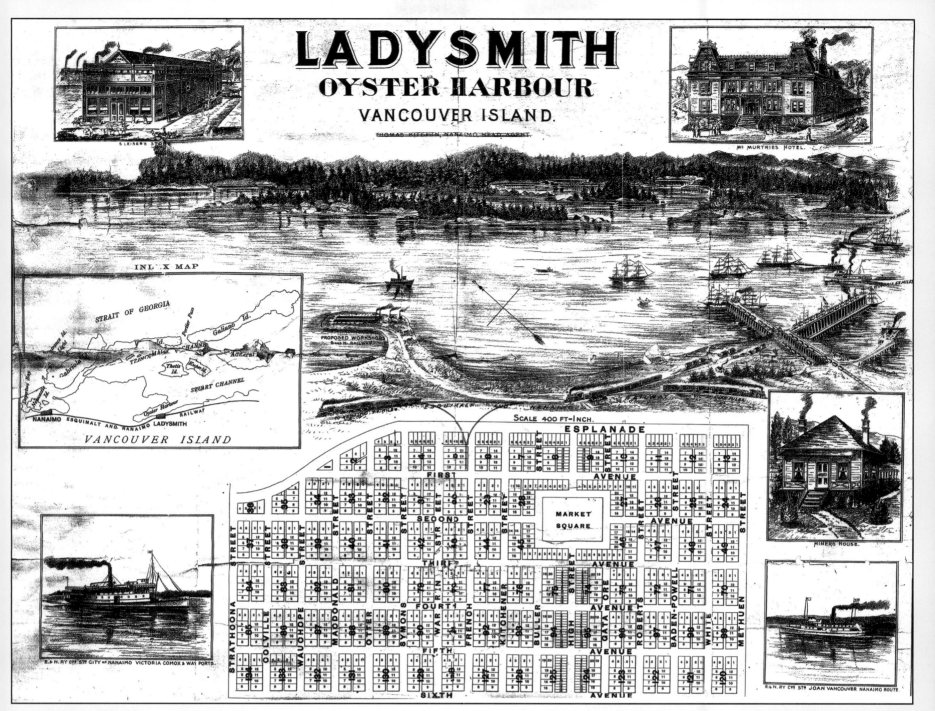

LADYSMITH
OYSTER HARBOUR
VANCOUVER ISLAND.

Thomas Kitchin, Nanaimo, Head Agent.

Map 546 (*above*).

This rather superb map was produced in 1901 to sell lots—presumably to non-miners—in the new town of Ladysmith, created by James Dunsmuir on his Esquimalt & Nanaimo land grant and named during the Boer War. Some nine hundred miners worked at the Extension Mines, and most lived in Ladysmith. The *Proposed Workshops E & N Railway* did not materialize, but in its place was a smelter, which began work in December 1902 using copper ore from Mount Sicker (see page 164). Dunsmuir had joined the Tyee Mining Company and provided land on the Ladysmith waterfront for the company's smelter. By 1905 there was also a shingle mill on the Ladysmith waterfront.

Map 547 (*right*).

This map dated 1905 shows the extensive coal *Loading Wharves* and the *Tyee Smelter Reserve*. It also depicts a rail spur entering the town, which is not likely to be accurate since there is a steep slope here up from the waterfront. *Below*. The massive coal wharves at Ladysmith, probably about 1905.

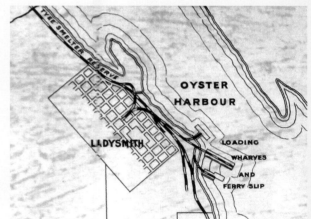

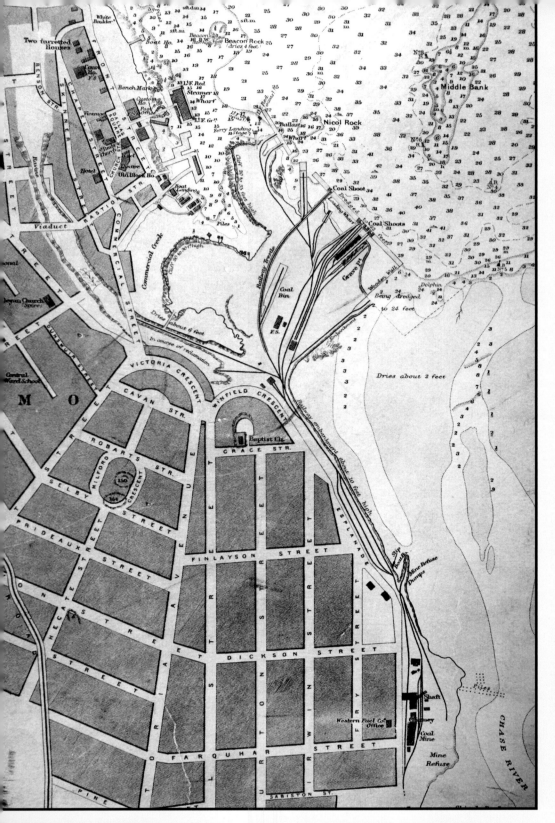

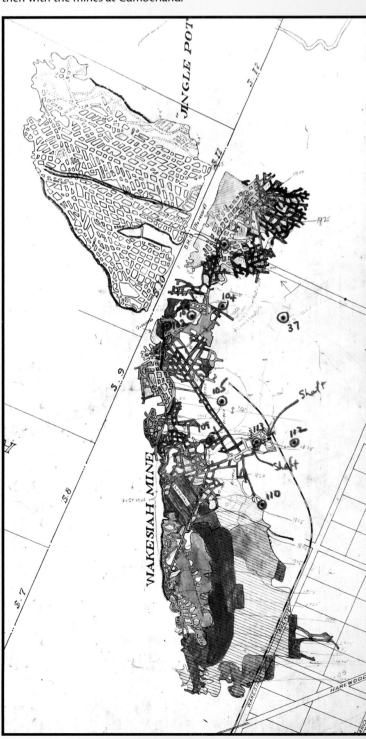

MAP 548 (*left*).
Detail of coal-loading facilities at Nanaimo in 1899, mapped by the surveying ship HMS *Egeria* and published in a British Admiralty chart in 1912. The Bastion is the *Old Block Ho.* At the end of the railway tracks are *Coal Shoots*, allowing coal to be loaded directly into ships' holds. The tracks are on a *Railway Trestle.* At bottom is a *Coal Mine*, shown owned by *Western Fuel Co.* (1902–27); note how the area behind the name appears slightly lighter, the sign of an alteration of the plate that would have been done after Western Fuel Company purchased the mine from the New Vancouver Mining & Land Company in 1902. This was the mine (VCMLC No. 1) in which a massive explosion killed 157 miners in 1887, the worst mining disaster in British Columbia's history. There are several piles of *Mine Refuse* shown. The original area of Colviletown, around the Bastion, was an island (Cameron Island), as evidenced by the maps on the previous page. To the west was a ravine, and part of it is shown as in course of reclamation; the entire ravine was eventually filled with mine waste and garbage.

MAP 549 (*left, bottom*) and **MAP 550** (*below*).
The coal mine map as art? Two particularly attractive maps of the intricate pattern of the underground labyrinth that is a coal mine, both from the Nanaimo area: the Northfield Mine, near Departure Bay; and Wakesiah and Jingle Pot Mines, to its south. All the maps come from the collection of Alexander Buckham, a coal mine geologist with the federal government until 1949 and then with the mines at Cumberland.

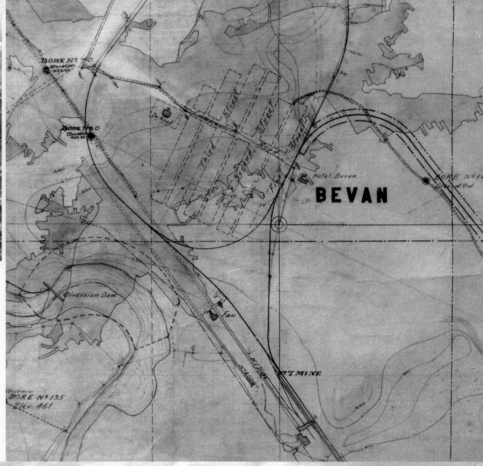

it. Coal was shipped through Departure Bay. By 1883 Dunsmuir had bought out his partners and was in sole control; the company was now R. Dunsmuir & Sons. The same year Dunsmuir signed a contract to build the Esquimalt & Nanaimo Railway (see page 164), which was completed in 1886 and extended to the Wellington Mine the following year.

After Robert Dunsmuir died in 1889 the company was effectively run by his son James (though voting control was in the hands of Robert's wife, much to James's chagrin). About 1895 coal was discovered to the west of the Wellington Mine, and since the latter was approaching exhaustion by this time, plans were made to develop this new field, at first called Wellington Extension and then simply Extension. A new railway line surveyed from Extension to Departure Bay, however, needed permission to cross land owned by the NVCML, which, likely because James Dunsmuir had refused a similar request from the NVCML to cross his E&N line in 1888, was refused. Not to be outdone, Dunsmuir built a rail line from Extension to Oyster Harbour south of Nanaimo (Map 545, page 226), where he built extensive wharves and bunkers. The first coal was shipped from Extension in 1899. In 1897 Dunsmuir laid out a townsite on land adjacent to Oyster Harbour—land he owned by virtue of the E&N land grant—and decreed that miners at the Extension Mine had to live there. A miners' train would take them to work each day.

When, on 1 March 1900, word reached Dunsmuir—who happened to be at Oyster Harbour that day—that the South African city of Ladysmith, besieged for several months during the Boer War, had been relieved, he is reputed to have announced that henceforth the town would be known as Ladysmith.

Dunsmuir sold the E&N land grant to the Canadian Pacific Railway in 1905, and all his Vancouver Island mining interests were sold in 1910 to Canadian Collieries (Dunsmuir) Limited. Ladysmith and Extension were the scene of one of the longest and most grievous strikes in British Columbia's history in 1912–14—a strike about safety procedures and attempts at unionization, a strike that had spread from the Comox Valley fields over concern about gas explosions. Miners were evicted from company housing, and their

Map 551 (above).

Bevan, also shown in the photo, above, was a company town created close to No 7. Mine a few kilometres north of Cumberland, in the Comox Valley coalfields. The plain frame houses were rented to miners for $11 or $12 per month and when mining ceased were moved to other locations, many ending up in logging camps. Underground lower seam workings are shown in red.

Above, left, is a view of the entrance to No. 7 Mine about 1910, when Bevan was first constructed.

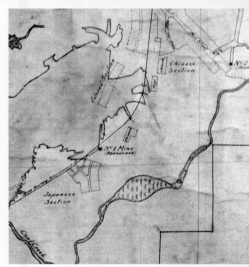

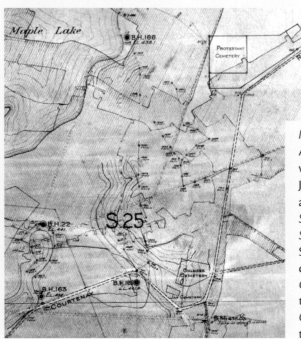

Map 552 (above) and Map 553 (left).

At its peak between 1900 and 1914, Cumberland was home to about 3,000 Chinese and 1,500 Japanese mine workers, housed in segregated accommodation. Map 552 shows the Japanese Section and the Chinese Section—and a Colored Section—a short distance west of Cumberland. Segregated in life, they were also segregated in death: Map 553 shows the location of the Jap Cemetery and the Chinese Cemetery at a distance from the Protestant Cemetery and R.C. Cemetery athwart the road from Cumberland to Courtenay, just south of Maple Lake.

Miners from Comox No. 5 Mine posed for this photograph after extracting a record 524 tons of coal during a single eight-hour shift. These were brave men engaged daily in a dangerous occupation to earn a living for themselves and their families. The Cumberland Museum holds a list of the fatalities in the Cumberland mine alone; between 1888 and 1964 some 289 miners were killed, 69 in 1901, and 33 in 1923. Most were the victims of underground explosions.

MAP 554 (*left*).
Comox *No. 4 Mine*, on the north side of *Comox Lake*, about 1915.
Right.
A view of No. 4 Mine.

MAP 555 (*right*).
The underground map was entirely different. This is part of No. 4 Mine showing the intricate pattern of pillars and stalls.

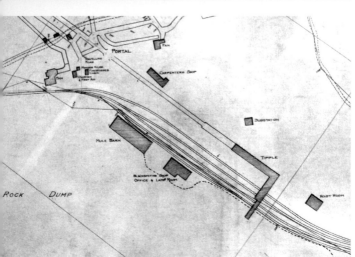

MAP 556 (*left*).
Many of the maps of the coalfields are huge, being working documents. This is a view of the map of Comox No. 4 Mine as displayed in the Cumberland Museum today; it is perhaps six metres long and two high, in a narrow passage—next to impossible to photograph!

strike pay had run out, when in August 1912 riots broke out, bombs were thrown, strike breakers and Chinese workers attacked, and the militia called in to restore order. The strike was never properly resolved and ended because of the start of World War I in August 1914.

The Extension Mine closed in 1931, but other coal mines in the Nanaimo area continued production until the 1950s.

The Dunsmuirs were also responsible for developing the coal resources of the Comox Valley, originally found in 1869 by Sam Cliffe, a Nanaimo miner. He did not have the capital to develop the coal deposits here and so sold out to Robert Dunsmuir.

MAP 557 (*far left, bottom*).
Detail of No. 4 Mine about 1922 shows the *Mule Barn*, the mine *Portal*, and the *Tipple* across the rails. Coal would be hauled on rails by the mules to the top of the tipple, where it could be loaded into railcars by gravity.

MAP 558 (*left*).
This 1924 map shows *Cumberland*. The ethnically segregated areas are those strung out to the west of the grid of roads. To the north is *Puntledge,* another satellite settlement, like *Bevan* built to house miners nearer their work. Comox Lake is at left.

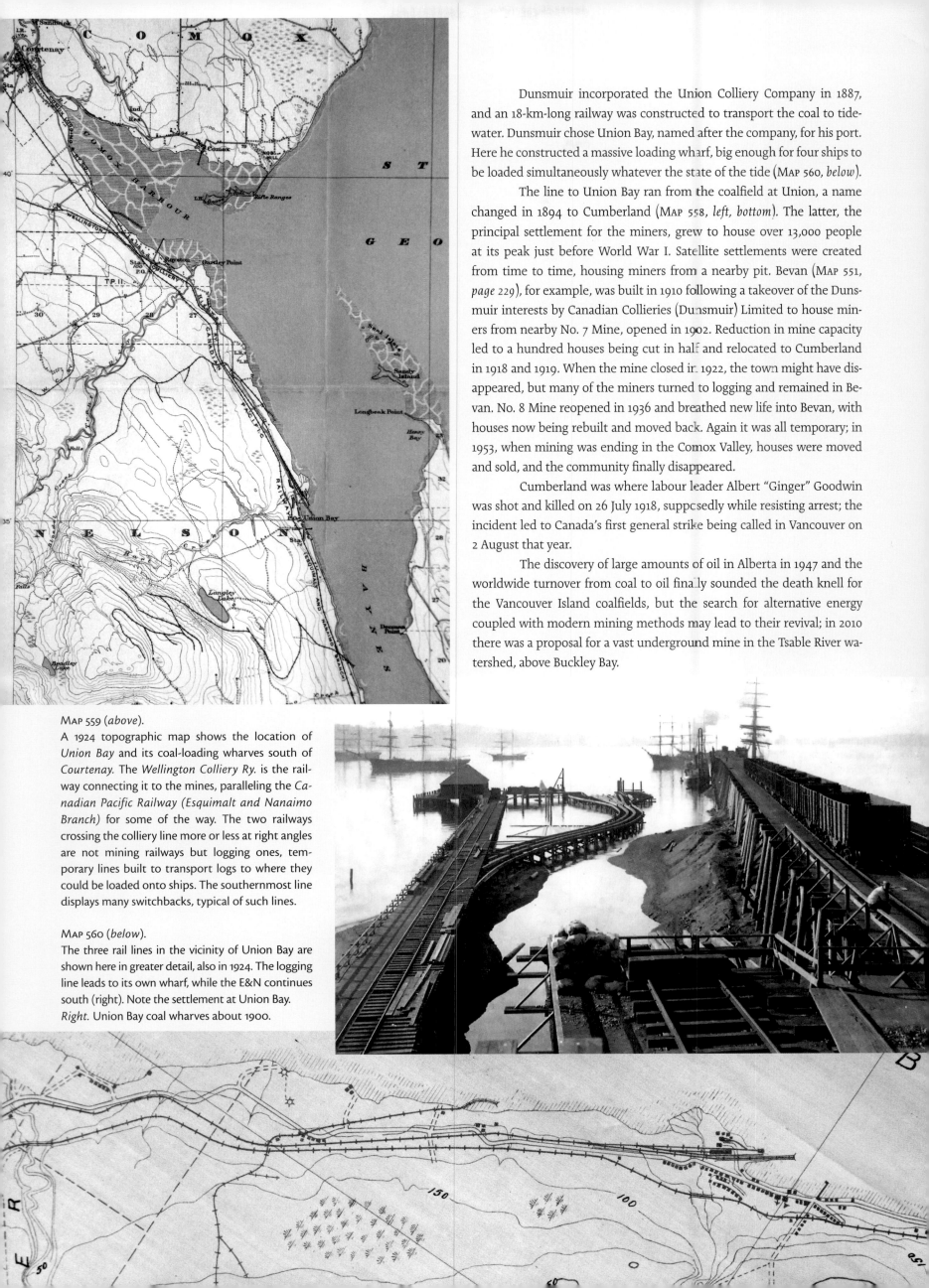

Dunsmuir incorporated the Union Colliery Company in 1887, and an 18-km-long railway was constructed to transport the coal to tidewater. Dunsmuir chose Union Bay, named after the company, for his port. Here he constructed a massive loading wharf, big enough for four ships to be loaded simultaneously whatever the state of the tide (MAP 560, *below*).

The line to Union Bay ran from the coalfield at Union, a name changed in 1894 to Cumberland (MAP 558, *left, bottom*). The latter, the principal settlement for the miners, grew to house over 13,000 people at its peak just before World War I. Satellite settlements were created from time to time, housing miners from a nearby pit. Bevan (MAP 551, *page 229*), for example, was built in 1910 following a takeover of the Dunsmuir interests by Canadian Collieries (Dunsmuir) Limited to house miners from nearby No. 7 Mine, opened in 1902. Reduction in mine capacity led to a hundred houses being cut in half and relocated to Cumberland in 1918 and 1919. When the mine closed in 1922, the town might have disappeared, but many of the miners turned to logging and remained in Bevan. No. 8 Mine reopened in 1936 and breathed new life into Bevan, with houses now being rebuilt and moved back. Again it was all temporary; in 1953, when mining was ending in the Comox Valley, houses were moved and sold, and the community finally disappeared.

Cumberland was where labour leader Albert "Ginger" Goodwin was shot and killed on 26 July 1918, supposedly while resisting arrest; the incident led to Canada's first general strike being called in Vancouver on 2 August that year.

The discovery of large amounts of oil in Alberta in 1947 and the worldwide turnover from coal to oil finally sounded the death knell for the Vancouver Island coalfields, but the search for alternative energy coupled with modern mining methods may lead to their revival; in 2010 there was a proposal for a vast underground mine in the Tsable River watershed, above Buckley Bay.

MAP 559 (*above*).
A 1924 topographic map shows the location of *Union Bay* and its coal-loading wharves south of *Courtenay*. The *Wellington Colliery Ry.* is the railway connecting it to the mines, paralleling the *Canadian Pacific Railway (Esquimalt and Nanaimo Branch)* for some of the way. The two railways crossing the colliery line more or less at right angles are not mining railways but logging ones, temporary lines built to transport logs to where they could be loaded onto ships. The southernmost line displays many switchbacks, typical of such lines.

MAP 560 (*below*).
The three rail lines in the vicinity of Union Bay are shown here in greater detail, also in 1924. The logging line leads to its own wharf, while the E&N continues south (right). Note the settlement at Union Bay.
Right. Union Bay coal wharves about 1900.

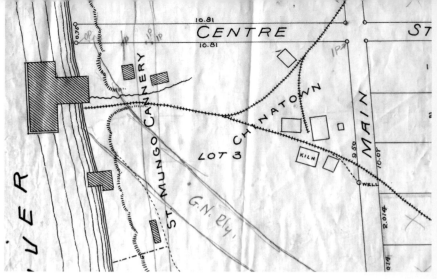

Fishing for Salmon

Arguably British Columbia's oldest industry, the coastal salmon fishery was originally responsible in no small part for the high densities of Aboriginal peoples along the coast, and the relatively easy availability of salmon was soon exploited by EuroCanadians.

The Hudson's Bay Company exported salmon from Fort Langley salted and packed in barrels. Although invented during the Napoleonic Wars, the canning process was not applied to salmon until the 1860s, and canneries did not proliferate in British Columbia until the 1880s. In 1882 the Phoenix Cannery in Steveston and the Richmond Cannery on Sea Island began operations. The North Pacific Cannery at what became Prince Rupert opened in 1889. The arrival of the transcontinental railway in 1885 enabled salmon to be more easily shipped to the east and on to Europe, and the industry grew apace.

Salmon processing and canning was a labour-intensive process and resulted in the employment of many Aboriginal or Chinese and Japanese workers, who would work for lower wages. The development of a mechanical processor known as the "Iron Chink," patented in 1905, reduced employment, especially after the Chinese Immigration Act of 1923, which halted Chinese immigration into Canada.

MAP 561 (*above*).
This is the *St. Mungo Cannery*, situated on the Fraser exactly where the southern end of the Alex Fraser Bridge is now located. The map seems to have been drawn about 1908, as the line of the *G. N. Rly.* (Great Northern) has been added in pencil. Living quarters for the Chinese cannery workers are at *Chinatown*. Two tramways and a *kiln* are depicted. The railway transported firewood to cook the salmon and to make charcoal, used to solder cans. The cannery closed in 1957 and the land sold in 1981 for the bridge's construction.

Above, left. A cannery collector boat towing the gillnet fleet to the fishing grounds. The photo was taken in the 1920s.

MAP 562 (*below, left, in two parts*).
Details from an 1897 fire insurance index map show the dozens of salmon canneries at the mouth of the Fraser River, especially in *Steveston*—enough to require an additional map (*left, bottom*) at a larger scale.

MAP 563 (*below*).
Typical of canneries along the British Columbia coast was the Wadhams Cannery on Rivers Inlet, on the mainland just north of Vancouver Island. This 1923 fire insurance map displays many huts used for living quarters on either side of the cannery indicating cannery workers were Chinese (*Chinese Bunk Ho.*) and *Japanese*. The photo shows the same cannery about 1902.

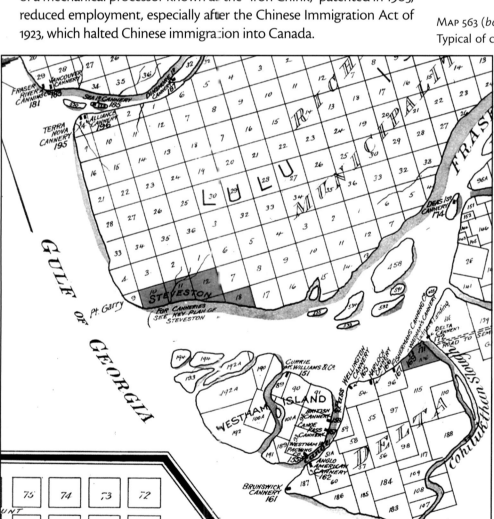

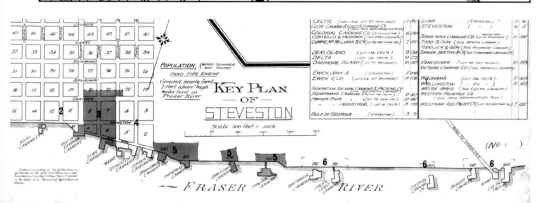

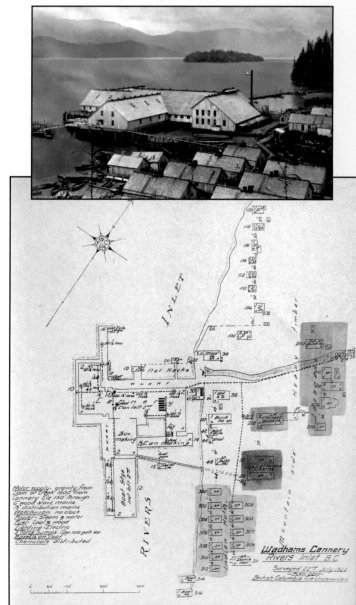

A Northern Treaty

In 1763, after the British defeat of the French in North America, King George III issued a Royal Proclamation that defined the basis for British governance: outside of Québec the land was defined as Aboriginal territory and required the negotiation of a treaty with the Crown before it could be settled by EuroAmericans. This was one of the factors behind the American Revolution, but the rule continued in force in the remaining British Territory.

In 1869, following the creation of Canada two years before, Rupert's Land and its northwesterly extensions—the land claimed by the Hudson's Bay Company—was transferred to Canada. In order to open this area legally for settlement, the Canadian government embarked on a long series of negotiations with Aboriginal peoples to obtain title to the land, resulting in the so-called Numbered Treaties, 1–11, negotiated between 1871 and 1921. One of these treaties, known as Treaty 8, affected the entire northeastern part of British Columbia.

After the Colony of Vancouver Island had been created in 1849 (see page 53), Governor James Douglas had made a series of purchases of limited amounts of land for areas around Victoria, Nanaimo, and Fort Rupert. The fourteen purchases, made between 1850 and 1854, are known as the Douglas Treaties. They were informally negotiated; several did not even state the area that they covered or the amount of money involved, and no maps were made to define their boundaries. Apart from these early treaties, Treaty 8 remained the only one signed with British Columbia Aboriginal peoples until the Nisga'a Treaty of 1998 (see page 335).

There is, however, confusion as to where the western boundary of Treaty 8 was intended to be drawn. No map was ever presented to Aboriginal negotiators. It seems clear from the government map drawn up before the signing that the boundary was intended to be the watershed between Pacific and Arctic waters (MAP 564, *above*), yet where the line was drawn was not the watershed. This was rectified on a map drawn by the government in 1900 (MAP 565, *left*), which shows the boundary line more exactly along the watershed, yet a later map (MAP 566, *right*), shows the boundary closer to where it was originally drawn, but not on the watershed. To add to the confusion, the text of the treaty refers to the western boundary only as the "central range of the Rocky Mountains." The imperfectly known and mapped geography of northern British Columbia at the time was largely to blame for this confusion, and the issue continues to be debated today.

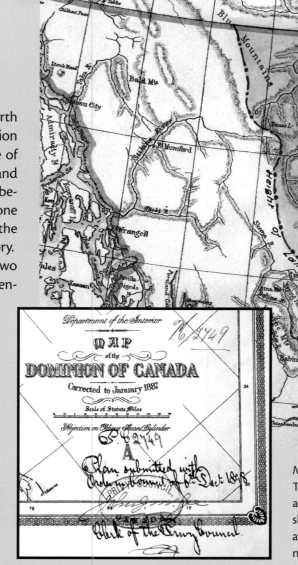

MAP 564 (*above*), with title (*inset, left*). The original map from the Privy Council authorizing the negotiation of Treaty 8. It shows the western boundary of the treaty area (A), extending to about 129°W at the northern boundary of British Columbia. *Height of Land*—usual terminology for a watershed—is written along the boundary line. The line is east of the actual watershed.

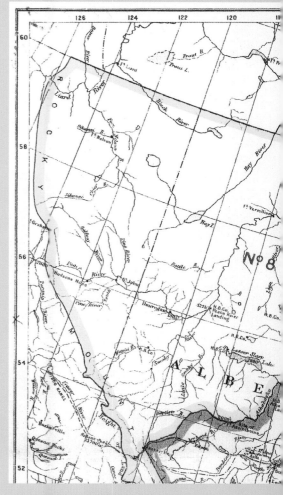

MAP 565 (*left*).
The map produced by the Department of Indian Affairs in 1900 shows the western boundary of the treaty extending beyond 130°W, which more exactly reflects a line drawn on the watershed.

MAP 566 (*right*).
In 1912 James White, geographer of the Department of the Interior, produced an official map showing all the treaties to that date. On this map the western boundary of Treaty 8 has been depicted extending only to 126°W, closer to where it was shown on the 1898 map but well east of the watershed line in the north. The *Height of Land* at the southern end is accurate, however.

The Alaska Boundary Dispute

When Russia signed a treaty with Britain and the United States in 1824 and 1825 setting a boundary to the countries' territories, no one cared that it was ill-defined. But after the United States took over the Russian claim by purchasing Alaska in 1867, and especially during the Klondike Gold Rush beginning in 1897, the exact position of the boundary suddenly assumed much greater importance.

The original treaty wording had specified the panhandle boundary as the summit of the mountains parallel to the coast, and where the mountains were more than ten marine leagues (56 km) from the coast, the boundary was to be drawn no more than ten marine leagues from the coast. "The coast" in a region such as the island-strewn and highly indented panhandle was in itself ambiguous, and if the mountains were not an obliging 56 km away from it, the exact position of a boundary line was undefined.

The Klondike gold discovery could have caused a war. The Canadian government sent soldiers to the region, and the redoubtable Sam Steele of the North-West Mounted Police set up a post at the top of Chilkoot Pass to enforce Canadian customs regulations. In 1898 the American and Canadian governments attempted to negotiate a mutually agreeable boundary but failed. They did agree to a temporary boundary in the area of the head of Lynn Canal, however. It was a start.

On 24 January 1903 U.S. secretary of state John Hay and British ambassador to the U.S. Michael Herbert signed a treaty (the Hay-Herbert Treaty, usually referred to as the Alaska Boundary Treaty) to settle the boundary issue. They agreed to set up a tribunal of six to decide where the boundary lay. Three Americans, two Canadians, and a British judge, Lord Alverstone, were appointed to the tribunal. Ominously, the Americans included Elihu Root, the U.S. secretary of war. A majority was required to agree to any settlement, but the tribunal's decision would be final.

The Americans wanted the easternmost boundary shown on these maps. This did seem to approximate the roughly mapped boundary marked on both the last Russian maps (Map 567, *left, top*) and the map showing the territory purchased by the Americans from Russia in 1869 (Map 568, *left*). This boundary, however, did not always follow mountaintops and was sometimes more than 56 km from the coast—however defined. The British and Canadian governments pushed for a boundary much closer to the sea; rightly or wrongly they were attempting to at least preserve access to the heads of the longer fiords. On 20 October a decision was handed down. The new boundary was to be drawn between those the two governments wanted, but generally much closer to the American claim—and farther from the coast—than the Canadian line. Critically, Canada lost all access to any of the inlets, thus effectively cutting off half of British Columbia from the sea.

The decision generated a great deal of anger in Canada, not so much against the Americans but against the British, for the British tribunal member, Lord Alverstone, chief justice of England, sided with the Americans to allow them to get their majority. It is possible that he was instructed to do so because Britain feared American president Theodore Roosevelt would use force. Five years earlier Roosevelt had taken part in the American invasion of Cuba, and the United States seemed in an empire-building mood.

The Alaska boundary decision was carefully documented, with a series of detailed map sheets showing the exact position of the final boundary. One is shown below (Map 570).

Map 567 (*above, top*).
This 1863 map appears to be the last Russian map to show a panhandle boundary, so it presumably shows what the Russians thought they were selling.

Map 568 (*above*).
This was the territory in the panhandle that the United States thought it was purchasing from Russia in 1867; it is part of a map documenting the purchase.

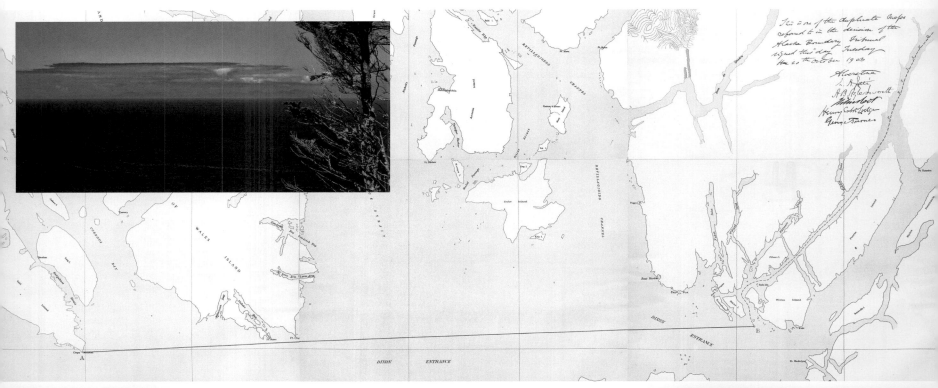

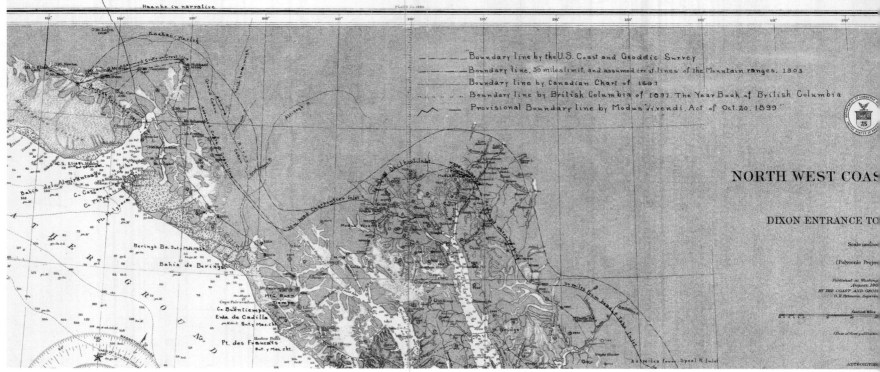

MAP 569 (*left, top*).

The boundary as first claimed by Canada and British Columbia was very close to the coast, as can be seen on this 1897 map, which also shows the *boundary claimed by United States*. The *Old Canadian Customs House* on the *Stikine R.* was in what is now Alaska.

MAP 570 (*left, bottom*).

The boundary as agreed in October 1903 was documented by a series of detailed maps, each signed by the six boundary commissioners. *Alverstone*, the British representative who sided with the Americans, heads the list.
Inset. Cape Muzon, Alaska, is just visible in this photo taken from Haida Gwaii north across Dixon Entrance.

MAP 571 (*above*).

The details of the Alaska boundary dispute were more complicated than they appear on many maps; this one gives some idea of the complexity. The detail is of the head of Lynn Canal and around *Skagway*, the principal route of gold seekers trekking to the Klondike after 1897 and the cause of much of the boundary frictions. The *Modus Vivendi* line was a temporary boundary agreed in 1899 around this critical area. Several boundaries are marked as 30 miles from the head of an inlet; this (56 km) was the interpretation of 10 leagues from the coast as stated in the 1825 treaty between Russia and the United States.

MAP 572 (*right*).

This American map drawn up just three days after the boundary agreement was signed is a copy of the map attached to the treaty. The newly agreed-upon boundary, in red, sits between that claimed by the Americans and that first claimed by the British, near the coast.

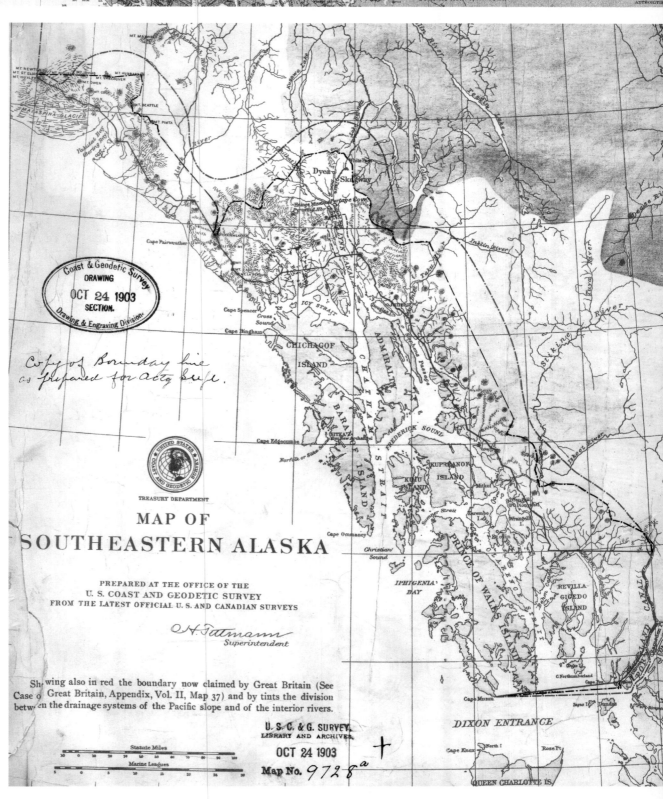

MAP OF
SOUTHEASTERN ALASKA

PREPARED AT THE OFFICE OF THE
U. S. COAST AND GEODETIC SURVEY
FROM THE LATEST OFFICIAL U. S. AND CANADIAN SURVEYS

O.H. Tittmann
Superintendent

Showing also in red the boundary now claimed by Great Britain (See Case of Great Britain, Appendix, Vol. II, Map 37) and by tints the division between the drainage systems of the Pacific slope and of the interior rivers.

THE BOOMING LOWER MAINLAND

The years just before World War I were boom times for Vancouver, as they were for all the major West Coast cities. Vancouver's population grew from 26,000 in 1901 to 100,000 in 1911, and continued growth was expected because of the building of the Panama Canal. Construction on the canal had actually begun in 1880 under French direction but had been abandoned in 1893 after many engineering difficulties. A new effort, led by the United States, began in 1904—after Panama was created from Colombian territory the year before—and now looked on track to be completed by 1916.

Port expansion schemes and even the creation of vast new ports such as one that would have covered the west coast of Richmond's Lulu Island (MAP 608, *page 245*) sprang into life. Three new railways added to the promise of real estate profits—the Grand Trunk Pacific, to Prince Rupert (see page 250); the Canadian Northern, another line to Vancouver (page 272); and the Pacific Great Eastern, which was not completed to its originally planned destination until the 1950s but did run from the British Columbia interior to Squamish (page 278). In addition, in 1910 an electric interurban line from New Westminster to Chilliwack opened up the Fraser Valley (page 270). And new municipalities emerged: the City of North Vancouver in 1907, West Vancouver in 1912, Port Coquitlam in 1913 (the name deliberately chosen to suggest a port despite its relatively inland position), and Pitt Meadows in 1914.

The Panama Canal was completed earlier than expected, opening in August 1914. But the timing could hardly have been worse. For that very same month World War I began in Europe. Commercial shipping volumes in the Pacific plummeted, and thousands of would-be employees and erstwhile investors left Canada for the war in Europe. With them demand for real estate in any form dropped dramatically, with the result that many of the schemes promoted to sell real estate and prepare for the coming boom found themselves without financial backing and without customers. Lots that had changed hands for hundreds of dollars just a couple of years before were being seized by municipalities for non-payment of taxes. Vancouver would survive, of course, but would henceforth grow at a more sustainable rate.

The boom had been a real estate promoters' paradise. In 1913 there were 1,053 real estate salesmen in the city; by comparison, there were only 325 grocers. Almost anything could be sold, it seemed, if it promised vast profits—and they all did. The newspapers and magazines of the period are full of wonderful advertisements that more often than not used maps to sell their subdivisions and schemes. The advertising—if not the projects they promoted—live on to amuse. But many, of course, were solid projects and promoted growth in the metropolis that was going to happen in any case—later if not at the time. On the next few pages are some examples of the real estate advertising and other promotional literature of the period.

Above, top. This 1912 collage of recent Vancouver construction demonstrates the burgeoning economic position of Vancouver.

MAP 574 (*above*).
The Klondike Gold Rush in 1897 put Vancouver on the map as a major supply base.

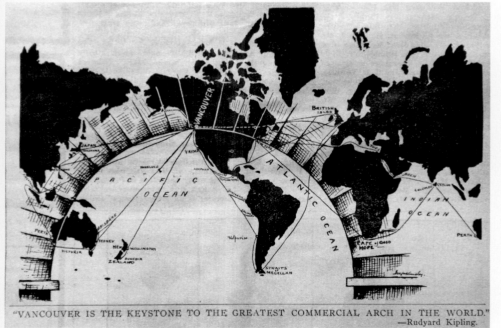

"VANCOUVER IS THE KEYSTONE TO THE GREATEST COMMERCIAL ARCH IN THE WORLD."
—Rudyard Kipling.

MAP 573 (*below, left*).
This map purporting to illustrate the "keystone" position of Vancouver—complete with a Rudyard Kipling quotation, no less—was published in April 1909.

MAP 575 (*right*).
A more reasonable map illustrating the key position of Vancouver.

Above, left.
This cartoon from a 1912 ad urges investors to ignore the suburbs—the "sideshows"—for another development that was not itself in the city. "Broadview" was adjacent to Hastings Townsite, annexed the year before.

Vancouver's Strategic Position On World Trade Routes

VANCOUVER--
The Distributing Base for Western Canada.

THE opening of the Panama Canal transformed the transportation map of North America and definitely established Vancouver as THE Distributing Base for Western Canada.

Western Canada's huge diversified production now moves to world markets via Vancouver, creating at this city an industrial and mercantile centre destined to dominate Pacific trade and commerce.

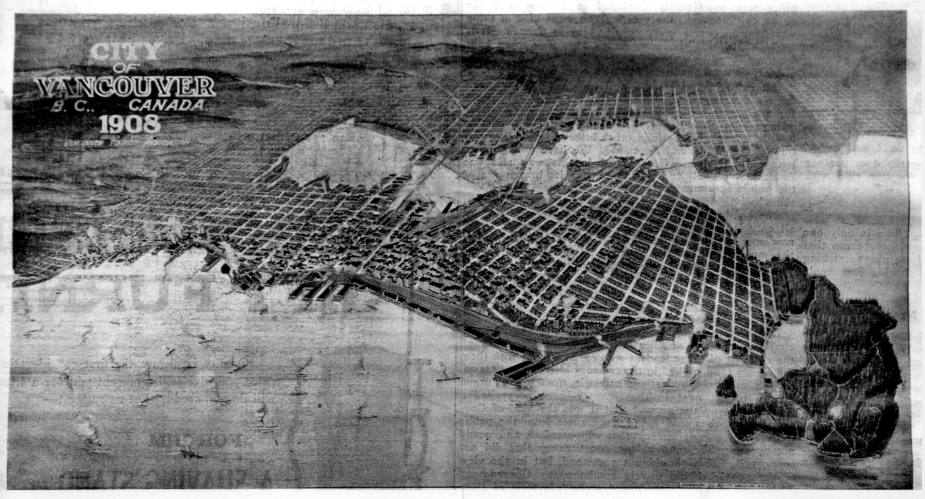

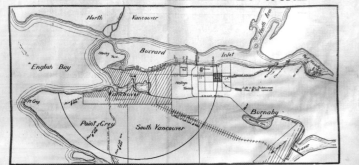

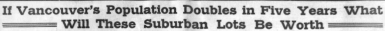
MAP 576 (*above, top*).
A nicely executed 1908 bird's-eye tourist map of Vancouver viewed from the north.

MAP 577 (*above, left*) and MAP 578 (*above*).
MAP 578 is another excellent bird's-eye map of Vancouver, this time viewed from the east. It was used to illustrate the proximity of a developer's subdivision, East Vancouver Heights, to the city. The map contains a number of features then only proposals, such as a Second Narrows bridge. This is clearly the same map used in the ad (MAP 577) promoting these kinds of artistic maps for real estate advertisers.

MAP 579 (*left*).
The case for purchasing lots, such as those in East Vancouver Heights, for investment. Indeed, lots bought in 1909, the date of this ad, would have been worth considerably more in five years, but then would have lost much of that value in the five thereafter.

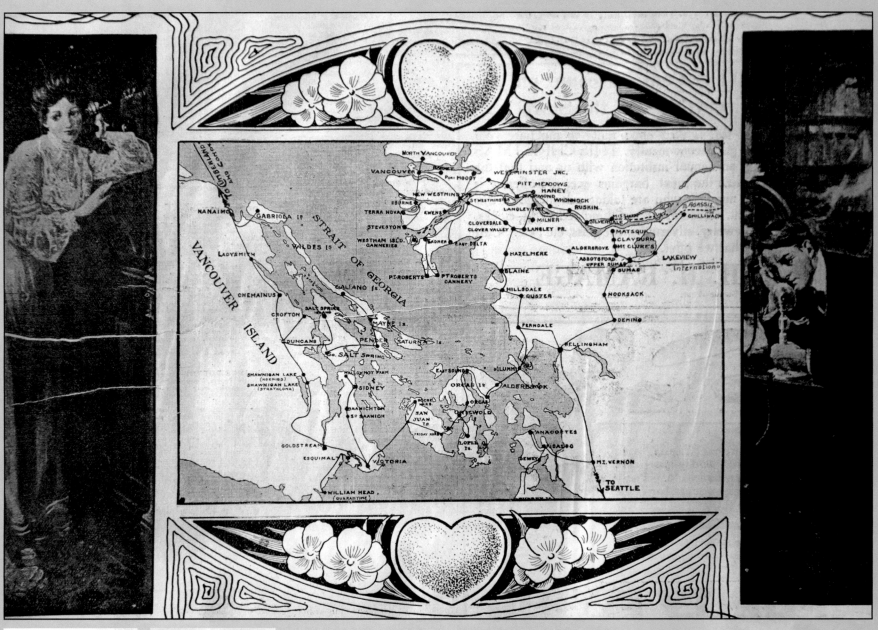

MAP 580 (*above*).

By December 1909, the date of this map, Vancouver had over eight thousand telephones, and 350 new phones were being installed every month. Vancouver had more private phones per capita than any other Canadian city. This map shows the network of main circuits—heavy copper cable trunk lines—then in operation, and the lines reach *Vancouver Island* via the American San Juan Islands. A new direct submarine cable was to be laid the following year. The map was part of what today we would probably call an infomercial on television or an advertorial in a magazine—lots of information but essentially geared to selling more phones. People apparently had to learn how to use the phone correctly—hence the pair of photos at left.

MAP 581 (*left*) and MAP 582 (*right*).

The City Beautiful planning movement was alive in Vancouver at the beginning of the twentieth century. It espoused the marriage of town and country to beautify the city. Eminent British landscape architect Thomas Mawson produced plans for Vancouver in 1913, which included making Georgia Street a pedestrian-only grand boulevard linking a civic centre with Stanley Park (MAP 581). This was to end at a *Round Pond* shown on his plan for the entrance to Stanley Park (MAP 582). Although the ideas appealed to many, with the onset of war they were not implemented. They did lead to the passing of the Town Planning Act in 1925 and the creation of the Vancouver Town Planning Commission the following year. It was this commission that hired St. Louis planning consultant Harland Bartholomew in 1928, and a number of features from his 1929 plan were incorporated into Vancouver. Similar ideas for a grand boulevard were used in North Vancouver (see MAP 595 and MAP 596, *page 242*).

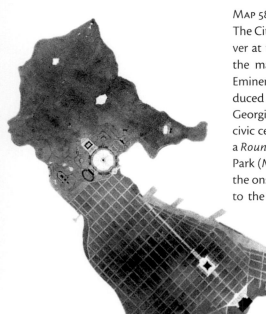

The above sketch of EBURNE TOWNSITE

BY A WELL KNOWN ARTIST, AN R. A., SPEAKS VOLUMES

IT SHOWS that EBURNE TOWNSITE is in the midst of a settled community and NOT in the backwoods.

IT SHOWS one mill, but four others are in course of erection, besides a flour mill, all a short distance from the Electric Railway Station.

It shows that settlers do not have to wait a YEAR OR TWO for transportation, as the B. C. Electric Company has a splendid service NOW, hourly at present, but half-hourly soon.

The run from Vancouver to Eburne is made in fifteen minutes, but when the car is in a hurry it does it in ten, and we expect it will be named THE SOUTHERN FLIER.

The junction of the THREE Electric Railways is north of the general store at the foot of Fourth St. while there are stations at foot of Granville Street on Townsend Road—TRAVELING ACCOMMODATION galore. If the SOUTHERN FLIER be too speedy for you the Mail Stage can land you at Vancouver Post-office in sixty minutes prompt, while the driving, wheeling and walking is good, and the distance on Granville Street is less than six miles.

AT EBURNE there is abundance of good water; there is electric light, telephone, postoffice, church, school, Odd Fellows' Hall, stores, etc., etc.

EBURNE TOWNSITE is one of the very few subdivisions that has offered buyers NO INDUCEMENTS in the shape of presents, prizes or dollar a week payments, and that has NOT OFFERED lots by auction to the highest bidder (if high enough), the owners being satisfied that they have a GOOD property and they offer sterling value in realty for sterling money, in fact a "Roland for an Oliver."

WE MAKE THE POSITIVE ASSERTION that NO Suburban Townsite around Vancouver can offer such advantages as EBURNE TOWNSITE does.

TAKING these advantages into consideration, there is no other Subdivision or Townsite offers such value in lots as we do at EBURNE; no place where you can buy SUCH LARGE LOTS—all 50 feet wide by 120, 130 and some even 140 feet deep—at such LOW PRICES and SUCH GOOD TERMS as you can in EBURNE TOWNSITE.

Just notice how many properties are advertised as being "near Eburne Townsite" close to Eburne" and how advice is given to buy in the "Eburne district." These prove our contention that Eburne Townsite is the Hub, the Crux, the Centre, of a live and rising district where values are steadily increasing. The nearer the Hub you buy, the quicker and greater the increase. THEREFORE, WHY NOT BUY RIGHT IN THE HUB ? Which is Eburne Townsite.

Present Prices for lots are from $250 up--1-4 cash ; balance over three years.

THE CENTRAL REAL ESTATE COMPANY

JOHN LOVE, Secretary. **Phone 203** **737 PENDER ST., VANCOUVER**
(60 Yards West from Granville Street)

MAP 583 (*above*).

This fine illustration, a hybrid view and map, was an ad for Eburne Townsite published in 1907 when it was part of South Vancouver municipality. The following year Point Grey was created, and Eburne was included within its boundaries. Note the interurban car at right; the Canadian Pacific line to Steveston had begun steam operation in 1902, and the line had been leased to B.C. Electric in 1905 for electric interurban operation (see page 270).

MAP 584 (*right*).

Eburne was still being advertised in 1911. Lots were priced higher on Oak Street than elsewhere owing to the imminent arrival (January 1913) of a streetcar line, interestingly the exact opposite of today—prices are lower on Oak Street because of the traffic noise. Note also on this map two interurban lines; the one to New Westminster from Eburne had just begun operation. Originally both sides of the river were called Eburne, and so, in 1916, the north bank community was renamed Marpole after Canadian Pacific general superintendent Richard Marpole. The view *below* shows Eburne across the road bridge from Sea Island, the left-hand bridge on the map, and reveals the area as quite heavily industrialized, with a number of sawmills. The photo was taken in 1911.

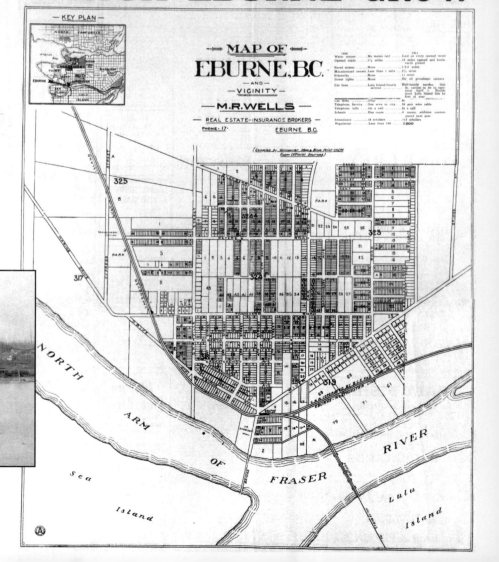

"VANCOUVER REPORTS PROGRESS."

Above.
This illustration appeared in a magazine in 1910, reporting on the construction boom in downtown Vancouver.

MAP 585 (*above, right*).
Point Grey was created from the western part of South Vancouver in 1908. Sometimes huge tracts of land were offered for sale. This map shows one such offering, for an auction in 1909. Note the planned streets for the University Endowment Lands.

TRITES & LESLIE
REAL ESTATE AND AUCTIONEERS
659 GRANVILLE ST.
VANCOUVER, B. C.

LANGARA
POINT GREY
BEING SUBDIVISION OF LOTS 139, 140, 540 AND 2027
NEW WESTMINSTER DISTRICT, B.C.
1909
LOTS COLORED RED ARE TO BE SOLD BY AUCTION

BROADWAY'S GLORIOUS FUTURE

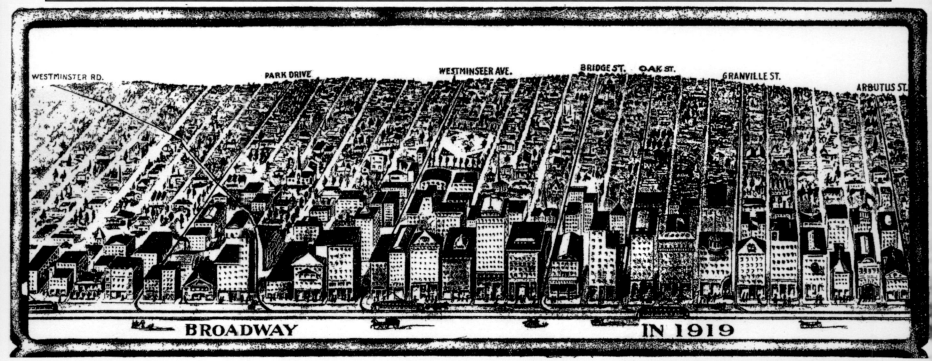

BROADWAY   IN 1919

MAP 586 (*above*).
In 1909 one imaginative real estate broker selling property along Broadway created this bird's-eye-type map showing what it was expected to look like ten years later. Unfortunately the broker had not allowed for the intervention of World War I.

MAP 587 (*left*).
Another real estate promoter created this map in 1911 to illustrate the position of its "Marine View" subdivision (at the red arrow).

MAP 588 (*right*).
Vancouver is shown as the commercial centre of the world, or at least the Pacific, on this little map published in 1908 by the Vancouver Tourist Association.

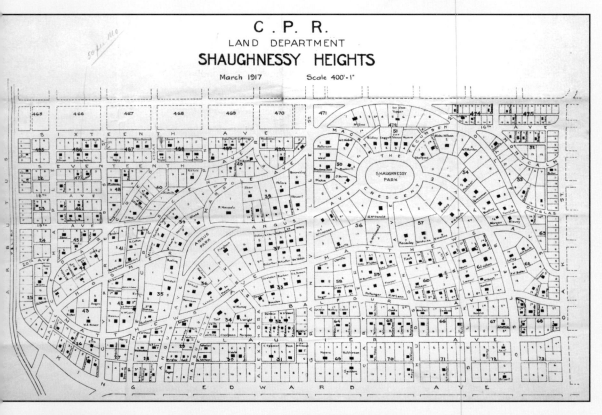

C. P. R.
LAND DEPARTMENT
SHAUGHNESSY HEIGHTS
March 1917   Scale 400'·1'

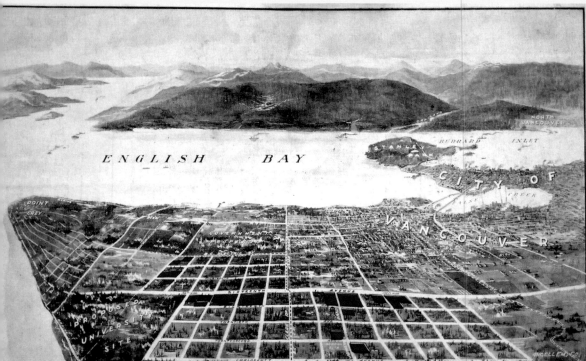

Map 589 (*left, top*).
In the early years of the twentieth century a number of wealthy landowners attempted to create an upscale alternative residential area to the West End in Grandview, then at the eastern end of False Creek. The Canadian Pacific, not to be outdone and not wishing to lose its control of land development around Vancouver, in 1907 began planning its own upscale area on its land grant. The result was Shaughnessy Heights. Before a single house was built the company spent $2 million preparing the site. This is the first part of the development, shown in 1917, by which time a goodly number of fine houses had been built. Owners' names are also shown, and they include *R. Marpole*, Richard Marpole, general superintendent of the railway's Pacific Division. Restrictive zoning has ensured that the area remains one of large houses and high incomes.

Map 590 (*left, centre*).
West-side residents had petitioned the provincial government to create Point Grey municipality in 1908 so that it could remain mainly residential in nature. Although much of the area at the time was forest (apart from the sawmilling centre of Eburne), most was generally subdivided for residential development, and some large areas were sold for this purpose (**Map 585**, *far left*). Few houses were built until the 1920s, though this did not stop speculators from buying lots. One of the most notorious pre–World War I real estate brokers was Alvo von Alvensleben, the advertiser of these lots shown in red in Point Grey but right on the boundary with Vancouver (16th Avenue). This well-prepared 1910 bird's-eye map names the *City of Vancouver* at right. Dashed lines show existing and planned streetcar lines.

Von Alvensleben was a German national involved in many real estate ventures, bringing much German and other European capital to British Columbia. He is perhaps best known for his ownership of the Wigwam Inn at the head of Indian Arm. He was widely assumed to be a German spy, and von Alvensleben's assets were seized at the beginning of the war. He spent the war years—until 1917—in Seattle. It did him no good, for when the United States entered the war that year, he was interned anyway.

Map 591 (*right*).
This very fine bird's-eye map was used to sell North Shore real estate in 1911—property that would the following year be part of the new municipality of West Vancouver. A ferry had begun operating between *Hollyburn* (today Ambleside) and the West End of Vancouver in 1909, serving land being developed by John Lawson. This map shows a rival subdivision by real estate brokers Irwin and Billings at *Dundarave*, complete with its own proposed ferries shown. (The Hollyburn ferry has deliberately been omitted.) A *Proposed Tunnel* is shown across the First Narrows. Lots were already being offered at high prices—$400 to $1,200.

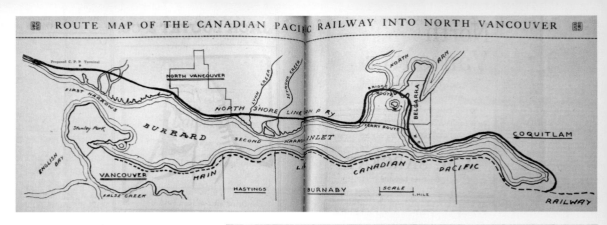

MAP 592 (*right*).
A planned route for the Canadian Pacific Railway onto the North Shore, published in 1912. From Port Moody the line was to cross Indian Arm either by ferry or by a bridge from Belcarra, where the inlet narrows and is shorter than the route ultimately chosen at Second Narrows. The latter was completed in 1925.

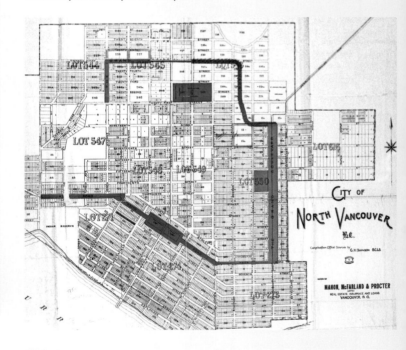

MAP 594 (*above*).
This is a map of the City of North Vancouver's townsite produced in 1907, the year the city was incorporated.

MAP 593 (*above*).
The City of North Vancouver was a real estate promoter's delight. Lots with views, a ferry connection with Vancouver, and—after 1906—streetcars and electric lights. Indeed, the coming of the streetcar was one of the forces behind the idea of carving off the city from the district—a compact city would be easier and cheaper to service. A map of the routes by 1911 is shown as MAP 362, *page 141*. This bird's-eye map and view combo dates from 1907. Note *Moodyville,* not included in the city until 1925, at bottom right.

MAP 595 (*right*) and MAP 596 (*right, bottom*).
The City Beautiful movement came to North Vancouver when a group of real estate promoters proposed a boulevard circumnavigating the city as part of a plan to create high-value lots, an idea that failed when the Canadian Pacific developed its Shaughnessy Heights (MAP 589, *previous page*). Grand Boulevard and Victoria Park remain as substantial remnants of this concept.

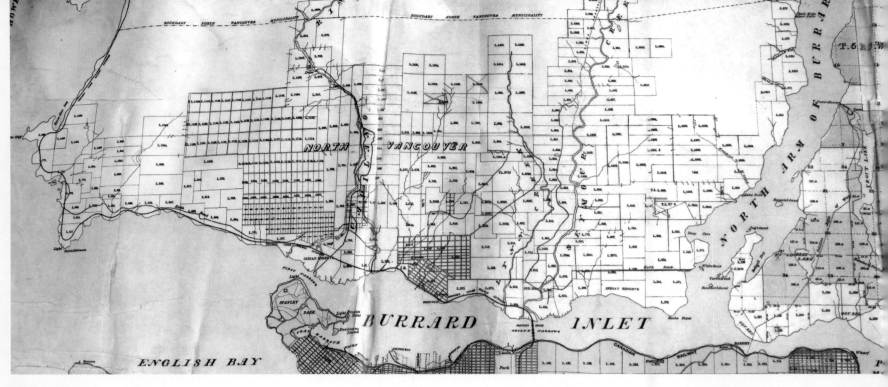

Map 597 (*above*).

The District of North Vancouver was incorporated in 1891 and covered the entire North Shore. As this 1905 map suggests, most of the population lived in what would become the City of North Vancouver in 1907. The district was split in half when West Vancouver was created in 1912.

Stanley Heights The Beauty Spot of North Vancouver....

$100 Each
$5.00 Down
$5.00 Each
Month.

No Interest
No Taxes
No Gullies
No Bedrock

Large
High
Dry
View
Lots
To
Good
Lane.

COME WITH US TO STANLEY HEIGHTS

Then you will know that any lot is a good one. Our boat leaves Billingsgate Wharf (just east of North Vancouver Ferry Wharf) each day at 2 o'clock.

PHONE 970. Jones & Franklin AND S. F. Munson
OPEN EVENINGS. 449½ PENDER ST. West.

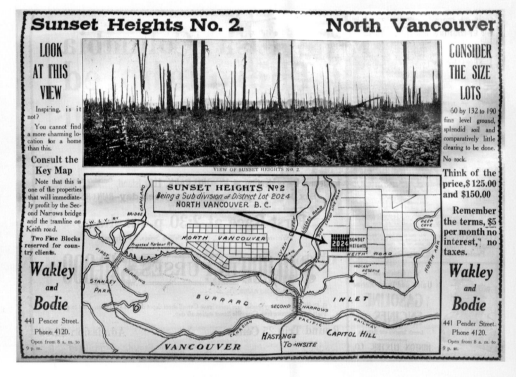

Sunset Heights No. 2. North Vancouver

LOOK AT THIS VIEW

Inspiring, is it not?

You cannot find a more charming location for a home than this.

Consult the Key Map

Note that this is one of the properties that will immediately profit by the Second Narrows bridge and the tramline on Keith road.

Two Fine Blocks reserved for country clients.

Wakley and Bodie

441 Pender Street.
Phone 4120.

Open from 8 a. m. to 9 p.m.

VIEW OF SUNSET HEIGHTS NO. 2.

SUNSET HEIGHTS No 2
Being a Sub division of District Lot 2024
NORTH VANCOUVER B. C.

CONSIDER THE SIZE LOTS

50 by 132 to 190 fine level ground, splendid soil and comparatively little clearing to be done. No rock.

Think of the price, $125.00 and $150.00

Remember the terms, $5 per month no interest," no taxes.

Wakley and Bodie

441 Pender Street.
Phone 4120.

Open from 8 a. m. to 9 p. m.

Map 598 (*above*),
Map 599 (*above, right*),
and Map 600 (*right*).

These three ads promote subdivision in the District of North Vancouver, all in 1909. Map 598, *Stanley Heights*, also shows the innovative *Roslyn* circular subdivision (see Map 602, *overleaf*). It seems that *Sunset Heights* (Map 600, published in July) must have sold satisfactorily, as *Sunset Heights No. 2* appeared in September of the same year (Map 599). The latter adds a view, showing a logged and burned-over landscape but showing that the location had a view. The *Surveyed Electric Ry* of July is gone, however. Also gone is the *Surveyed R.R. Line* towards *Deep Cove Summer Resort,* having been replaced with a *Proposed Harbour Ry* across the Second Narrows.

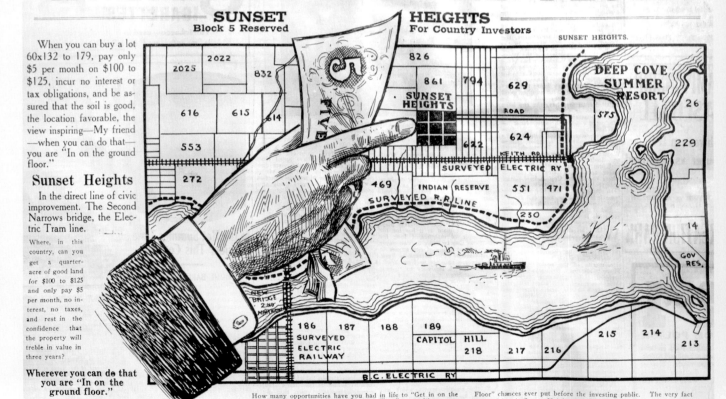

"IN ON THE GROUND FLOOR"

SUNSET HEIGHTS
Block 5 Reserved For Country Investors

When you can buy a lot 60x132 to 179, pay only $5 per month on $100 to $125, incur no interest or tax obligations, and be assured that the soil is good, the location favorable, the view inspiring—My friend —when you can do that— you are "In on the ground floor."

Sunset Heights

In the direct line of civic improvement. The Second Narrows bridge, the Electric Tram line.

Where, in this country, can you get a quarter-acre of good land for $100 to $125 and only pay $5 per month, no interest, no taxes, and rest in the confidence that the property will treble in value in three years?

Wherever you can do that you are "In on the ground floor."

Sunset Heights

441 PENDER STREET

441 Pender Street.

How many opportunities have you had in life to "Get in on the Ground Floor?" Ten years ago "Ground Floor" propositions were offered that today requires an elevator to reach.

You have often said "I'd go in on that if I could 'get in on the Ground Floor.'" Sunset Heights is one of the best "Ground

Floor" chances ever put before the investing public. The very fact that 400 lots in Sunset Heights have been sold in seven days is proof that the people know, appreciate and buy a "Ground Floor" offering. Are you one of the far seeing ones? You are—So don't neglect this for one more day or your opportunity may be gone.

WAKLEY & BODIE 441 Pender Street.

Our telephone number is 4120. Remember we will be open Saturday afternoon and evening.

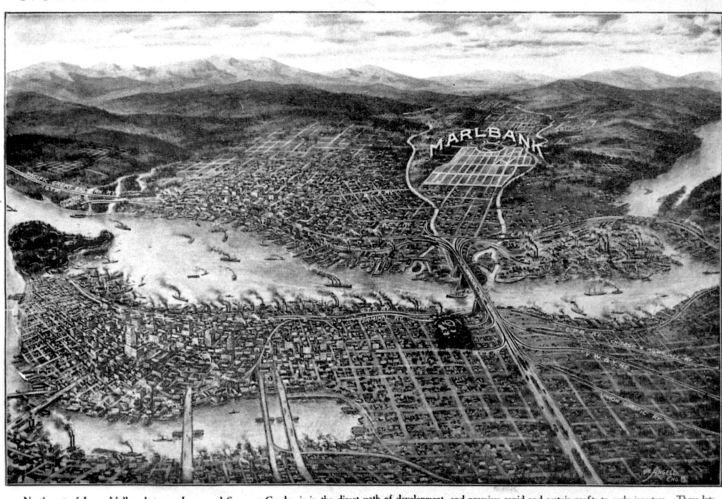
Map 601 (*above*).

This 1912 map is one of the finest bird's-eye real estate ads of the period, offering lots for $200 and $250 at *Marlbank*, a North Vancouver subdivision between Lynn and Seymour Creeks. The artistic licence taken is considerable. A multi-railway and virtual freeway road bridge crosses Second Narrows where none existed, and there is a railway bridge across Indian Arm. Numerous bridges cross False Creek, and Vancouver is depicted in the smoky way that was supposed to indicate success. The idea was to show the city in the future—with Marlbank as a safe investment among all that prosperity.

Map 602 (*below*).

Rosslyn was an innovative 1910 subdivision laid out with circular streets. Intended to attract an upscale clientele, the streets were named in order of supposed rank: Imperial Crescent as the inner circle, then Royal, King, Queen, Prince, Princess, and Duke. The developer built a wharf but could not sell the lots, and the land was later used for a cedar mill.

Map 603 (*right*).

This 1911 ad advertised the Grouse Mountain Scenic Incline Railway, a funicular railway planned for Grouse Mountain that was not built. A resort on the mountain, reached by a toll highway from the top of Mountain Highway, was completed in 1926. In 1951 a chairlift allowed access from Skyline Drive in North Vancouver, and a cable tramway was opened 2 km to the northwest in 1966, the predecessor of today's Grouse Mountain Skyride.

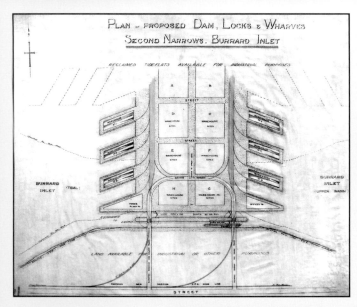

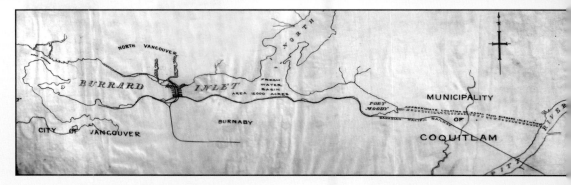

MAP 604 (*left*) and MAP 605 (*above*).

A 1910 proposal to dam the Second Narrows and dig a canal from *Port Moody* to the *Pitt River*, creating a freshwater basin east of the dam. The dam was supposedly cheaper to build than a bridge. The canal would have allowed large ships to reach what became Port Coquitlam without having to negotiate the difficult entrance to the Fraser River. The idea died when Vancouver city council refused to finance the dam, and a bridge, at the time slated to be complete by 1912, was finally opened in 1925.

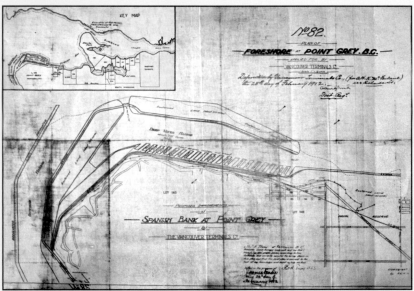

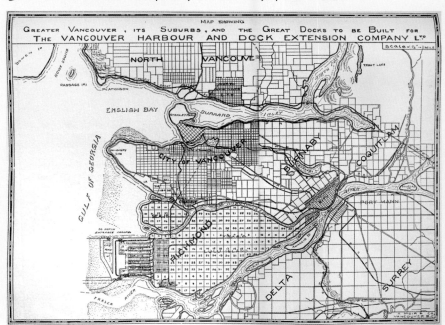

MAP 606 (*above*).

The anticipated boom in West Coast shipping with the opening of the Panama Canal spawned many schemes for new harbour facilities. One that would have changed the face of Vancouver forever was this 1912 proposal to convert Point Grey and Spanish Banks into a major harbour, thus avoiding the necessity of passing through the tide-controlled First Narrows. Spanish Banks would have been dredged to provide the fill for a massive breakwater. This plan was again considered in a 1929 planning report by consultant Harland Bartholomew but was rejected principally because of its high cost.

MAP 607 (*above*) and MAP 608 (*below*).

In 1911 industrialist Charles Pretty and other investors formed the Vancouver Harbour and Dock Extension Company and came up with this amazing proposal to create a vast dockland on Sturgeon Bank off *Richmond*. The cost and declining economic conditions killed this scheme, but others similar to it followed in the 1950s and in 1970, when what was intended to be the first part of the Roberts Bank Superport was opened (see page 324), for, if one ignores ecological considerations, the idea is entirely practicable. Pretty's scheme did leave us with this wonderful bird's-eye map, published in November 1912.

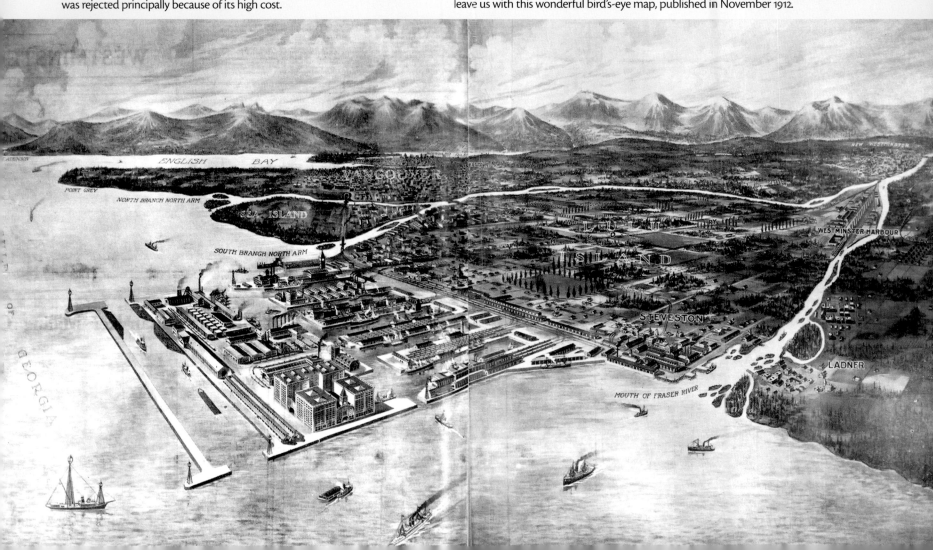

Map 609 (*left*).
A dock complex at Kits Point proposed in 1917 by the Chicago, Milwaukee, St. Paul & Pacific Railroad, which was contemplating a Vancouver presence following completion of its transcontinental line to Seattle in 1909. This scheme would have been built on an Indian reserve and thus circumvented the Canadian Pacific land grant. The Canadian Pacific had itself considered a similar scheme in the 1890s to avoid First Narrows (Map 346, *page 135*).

Map 610 (*below, centre*).
New Westminster, not to be left behind, commissioned this plan in 1912 to expand its own docks by cutting off the eastern end of Annacis Channel and using it as a harbour basin. The city was hoping to entice the Chicago, Milwaukee, St. Paul & Pacific or even the Union Pacific, as both had announced an intention to extend to Vancouver. But ships would still have had to enter via the shallow Fraser delta.

Map 611 (*below, bottom*).
This fine, unusually angled bird's-eye map was used in 1912 to promote *Industrial Terminals* on the Fraser after the New Westminster Bridge was completed in 1904.

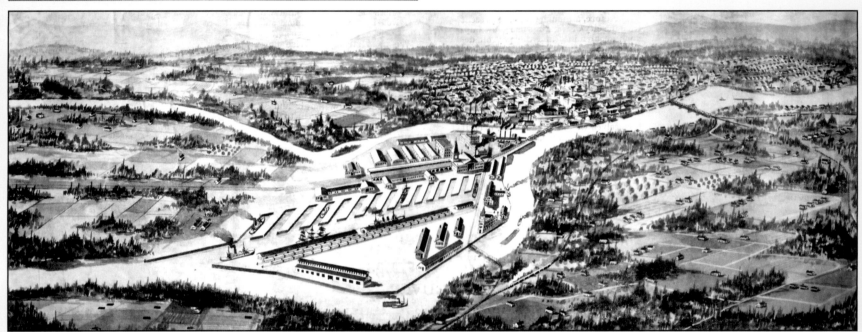

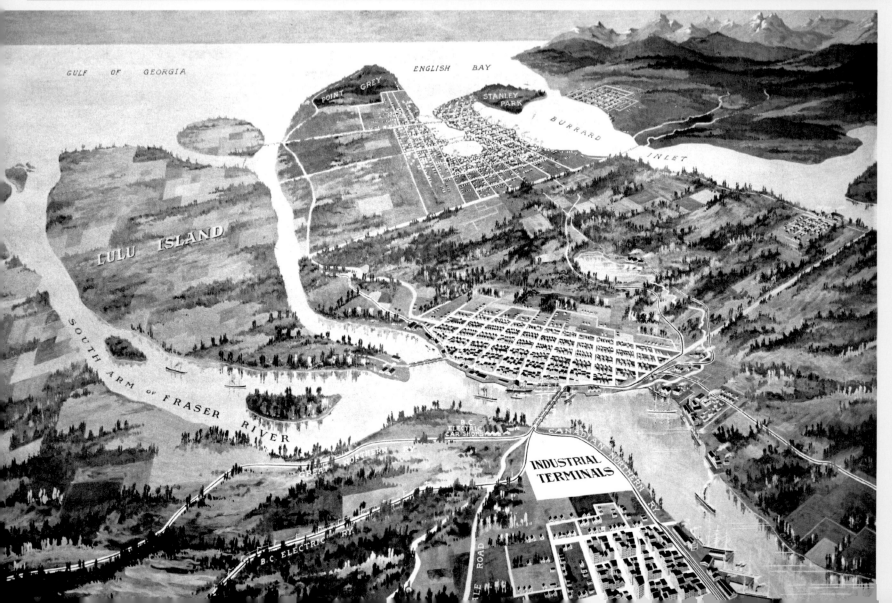

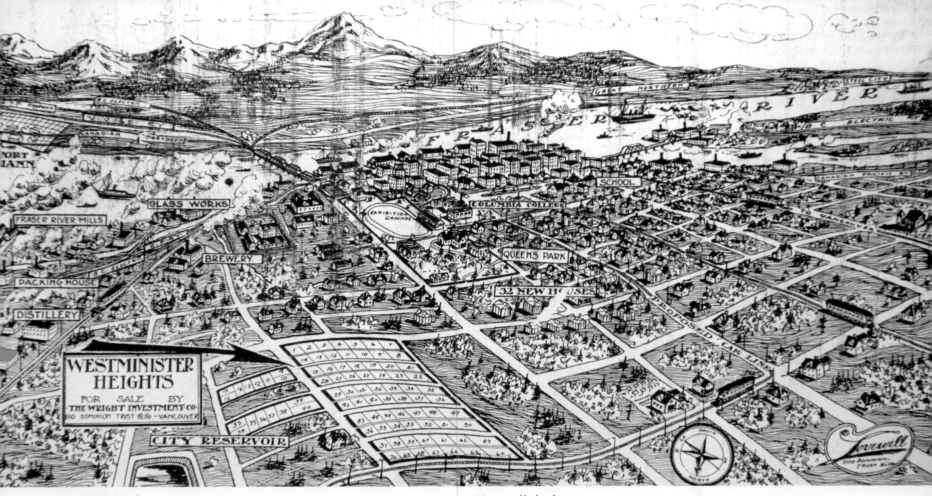

Map 612 (*above*).

This interesting map, part bird's-eye and part pictorial, was used to advertise a residential development in New Westminster in 1911, for sale, appropriately perhaps, by the *Wright Investment Co.*! Streetcar lines, existing and proposed, are shown complete with streetcars, and the interurban line to Vancouver is at right. Many railways, again both existing and proposed, converge onto the New Westminster Bridge.

Map 613 (*below*).

This detailed pictorial map appeared in 1910 to promote the "Twin City Industrial Subdivision" east of New Westminster. Lots could be purchased for $275, or $50 down and $10 per month. Many rail ines that did not exist are shown. Notice the attempt to depict the hilly nature of New Westminster with a three-dimensional "bulge," perhaps to contrast with the subdivision offered for sale.

Map 614 (*above*).

Many railways converge onto the New Westminster Bridge on this very detailed bird's-eye map drawn about 1912 to emphasize the proximity to both New Westminster and Vancouver of two subdivisions, Buena Vista and Burnaby Heights. This map, unusually, was not a newspaper ad but a huge production likely intended to be hung on a wall in the developer's office.

Map 615 (*below, right*), with photo (*right*) and ad (*below*).

White Rock developed as a summer resort after the *Great Northern Railway* was diverted around the Semiahmoo Peninsula in 1909. The map dates from 1912, the photo from the 1920s, and the ad from December 1912, when sales must have been hard to come by, for lots farther from the beach are now being offered free with a magazine subscription! The *Pier* shown here is in a different location from the present one. Note the *Auto Drive to Beach* to allow cars to drive on the extensive sands. The *New Station* for the railway is at left on the map and is also shown in the photo.

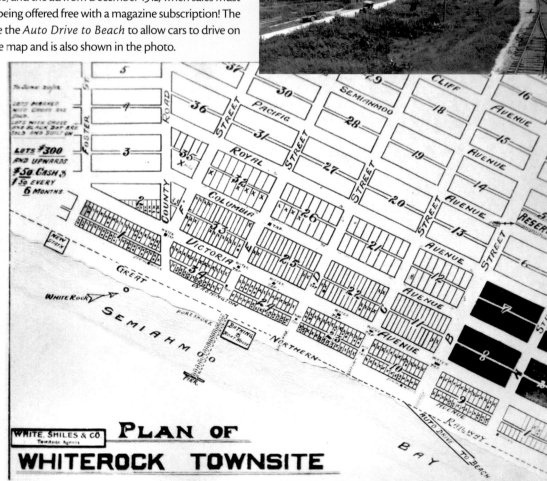

PLAN OF
WHITEROCK TOWNSITE

WHITE, SHILES & CO
Townsite Agents

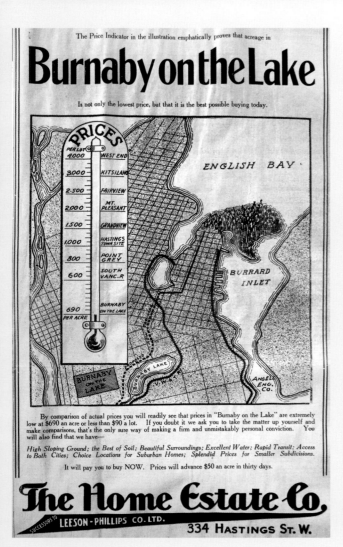

Burnaby on the Lake

Is not only the lowest price, but that it is the best possible buying today.

By comparison of actual prices you will readily see that prices in "Burnaby on the Lake" are extremely low at $690 an acre or less than $90 a lot. If you doubt it we ask you to take the matter up yourself and make comparisons, that's the only sure way of making a firm and unmistakably personal conviction. You will also find that we have—

High Sloping Ground; the Best of Soil; Beautiful Surroundings; Excellent Water; Rapid Transit; Access to Both Cities; Choice Locations for Suburban Homes; Splendid Prices for Smaller Subdivisions.

It will pay you to buy NOW. Prices will advance $50 an acre in thirty days.

The Home Estate Co.
SUCCESSORS TO LEESON-PHILLIPS CO. LTD.
334 HASTINGS ST. W.

Map 616 (*above*).

A "price thermometer" dominates this 1909 ad for a Burnaby subdivision. The ad also has a perspective map illustrating the development's proximity to the City of Vancouver.

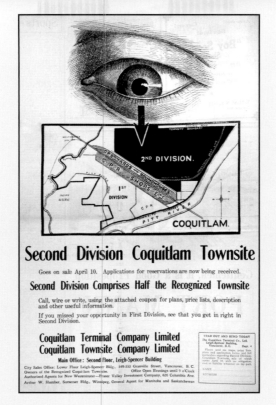

Second Division Coquitlam Townsite

Goes on sale April 10. Applications for reservations are now being received.

Second Division Comprises Half the Recognized Townsite

Call, wire or write, using the attached coupon for plans, price lists, description and other useful information.

If you missed your opportunity in First Division, see that you get in right in Second Division.

Coquitlam Terminal Company Limited
Coquitlam Townsite Company Limited
Main Office: Second Floor, Leigh-Spencer Building

Map 617 (*above*), Map 618 (*right*), Map 619 (*below, centre left*), Map 620 (*below, centre right*), Map 621 (*below, bottom left*), and Map 622 (*below, bottom right*).

Canadian Pacific established new marshalling yards at Westminster Junction in 1911, leading immediately to many attempts to sell surrounding land. Some of the ads were quite clever, as the selection shown here illustrates. The new municipality of Port Coquitlam was incorporated in 1913 as a result of all the new activity, but by 1914 it was seizing lots in large quantities for non-payment of taxes. The September 1915 issue of the *Coquitlam Star*, one of the last before the paper went out of business, carried twenty-one full pages of lists of properties for tax sale. Although offered at prices ranging from $5 to $7 of unpaid taxes, there were few buyers.

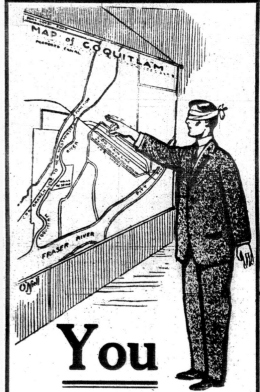

You

can't afford to pick blindfolded at **COQUITLAM**

We Have **3** of the Best Buys on INSIDE Property on the Market Today

These Will Not Be on the Market for Many Hours

$50 Cash Will Handle One.
$100 Cash Will Handle Another.
$125 Cash Will Handle the Other One.

Two of these are absurdly low in comparison with anything that is offered today. These are not so-called inside lots. THEY ARE THE CREAM OF THE MARKET, and the man fortunate in getting any one of them has a good fat profit coming to him NOT in the dim future, but right off the reel. Come in and go into the situation THOROUGHLY. Compare prices and locations on the map. You'll agree with our verdict.

Yours for quick profits

COX & STEPHENSON
47A Hastings St. West

Buy in the Manufacturers' Subdivision
THE CREAM OF THE TOWNSITE

SECTIONS 9, 16 AND 17, the only property in Coquitlam, onto which you can step from the present main line tracks, that has deep-water facilities. The ideal subdivision of Coquitlam is then before you. You see a property devoid of trees, brush or stumps, ground dry, level and ready for the manufacturer. Industries are buying here now to take advantage of main line facilities. The Western Canada Power Company's locations are in the midst of our lands. The Dewdney Trunk Road intersects our property. The traffic of Coquitlam will pass through the streets of this subdivision.

RIGHT BETWEEN THE MAIN LINE AND DEEP WATER

Here's where the manufacturers will locate, and here's where ocean-going vessels will dock. With the erection of factories, the building of stores, banks, hotels, etc., on which the adjoining lots will be simultaneous. Back from the waterfront the town of Coquitlam will grow, like any other town.

Get in now on the ground floor on the Power and Waterfrontage Lots
that capital and industry will develop.

Listen: A vacant corner lot in the townsite, purchased in the fall of 1910 for $450.00, sold this year for $6,000.00. Do you know what this means to you? It means that if you purchase lots in this subdivision you will make big money this summer. If you hold them, your profit possibilities cannot now adequately be estimated, for each week more evidence is forthcoming of the quick and sure increase in values.

PRICES $600 UP, AND ON VERY REASONABLE TERMS

It will cost you no more to buy here, where the increase in values will be most pronounced, than where development will be slower. But the lots are few and going quickly. Your biggest chance is here and now.

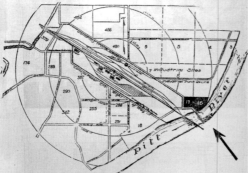

North American Lumber Co. Limited

Vancouver Offices: 614 Pender Street West   134 Hastings Street West   Branch Office: Coquitlam, B.C.
Phone Seymour 8183   Phone Seymour 2274   Phone Seymour 5382
OPEN NIGHTS

If you are one of those who have not yet profited in Coquitlam, here's the biggest chance yet

Don't wait! Read this important opening announcement

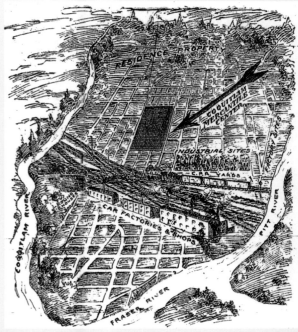

GRAND TRUNK PACIFIC

Even before the Canadian Pacific constructed the first transcontinental railway, the Grand Trunk Railway, Canada's earliest large railway network, had been asked by the federal government to build one and, at least twice, had turned the request down. The idea of a second transcontinental line persisted—one to travel the Prairies farther north than the Canadian Pacific, more or less along the route Sandford Fleming first proposed in the 1870s (see page 114). In 1895 Québec interests emerged behind a transcontinental scheme they thought would develop the northern part of that province (which then ended at James Bay) called the Trans-Canada Railway, and by 1902 the proposal had been finalized. It envisaged a line terminating on the British Columbia coast at Port Simpson.

But the scheme was sidelined by another, similar project proposed by Charles Melville Hays, an American railroad president hired by the Grand Trunk to revitalize its affairs. Here was a project of immense scope, it was thought, bound to elevate the railway to the level of the upstart Canadian Pacific. Prime Minister Wilfrid Laurier thought this would satisfy his Québec supporters, but the Grand Trunk Pacific Railway (GTPR), the subsidiary the Grand Trunk incorporated in 1903, was not interested in the eastern portion. Here the government agreed to build the line, called the National Transcontinental, which would connect with the GTPR line and also be leased by the company. In the end, the GTPR refused to carry through with that promise, but the western portion was built, pressed onward by Hays until his untimely demise in 1912 as a passenger on the *Titanic.* An agreement was reached

MAP 623 (*left, top*).
This fine map of the proposed route of the Grand Trunk Pacific appeared—appropriately enough, given the way the map is portrayed—in the *Christmas Globe* in 1906.

MAP 624 (*left*).
Initial proposals included extensions from the main line up the Skeena Valley and right to *Dawson*, in the Klondike region of the *Yukon.* Also included was a route south from *Ft. George* to connect with *Vancouver*, a proposal that became the Pacific Great Eastern, following a different route (see page 278).

MAP 625 (*above*).
The extent to which the Liberal Party of Wilfrid Laurier was bound up with the Grand Trunk Pacific is illustrated by this banquet menu header from a party fundraiser in 1906. The new railway crosses the continent anchored at each end by classical figures.

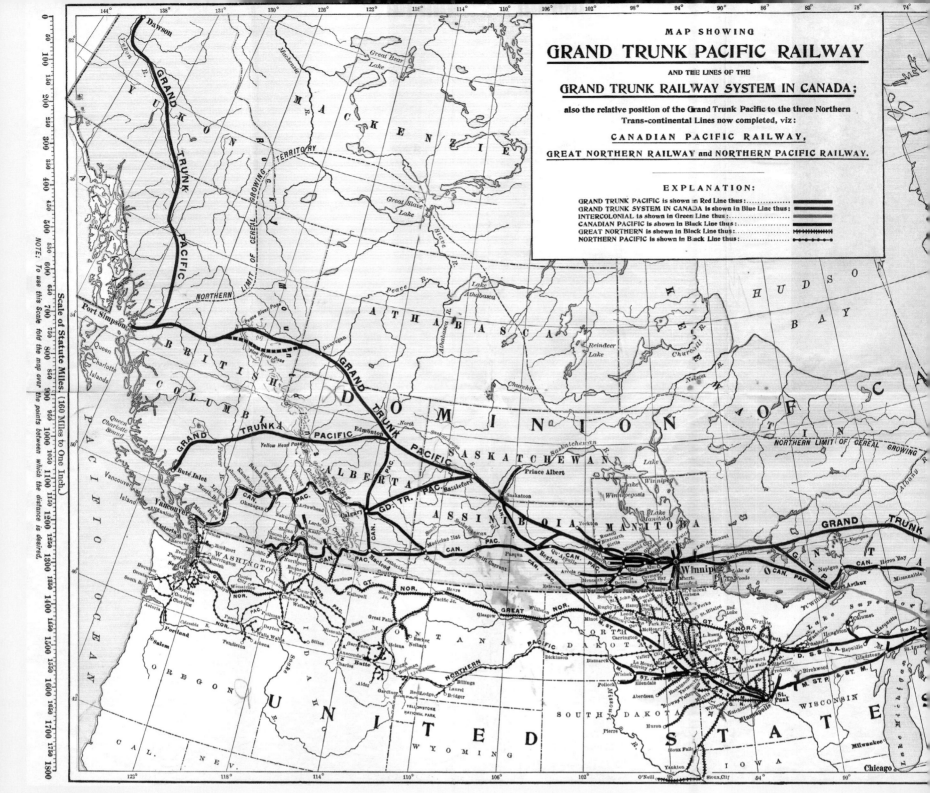

Map 626 (*above*).

This is the western part of a map that appeared in a 1904 booklet, which seems to have been a Liberal Party document, outlining the rationale for government support of the Grand Trunk Pacific and the National Transcontinental as a second transcontinental link. The line terminates here at *Port Simpson*, while another line to the Pacific terminates at Bute Inlet, also a Fleming-surveyed route.

Map 627 (*below*).

A more detailed route, now to *Prince Rupert*, appeared in the *Victoria Colonist* in 1909. Few towns are located along the line, a situation soon to change.

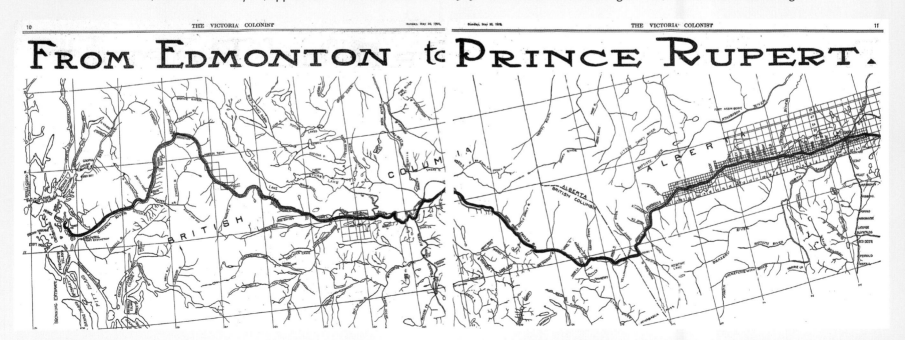

with the GTPR in April 1903, and a bill ratifying the agreement and allowing for government financial aid passed through Parliament on 24 October.

Much of the railway's route, on the Prairies, through the Rockies using Yellowhead Pass, and in British Columbia, was the same as one Fleming's railway surveyors first surveyed. There was one important exception: the terminus was not to be Port Simpson but an entirely new port created on Kaien Island to the south, where a new railway-inspired city would rise—Prince Rupert.

The railway had in 1906 incorporated a subsidiary, the Grand Trunk Pacific Town and Development Company, to acquire townsites and develop them to maximize profits accruing from the coming of the railway. In Prince Rupert it would have no real competition, in contrast to just about everywhere else on the railway's route.

Most of the land selected for the railway townsite was an Indian reserve, and it took much negotiation and backroom wrangling to acquire it. The federal government was willing to grant the land to the Grand Trunk Pacific, but if it did so, the province was entitled by law to what were called the reversionary rights—the transfer to it of all land taken out of the reserve. And so the railway had to agree to pay $2.50 per acre (about $6.18/ha) to the province for the reversionary rights and agree to reconvey a quarter of the waterfront back to the province. It also agreed to begin constructing the railway east from Prince Rupert.

In 1907 Brett & Hall, a town planning consulting firm from Boston, was commissioned to produce a plan for the new model city worthy of its coming role as a major railway terminus. George D. Hall, senior partner of the firm, was GTPR president Charles Hays's son-in-law and a student of famous American town planner Frederick Law Olmsted. Based on emerging principles of the "City Beautiful" movement, the plan

Map 628 (*left*) and Map 629 (*below*).
A fine promotional pamphlet titled *Prince Rupert and a Dawning Empire*, published by the Prince Rupert Publicity Club and Board of Trade in 1910. This local body produced several promotional pieces with a quality that far outstripped the railway's effort to promote its own creation. The brochure cover (Map 628) showed Prince Rupert and a train approaching from the east—with nary a mountain in sight, except the Rockies in the background. Inside the brochure was this excellent bird's-eye map (Map 629) of the entire British Columbia coast, which also showed the whole of the Grand Trunk Pacific line within the province. The map is a rarely seen perspective, designed, of course, to make it look as though Prince Rupert was the natural focus point of the West Coast.

Over 1,000 Miles of Coast Line, Through the Myriads of Beautiful Islands Forming th

A VAST NEW EMPIRE, OF WHICH PRINCE RUPERT IS THE TH

S. S. Prince George
of the Grand Trunk Pacific Line.

MAP 630 (*right*).

This map, published in 1910, shows the Grand Trunk Pacific route in British Columbia, with inset maps showing *Prince Rupert* and a hemispheric map showing the positional advantage of the city for shipping to the Orient. The coastal route planned for the railway's steamships is shown with a dashed line, as is the proposed route of the connector line between Vancouver and *Fort George*, eventually built as a separate venture, the Pacific Great Eastern. Note the routing of this line into Vancouver, different from the one ultimately taken down Howe Sound.

Above. The impressive steamship *Prince George*, which along with its sister ship, the *Prince Rupert*, was brought to the coast in 1910 to link Prince Rupert with Vancouver, Victoria, Seattle, and Stewart, Alaska. The ships were named "princes" as a counterpoint to the Canadian Pacific's "princesses."

"Inland Passage"

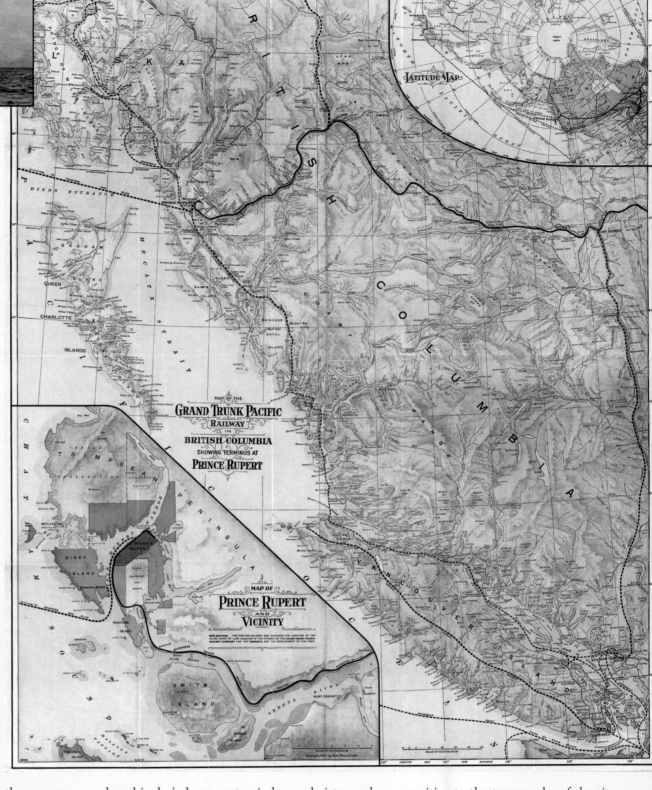

the company produced included crescents, circles, and vistas and was sensitive to the topography of the site (MAP 631, *overleaf*). By early 1909 the townsite had been cleared and 2,400 lots had been staked. In May sales began with auctions in Vancouver and Victoria. By the end of the year the population was three thousand.

In accordance with the GTPR's agreement with the provincial government, construction began from Prince Rupert eastward in 1908, but progress along the difficult Skeena Valley was slow, requiring much rock blasting and excavation. And labour was in short supply. Not until July 1910 was a bridge completed to connect Kaien Island to the mainland. By the end of 1912 the end of steel had only reached the vicinity of Hazelton, some 285 km from the coast. Faster progress had been made working westward on the Prairies, however, and the rails crossed into British Columbia in November 1911. The rails from west and east finally met near Fort Fraser, 600 km east of Prince Rupert, in April 1914,

Right.
The first locomotives in Prince Rupert arrived by sea. Here the first two are unloaded in 1910 onto temporary tracks laid to the edge of the shoreline. They would be used in the construction of the line eastward.

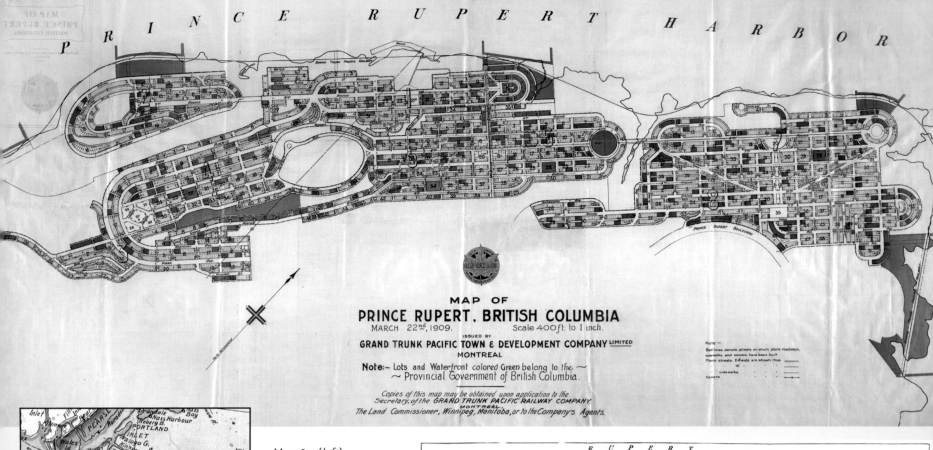

MAP OF
PRINCE RUPERT, BRITISH COLUMBIA
MARCH 22ND, 1909.   Scale 400ft. to 1 inch.

ISSUED BY

GRAND TRUNK PACIFIC TOWN & DEVELOPMENT COMPANY LIMITED
MONTREAL

Note:- Lots and Waterfront colored Green belong to the ~
~ Provincial Government of British Columbia.

Copies of this map may be obtained upon application to the
Secretary, of the *GRAND TRUNK PACIFIC RAILWAY COMPANY*
MONTREAL,
The Land Commissioner, Winnipeg, Manitoba, or to the Company's Agents.

Map 632 (*left*).
The location of *Prince Rupert* on *Kaien* Island at the terminus of the line down the *Skeena* is shown on this 1914 map. *Metlakatla* (see page 110) is across the harbour, and *Port Simpson*, the originally planned terminus, is at the north end of the *Tsimpsean Penn.*

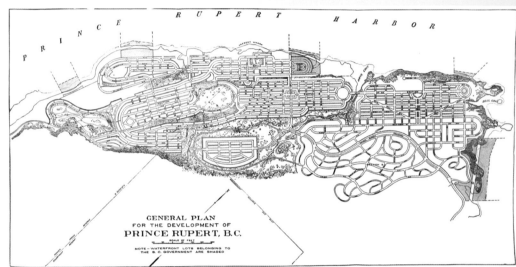

Map 631 (*above, top*) and Map 633 (*above*).
The GTPR hired landscape architects Brett & Hall of Boston to lay out the townsite for Prince Rupert. The firm produced a plan that tried to accommodate the topography of the site and generally succeeded. Most of today's Prince Rupert is based on this plan. Map 631 was contained in a 1909 promotional brochure and shows only the part being sold at the time. The blue-green parcels are provincial government properties—including those agreed to with the purchase of the reversionary rights, and a quarter of the town lots, which since 1896 were required to be allocated to the province in any new townsite. Map 633 shows the entire Brett & Hall plan.

Map 634 (*below, right*).
The Grand Trunk Pacific had grand plans for its western terminus. In 1911 the railway commissioned Francis Mawson Rattenbury, architect of the provincial legislative buildings in Victoria, to design a fine hotel, a railway station, and a steamship terminal in Prince Rupert, all shown here on his 1913 perspective drawing of the project.

Left. This ad for Prince Rupert, with photos showing the city's five-year progress, appeared in a GTPR brochure for Smithers, published in 1914.

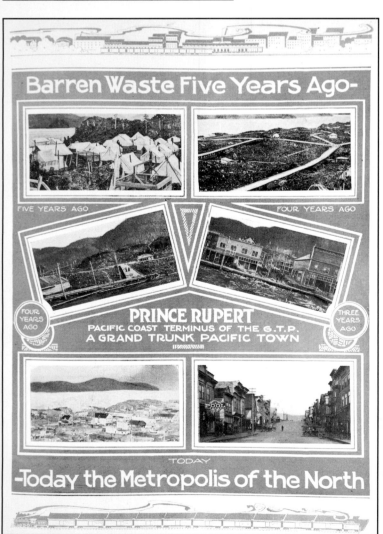

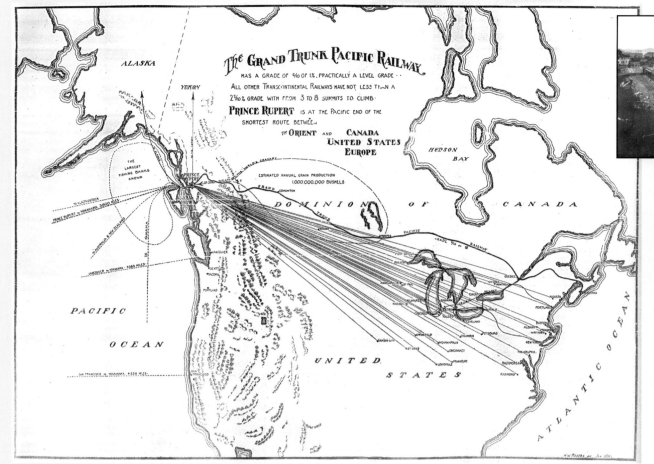

although service did not commence until August—the very month that saw the start of World War I. The railway would never be able to recoup its investment in the high-quality, gently graded line it had built.

Speculators had, as with any railway, tried to anticipate where the line might be routed and, more importantly, where stations and divisional points (where crews and locomotives were changed) would be. The railway had constantly attempted to outfox the real estate men and establish townsites for itself, where it could keep the profits to be made from the increase in values.

At no place was this battle more marked than where the line was to cross the Fraser and proceed west up the valley of the Nechako. Here the Hudson's Bay Company had built Fort George in 1807, and here two rival communities sprang up, separated by an Indian reserve between them. George Hammond, with his company, Natural Resources Security, bought up land to the west, subdividing it into a townsite he called Fort George, adding to it several times.

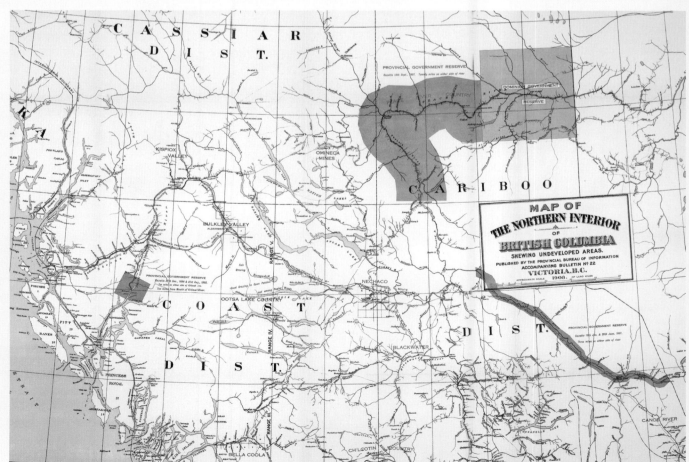

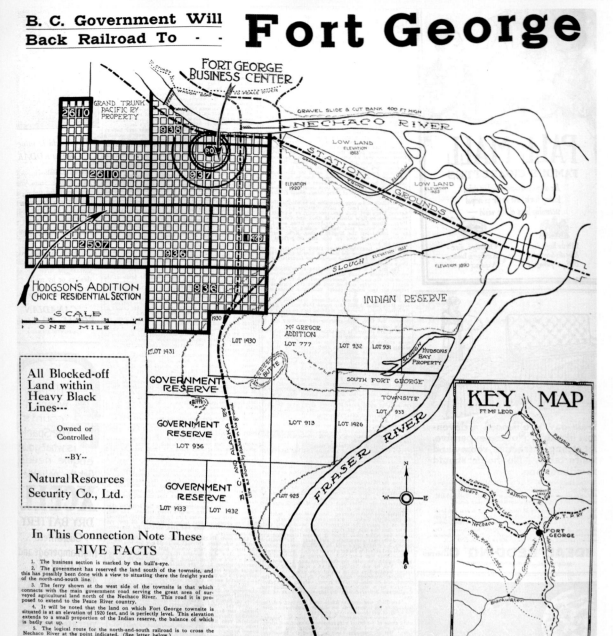

MAP 638 (*left*).

A fine ad, published in March 1912, for the Fort George Townsite of the Natural Resources Security Company owned by seasoned promoter George J. Hammond. The competing townsite of *South Fort George* (see MAP 645, *page 258*) is also shown south of the *Hudsons Bay Property* on the *Fraser River*. The Grand Trunk Pacific outmanoeuvred both by obtaining title to the *Indian Reserve* between the two for its townsite, which it named Prince George. The *Station Grounds* are already shown on the reserve. This map not only shows the Grand Trunk Pacific but also another line south to north labelled the *B.C. and Alaska Ry.*, which had been chartered in 1910 to build from the Canadian Pacific line at Lytton to Teslin Lake far to the north via Lillooet and Fort George.

MAP 639 (*above*).

By 1919, the date of this map, the British Columbia & Alaska Railway has been replaced by the *P.G.E. Ry.* (Pacific Great Eastern) on this map illustrating the central position of Prince George, but that railway did not make it to the new city either—at least not until 1952 (see page 320). A similar map, showing the central location of competing private townsite Fort Salmon, is MAP 725, *page 283*.

MAP 640 (*below*).

Another Hammond ad for his *Fort George Townsite*, this one published in May 1913. The Indian reserve is now *G.T.P. Property*, and Hammond is trying to show how near the *Main Passenger Station* would be to his development. In fact the railway kept it as far away from him as possible.

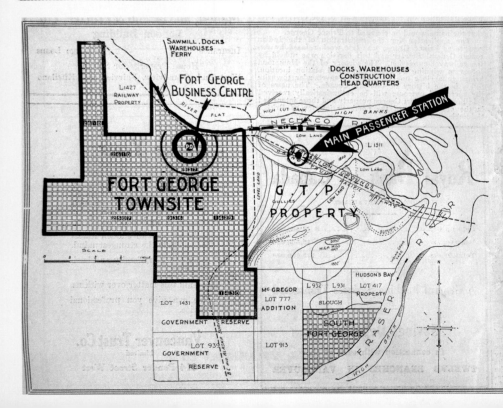

THE GRAND TRUNK PACIFIC RAILWAY CO.'S TOWNSITE OF PRINCE GEORGE

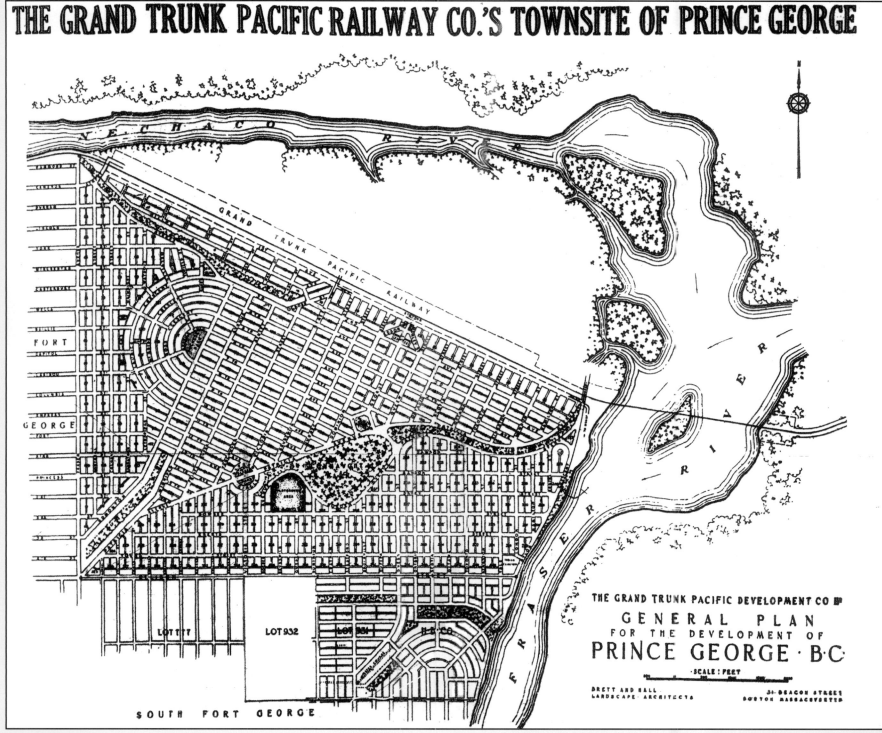

THE GRAND TRUNK PACIFIC DEVELOPMENT CO.

GENERAL PLAN
FOR THE DEVELOPMENT OF
PRINCE GEORGE · B·C·

·SCALE : FEET·

BRETT AND HALL
LANDSCAPE ARCHITECTS

31 BEACON STREET
BOSTON MASSACHUSETTS

Map 641 (*above*).
The Brett & Hall townsite plan for the Grand Trunk Pacific's site on the Indian reserve, first published in the *Fort George Herald* in April 1913. It followed "City Beautiful" princi-ples similar to their plan for Prince Rupert. The crescents in the western part may have been created to make it more difficult to travel between the Fort George townsite and the railway station. The main street, *George St.*, connected the *Station* in the north with *Princess Square* in the south (the octagonal block), the (first) location of city hall.

Map 642 (*right*).
A Hammond map from late 1912 shows the location of *G.T.P. Termi-nals*, again nearer to his *Fort George Townsite* than they would be built. In addition a station for the *B.C. & Alaska* is shown on the edge of his townsite. The *G.T.R. Ry. Fort George-Vancouver Line*, now shown entering the *Indian Reservation* from the south, is the intended route of the Pacific Great Eastern, although that eventually ended up on the other side of the river so as to utilize the Grand Trunk Pacific's bridge. *Below* is the *Central Townsite* in 1913.

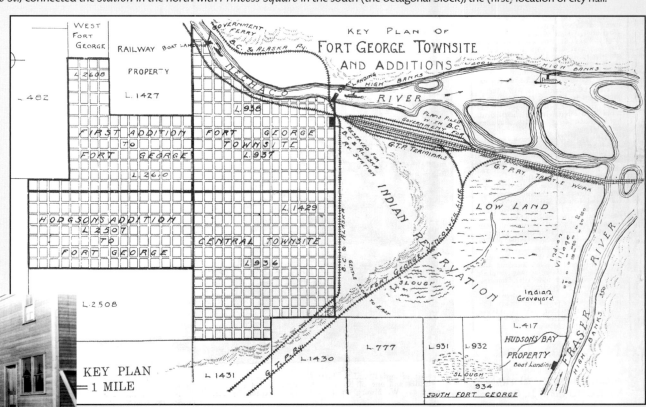

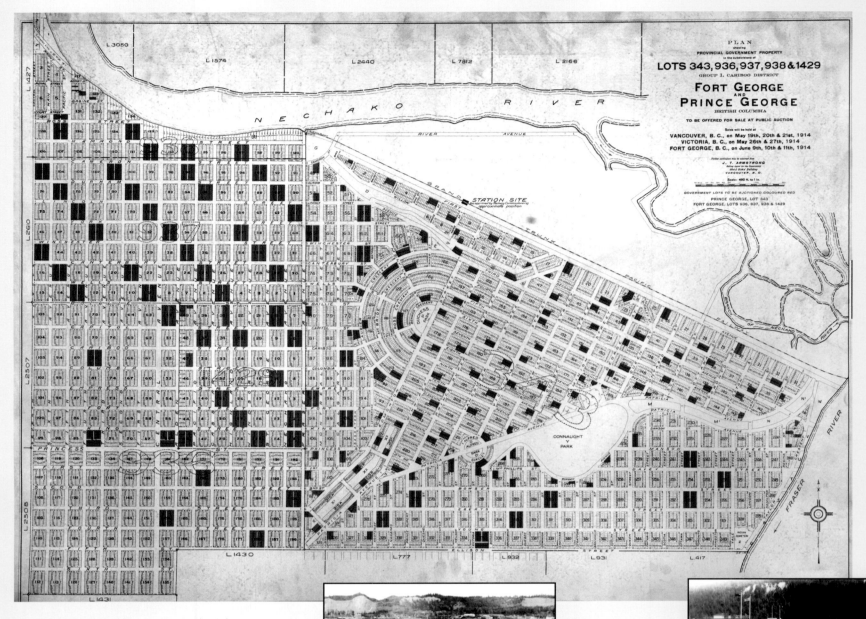

Plan
showing
PROVINCIAL GOVERNMENT PROPERTY
in the subdivisions of
LOTS 343, 936, 937, 938 & 1429
GROUP 1. CARIBOO DISTRICT
FORT GEORGE
AND
PRINCE GEORGE
BRITISH COLUMBIA
TO BE OFFERED FOR SALE AT PUBLIC AUCTION
Sales will be held at
VANCOUVER, B. C., on May 19th, 20th & 21st, 1914
VICTORIA, B. C., on May 26th & 27th, 1914
FORT GEORGE, B. C., on June 9th, 10th & 11th, 1914

J. T. ARMSTRONG

GOVERNMENT LOTS TO BE AUCTIONED COLOURED RED
PRINCE GEORGE, LOTS 343
FORT GEORGE, LOTS 936, 937, 938 & 1429

Map 643 (*above*) and Map 645 (*right*).
Since 1896 the law had required townsite developers to give 25 per cent of their lots to the provincial
government. Here the government is offering its
lots for sale (coloured red on the map) at auctions
in Vancouver, Victoria, and Fort George in May 1914.
Map 643 shows those in Fort George and Prince George, while Map 645 shows those in South Fort
George. The photos show Prince George (*inset, above*) and South Fort George (*inset, right*) at about
the time of the auction.

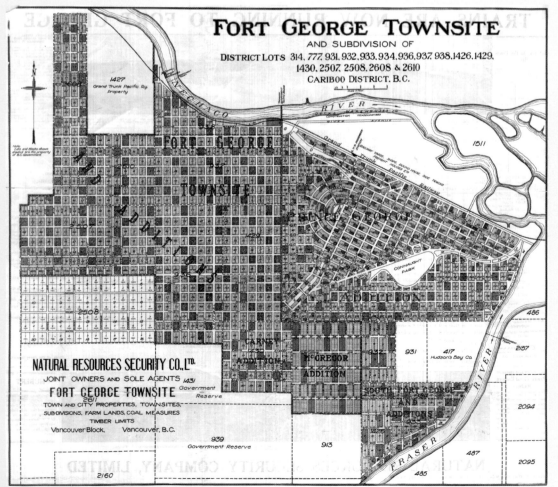

FORT GEORGE TOWNSITE
AND SUBDIVISION OF
DISTRICT LOTS 314, 777, 931, 932, 933, 934, 936, 937, 938, 1426, 1429,
1430, 2507, 2508, 2608 & 2610
CARIBOO DISTRICT. B.C.

NATURAL RESOURCES SECURITY CO. LTD.
JOINT OWNERS AND SOLE AGENTS
FORT GEORGE TOWNSITE
TOWN AND CITY PROPERTIES, TOWNSITES,
SUBDIVISIONS, FARM LANDS, COAL MEASURES
TIMBER LIMITS
Vancouver Block.   Vancouver, B.C.

PLAN
showing
PROVINCIAL GOVERNMENT PROPERTY
in the subdivisions of
LOTS 933, & 934,
GROUP 1. CARIBOO DISTRICT
SOUTH FORT GEORGE
BRITISH COLUMBIA
TO BE OFFERED FOR SALE AT PUBLIC AUCTION
Sales will be held at
VANCOUVER, B. C., on May 19th, 20th & 21st, 1914
VICTORIA, B. C., on May 26th & 27th, 1914
FORT GEORGE, B. C., on June 9th, 10th & 11th, 1914

J. T. ARMSTRONG

Map 644 (*left*).
Not for the first time did George Hammond appropri-
ate for himself another's development (see also Map 668,
page 264). Here the railway's townsite has audaciously
been portrayed as the *Prince George Addition*, the im-
plication being that it was merely another expansion of
his own Fort George townsite.

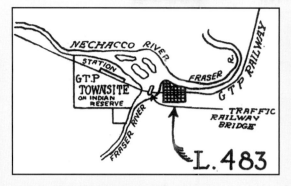

Map 646 (left) and Map 647 (left, inset).
Private townsite promotions always followed and often preceded railway building, and the Grand Trunk Pacific was no exception. *Willow River* (*Willow City* on Map 647), north of Prince George, was one such development, though a "railway authorized" one. Map 646, from a promotional brochure, emphasizes the agricultural potential of the region. The railway did go through the town, but the site had no advantage over Prince George; today only a small settlement remains.

Map 648 (above, centre).
This ad for land competitive to the GTPR's townsite appeared in April 1913. "A subdivision showing promise of quick profit at prices within the reach of all," said the ad, " why buy 25-foot lots when you can get 1½ acres for half the price and within closer radius of the G.T.P. depot."

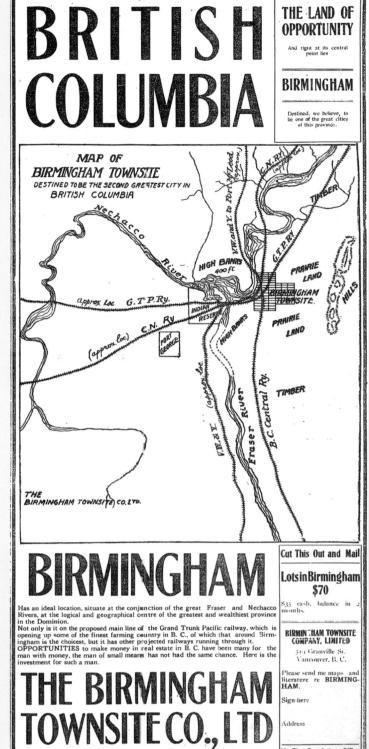

Immediately south of the Hudson's Bay Company property on the Fraser was South Fort George, a subdivision laid out at the steamboat landing (Map 645, left, bottom). Grand Trunk Pacific had intended to create a railway terminal and yards on the northern part of the Indian reserve and include land to the west for a townsite, but lot owners there demanded $75 per acre ($185/ha), which the company was unwilling to pay. And so it determined to acquire the entire reserve for its townsite. After an enormous amount of negotiation and manoeuvring, in November 1911 the company purchased the land for $125,000, or $91.51 per acre ($226/ha). It also acquired the reversionary interest from the province, a much easier process than in Prince Rupert—there the premier, Richard McBride, had wanted to ensure the railway was built, but here the railway was well under construction. In the period before the purchase was finalized, the company had been opposed by other businesses, including Charles Millar of the B.C. Express Company, who had wanted to acquire the eastern half of the reserve for docks and warehouses to service his steamboats. The GTPR agreed to sell them 200 acres (81 ha) in the southern part of the reserve for $59,296, or $300 per acre ($740/ha).

Hammond was furious but could do nothing. He offered the railway $1 million for the right to market reserve lots if he could retain revenue over and above this amount. His offer was rejected. This did not stop Hammond from producing ads that suggested the new townsite was just an extension of his Fort George (Map 644, far left). He also pressed a drawn-out battle to have the railway build its terminal as far west as possible, but despite being served Railway Commission orders for a western site, the GTPR ended up building its terminal as far away from Hammond's lands as it could.

And the Grand Trunk Pacific was going to sell its new townsite itself. Brett & Hall were again commissioned to design an elegant townsite, which they completed early in 1912 (Map 641, page 257). Using "City Beautiful" principles once more, the plan showed most of the streets running at right angles to the railway and included some crescents that made the journey from the station to Hammond's townsite longer than it needed to be; whether this was deliberate or not has never been proven, but it seems likely, for the crescents are located only in one specific area where this would be achieved. A 2009 planning study showed "commuter desire lines" across the crescents, indicating that even today the street morphology is less than optimal for residents just to the west.

Map 649 (right, top) and Map 650 (right).
One townsite promotion that seems to have had a lot of effort put into it, but that went absolutely nowhere, was the proposed city of *Birmingham*, a rival for Fort George on the eastern bank of the Fraser. Indeed, this site seemed as good as that of Hammond's Fort George; the railway did go through the location. But the Grand Trunk Pacific was not going to obligingly stop in a townsite where it would make no money, especially when the possibilities of the Indian reserve across the river seemed wide open. Note how, on Map 649, *Fort George* has been deliberately misplaced to seemingly be far from the railway, while the *Indian Reserve* has been made subtly larger to exclude possible competitors there. Birmingham's promoters, as those of Fort George, likely did not think the railway would take over the reserve. These ads are from 1910 newspapers.

FORT FRASER, B.C.

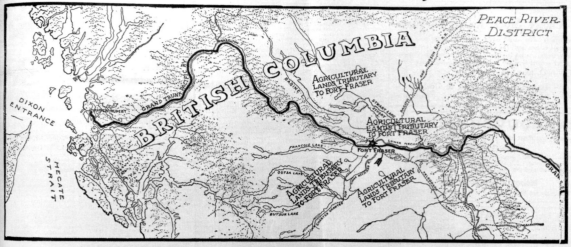

PEACE RIVER DISTRICT

Examine the Map and See Why Fort Fraser has the Greatest Location of any Town on the Grand Trunk Pacific Railway in B.C.

Of late, much has been written regarding the wonderful agricultural districts surrounding Fort Fraser. A casual observance of the above map will prove conclusively that Fort Fraser has the geographical location to control all the trade of the vast agricultural areas of the interior.

Fort Fraser Today Offers More Inducements to Either the Large or Small Investor than Any Western Canadian Town Ever Offered

Fort Fraser has the location, farming land and is an old established trade centre. No better investment can be found in Western Canada than either a large or a small investment in Fort Fraser property today. Look to Fort Fraser for your fortune and future happiness.

Dominion Stock & Bond Corporation Ltd.
Winch Building   Vancouver, Canada   **Dominion Building**

FORT FRASER (THE HUB OF B. C. ON THE G. T. P.)

A COMMERCIAL and industrial centre surrounded by a vast area of the richest agricultural land in all the West, embracing the famous Nechaco Valley, the Stuart Lake Country, Bulkley Valley, Blackwater Valley, Ootsa Lake Country and part of the great Peace River Valley; also within 125 miles just west of the Bulkley Valley, the large forests of the Coast Range Slope and the well-known Chilcotin grazing and stock-raising country.

Directly on the main line of the Grand Trunk Pacific Railway, at the east end of Fraser Lake, on the Nechaco River, commanding over 1,000 miles of navigable waterways; a natural rail, lake and river distributing point, affording an easy grade for railroad building in every direction.

A townsite possessing a wealth of natural resources, which can neither be taken away by force nor stolen. No combination of forces on earth can defeat or retard the growth and ultimate supremacy of Fort Fraser when the Grand Trunk Railway is completed in 1913.

Be your resources great or small, there is here a genuine and exceptional opportunity for investment. Fort Fraser is owned and vouched for by British Columbia men of high social and financial standing, who invite and recommend a searching investigation by those who may be interested.

If you will act quickly you may now secure choice business lots at first prices; inside lots, $100 and up; double corners, $250 and up; terms easy enough for anyone.

Literature containing facts, figures and proofs will be mailed to any address on request.

FORT FRASER TOWNSITE   113 Hastings Street East, VANCOUVER, B. C.
A. C. HIRSCHFELD, Manager

FORT FRASER
The Heart of British Columbia

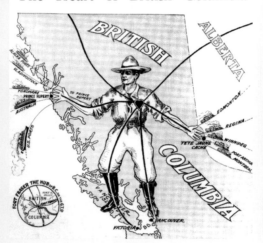

Undoubtedly the most advertised townsite along the route of the GTPR in British Columbia was Fort Fraser, at the eastern end of Fraser Lake, the result of the aggressive promotional policy of its principal owners, Dominion Stock & Bond Corporation. Despite all the alluring advertising, Fort Fraser never grew beyond the small village that remains today. Many of the ads incorporated innovative maps such as those shown here.

MAP 651 (*above, top left*).
This 1913 ad emphasizes the position of *Fort Fraser* as surrounded by agricultural lands.
MAP 652 (*above, top right*).
Fort Fraser Nechaco Valley is at the hub of many economic regions in this innovative 1911 ad, where wheels and cogs form what can be called a map nonetheless.
MAP 653 (*above, left*).
Fort Fraser is now depicted not as a hub but as a heart in this ad from about 1912. The hub also appears at bottom left.
MAP 654 (*above, centre*).
Half the population of Canada appears to be trekking along the *Sure Road to Wealth* to the *Land of Big Crops* in another innovative Fort Fraser ad, this one published in May 1913.
MAP 655 (*above, right*).
A last spike looms large over an enormous townsite in this ad from 1912.
MAP 656 (*right*).
The hub concept shows up again in this 1912 ad, this time with unlikely railways, overly generalized, converging on *Fort Fraser*, with, as usual, no rival settlements shown.

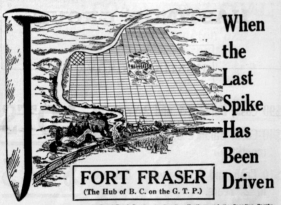

When the Last Spike Has Been Driven

FORT FRASER
(The Hub of B. C. on the G. T. P.)

In the summer of 1913, when the Grand Trunk Pacific, the Canadian Northern and the Canadian Pacific, three great steel highways, in majestic potency, trail from ocean to ocean, 4000 miles across the Dominion of Canada; when the west-coast province of British Columbia has been brought into competitive commercial communication with almost the entire world; when millions of capital have been invested within her borders in lumbering, mining, fisheries and public utilities, and when the "black banners of industry" are to be seen floating over thousands of cities, towns and villages all over her agricultural, mineral and timber areas; what then will be the value of property in a trade centre like

FORT FRASER?

1. In the heart of British Columbia, on the main line of the G. T. P., on Fraser Lake and the Nechaco River, commanding over 1000 miles of navigable waterway; a natural rail, lake and river distributing point, affording an easy grade for railroad building in every direction.

2. Surrounded by ten million acres of the richest agricultural land in all the west, embracing the famous Nechaco, Stuart Lake, Bulkley, Blackwater, Ootsa Lake, and Peace River Valleys; also within one hundred and twenty-five miles of the big Omineca, Finlay River and Cariboo mining districts, as well as the immense coal fields just west of the Bulkley Valley.

Fortunes were made by the early investors in Prince Rupert, Calgary, Edmonton, Regina, Saskatoon, Moose Jaw and Lethbridge.

Now, in its infancy, Fort Fraser has a future brighter than any other new townsite that ever came into existence in the whole of Western Canada.

More than forty entire blocks in this townsite have been reserved by the British Columbia Government and the title to Fort Fraser lots is guaranteed by the government and deposited in trust with the Dominion Stock & Bond Corporation, Limited, Vancouver, B. C. Capital $2,000,000.

For awhile, lots $150 and up, on terms of one-tenth cash and balance in eighteen equal monthly payments, without interest or taxes.

Have you the business acumen to grasp a genuine opportunity? Are you possessed of that essential qualification for success—quick decisive action? Then get particulars immediately. Call or write.

Dominion Stock & Bond Corporation, Ltd.
Capital $2,000,000

Ground Floor, Winch Building, Hastings St., Vancouver, B. C. | Telephone Seymour 7670

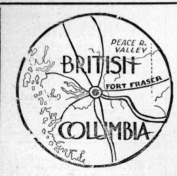

PEACE R. VALLEY — BRITISH COLUMBIA — FORT FRASER

FORT FRASER
A City Made Necessary by Nature

Can you call to mind a large city anywhere that was not favored by nature before it was developed by man?

From Halifax to Vancouver, practically every Canadian city of size owes its growth to some natural advantage.

So with Fort Fraser—the coming Commercial Capital of Interior British Columbia.

Situated at the head of 1000 miles of navigable waterways, with many rich valleys radiating from it, and on the line of the newest transcontinental Empire-builder, the Grand Trunk Pacific Railway.

Fort Fraser can truly be said to be a
City Made Necessary by Nature

When the railroad arrives next year, with thousands of thrifty settlers, Fort Fraser will jump into prominence that will be universal.

Prices are sure to rise—values will double and treble—a "boom" will ensue.

Are you going to get your share? Not unless you own a lot in Fort Fraser. You can get one today for $200.00, on the easiest of terms.

Don't wait—look this up at once.

DOMINION STOCK & BOND CORPORATION, LIMITED
Winch Building   Vancouver, B. C.

Canberra North

Booster magazine *Canada West* publisher and railway publicist Herbert Vanderhoof wanted his namesake town to be something special. In response to his request, in 1914 or 1915, Walter Burley Griffin and Francis Barry Byrne, both "flat roof" Prairie School architects who had apprenticed under Frank Lloyd Wright, produced a plan for an extension to Vanderhoof, an extension that was quite different from the original plan. The latter was a traditional grid pattern plan, laid out, in Griffin's words, "to squeeze the last penny of profits out of the sale of lots" by "cutting the property to pieces with cross streets." The extension plan, however, was quite different, with curved streets conforming to the topography along the lines of the "City Beautiful" movement incorporated into the Prince Rupert and Prince George plans.

Griffin was on his way to becoming one of the most famous town planners in the world, for in 1912 he won an international design competition for the new capital city of Australia—Canberra. One of the reasons Griffin won the Canberra competition was the superb accompanying drawings in his submission—some included gold leaf—by his wife and fellow architect Marion Mahony Griffin; unfortunately nothing similar accompanied the Vanderhoof plan!

Griffin never came to Vanderhoof, merely making use of a survey of the topography to draw up his scheme. But he had done just that for Canberra at first.

Unfortunately Griffin's plan was never used, and Herbert Vanderhoof's ideas for a model town and a literary centre fell to the realities of economics. Indeed, because of flooding problems on the south side of the river, in 1919 much of the community simply moved over to a competing townsite on the north side, and south side development—on the land covered by Griffin's plan—only resumed after World War II.

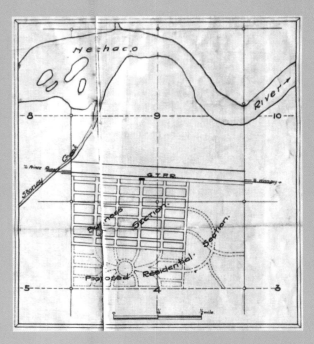

MAP 657 (*above*).
This is an inset map from an undated blueprint of Vanderhoof, thought to be from about 1914, and shows the proposed extension to the town strangely with another configuration—one closer to the circles and radial streets that Griffin designed for Canberra. It could be a representation of an earlier attempt made before details of the topography were known to Griffin.

MAP 658 (*below*).
Though only this poor copy appears to have survived, this 1914 or 1915 plan shows the Griffin and Byrne street layout of Vanderhoof quite well. The plan seems to have been initiated by Griffin but finished by Byrne, who managed Griffin's Chicago practice at this time, when Griffin had gone to Canberra. This plan is listed by some sources as Byrne's and in others as by Griffin; it is the only North American town plan of half a dozen or so by Griffin that had both their names on it. Byrne is also said to have prepared a similar plan for Giscome, also on the Grand Trunk Pacific, but this plan—like the settlement itself—has not survived.

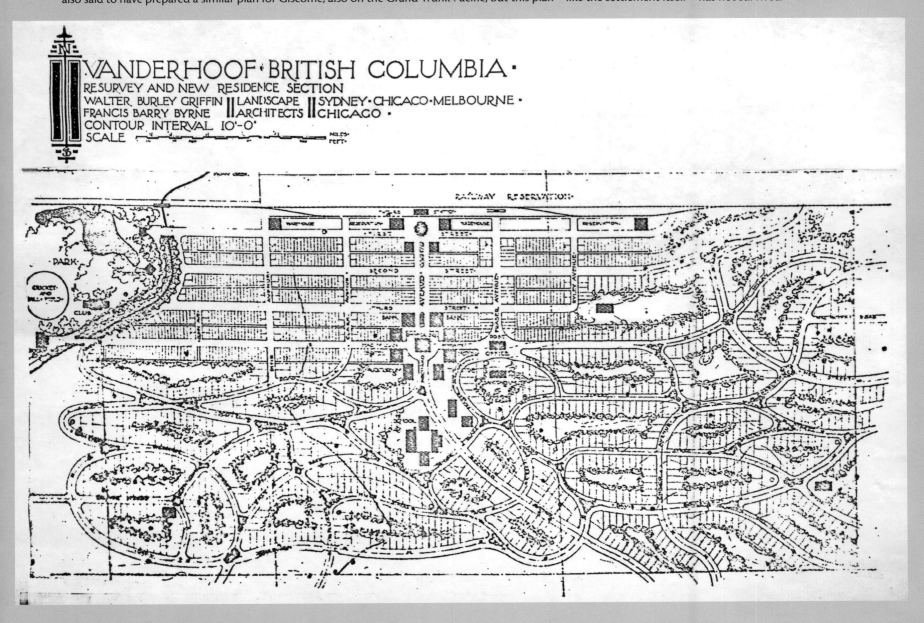

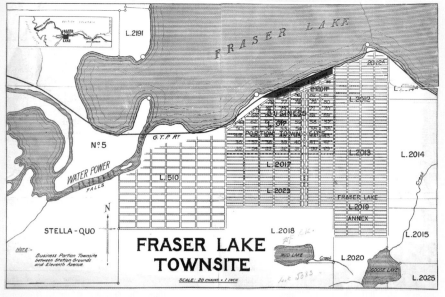

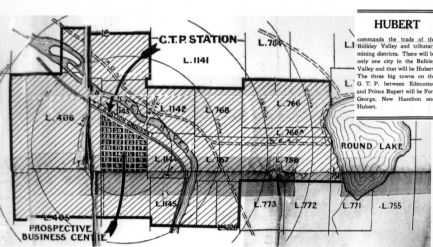

Map 659 (*left, top*).
Fraser Lake had less promotion than Fort Fraser but today is the larger of the two. This is the town plan from about 1912.

Map 660 (*left, second from top*).
The townsite at Hubert (the "only one city in the Bulkley Valley") lost out to nearby Telkwa. This is a map from an ad published about 1911.

Map 661 (*left, second from bottom*).
Amundsen probably appeared to be a sure bet, as it was at Pacific, the assumed requisite 120 miles (200 km) from Prince Rupert for a divisional point, but it was bypassed by the railway and never developed. The 1912 map shows fanciful railway yards.

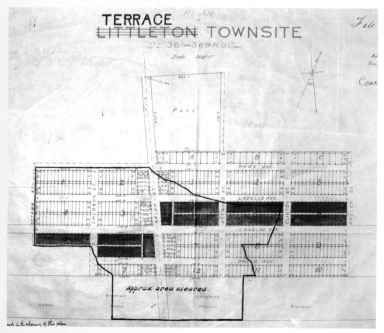

Map 662 (*left, bottom*).
A townsite was laid out at Kitsumkalum in 1910 on land originally pre-empted by George Little, who offered the Grand Trunk Pacific free land for its operations. The name of Little's townsite, Littleton, was refused by the post office because of duplication elsewhere, and the name was changed to Terrace. The red-coloured lots are government owned. Note the outline, in black, of the *Approx area cleared*.

Map 663 (*above, top right*).
The official town plan for the divisional point of Smithers included some features of the Brett & Hall plan seen in Prince George. At right angles to the station, *Main Street* is anchored at the other end by *City Hall Park*. Lots in black are provincial government blocks (not white as noted on the map; it was a blueprint that has been inverted for clarity).

Map 664 (*above*).
This ad for *Smithers* appeared in September 1913 and shows why it was a divisional point, more or less equidistant from *Prince Rupert* and *Fort George*.

Companies, real estate syndicates, and the like were forever trying to guess where the railway would locate its divisional points and its stations so that they could stake claims and buy land that would appreciate in value after the railway arrived. Some, such as the Dominion Stock & Bond Corporation at Fort Fraser (page 260), spent a lot of money advertising their site. Many lost out when they were deliberately bypassed. The townsites of Aldermere, Hubert (MAP 660, *far left*), and Telkwa lost out when the railway selected another nearby site for its divisional point, naming it after its chairman, Alfred W. Smithers (MAP 664, *left, top*). Aldermere and Hubert are today but fields, although Telkwa survives as a small settlement. The Grand Trunk Pacific had all the cards. In the Skeena Valley, George Little agreed to give the railway land for a station and yards and was allowed to add his own subdivision close by. Littleton, however, was not an acceptable name to the post office, owing to duplication, and so he changed the name to Terrace (MAP 662, *far left, bottom*). Others, such as Amundsen, named after the Arctic explorer and at the start of the railway's Pacific subdivision, were simply bypassed (MAP 661, *far left*).

The railway lost out, however, at Hazelton, where the Bulkley River meets the Skeena. The original settlement was on the other side of the river, shown on several of the maps as Old Hazelton. The railway decided to establish a station at South Hazelton because the landowners

MAP 665 (*above*).
This fine panoramic map was the centrepiece of a promotional brochure the GTPR published for its townsite of *Smithers*, complete with a more detailed map of the town as an inset.

MAP 666 (*right*).
This announcement was published on 23 November 1911 and was one of the opening salvos of a battle over the competing townsites of South Hazelton (on *Lot 851*) and New Hazelton (on *Lot 882*). The GTPR expended much time and money supporting South Hazelton because the owners there had agreed to give the railway half of its land, but New Hazelton was in a far better location and would emerge as the enduring settlement.

MAP 667 (*below*).
This 1913 provincial government map shows the line east of *Hazelton* as under construction (dashed red line). *Aldermere* is shown, near *Telkwa*, though not Hubert, which was also close. *Littleton*, soon to become Terrace, is shown, on the *Skeena River*.

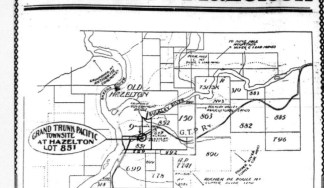

Grand Trunk Pacific Railway
Townsite at Hazelton

SCALE OF MAP:—ONE INCH TO THE MILE

The Grand Trunk Pacific Railway Townsite at Hazelton is situated on Lot 851, as per above plan. It will be registered as SOUTH HAZELTON and it is the intention of the company to build a station on this townsite in the spring of 1912.

There will be no station at Ellison,---one mile west.

Purchasers of lots at Ellison will be fully protected.

Surveys are completed and plans will be published just as soon as the government makes selections of lots.

The Land Commissioner of the Grand Trunk Pacific Railway will issue all agreements and deeds.

Notice to the Public.

The Grand Trunk Pacific Townsite of South Hazelton is situated on Lot 851, and it is the intention of the Company to build a station on this Townsite in the Spring of 1912.

Land Commissioner.

Winnipeg.
Nov. 6th, 1911.

G. U. Ryley
Grand Trunk Pacific Railway Land Commissioner

GRAND TRUNK PACIFIC TOWNSITE AT HAZELTON

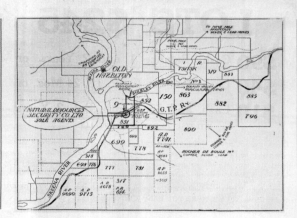

Citizens of Hazelton Enthusiastic

15. H. P.   11 Ash

Hazelton, B.C.
Oct. 20th, 1911

Natural Resources
Security Co. Ltd.
Bower Bldg.
Vancouver, B.C.

Town enthusiastic over South Hazelton.

Keep us informed concerning new developments.

Aldous & Murray.

COPY OF TELEGRAM
From Grand Trunk
Pacific Railway Co.

87aef   26 Dh & Dh

Winnipeg, Man.
Oct. 19th, 1911

W. J. Sanders
Natural Resources Co.
Vancouver, B.C.

Agreement accepted.
Company will build station soon as possible.

11 15 a.m.   G. U. Ryley

ANNOUNCEMENT:

The Natural Resources Security Company Limited begs to announce to the public that the Grand Trunk Pacific Railway has just signed an agreement whereby the official and only railroad townsite at Hazelton will be located on Lot 851 as per above plan.

There will be no station at Ellison, one mile west. Purchasers of lots at Ellison will be fully protected. Surveys are completed and plans will be published just as soon as the Government and Railroad make selection of lots.

The Land Commissioner of the Grand Trunk Pacific Railway will issue all agreements and deeds and the Natural Resources Security Company Limited will have full charge of all sales. Yours truly, GEORGE J. HAMMOND, President Natural Resources Security Co. Limited, Bower Block, Vancouver, B. C.

Only One Hazelton

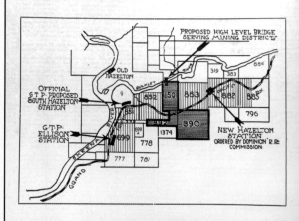

The above sketch plan, drawn to scale, gives an exact portrayal of the situation at Hazelton. Ellison station is impracticable and has been abandoned by the railway company. South Hazelton has been turned down by the Dominion Railway Commission, which body has ordered a station at New Hazelton. Lots 159, 890 and 892 are owned by us, and the east half of Lot 863, which is Section One, New Hazelton, has been practically purchased by us.

MAP 669 (*above*).
Published on 30 December 1911, this ad, also by Hammond, shows (dark shading) the land owned by his company. *New Hazelton Station Ordered by Dominion R.R. Commission* was so ordered, but the GTPR refused to comply for another two years. The map also shows another early competing townsite, *Ellison*, farther down the *Skeena River*.

MAP 668 (*above*).
Adding to the confusion about townsites at Hazelton was George Hammond's Natural Resources Security Company, which placed this ad on 28 October 1911 showing it as the sole agent for the GTPR townsite of South Hazelton. This was a fabrication that seems only to have been made to bolster the prospects of the land Hammond owned nearby (see MAP 669, *above, right*). The telegram has been altered: syndicate leader W.J. Sanders is shown as part of Hammond's Natural Resources Company.

MAP 670 (*left*).
An ad for *New Hazelton Heights*, a subdivision adjacent to Robert Kelly's townsite of New Hazelton; everyone seemed to be trying to get a piece of the action! It seemed that Kelly's site would be the winner in the battle of the townsites simply because it was a practical location for a station, whereas the GTPR's site was not.

MAP 671 (*below*).
This misleading bird's-eye map was published by the GTPR to promote its *South Hazelton* townsite. It is shown, incorrectly, as more or less on flat land across the river from *Old Hazelton*; New Hazelton, off in the distance here, is not named. *Silver Standard Mines* close by in fact loaded its ore at New Hazelton once the railway commissioners ordered the railway to place a siding there; South Hazelton was too steep for loaded ore wagons.

—put ten dollars into
the new city of
HAZELTON

"the city born in a silver bowl"

and start yourself on the road to wealth

---you men who are tired working your head and hands off, with nothing to show for it at the end of the year, put $10 a month into a lot in this young giant city of the north

Our property is

New Hazelton Heights

four blocks south of the financial and business section, and the growth of New Hazelton will be south.

New Hazelton has a firmer foundation for greatness than Calgary, Saskatoon or Edmonton had when they started.

Just think—the wealth that would have been yours now had you put $10 a month into a lot close in any of those places a few years ago.

Lots are regular city size, 33 ft. by 120 ft.

Prices Today Are $100.00 to $300.00

They will be advanced February 1

Terms $10 cash and $10 monthly

No Interest—no Taxes

FREE Title to the lots is guaranteed by B. C. Government. Information of business openings and positions, also Maps, Plans, etc.

Standard Securities Limited

SOLE SELLING AGENTS

515-519 Pacific Building   Vancouver, B.C.

BANKERS, IMPERIAL BANK

Associate Selling Agents with Clements & Heyward
882 New Hazelton, Section 2, "The Railroad Centre"

Study the plan

---read what the editor of the Omineca Herald says: *The New Hazelton Paper*, says:

"Standard Securities have the section of land tieing on to New Hazelton, only four blocks from the main business street (14th) and the property is all high, dry, and level, in fact it is some of the choicest land on the whole townsite."

"By putting this property on the market the town will naturally grow towards the high level ground rather than down towards the low, marshy ground."

"The Standard Securities have a good property, and it will be a good speculation when the lots are on the market."

Right.
The GTPR repudiation of the October Hammond ad, published in a Winnipeg newspaper and repeated in the *Fort George Herald* on 30 December 1911. An accompanying editorial suggested that Hammond did it to "bull the lot market up the Skeena River."

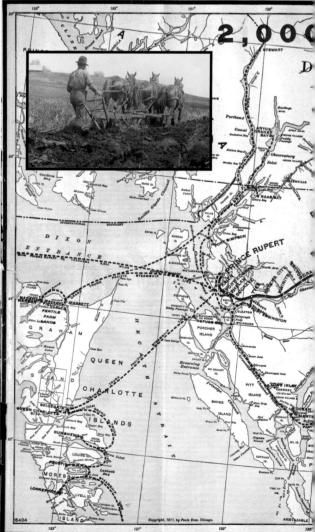

W. J. SANDERS

GENERAL AGENT FOR THE

GRAND TRUNK PACIFIC TOWNSITE
OF SOUTH HAZELTON

LEIGH SPENCER BUILDING   VANCOUVER, CANADA

Above.
The first transcontinental Grand Trunk Pacific Railway train leaves Prince Rupert on 9 April 1914. The first passenger train had left on 14 June 1911 but had gone only as far as Copper River (Copper City, shown on Map 667, *page 263*).

there agreed to its demand for a share of the profits. But the location was steep and unsuitable for loading ores from the local mines, and it competed with another townsite a short distance away called New Hazelton, owned after 1911 by Robert Kelly, of wholesale grocer Kelly, Douglas and Company. Despite clear evidence that New Hazelton was the better location, obstinate railway officials went to extraordinary lengths to support its original choice at the expense of Kelly's town. Exploits such as this increased costs unnecessarily.

World War I broke out only four months after the Grand Trunk Pacific was completed, decimating traffic. The railway was almost immediately in financial trouble. In 1919 it defaulted on its bonds and was taken over by the government; by 1923 it had been integrated, along with its fellow indigent railways, into Canadian National.

Map 673 (*below*).
Arable land and agricultural land abound on this promotional map published by the GTPR in 1911. The agricultural potential would never be enough to provide significant traffic for the railway. All the townsites mentioned elsewhere in the text can be located on this map.
Inset, left. Plowing with horses at Hubert, near Telkwa.

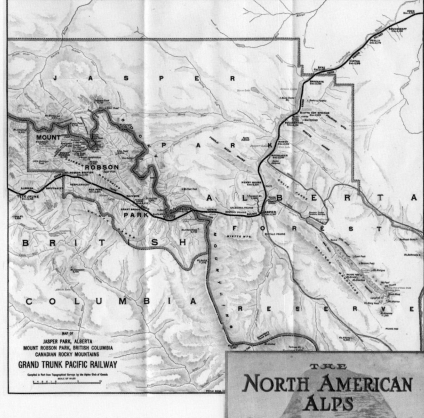

Map 672 (*above*).
The Grand Trunk Pacific needed any revenue it could get once it began operations, and, like the Canadian Pacific, turned to promoting tourism. This map and brochure, aimed no doubt mainly at Americans (the U.S. not yet having entered the war) marketed the Rockies as the North American Alps. The line through the Rockies duplicated that of the Canadian Northern, running parallel to it in some locations, and in 1917 the company was ordered to rip it up, with the steel going to the war effort. The two railway companies would soon be one and the same in any case.

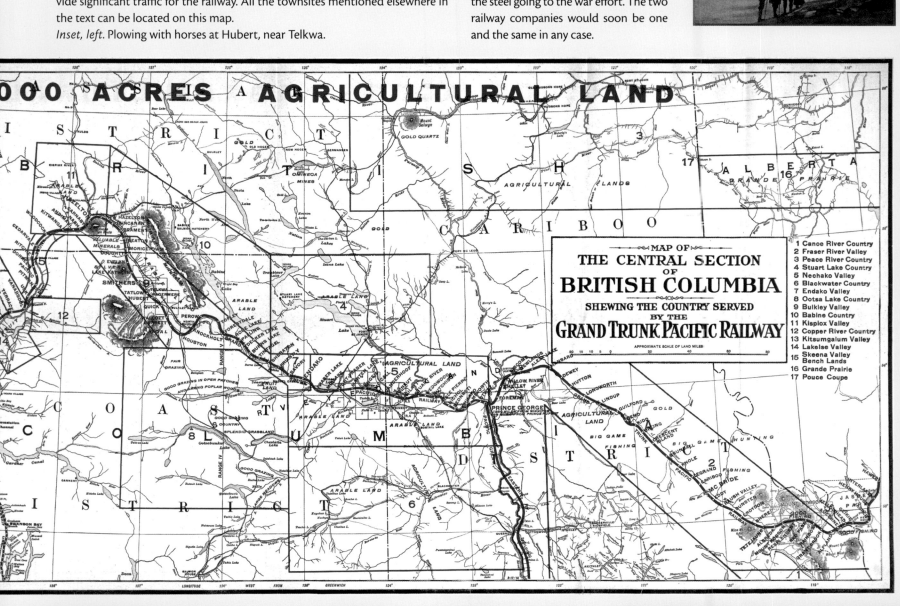

The Mapmaking Missionary

We owe many of the early details of the northern part of British Columbia's geography to an Oblate missionary, Father Adrien-Gabriel Morice, who came to the region in 1880 and stayed until 1904.

Morice is perhaps most famous for his creation of a written form of the Carrier language, which helped him in his work, but he is also well known for his maps. As he travelled, he made a habit of noting where he was, and the features around him, although he possessed only elementary instruments. He later used this information to compile the first comprehensive map of northern British Columbia, which was published in 1907 (MAP 674, *right*).

MAP 674 (*right*).
Adrien-Gabriel Morice's 1907 map of northern British Columbia, the first published detailed map of the region.

MAP 675 (*below*).
Morice's map of the Nechako basin, produced in 1904. The red lines are routes that he travelled; one ends on a mountaintop (*M*^{t}. *Glacier*, at left). He has carefully noted both villages with and without a church, a detail no doubt important to him. The map also notes Indian trails (*Sentier indien*) and vegetational zones, such as *Zone du pin Douglas* (Douglas fir). The Gardner Canal (*Fjord Gardner*), location of Kemano (just off map; see page 322) is at left. *F*^{t}. *S*^{t}. *James* is at top right.

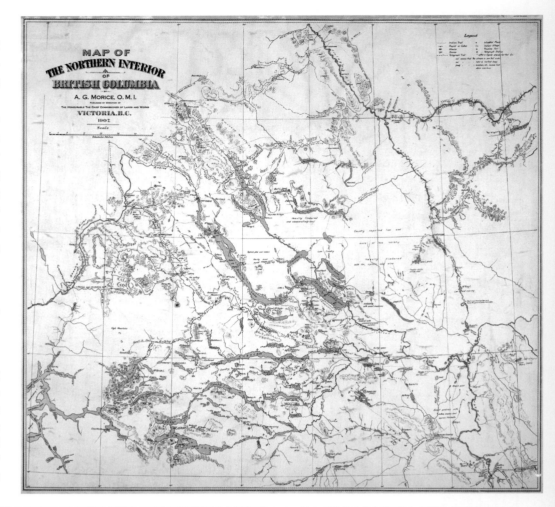

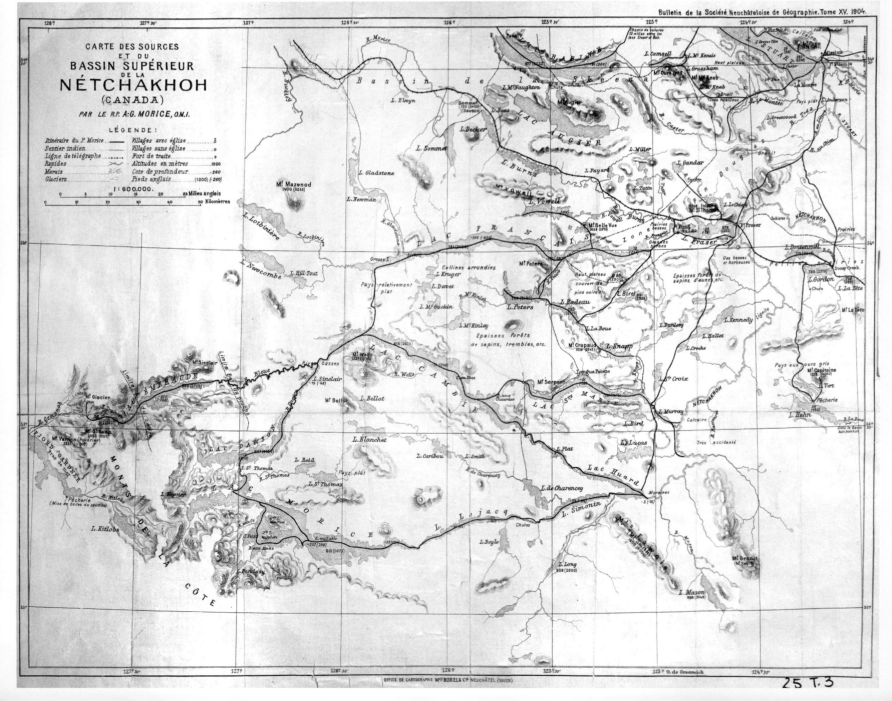

The Doukhobors

The Doukhobors were a Russian religious group notable for their pacifism. Their refusal to be conscripted led to considerable persecution in Russia, and in 1897 the Russian government, wishing to be rid of them, allowed them to emigrate. In 1898 and 1899 about 7,400 Doukhobors came to Canada because federal immigration minister Clifford Sifton, looking to populate the Prairies, welcomed them.

Internal disputes and problems resulting from the Doukhobors' refusal to register land individually led a group headed by Peter Verigin, the Christian Community of Universal Brotherhood, to move to British Columbia. Between 1908 and 1912 Verigin purchased land around Grand Forks, in the Slocan Valley, and around Castlegar, registering the land in his name but deeding it to the community.

Some eight thousand Doukhobors moved to British Columbia and began planting orchards, becoming well known for their excellent preserves (MAP 676, *right, top*). For nearly thirty years their utopia flourished in the Kootenays, though there were periodic internal conflicts and conflicts with government. They refused, for example, to register births, deaths, and marriages or send their children to school.

One group, known as the Sons of Freedom, especially protested what they saw as government interference in their lives and also wanted to remind other Doukhobors of their principles. They became well known for their unique form of protest—mass nudity, an act punishable by three years in prison after 1931. Later the group set fire to public facilities such as courthouses and schools. Peter Verigin was killed in an unexplained explosion in 1924. In 1938 the Depression led the banks to foreclose on the mortgage on their lands.

About twenty thousand Doukhobors still live in British Columbia, principally in the Kootenays and especially around Grand Forks.

MAP 677 (*right*).
The *Lands of the Doukhobor Society* cover the east bank of the Columbia River near *Castlegar* on this 1916 map, and other tracts labelled *Doukhobors* are just south of Castlegar. *Brilliant*, a Doukhobor name, is noted in two places. Today Brilliant is across the river from Castlegar.

Inset, right, and *below.*
Doukhobor children and a group engaged in communal farm work near Grand Forks about 1917. The photos were taken by Charles McKinnon Campbell, mine superintendent at Phoenix (see page 195).

MAP 676 (*above*).
A map-view of a Doukhobor orchard at Brilliant, near Castlegar, on a jam label from the Christian Community of Universal Brotherhood about 1920. The jam factory at Brilliant began producing in 1915.

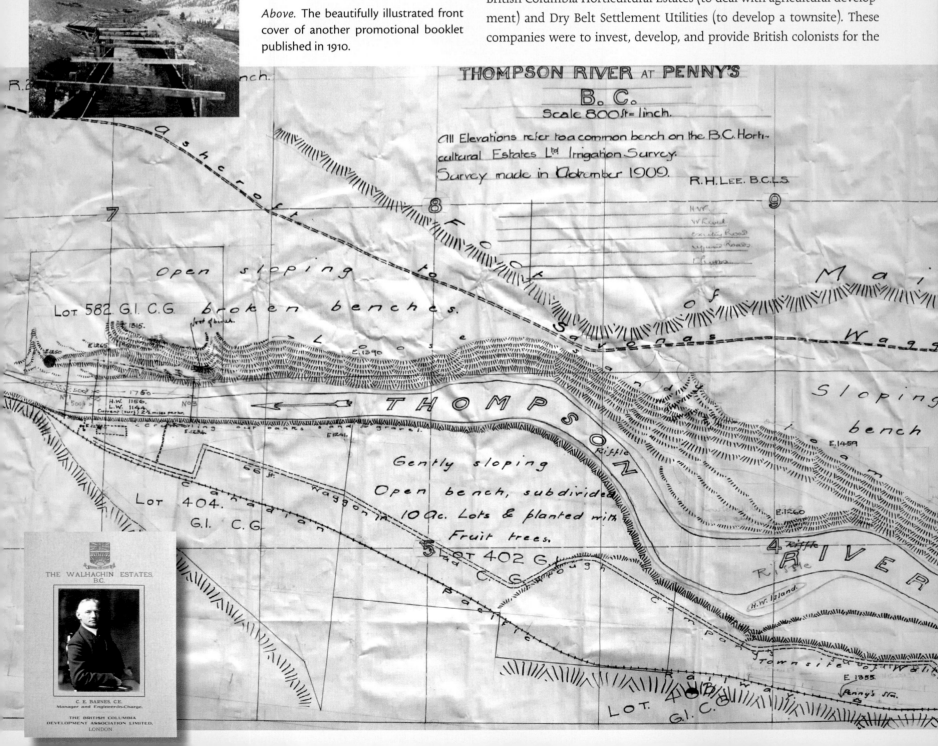

GENTLEMEN FARMERS

As fruit growing in the Okanagan Valley was proven successful in the early twentieth century, it was only natural that some would look beyond its boundaries for similar areas where orchards might be established. One attempt of special note was in the Thompson Valley west of Kamloops at a place called Walhachin.

But the Thompson Valley here is narrow, leading to greater extremes in temperature. In addition the river is entrenched, making the delivery of water to the benchlands above commercially infeasible with the technology of the day. Nevertheless, these difficulties did not discourage a group of mainly British investors.

An American surveyor, Charles Barnes, saw that fruit trees were growing on a small patch of land at the ranch of Charles Pennie and assumed that the entire area would support such a venture. He was wrong; Pennie's orchard had been carefully created on just about the only patch of land on which the trees might grow, being stone free and protected from killing winds that blew all winter off the river.

Barnes interested a British company, the British Columbia Development Association, which had already invested in other British Columbia projects and which was principally composed of aristocratic investors listing themselves as "gentlemen." Two other companies were created—British Columbia Horticultural Estates (to deal with agricultural development) and Dry Belt Settlement Utilities (to develop a townsite). These companies were to invest, develop, and provide British colonists for the

MAP 678 (below).
A survey of Walhachin carried out in 1909. The *Townsite of Walhachin* is at bottom right, as is the Canadian Pacific Railway station *Penny's Stn.* At centre left on the south bank is land *planted with Fruit trees.* The main irrigation flume from Deadman's Creek ran along the edge of the *Foot of Main Hillside* so as to distribute water by gravity. *Inset, top,* is a view of that flume as it rounds the hill from the Deadman Valley into that of the Thompson, and *inset, bottom,* is the cover of a booklet produced for potential investor-colonists in 1910.

Above. The beautifully illustrated front cover of another promotional booklet published in 1910.

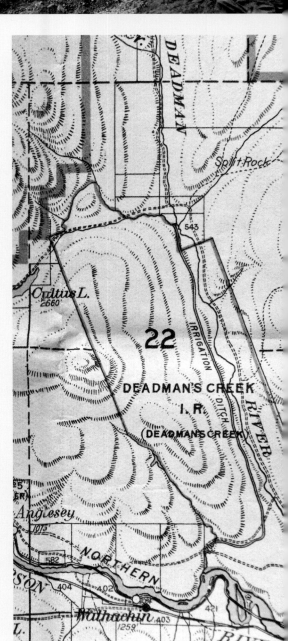

new Eden that was to be created here. In 1908 the Pennie ranch was purchased, along with adjacent land, and more land across the river, the latter purchased from the federal government for one dollar an acre ($2.48/ha), bringing the total project size to over 2,000 ha.

To provide the essential water for the orchards, a scheme was devised to bring water from Deadman's Creek, some 12 km to the north (MAP 679, *right*), and Barnes oversaw the construction of a two-metre-wide wooden flume, which brought water by gravity to a series of irrigation ditches dug between the trees. Some water was channelled to the south bank orchards from local streams, but in 1911 a steel pipe was suspended across the Thompson to bring additional water from Deadman's Creek to the south bank.

It was made clear from the start that only upper-class Englishmen would be welcome. A luxurious hotel was built in 1910 where dinner guests were required to wear formal evening clothes. English-style gardens were laid out, and everything possible was done to maintain upper-class English "standards." Class-consciousness was rife. A little bit of Old England had been created on the banks of the Thompson.

It could not last. The climate was just too extreme for effective fruit growing, and when trees were killed by the cold, it took four years or more to replace them. As an economy measure low-grade lumber had been used to construct flumes, and it proved too thin to prevent warping or to properly seal, resulting in the loss of perhaps 40 per cent of the water. The final straw was World War I. Since most of the settlers were patriotic Englishmen, they immediately answered the call of the Mother Country and by 1916 no English orchardists of military age remained. It was left to the older men, and to hired help, to maintain the irrigation system and fruit trees. Gales and washouts did not help, and when flumes broke they often were not repaired. And then, after the war, many of the colonists found they could get good jobs in Britain and chose not to return.

By 1922 the main flume was no longer bringing water to Walhachin, and the last of the British settlers moved away. The bold English experiment on the banks of the Thompson reverted to grazing land.

MAP 679 (*right*).
This 1916 map shows how the main flume, labelled as an *Irrigation Ditch*, supplied water from the *Deadman River* (Creek) to *Walhachin* on the Thompson, at bottom. By this date the Canadian Northern Railway had built its line through Walhachin.

Left, top; below; and *top right.*
Photos from a presentation album depicting the progress at Walhachin in 1910. At *left, top* is a view of the townsite with the Canadian Pacific line in the foreground; *below*, newly planted fruit trees are being provided with water via a network of shallow distribution channels also used to grow ground crops such as potatoes and onions between the trees to provide a cash crop while the trees matured. *Above*, men digging part of the main ditch along the *Foot of Main hillside* shown on MAP 678.

Left, centre.
The steel pipe strung across the Thompson in 1911 to bring water from the main flume on the north bank to the south bank fields. The following year it was replaced with a larger-diameter but wooden pipe to supply a new estate, that of the Marquis of Anglesey. However, the new pipe leaked and succeeded in delivering less water than the first.

Interurbans

Electric interurban lines were essentially longer-distance streetcar routes that used variable-length trains of multiple cars.

The first interurban began operating in October 1891 between Vancouver and New Westminster (see page 140). The next really began operation in 1905 and ran through Vancouver's Kerrisdale and on to Steveston, though the line had been built three years before as a steam-operated line by Canadian Pacific, trying to head off a possible invasion of its Vancouver territory by the Great Northern. The line was leased to British Columbia Electric Railway Company (BCER), which ran "Sockeye Express" interurbans every hour to service the canneries at Steveston.

But the completion of the New Westminster Bridge over the Fraser at its namesake city in 1904 opened up the Fraser Valley, and BCER began constructing a line from New Westminster to Chilliwack, using a 1906 charter for the Vancouver, Fraser Valley & Southern Railway Company. The interurban line was completed in 1910. An inaugural train with Premier Richard McBride aboard ran to Chilliwack on 3 October 1910, albeit with the necessary substitution of steam power for one section where a power pole had come down.

The 103-km-long line was the longest interurban in Canada and brought passenger and freight service to the valley along with

Map 680 (above).
A bird's-eye map, looking north-northwest, from an ad for land at Chilliwack, reached by the interurban line in 1910. In front of *B.C. Electric Railway* (with electric unit) as it runs into the *City of Chilliwack* is the *Canadian Northern Railway* (with steam locomotive). However, only the Canadian Northern (now Canadian National) line is roughly correctly positioned (though track would not be laid here until the following year); the interurban line approached Chilliwack from the south and ended at a loop; it never ran east of Chilliwack as shown here. The striking graphic (*left*) is from a 1930s tourist brochure. Car 1334 did not exist!

electricity. It signalled the end of river steamboats and encouraged settlement away from the river, for example at Langley Prairie—today the City of Langley. Fresh milk and vegetables from the valley could now be sold in Vancouver.

In the Saanich Peninsula BCER began an interurban service two years later (Map 685, *right, bottom*). It was essentially a real estate speculation: the company bought up land and hoped to make money when it appreciated in value because of the interurban. But it was 1913, and the hoped-for increases in value did not materialize. The Saanich interurban lasted only until 1924.

The Fraser Valley interurban ended its passenger service on 30 September 1950, and the route to Steveston was shut down in February 1958, both victims of the automobile and truck.

Map 681 (*left*) and Map 682 (*below*).
The interurban opened up the Fraser Valley to its Vancouver markets, and real estate salespeople saw great potential in it. Both of these ads, which ran in November and September 1910, respectively, feature the interurban prominently. Map 681 notes *1 hour by Chilliwack Tram* and promises to "have you back for lunch." Map 682 advertises *10,000 acres* on *Matsqui Prairie* and another *30,000* on *Sumas Prairie*, location of the soon-to-be-reclaimed Sumas Lake, and the fruit and vegetables adorning the ad's perimeter emphasize agricultural abundance. A little map of the *Vancouver to Abbotsford* line, complete with an interurban car, shows it to be *40 miles* (64 km).

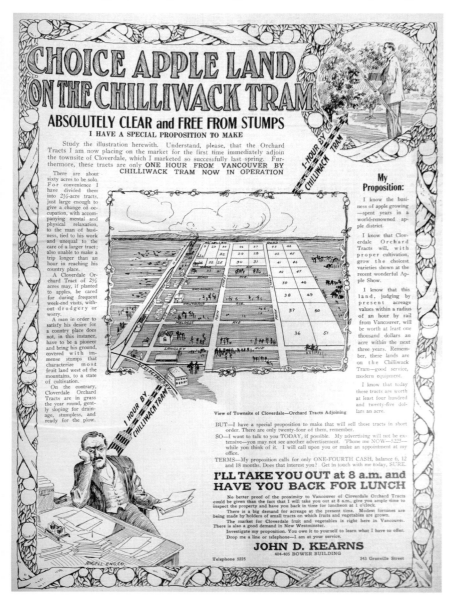

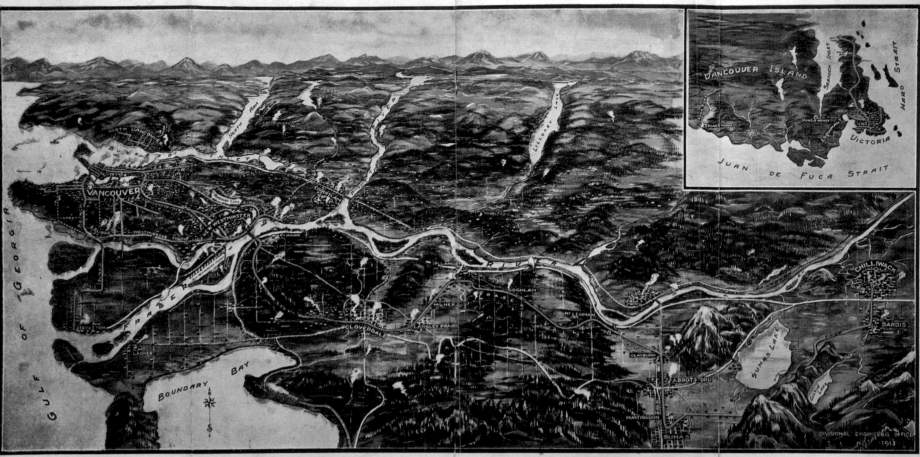

MAINLAND LINES—Vancouver, South Vancouver, Point Grey, North Vancouver, New Westminster. Vancouver and Chilliwack. Vancouver and New Westminster (3-lines). Vancouver and Steveston.

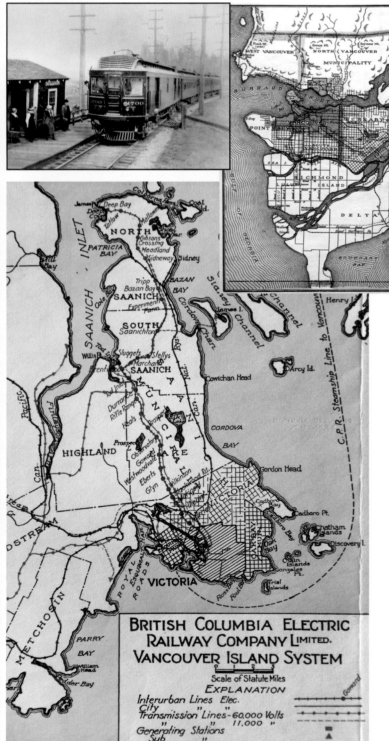

Map 683 (*above, top*) and **Map 684** (*above*).
This superb bird's-eye map of the Fraser Valley interurban lines contrasts with the more businesslike, but still attractive, regular map. The bird's-eye, which includes an inset view of the Saanich line, was part of a tourist brochure published in 1913 and, one would imagine, aimed to attract tourists to the scenic valley. The regular map shows and names all the stations. It was published in 1923; note *Sumas Lake (Reclaimed)* (see page 304).

Map 685 (*left*).
From the same brochure as **Map 684** comes this map of the Saanich Peninsula interurban from *Victoria* to *Deep Bay*. A year after the map was published, passenger service ceased.

Map 686 (*right*).
Details of the Steveston interurban are shown on this 1929 map. A complex of short freight lines service industrial areas in *Marpole*.

Left, centre. One of the finest photographs of the Fraser Valley interurban is this one showing a four-car westbound train pulling into Sullivan Station, at today's 152 Street and 64 Avenue in Surrey. The motor car at front is also the baggage car.

CANADIAN NORTHERN

The Canadian Northern Railway was the last North American transcontinental line and was built to Vancouver in 1915, in direct competition with Canadian Pacific, by two railway empire builders, William Mackenzie and Donald Mann. The pair had created a network of lines on the Prairies by buying up smaller railways, but when the Grand Trunk Pacific began building to Prince Rupert they were determined not to be outflanked. A subsidiary, the Canadian Northern Pacific Railway, was chartered in 1910, and construction began.

The route selected was through the Yellowhead Pass, cheek by jowl with the Grand Trunk in many places, and then down the North Thompson; from there the line passed through the Fraser Canyon, which gave the railway much trouble, for the Canadian Pacific had selected the better side to lay track in the narrowly defined canyon, and when it crossed to the other side, the Canadian Northern was forced to do the opposite.

And getting into Vancouver was yet another problem, for Canadian Pacific with its land grant seemed to have tied up all possible routes. But this was the sort of problem Mackenzie and Mann were good at solving. At first they had intended that the terminus would be on the south side of the Fraser, and a townsite was registered there in 1909. Land was sold from 1910 for a new city to be called Port Mann. But, no doubt to the investors' chagrin, it soon became clear that Mackenzie and Mann did not in fact intend to stop at Port Mann. In April 1912 they arranged with the Great Northern for running rights over the VV&E line to False Creek (see page 209), and then—as the pièce de résistance—proposed to fill the remaining mudflats at the eastern end of False Creek for a grand terminus adjacent to the new one being built for the Great Northern.

Vancouver city council agreed to the proposal in February 1913. The railway was to reclaim the land, return 35 acres (14 ha) to the city, and purchase the remaining 129 acres (52 ha) for one dollar. But there were

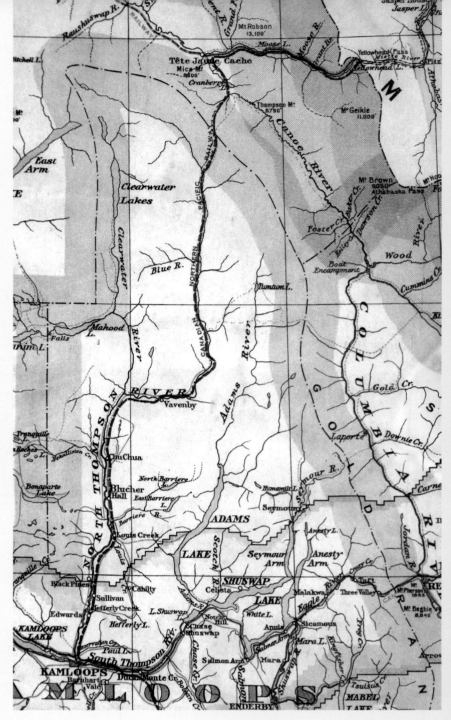

MAP 687 (*above*).
This 1912 map shows the proposed line of the *Canadian Northern Pacific Railway* (dashed line) down the valley of the *North Thompson River* to *Kamloops*.

MAP 688 (*left*).
This grand plan for Port Mann—"official," no less—appeared in the *Vancouver Province* newspaper on 20 February 1912.

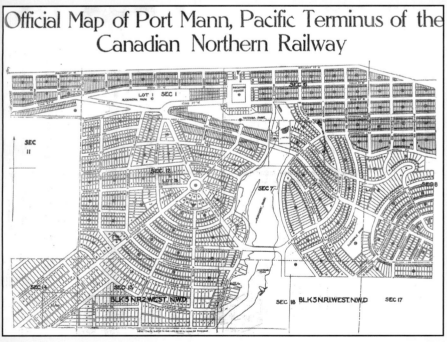

Official Map of Port Mann, Pacific Terminus of the Canadian Northern Railway

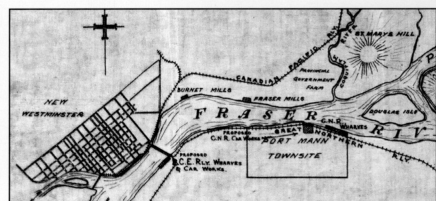

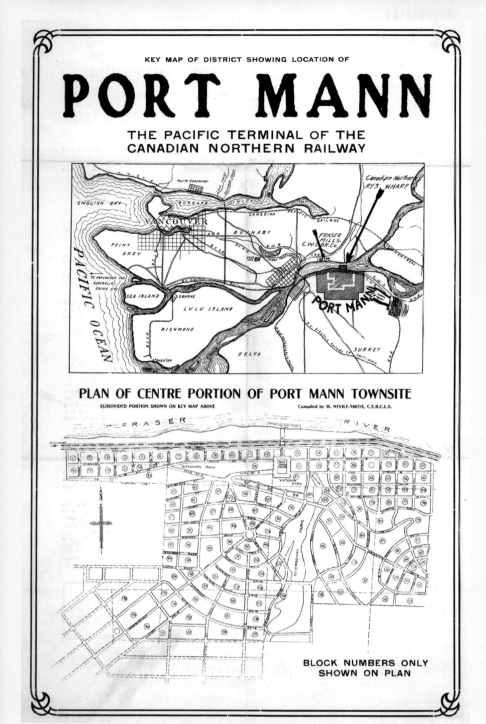

KEY MAP OF DISTRICT SHOWING LOCATION OF

PORT MANN

THE PACIFIC TERMINAL OF THE CANADIAN NORTHERN RAILWAY

PLAN OF CENTRE PORTION OF PORT MANN TOWNSITE

SUBDIVIDED PORTION SHOWN ON KEY MAP ABOVE

Compiled by H. NEVILE-SMITH, C.E.B.C.L.S.

BLOCK NUMBERS ONLY
SHOWN ON PLAN

Map 689 (*far left, centre*).
Port Mann Townsite is located on this map from about 1909.

Map 690 (*far left, bottom*).
All that had been surveyed and cleared by 1912 was this area at the base of the escarpment. This was the area in which the railway would build its freight yards, and they remain there today. Today the Port Mann Bridge crosses approximately in the position of the "F" in the word "Fraser."

Map 691 (*left*).
Port Mann looms large, rivalling *Vancouver* and *New Westminster*, in this fine ad published in 1912.

Map 692 (*above*).
The new city of *Port Mann* appears as though it already existed on this 1913 map, an updated edition of Map 500 (*pages 208–09*).

Map 693 (*below*).
An imaginative ad that appeared in a magazine in March 1912, making *Port Mann Official Townsite* "A Dog-on-Good Proposition."

Right
The first real estate office—a tent—appeared on the Port Mann site in the summer of 1910.

further terms in the contract: the railway was to build a large terminal and a hotel, make Vancouver its headquarters, link with a trans-Pacific shipping service, and lay an electrified double-track line in a tunnel from the Fraser to False Creek. In fact Mackenzie and Mann were running out of money and stalled with the reclamation work, and little else was done. An ultimatum from the city in 1915 hurriedly brought Mackenzie from Toronto to calm the situation; work began once again, and by 1917 eastern False Creek had been completely filled in. A grand terminus was built in 1917, but all the other projects in the city contract fell by the wayside.

Two years earlier, on 23 January 1915, the last spike on the Canadian Northern line was driven at Basque, just south of Ashcroft, and the transcontinental line was complete. The timing was atrocious. Five months before, World War I had begun. Two years after the line's completion it was nationalized by the federal government, and, along with the Grand Trunk Pacific it became Canadian National Railways.

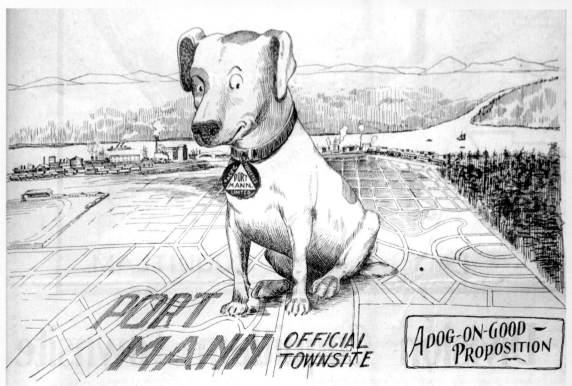

Gentlemen, it is a Good Proposition

PORT MANN

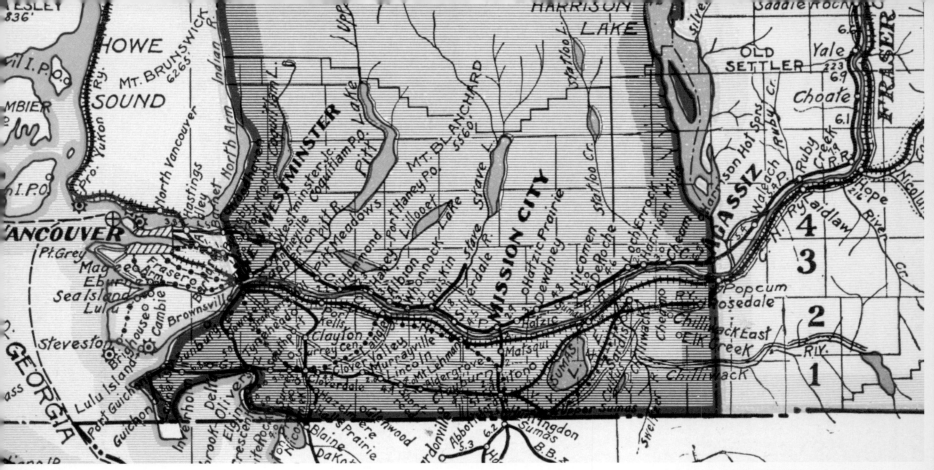

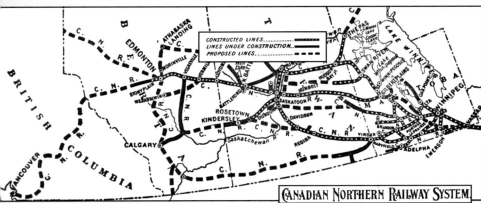

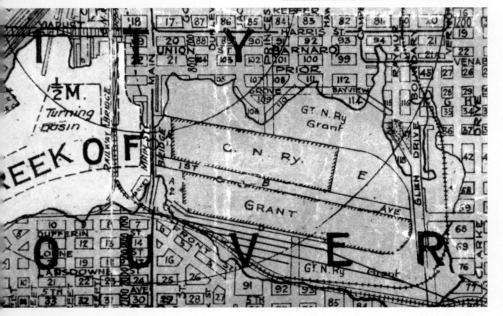

On Vancouver Island a line was planned between Victoria and Port Alberni (MAP 700, *far right, bottom*), and a line between Victoria and Patricia Bay, at the northern end of the Saanich Peninsula, was to connect with a train ferry to the mainland. Although work began on the Port Alberni line in 1911, the war disrupted the project, and by September 1918 less than 7 km of track had been laid. A renewed effort begun at that time to access spruce forests needed for aircraft was cut short when the war ended two months later. All work ceased in 1928, when the line had still only reached the western end of Cowichan Lake.

The 26-km-long line to Patricia Bay was also delayed, opening in April 1917. A ferry, the SS *Canora*, arrived in December and provided service, though only for freight, until 1968. Austerity was the byword, however, once the government took over the Canadian Northern. Canadian National often saw no need to provide passenger service. The truncated Port Alberni line did help the lumber industry develop for many years, but all Canadian National service on Vancouver Island had ceased by the early 1990s.

MAP 697 (*right*).
This 1912 Canadian Northern Pacific Railway blueprint shows how its planned station and freight offices are to fit between the *Passenger Depot G.N.R.* and that company's lines to the south. *False Creek* is across *Main Street* and separated from it by a *Sea Wall*.

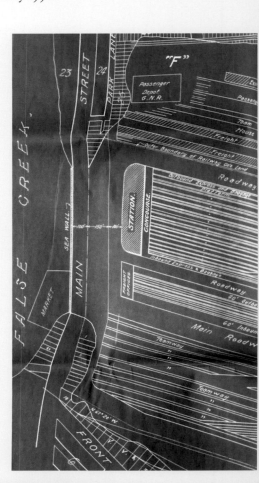

MAP 694 (*above, top*).
The path of the *CN Ry* (black line with cross strokes) can be carefully followed along the south bank of the Fraser on this distinctly overcrowded 1914 map showing all the existing and proposed railway lines of the region. The *V.V.&E. Ry.* is incorrectly shown south of *Sumas L.*; it was the line on the north side of the lake and joined the CN line at Cannor, near the Vedder Canal (see page 220); it did not continue to *Rosedale*, as shown on this map.

MAP 695 (*above, centre*).
The 1910 map of the Canadian Northern system showed merely an ill-defined proposed line into British Columbia compared with a dense network of lines elsewhere. The western part of the map is shown here.

MAP 696 (*above*).
The *C.N. Ry. Grant* is shown in the central part of False Greek on this 1914 map. It is flanked on both sides by the horseshoe-shaped *Gt. N. Ry. Grant* (see MAP 502, *page 209*). The original VV&E line crosses False Creek on a trestle to its depot.

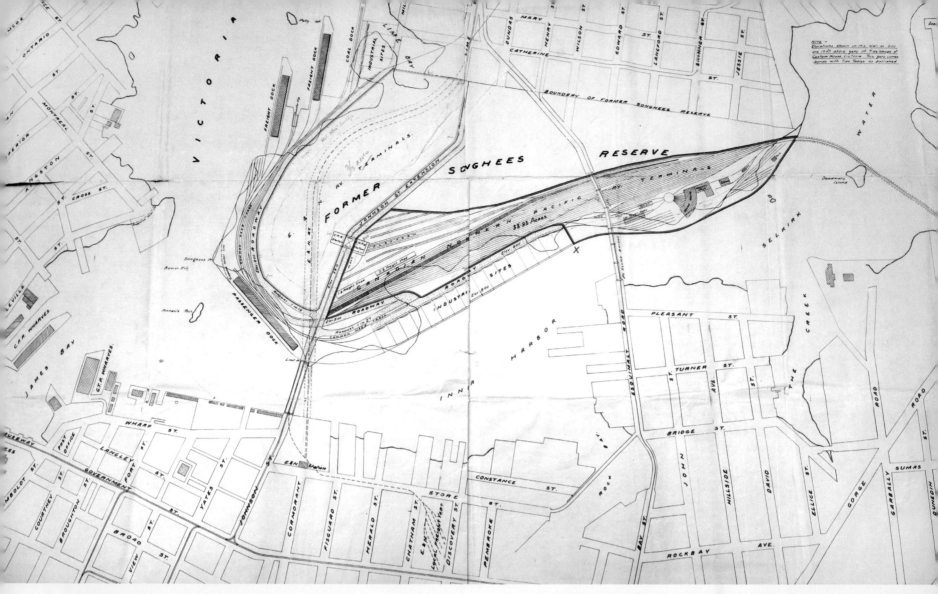

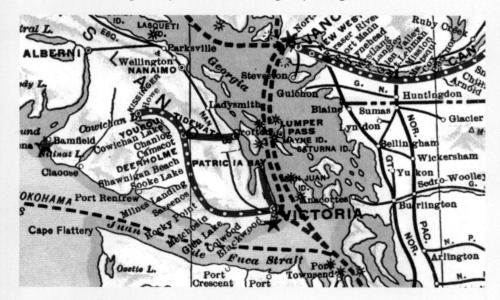

MAP 698 (above).
To find land for terminal facilities in Victoria, the Canadian Northern gained the use of the former Indian reserve land that had been allocated to the bands collectively known as the Songhees by James Douglas in 1850–54. This blueprint likely dates from just before the building of the yards there, which began operating in 1920.

MAP 699 (above).
Part of a Canadian National system map published in 1929 depicts the operations of that railway on the old Canadian Northern routes and its own extensions more or less at its peak. The line to Port Alberni reaches only the western end of *Cowichan L.*, where construction was halted and the grade beyond abandoned in 1928. A branch to Cowichan Bay is shown; this was built in 1924 to eliminate the need to transport lumber from the Cowichan Lake area to *Victoria* and then north again to *Patricia Bay*. At Cowichan Bay facilities were built to allow lumber to be transferred to ocean-going freighters from trains arriving at tidewater. The line from Victoria to Patricia Bay, opened in 1917, was used for some time by a gas railcar to provide passenger service but was principally intended for freight, especially after nationalization. The railway ferry SS *Canora* (the name a composite of <u>Ca</u>nadian <u>No</u>rthern <u>Ra</u>ilway) ran from Patricia Bay to the mainland, though, as shown here, also used facilities at Victoria. The Canadian National line sweeping into Vancouver is also shown, as are steamship connections to other coastal ports.

MAP 700 (above).
The complete intended route of the Canadian Northern line from *Victoria* to *Port Alberni* is shown as a dashed blue line (marked *C.N.P. Ry.*) on this map published in 1912. The line was not built beyond the *Nitinat R.*, but the entire route to Port Alberni, and beyond to the Cumberland coalfields, was surveyed, and most of the section to Port Alberni was even graded. Much of the abandoned grade was soon utilized by logging company railways. The branch from Victoria to Patricia Bay, at the northern end of the Saanich Peninsula, is also shown, as is the Great Northern's line (in green) to Sidney (see page 169).

Tunnels through the Mountains

Canadian Pacific had built its transcontinental line in a hurry, leaving two bottlenecks in the mountains. One was in Kicking Horse Canyon, where the temporary expedient had been the "Big Hill," with its safety switches and 4.4 per cent grade (see page 123), and the other was in Rogers Pass, at the summit of the route through the Selkirk Mountains. These stretches of track were expensive to operate, typically requiring helper locomotives, and prone to disasters such as a snow slide in Rogers Pass in 1910 that killed sixty-two men.

Facing increased competition, the railway determined to deal with these problems. First it tackled the Kicking Horse Canyon. Chief engineer John E. Schwitzer came up with an ingenious solution: reduce the grade by lengthening the path with two circular tunnels. Called the Spiral Tunnels, they had arcs of 232 and 234 degrees but, importantly, did not lessen the scenic opportunities for passengers, an important consideration for the railway's promotion of the mountains to tourists (see page 290).

Work began in January 1908. Besides the two new Spiral Tunnels, considerable work was required to realign about 5 km of track on either side. Work was completed, and the new line opened in August the following year. The tunnels became a tourist attraction in their own right, hailed as they were as one of the engineering wonders of the modern world, and Canadian Pacific over the years produced a number of maps and other promotional material to inform its passengers (Map 705, *far right, top*).

The remaining significant bottleneck was the Selkirks. Here a more conventional solution was reached: an 8-km-long tunnel under Mount Macdonald that allowed the rail line to bypass Rogers Pass and emerge in the Beaver Valley, where the grade was again gentler down to Revelstoke. Construction began in the summer of 1913, and the tunnel was completed in July 1916, when it was opened by the governor general of Canada, the Duke of Connaught, after whom it was named. Despite being a conventional tunnel, it had its own claim to stardom—when completed, it was the longest railway tunnel in North America. A casualty, however, was the railway's Glacier House hotel, bypassed by the new route (see page 291).

A new and even longer tunnel was completed under Rogers Pass in 1988. The Mount Macdonald Tunnel eliminated the need to use helper locomotives on long trains such as coal unit trains by reducing the east slope grade. The tunnel passes 91 m below the Connaught Tunnel. At 14.7 km, the Mount Macdonald Tunnel is now the longest in the Americas. The Connaught is still used, thus creating multiple tracking, which increases capacity. Trains usually run west though the Mount Macdonald and east through the Connaught.

Map 701 (*right, top*).
This contemporary railway cartoon celebrates the completion of the Spiral Tunnels in 1909 and the achievement of chief engineer John Schwitzer, whose boots are visible at the upper tunnel portal.

Map 702 (*right, centre top*).
Dating from about 1906 or 1907, Schwitzer's blueprint shows possible alternative positions for the Spiral Tunnels and different track alignments.

Map 703 (*right, centre bottom*).
Published in 1908, when the Spiral Tunnels were under construction, this map displays the *Present Line* and *New Line* for the track through the tunnels.

Map 704 (*right, bottom*).
A 1950s bird's-eye map of the Spiral Tunnels produced by Canadian Pacific.

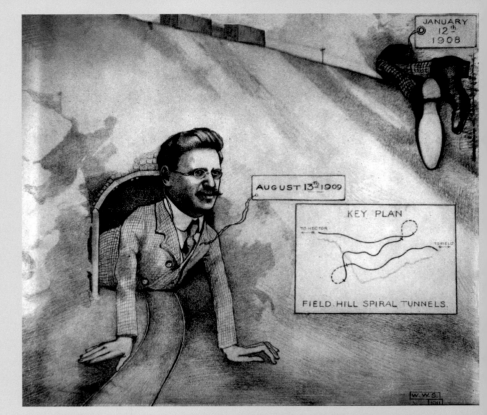

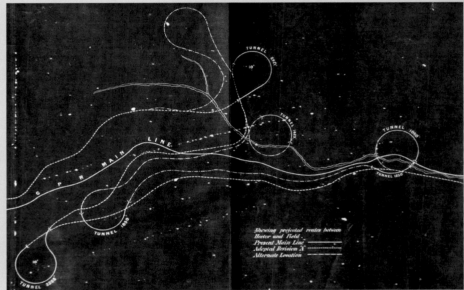

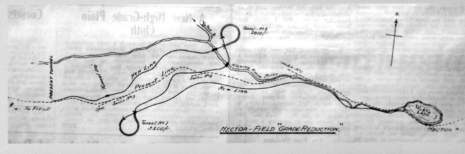

WHITE DOTTED LINES REPRESENT THE ROUTE TAKEN BY TRAIN THROUGH SPIRAL TUNNELS.

Map 705 (above).
Depicting the new route of the Canadian Pacific through the Spiral Tunnels far better than any text could ever do is this bird's-eye map from the 1930s that formed part of a menu for those dining on the train. The map was reproduced a number of times in several different guises, as it demonstrates quite well the way the spirals cut the grade.

Map 706 (below).
Part of a topographic map published in 1915, while the Connaught Tunnel was under construction. Here it is called by its original name, the *Selkirk Tunnel*.

Map 707 (below).
A promotional postcard likely published about the time the Connaught Tunnel was opened in 1916 includes a map of the new route and shows how Glacier House hotel was cut off by it. Views include Mount Macdonald and a train entering the tunnel—though one would not have thought that a tunnel would be appealing to tourists travelling through such spectacular scenery.

CONNAUGHT TUNNEL.—CANADIAN ROCKIES.

Through Mt. MacDonald. Canadian Pacific Railway.
LONGEST TUNNEL IN NORTH AMERICA.
Length 5 miles with double track.   Cost $5,500,000.00.   Completed in 23 Months and one year ahead of contract time, the fastest time ever recorded on this Continent.   Shortens the rail time 4.3 miles and reduces the elevation of track through Rogers Pass by 552 ft., eliminates 5 miles of snowsheds, also curvature to the amount of seven complete circles.

View showing Tunnel under Mt. MacDonald.
(Height 9,860 ft.)

View showing location of Tunnel and Old and New Lines.

Entrance to Connaught Tunnel.

PRINCE GEORGE—EVENTUALLY

There had been several proposals to build a railway line from Vancouver to the Yukon, a proposition that became more enticing after gold was discovered in the Klondike, but they did not get farther than the Lower Mainland. MAP 500, *pages 208–09,* shows the Vancouver, Westminster & Yukon line that became the Great Northern and Canadian Northern entrance into Vancouver. Maps such as MAP 694, *page 274,* show a proposed Yukon Railway up Howe Sound. But this was expensive country to build in unless you had a clearly remunerative destination.

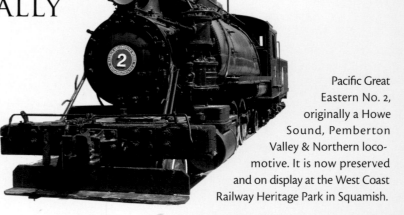

Pacific Great Eastern No. 2, originally a Howe Sound, Pemberton Valley & Northern locomotive. It is now preserved and on display at the West Coast Railway Heritage Park in Squamish.

A railway to the Peace River region had also been proposed, and the building of the Grand Trunk Pacific potentially allowed a rail line north from Vancouver to join with it and create another transcontinental line to Vancouver. The Grand Trunk Pacific itself initially had intended to build a line south to Vancouver and had surveyed a line routed down Harrison Lake before coming to an agreement with the promoters of the Pacific Great Eastern, chartered in February 1912 with its bonds guaranteed by the railway-friendly premier, Richard McBride (see page 283).

The Pacific Great Eastern Railway (so named because of its principal financial backer, the Great Eastern Railway of Britain) made an arrangement to take over the 15 km of line of the Howe Sound & Northern Railway (HS&NR). Previously the Howe Sound, Pemberton Valley & Northern, the railway had begun in 1909 to build a line north from the head of Howe Sound to tap the rich timber resources of the valley—but not before first buying up all the land for miles around. In 1909 the company laid out a townsite that was called Newport (MAP 709, *right, bottom*), a name changed in 1914 to Squamish, as the original settlement was known.

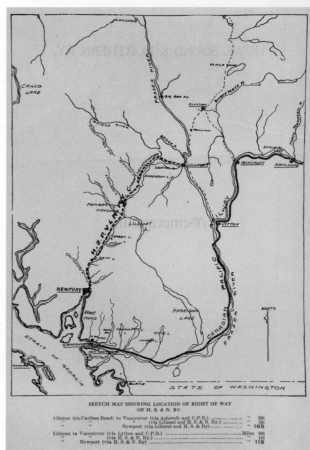

SKETCH MAP SHOWING LOCATION OF RIGHT OF WAY OF H. S. & N. RY.

The Birth of a New Seaport Occurs But Seldom
NEWPORT IS THE LATEST ARRIVAL
Watch Local Papers for Further Announcements.

The British American Trust Co. Limited
LAND AGENTS
431 Seymour Street. Vancouver, B. C.

MAP 708 (*right*).
This map, drawn in 1911, shows the intended track of the Howe Sound, Pemberton Valley & Northern Railway (*H.S.P.V. & N. Ry*) from *Newport* (Squamish) to *Lillooet.*

MAP 709 (*below, right*).
Although laid out in 1909, for some reason the townsite of Newport appears not to have been registered until 1912. This is a survey—note there is no name—completed on 30 May 1912 by Ernest A. Cleveland, who had been responsible for locating much of the railway's route. The *Howe Sound and Northern Railway* right-of-way runs down the eastern side of the town to docks at the *Head of Howe Sound.* The promoters of the HSPV&N bought up all the land before announcing what they were doing. The photo of the townsite sales office (*left, bottom*) was taken in October 1909; the photo of Cleveland Avenue (*right*) was taken in 1914. Comparison of this map with a modern one reveals that further land was later created south of that shown on this map by reclamation of the mud flats that formerly occupied the Squamish River estuary.

Above, left.
This ad announcing the "birth" of Newport invoked patriotism, suggesting that it was the latest of Britannia's settlements. The name "Newport" was considered more saleable than "Squamish."

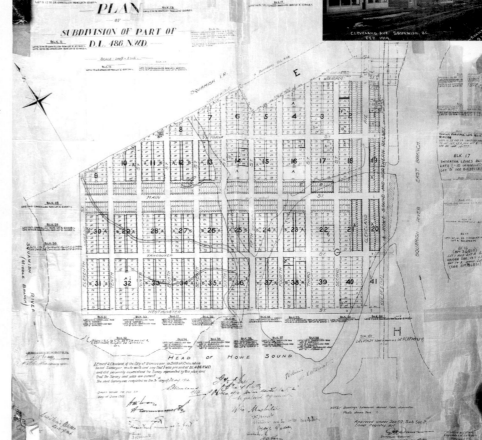

MAP 710 (*above*).
The proposed PGE line from *North Vancouver* to *Newport* and beyond, shown on a 1912 map.

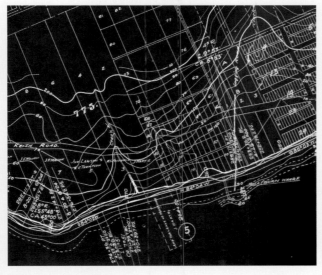

MAP 711 (*above*).
Surveyed in August 1912 and stamped "Sanctioned" and signed by the minister of railways on 3 September is a revised location plan for the Pacific Great Eastern in what became, later that same year, West Vancouver. This portion covers what is now Ambleside; note the property of John *Lawson* and *Hollyburn Wharf*. Lawson purchased land here in 1906 and with other investors began a ferry service from the wharf to Vancouver in 1909; Hollyburn was the name he bestowed on the project. Also shown is a *Logging Road* straight through the surveyed subdivision.

MAP 712 (*right*).
The proposed line of the *Pacific Great Eastern* from *North Vancouver* to *Newport*, shown on a 1914 map. The section in Howe Sound would not be completed until 1956.

MAP 713 (*left*).
Only the section north of Lillooet to Fort George is shown dashed as under construction on this map from about 1914.

MAP 714 (*right, bottom*).
White Cliff, a townsite laid out by Charles Tupper and Colonel Albert Whyte in 1909 at Horseshoe Bay and Whytecliff in what is today West Vancouver. The men backed another railway project that did not materialize, the Vancouver & Northern Railway. The location had originally been named White Cliff on an Admiralty chart. This map was created in 1910.

Construction of the northern part of the Pacific Great Eastern had reached Chasm, a few kilometres north of Clinton, in 1915, when the railway ran out of money and defaulted on its bonds. As with many grand projects of this period, the railway's timing was terrible. With the onset of World War I financial backing evaporated and potential customers disappeared.

In 1918 "Honest John" Oliver became premier, a man distinctly less enamoured of railways than McBride, and in 1918 the company, by then in dire financial straits, was taken over by the government.

Construction on the southern portion of the railway had begun in 1912 and by the end of the following year reached Dundarave in the new municipality of West Vancouver. Passenger service was inaugurated with due ceremony on 1 January 1914. Only six months later the line was completed through to White Cliff, a town platted a few years earlier at today's Horseshoe Bay (MAP 714, *right*). The station here was called Whytecliff (MAP 722, *overleaf*), the name changed at the request of Colonel Albert Whyte, one of the original townsite owners. Regular passenger service began on 21 August.

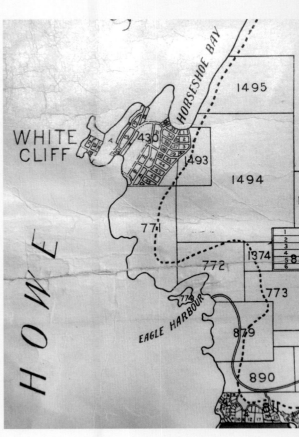

MAP 715 (*above*).
Grand Trunk Lands Co. land in central British Columbia across the path of an expected Grand Trunk Pacific extension to Vancouver instead found itself on the line of the associated Pacific Great Eastern. The map is from a sales booklet published in 1910 (*inset*). Note that the map title carefully states the company's lands are approximate. The Grand Trunk Lands Company was merely intending to facilitate—for a fee—individual claims to land from the government.

MAP 716 (*above, right*).
Only a short distance north of *Newport* is shown as complete on this 1913 map; the rest of the route to *Fort George* is projected; note also the projected Canadian Northern (*Can. Nor. Pac. Ry.*) line on the North Thompson.

Whytecliff was intended as a temporary terminus but became a permanent one. When the company was becoming short of funds, service on the southern section ceased for a while during 1918. After the government took over, service lasted until 1922, when it was again shut down. This time public pressure resulted in reinstatement of service. But it was only a temporary reprieve; in 1925 the opening of the Second Narrows Bridge increased competition from automobiles, and in 1928 the government reached an agreement with the municipality of West Vancouver whereby it received $150,000 to allow the line's closure—money that was spent paving Marine Drive all the way to Horseshoe Bay.

QUESNEL SOUTH   HIXON CREEK SOUTH

FORT GEORGE NORTH

FORT GEORGE SOUTH

SQUAMISH DOCK

Map 717 (*below, left*) and Map 718–Map 722 (*all on this page*).

Details of an enormous map of the completed and projected route of the Pacific Great Eastern produced by the railway's engineering department, dated 1915 but updated through 1920. The whole map is rolled and is 11 m long. Thickening of the lines represents railway lands.

Map 717 shows the line at *Williams Lake*. Map 718 (*above, top*) shows the line in the vicinity of *Quesnel Townsite*; at *Australian Creek* is the *End of Steel December 1920*, and at right is *Cottonwood River*, which required a long bridge that was not built until 1952. Note also the changed position of the *Kersley* station. Map 719 (*left, centre*) shows the planned junction with the *Grand Trunk Pacific (Right of Way)* near *Prince George*. Map 720 (*right, centre*) depicts the line at *Mons Y, Alta Lake*, later the site of Whistler. Map 721 (*above, left*) shows the southern end of the main section of the line at the *Govt Wharf, Squamish Dock*. And Map 722 (*below*) shows the completed southern section from *North Vancouver* to *Whytecliff*, run until 1929, after 1920 using gas-electric railcars such as the one shown, *inset*. At *right* is a railway ad for Horseshoe Bay, at the western end of the southern section.

Municipality of West Vancouver

NORTH VANCOUVER NORTH

WHYTECLIFF
CYPRESS Pk.
DUNDARAVE
WESTON
HOLLYBURN
NORTH VANCOUVER

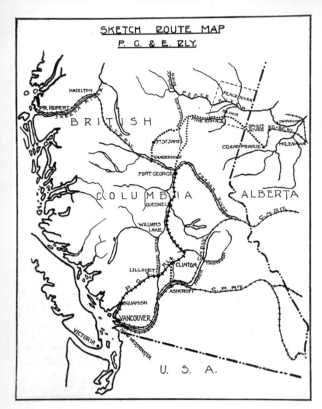

MAP 723 (left).
A sketch map from the Sullivan report of 1922 shows not only the line of the *P.G.E. Rly.* from *Squamish* to *Fort George* but also two possible routes farther north to the *Peace River Block*, the southeastern part of which the federal government had opened to settlement in 1912 (see page 286). Sullivan recommended that, because of a lack of anticipated traffic from the Grand Trunk Pacific and the undeveloped nature of the route itself, "the road from Quesnel to Prince George be abandoned and the track taken up and salvaged." The government agreed, and Quesnel became the northern terminus for the next thirty years. Sullivan also recommended not laying any track north of Prince George or along Howe Sound to Squamish, and the abandonment of the North Shore section. All his recommendations were followed, as was another suggesting the use of gas-powered railcars instead of steam locomotives for passenger trains.

MAP 724 (below).
This map from a government brochure advertising farmland along the route of the Pacific Great Eastern dates from the mid-1920s. The line is shown as complete (solid line) north from *Squamish* to the *Cottonwood River* and the rest of the line to *Prince George* as projected (dashed line). The southern portion of the railway is also shown as complete; this line was operated until 1928 after an attempt in late 1922 to close it was rebuffed by public outcry.

Inset is the cover of the government booklet; settlement was desperately needed, especially between Lillooet and Clinton, to provide business for the railway. One of the options Sullivan considered was a diversion of the line at Clinton to Ashcroft, where it could join Canadian Pacific or Canadian National and thus be linked to Vancouver. Both these lines are shown on the map, as is the one previously belonging to the Grand Trunk Pacific, but only Canadian Pacific has managed to remain privately owned.

On the northern section of the Pacific Great Eastern construction northward resumed after the government takeover in 1918. Williams Lake was reached in September 1919 and Quesnel in July 1921. By the end of 1922 track had reached the Cottonwood River, where a long and expensive bridge was required. Construction jumped the river, continuing northward, and the line was partially constructed to Red Rock Creek, only 30 km from Prince George, when the government gave up.

The government was fed up with the constant drain of construction costs on its coffers and in 1922 commissioned John G. Sullivan and two others to report on what should be done. Sullivan, a former Canadian Pacific consulting engineer, produced a scathing report in which he doubted the railway would ever be able to pay its way, especially north of Prince George (MAP 723, *above*).

The report was enough for premier John Oliver; the bridge over the Cottonwood, for which the steel had already been ordered, was cancelled. Quesnel was to be the "temporary" terminus.

In fact Quesnel remained the northern terminus for another thirty years; track already laid north of the city was taken up and sold for scrap. The railway was long derided for going "from nowhere to nowhere," and with its apparent leisurely operating style acquired monikers such as "Please Go Easy" and "Past God's Endurance."

The Cottonwood bridge and the connection to Prince George was completed in 1952, and four years after that track was relaid in West Vancouver and the section along Howe Sound built, finally connecting North Vancouver with Prince George in 1956 (see page 320). The railway had reached Prince George. Eventually.

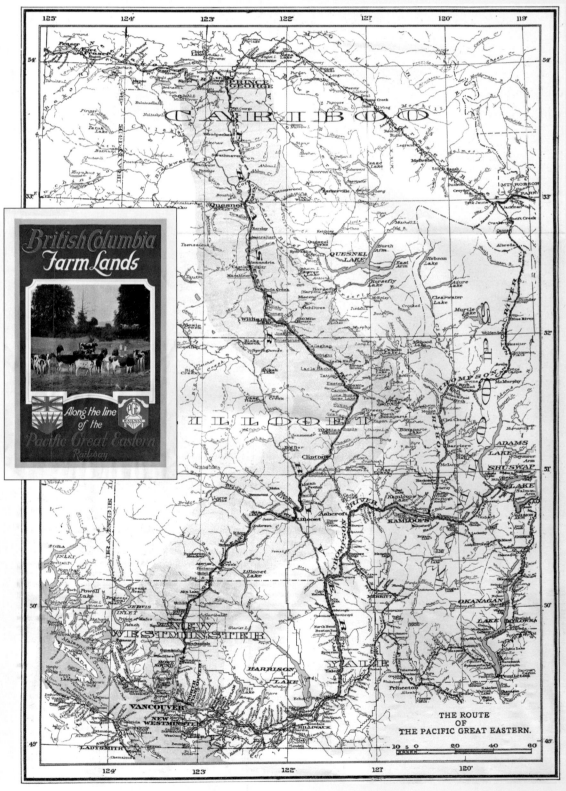

THE ROUTE OF THE PACIFIC GREAT EASTERN.

Railways Everywhere

Towards the end of the nineteenth century railways came to be seen as essential elements of growth and success, something no self-respecting town or aspiring city could afford to be without; a rail line was utterly critical for any townsite promoter. This was not surprising given the alternatives, and the sparsity and condition of roads and trails. Railways offered speed, convenience, and bulk transport.

British Columbia became railway mad: in the period between 1882 and 1913 no fewer than 212 railways were provincially chartered, and of course there were those federally chartered—the Canadian Pacific and Grand Trunk Pacific—as well.

Richard McBride, premier from 1903 to 1915, was perhaps the railways' greatest supporter at a time of maximum railway euphoria. Rising in the legislature on 20 February 1912, McBride pointed to 2,922 miles (4,675 km) of railways "built or under contract," comparing it with just 650 miles (1,040 km) of track in 1904 as he expounded upon his "railway policy"—which amounted to aiding railway construction any way he could—which would, he hoped, add another 845 miles (1,352 km). Included in the latter figure was the whole of the planned Pacific Great Eastern, "with an estimated length of 450 miles (720 km)."

War, and then the rise of the automobile and truck, and the construction of paved roads, ultimately ended the railway boom. In 1927 the provincial government passed the "Defunct Railway Companies Dissolution Act," which revoked the charters of 133 railways that had not built any lines.

MAP 725 (*above, right*).
It seems that just about every railway ever chartered that might pass anywhere near the promoter's townsite of *Fort Salmon*, just north of today's Prince George, is shown on this map, which appeared, of course, in an ad, on 14 December 1912. Despite the impressive railway display, the townsite, like most of the railways, never progressed beyond the planning stage.

MAP 726 (*right*).
Five months later the rival townsite of *Fort George* (see page 256) was featured in an ad with a similar railway map. Most of the railways were still mere possibilities, but the town this time, thanks to the *Grand Trunk Pacific*, was for real.

MAP 727 (*left, bottom*).
Premier Richard McBride is depicted in a March 1912 cartoon as the "Colossus of Roads" standing on a British Columbia with railway schemes mapped: the *C.N.P. Ry.* (Canadian Northern Pacific); the *E.&N. Ry* (Esquimalt & Nanaimo), with lines to *Hardy Bay* (Port Hardy); and the *G.T.P. Ry.* with a line to *Vancouver* via *Lillooet* and Howe Sound. The other cartoon (*left*) is from 1908 and shows McBride ready to build his own railways in British Columbia if the transcontinentals would not. Charles Melville *Hayes* (Hays) of Grand Trunk Pacific and Thomas *Shaughnessy* of Canadian Pacific peer westward over the Rockies.

MAP 728 (*below*).
Railways, projected in red (and existing in black), abound on this 1898 map. Notable is the *British Pacific Rly.*, which was to go from Vancouver Island up the *Bute Inlet* route (see page 114) to *Edmonton,* and the *Vancouver and Victoria Route to Klondike,* which hopefully intended to integrate steamship with rail, as it is shown hopping from island to island. The Great Northern–controlled *Vancouver, Victoria and Eastern* (VV&E) Coast-to-Kootenay line is projected across southern British Columbia (see page 216).

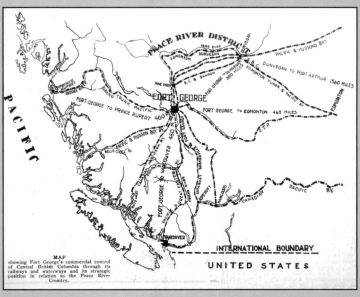

COLOSSUS OF ROADS

SURVEYING BRITISH COLUMBIA

Creating accurate maps of mountainous terrain has always been a challenge, yet the principal surveying method, triangulation, lends itself to such topography. Triangulation works on the fact that if you measure a base line and then measure angles from its ends to another object, you can work out the length of the other two sides of the triangle geometrically. Then you have a new line to take further angles from. But any error will accumulate, so extreme accuracy is essential. The best surveyors had two qualities: they did not mind hacking their way through remote forest, bush, and swamp to achieve a vantage point, and they were sticklers for accuracy. The maps shown here are a small sample of surveying work carried out in British Columbia. Some of the maps created were works of art in their own right. Today mapping still uses triangulation but measures distances, positions, and angles using sophisticated electronics and satellites, the latter triangulating vertically instead of horizontally.

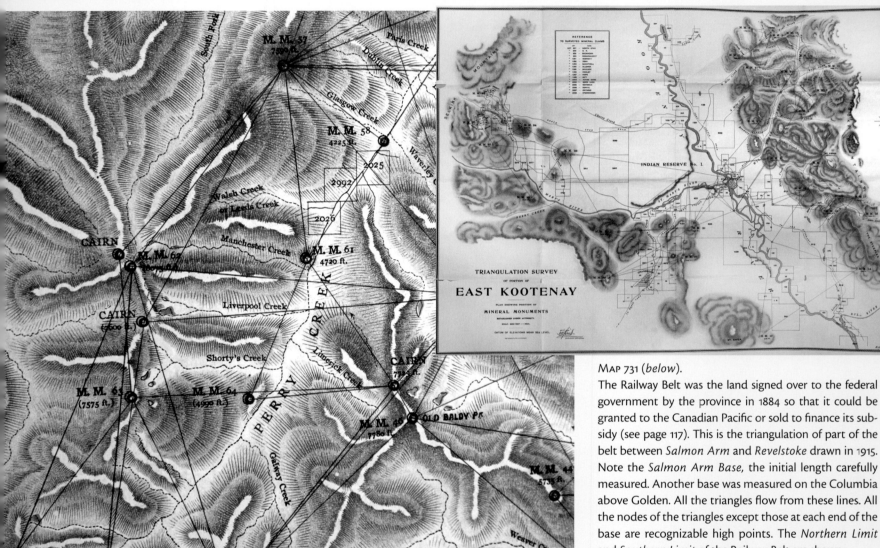

MAP 731 (*below*).
The Railway Belt was the land signed over to the federal government by the province in 1884 so that it could be granted to the Canadian Pacific or sold to finance its subsidy (see page 117). This is the triangulation of part of the belt between *Salmon Arm* and *Revelstoke* drawn in 1915. Note the *Salmon Arm Base,* the initial length carefully measured. Another base was measured on the Columbia above Golden. All the triangles flow from these lines. All the nodes of the triangles except those at each end of the base are recognizable high points. The *Northern Limit* and *Southern Limit* of the Railway Belt are shown.

MAP 729 (*above*).
Detail of part of a hand-drawn triangulation survey of the East Kootenay, surveyed in 1898. The area shown is about 25 km west of Cranbrook. *Perry Creek* flows into St. Mary's River and can also be seen on MAP 730. Topography is depicted using hachures, thin lines drawn facing down the slope, which, especially here, give an excellent idea of the relief. They are time-consuming to draw, however, and later maps used contours.

MAP 730 (*above, right*).
One of the maps produced during the 1898 survey of the East Kootenay showing *Fort Steele, Cranbrook,* and the *Kootenay River.* The triangulation has fixed the position of the *Sullivan* and *North Star* mineral claims on *Mark Creek,* at what became Kimberley (see page 190).

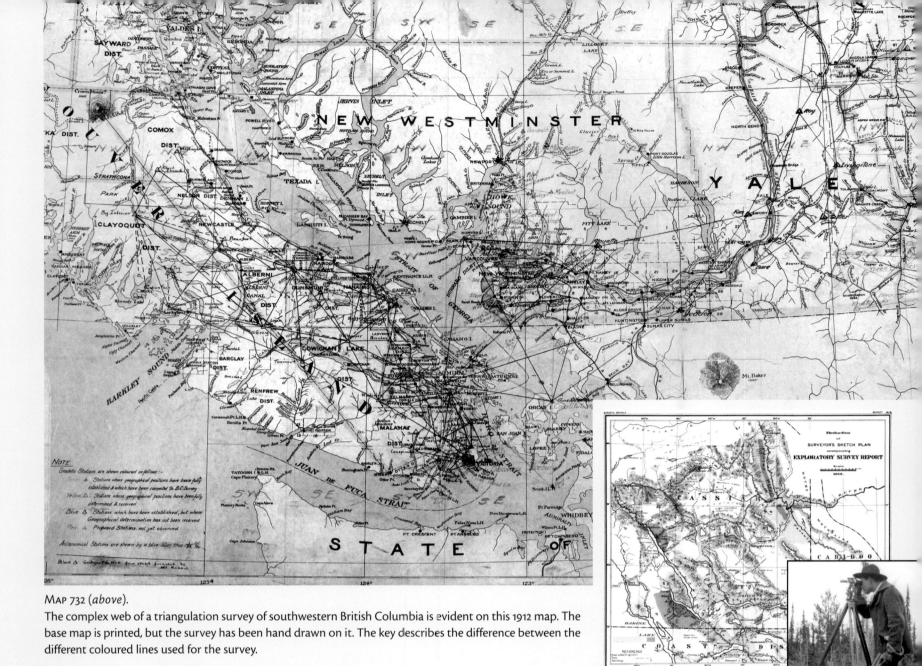

Map 732 (*above*).

The complex web of a triangulation survey of southwestern British Columbia is evident on this 1912 map. The base map is printed, but the survey has been hand drawn on it. The key describes the difference between the different coloured lines used for the survey.

Map 734 (*below, left*) and **Map 735** (*below, right*).

Marking a boundary almost always requires a survey. Before 1905, British Columbia's eastern boundary was with the Northwest Territories, and an approximate boundary line was acceptable. That year the province of Alberta was created, and especially given the coal-mining right on the boundary in Crowsnest Pass, the boundary between the two provinces needed to be better defined. An Interprovincial Boundary Commission was appointed in 1913 to survey and define the boundary. Surveyors were appointed as boundary commissioners for Alberta, for British Columbia, and for the federal government. The survey was carried out in 1913–17 and published in a boundary atlas, consisting of detailed map sheets. **Map 734** is an index map showing the arrangement of the sheets in one volume covering the boundary from the Yellowhead Pass north. **Map 735** shows a detail of one of the sheets at

Map 733 (*above*).

Frank Swannell was one of British Columbia's best-known surveyors. He was responsible for surveying and producing maps of vast areas of the north, such as the 1913 one shown here. Swannell also accompanied the 1934 Bedaux Expedition (see page 310). The photo shows Swannell at his theodolite at Lac La Hache in 1908.

Yellowhead Pass. The solid black line represents the boundary as surveyed, and the dashed black line is where the boundary was along a defined watershed and hence where it was thought unnecessary to survey. The *Canadian National Railways* line through Yellowhead Pass is marked. This is the old Canadian Northern Railway line. The trail (red dashed line north of *Yellowhead Lake*) is the bed of the Grand Trunk Pacific, from which the rails have been recently removed.

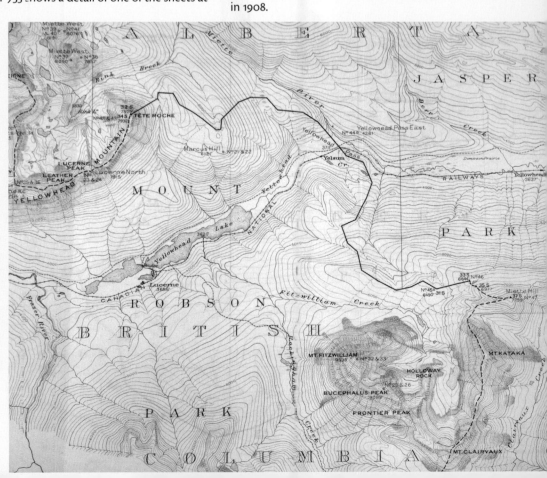

Population and Pre-emption

The federal government under Wilfrid Laurier (1896–1911) instigated a policy of encouraging immigration from European countries to Canada, specifically wanting to populate the Prairies, a policy implemented by his energetic minister of the interior, Clifford Sifton. Immigration levels that stood at 17,000 in 1896 had increased to over 400,000 per year when World War I began.

British Columbia did not have the vast amounts of agricultural land of the Prairies but nevertheless wanted to attract immigrants to create growth. The population of the province increased from less than 100,000 in 1891 to over 400,000 by World War I, mostly resulting from immigration.

Settlers were attracted by the promise of free land. Pre-emption of unsurveyed land was allowed and was conveyed to the pre-emptor after the latter met certain conditions, such as residence for four years and improvements to the property. Crown grants of surveyed land cost $2.50 per acre after 1880. In British Columbia in the forty years between 1873 and 1913 there were 30,495 Crown grants of land and 33,784 pre-emptions.

Above. "Canada's Land of Promise" on the cover of a 1913 government booklet.

Above.
Not quite the huddled masses yearning to breathe free, but close. People from all over the world, as well as all over Canada, were attracted by the promise of free or cheap land and a new life. Miss British Columbia, opportunities in hand, welcomes them all. This cartoon appeared in 1907.

Map 736 (*below*).
A map of Salmon River valley land north of Campbell River open for pre-emption in 1914 tops an immigrant's steamer trunk and other luggage and is flanked by government agriculture handbooks in this display at the Campbell River Museum. The logging community of Salmon River became Sayward in 1911.

Map 737 (*right, centre*).
Alienated land—that is, land pre-empted or in the process of pre-emption—is shown in orange-brown on this 1905 map of *Barkley Sound*. Timber leases are green.

Map 738 (*right*).
Around *Fort St. John* is the Peace River Block, shown here as a *Dominion Govt. Reserve* and stated to be for settlers only. The land was transferred to the federal government by the provincial government in 1884 to compensate it for alienated lands within the Railway Belt. It was opened to settlement beginning in 1912.

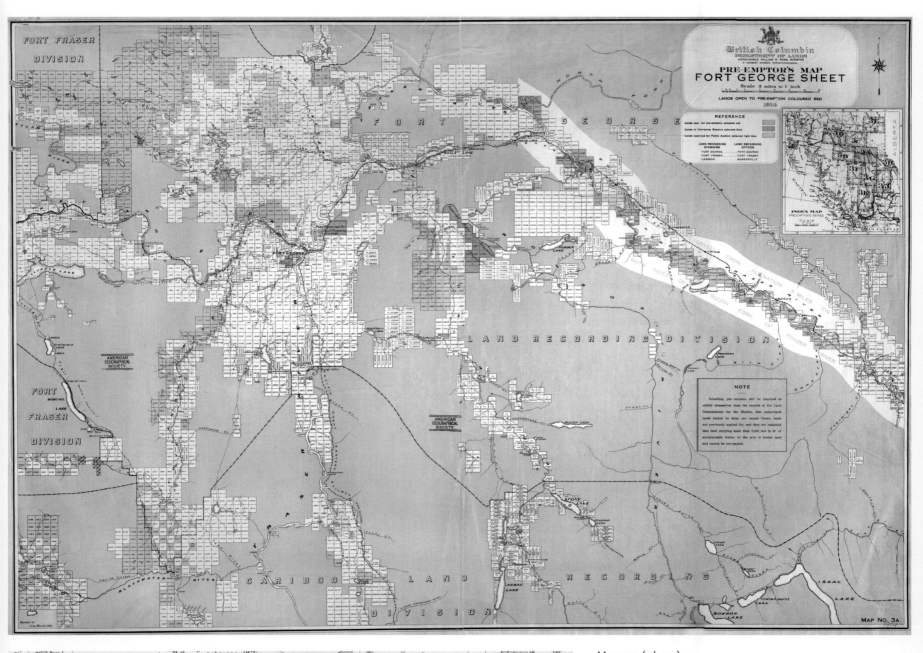

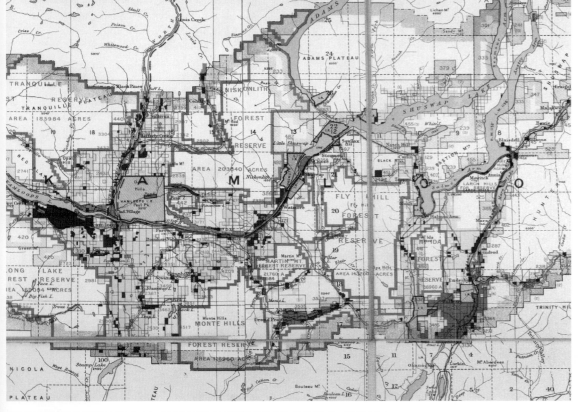

MAP 739 (*above*).
Land available for pre-emption in 1914 is shown in pink on this "Pre-emptor's Map" of land in the vicinity of Prince George. Although the majority of the area seems to still be available, most of the most desirable valley land has already been taken. The province published a number of map sheets like this to inform potential settlers.

Below. This rather splendid cartoon from 1908 reflects the economic optimism then sweeping the province. Miss British Columbia welcomes a ship swept in on a "tide of prosperity."

BRITISH COLUMBIA'S SHIP NOW ARRIVING

LEGEND

Homesteads (patented, entered for and unpatented)	Area 742,500	acres (approx.)
Sales, Special grants, Mining Lands sales	" 113,200	"
Lands disposed of by Provincial Government	" 295,190	"
Indian Reserves	" 183,800	"
Forest Reserves and Parks	" 2,476,700	"
Timber Berths	" 1,153,900	"
Grazing Leases	" 412,000	"
Total area of Belt	10,976,000 acres	

MAP 740 (*above*).
A colourful map of land status in the part of the Railway Belt between *Kamloops* and the North Okanagan, published in 1914. Pre-empted land—"homesteads"—is shown in yellow.

SETTLING SOLDIERS

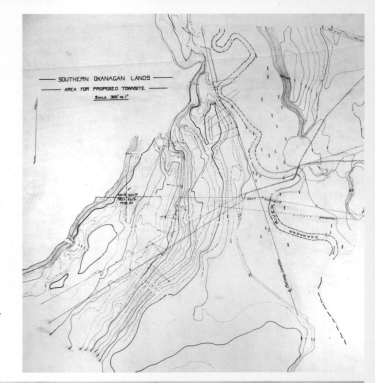

Premier John Oliver saw that soldiers returning from World War I could help populate and develop the province, and he set about devising a number of schemes for their resettlement. In 1917 a Land Settlement Board was created, and settlement areas were established at a number of locations around the province, including at Creston (Camp Lister), Merville on Vancouver Island near Courtenay, Fernie, Telkwa, and Vanderhoof. The government also passed the Soldiers' Land Act in 1918 and under it began the South Okanagan Lands Project (SOLP), which purchased 22,000 acres (8,900 ha) from the Southern Okanagan Land Company (see page 153) and began an ambitious scheme to reclaim some 3,200 ha from the dry valley.

Between 1919 and 1926 a dam was built on the Okanagan below McIntyre Bluff and a mainly concrete-lined main canal—known locally as the "Ditch"—was dug south to Osoyoos. A siphon was constructed: a 590-m-long, 2-m-diameter pipe that enabled water to temporarily flow uphill; it was made initially of wooden staves, like a barrel. This carried water from the east bank of the river to the benches on the west side below Oliver, a townsite named for the premier, laid out in 1920 to act as the regional centre of the irrigated district. Electrical power was supplied after 1922, allowing water to be pumped from the canal to uphill fields. Veterans, lured by easy payment terms, purchased five- and ten-acre parcels of land and planted orchards. Cantaloupes were grown between the rows to provide an income until the trees bore fruit.

The Kettle Valley Railway was extended south from Okanagan Falls to Haynes, just south of Oliver, in 1923, and was extended to Osoyoos in 1944. The South Okanagan Lands Irrigation District took over irrigation operations in 1964, and since 1989 it has been run by the towns of Oliver and Osoyoos. The system delivers nearly 100 billion litres of water a year to peach orchards, vineyards, and fields of other crops.

Map 741 (*right, top*).
A survey of the site of Oliver and irrigation possibilities from about 1916. Oliver was built on the west slope of the river bank; its position can be judged with reference to the townsite map (Map 742). Also shown, as a solid red line, is the *Syphon Line* and, as a dashed red line, the *(Abandoned) Syphon,* a previously surveyed route for the siphon to cross the *Okanagan River.*

Map 742 (*right, centre*).
The 1921 townsite plan for Oliver, named after Premier John Oliver. The siphon line follows *Johnstone St* (356 Avenue, now School Avenue) to a *Canal Lateral*; this is now all underground. Note that this map has north at the right.

Map 743 (*below*).
Also with north at the right, this map from about 1921 shows the *Townsite of Oliver* and shows how the *syphon* connected the two parts of the *Main Canal* on each side of the river. The surveyed line of the railway (dashed with crosshatches) runs through Oliver. The Kettle Valley Railway arrived in 1923, at first connecting at Okanagan Falls with a barge service from South Penticton. The railway was formally absorbed into the Canadian Pacific in 1931, and a rail line replaced barge service on Skaha Lake. Rail service was withdrawn in 1977 after refrigerated trucks had taken much of the fresh fruit transportation business.

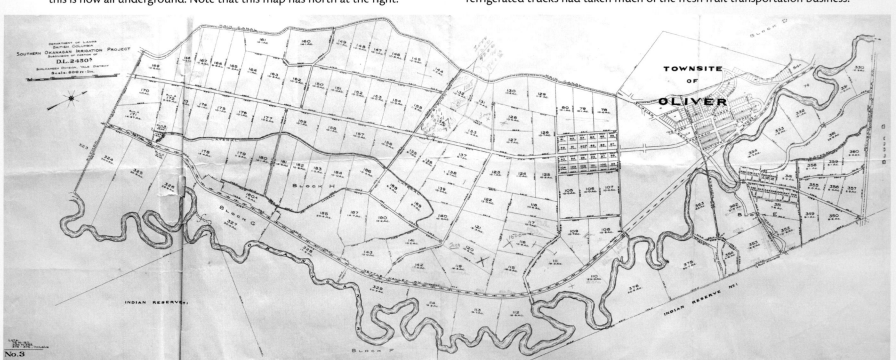

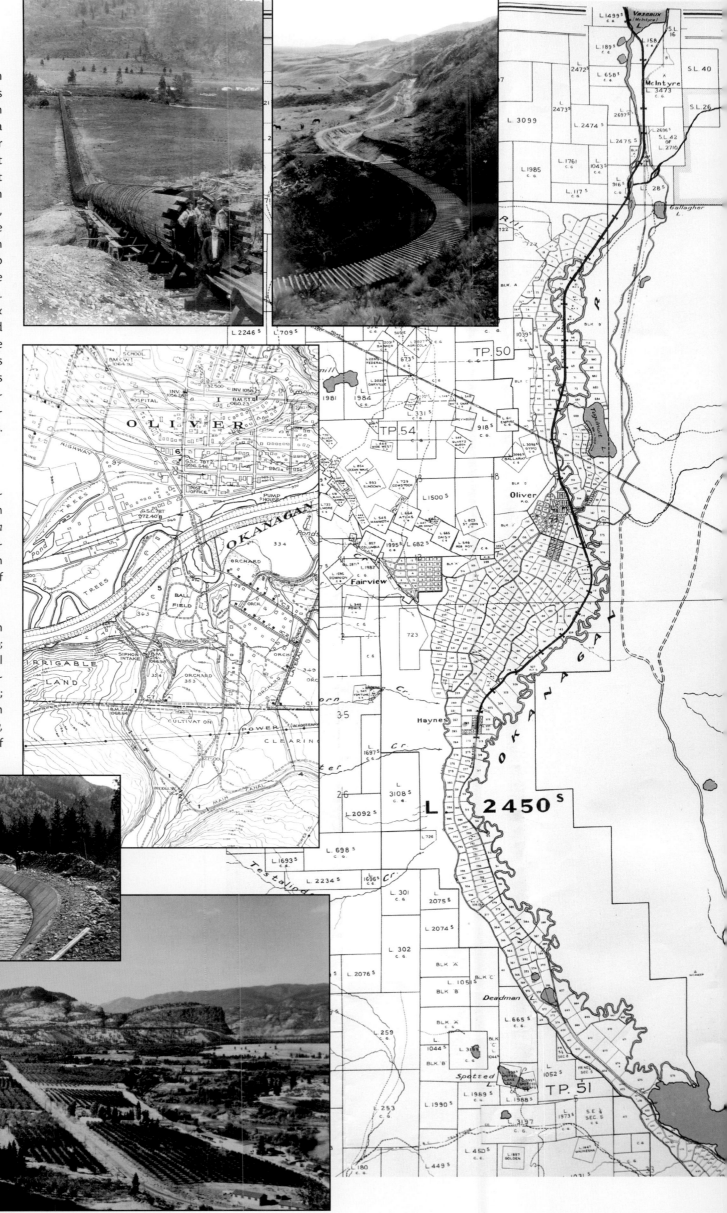

Map 744 (*far right*).

The length of the canal (in blue) feeding water to the fields around Oliver can be seen on this 1934 map. The intake is a dam on the *Okanagan R.* near *Vaseaux (McIntyre) Lake*; it crosses the Okanagan River at *Oliver* in the siphon, runs though the middle of fields to *Haynes*, and then stays on the uphill side of the west bank fields south from there. Also on this map are the transmission lines of the *Okanagan Water & Power Co.* (later West Kootenay Power & Light and now FortisBC), and the railway (thick black line with some cross-hatches) runs south to Haynes. This line was completed in 1923 and was extended to Osoyoos in 1944 to increase wartime fruit shipments. Osoyoos Lake is at bottom.

Map 745 (*right*).

Details of the canal, siphon intake, and railway line at *Oliver* on a 1955 map. Now the *Okanagan River* has been canalized, a flood-prevention project completed in 1958. The old meandering bed of the river is still visible.

Photos: *top left*, the stave siphon under construction about 1920; *top right*, the irrigation canal curves above what would become irrigated fields, about 1925; *below*, another view of the main canal, about 1923; and *below, bottom*, orchards just south of McIntyre Bluff, about 1928.

MARKETING THE B.C. ROCKIES

Canadian Pacific recognized the tourist potential of the Rocky Mountains even before its transcontinental line was complete. It was a bonus that would help build the traffic on the line so desperately needed to create revenue. The mountain sections of the railway were to immediately prove expensive and difficult to maintain because of the winter snows, and here the mountains were offering to give something back. Acting on advice from the railway, the government created national parks and reserves. What became Banff National Park was reserved in 1885 (see MAP 762, *page 298*), and Glacier and Yoho followed the next year; Mount Revelstoke National Park was created in 1914 and Kootenay in 1920. Railway hotels were built in a number of scenic spots along the line, aimed specifically at tourists, unlike, for example, the Hotel Vancouver, which was intended for businesspeople and those connecting with steamships. Most of the tourist hotels were accessible only by rail.

In the process of marketing the Rockies as a tourist destination, Canadian Pacific produced many fine brochures and maps, several of which are illustrated on these pages. It also commissioned photographers and artists to record images that could be used in its promotional material. Even places not directly accessed by the railway were promoted, since the only practical way of getting there, at least until the advent of the automobile and better roads, was by rail. The railway created five so-called bungalow camps, which provided a level of luxury below that of the resort hotels but above that of camping (MAPS 749 and 751, *page 292*). A string of various other camps, chalets, tea houses, and simpler lodges, such as the Lake Wapta Lodge at Hector, in the Kicking Horse Canyon, was established by the railway to cater to every pocketbook—as long as you could afford the train fare, of course. And the hot springs that emerged at a number of places in the mountains could also be a tourist attraction. Indeed, Banff itself began that way.

Above. Canadian Pacific's Mount Stephen House hotel at Field, at the bottom of the "Big Hill." Mount Stephen is in the background. Originally built as a stop for passengers so as to avoid the use of heavy dining cars, the hotel was expanded in 1901–02 as a destination resort hotel. The painting is by Edward Roper, about 1890.

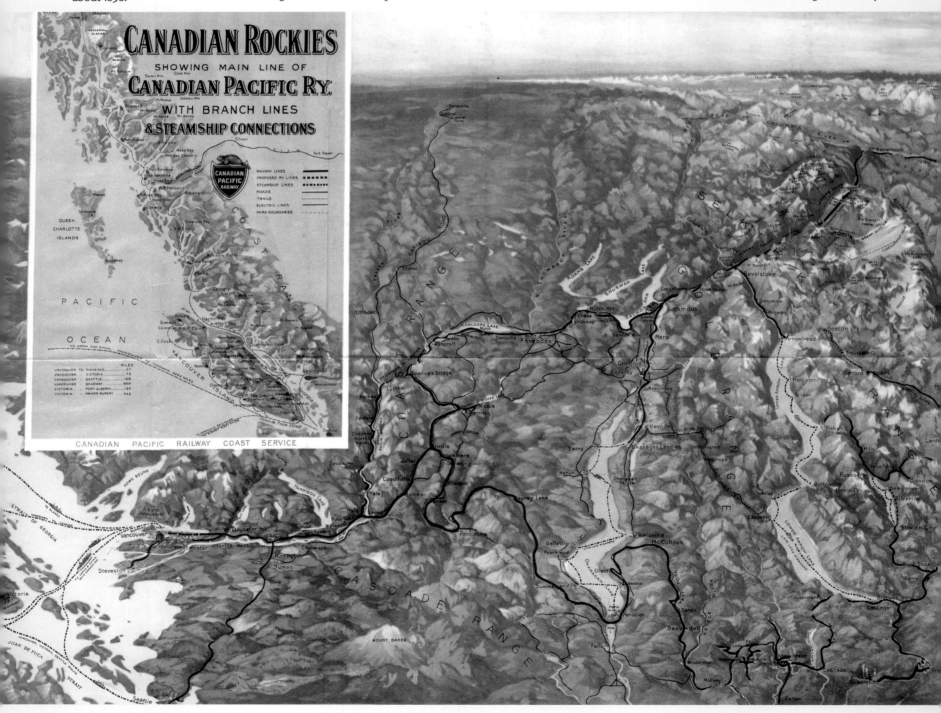

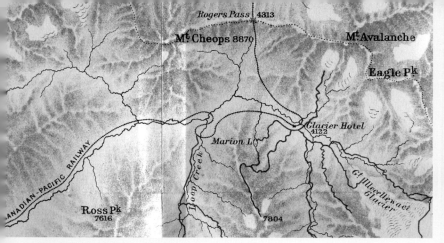

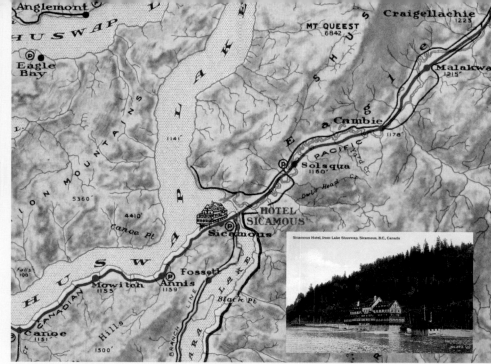

MAP 747 (*above*).
This 1888 map shows the railway's *Glacier Hotel* (Glacier House) at the extremity of an extensive loop of the track in the Illecillewaet River valley below the (misspelled) *Gt. Illecellewaet Glacier.* The hotel was closed soon after the railway diverted its line through the Connaught Tunnel in 1916 (see page 276).

MAP 748 (*above, right*).
Canadian Pacific's *Hotel Sicamous*, built in 1898 on the shores of Shuswap Lake, is depicted on this 1927 map. The hotel is also shown in a postcard view (*inset*).

MAP 746 (*below*).
This fine three-dimensional map was published by Canadian Pacific in 1917 as part of *Resorts in Canadian Pacific Rockies*, an extensive booklet for tourists; the cover is illustrated *right*. The map has company rail lines shown in solid red and lake steamship routes in dashed red. Some company hotels are shown, such as *Mt. Stephen House*, *Glacier House*, and the *Sicamous Hotel.* The inset map shows steamship connections.

LAKE WINDERMERE CAMP
In the Canadian Pacific Rockies

MAP 749 (*right*) and MAP 751 (*right, centre*). These two maps, dated 1923 and 1927, respectively, indicate the location of various Canadian Pacific bungalow camps and tea houses (in red on both maps). On the 1927 map roads have appeared—roads that would eventually rob the railway of its customers—though at this date most roads were of poor quality and not necessarily connected. MAP 751 shows the railway line in red, the road in black. MAP 749 also depicts *Sinclair Hot Springs*, since 1920 located in Kootenay National Park and soon to become Radium Hot Springs.

Left and *right*. Brochure covers for *Lake Windermere Camp* and *Wapta Bungalow Camp*, both in the *Canadian Pacific Rockies*.

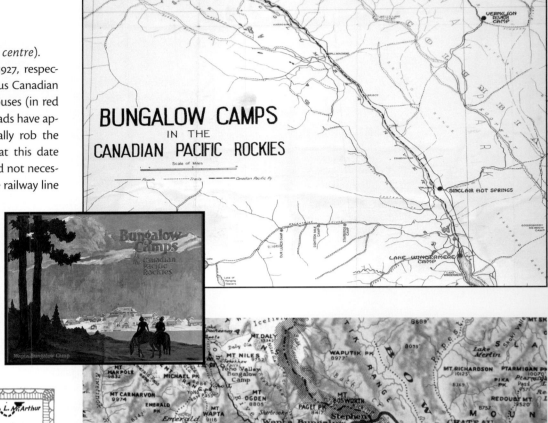

BUNGALOW CAMPS IN THE CANADIAN PACIFIC ROCKIES

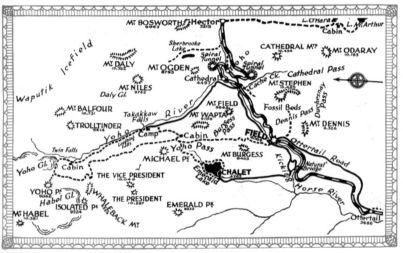

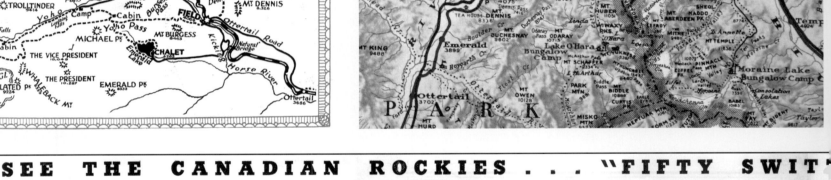

SEE THE CANADIAN ROCKIES . . . "FIFTY SWIT

PANORAMA OF THE CANADIAN ROCKIES

BANFF LAKE LOUISE
YOHO VALLEY
EMERALD LAKE
GOLDEN

— MOTOR ROAD
— CANADIAN PACIFIC RAILWAY

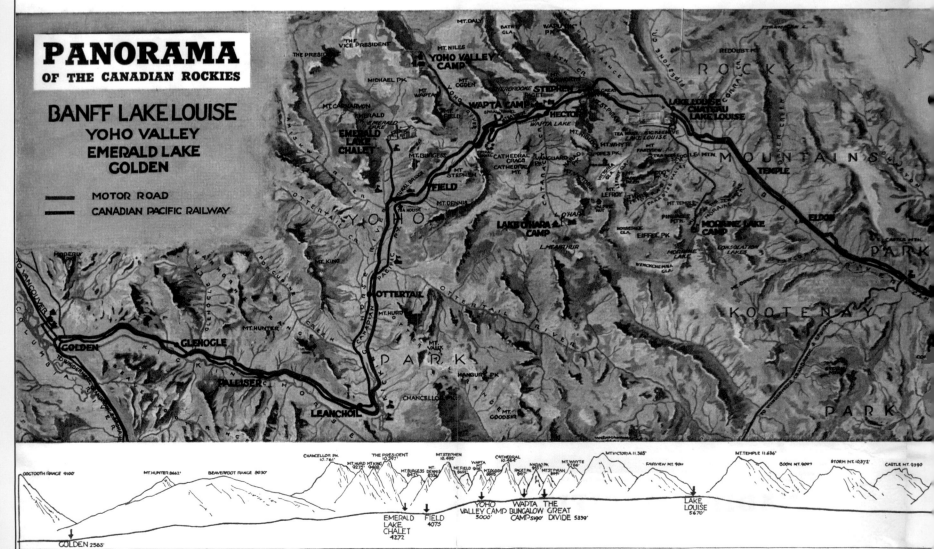

MAP 750 (*far left, centre*).
From the same Canadian Pacific 1917 booklet as MAP 746, *page 291*, this map details recreational facilities accessed from the railway hotel at *Field*, including *Camp[s]*, *Trail[s]*, *Cabin[s]*, a *Chalet*, and *Fossil Beds*. This was the location of the famous ancient fossil-laden Burgess Shale. There was lots for hotel guests to do. The tariff at Field offered a horse ride from Field to *Emerald Lake* via *Burgess Pass* for $3; a "tally-ho coach" to *Takakkaw Falls* was $2 per person, and a three-day horse trip to *L. McArthur* was $9.

MAP 752 (*below*).
Perhaps aimed at European tourists, Canadian Pacific advertised the Rockies as "Fifty Switzerlands in One" on this 1936 map called "panorama" rather than a map, though it clearly is the latter. The map focuses on the Rocky Mountain section of line between Banff and Golden. Of course, despite emphasizing the size of the Rockies, only the part accessible from the Canadian Pacific line is shown. A *Motor Road* now parallels the rail line. Hotels, camps, chalets, and tea houses are still shown, though the Depression was in full swing at this time. Below the map is a profile of the rail line with the mountains visible from it labelled.

MAP 753 (*above*).
This fine bird's-eye map of the Upper Kootenay and Upper Columbia Valleys from *Golden* to *Cranbrook* has a spectacularly exaggerated vertical scale, making the mountains appear very jagged. The map was included in a 1913 prospectus intended to attract investors in *Radium Hot Springs*, the position of which is prominently noted. The Kootenay Central, a Canadian Pacific subsidiary, is the black line up the valley; it was completed in 1913.

Radium Hot Springs had been known to the indigenous people of the area for hundreds of years, but the first historical mention of them was in 1841, when they were visited by George Simpson, governor of the Hudson's Bay Company. The springs were purchased from the government in 1888 by a Briton, G. Roland W. Stuart, for $160. In 1914 a syndicate he formed, the Radium Natural Springs Syndicate (for which this map and prospectus were issued), began constructing a pool. The syndicate sent water samples to scientists from McGill University who tested the waters and found small amounts of radon, a by-product of the decomposition of radium. The following year the spring was renamed Radium Hot Springs (it was previously Sinclair Hot Springs, as on MAP 749, *left, top*, on Sinclair Creek). The timing of the start of construction could not have been worse; the pool was not completed because of the onset of World War I. In 1920 the springs were included within the new Kootenay National Park and were expropriated two years later. Canadian Pacific built a lodge. This time, however, the railway was not to have any kind of monopoly, as other tourist accommodations were soon built once a road, the Banff–Windermere Highway, was completed in 1923 (see page 298). The same year a log bathhouse and pool were completed.
Inset. The photo shows the modern pool, alongside Sinclair Creek.

Top left. A brochure for the Sinclair Hot Springs Bungalow Camp, still retaining the old name for Radium Hot Springs despite its being published after 1923, when the road was completed.

THE FRASER RIVER NEAR LYTTON

BULKLEY GATE

S.S. PRINCE RUPERT ENTERING VANCOUVER HARBOUR

THE TRIANGLE TOUR
OF BRITISH COLUMBIA
JASPER NATIONAL PARK
MOUNT ROBSON PARK
CANADIAN ROCKIES
AND
THE SCENIC SEAS OF THE
NORTH PACIFIC COAST
CANADIAN NATIONAL
RAILWAYS
C.N.R. ■■■ ● S. SHIP LINES ▬▬▬ TRAILS -----
BOUNDARIES OF PARKS ▨▨▨

MAP 754 (*above*) and MAP 755 (*right, inset*).
Canadian National Railways had both the old Grand Trunk Pacific and Canadian Northern routes through the Rockies and Jasper National Park, created in 1907. Not to be outdone by Canadian Pacific, Canadian National published this

excellent three-dimensional map in 1929 to advertise what it called the "Triangle Tour," in which the three sides of the triangle were the Grand Trunk route, the Canadian Northern line, and a steamship connection between Prince Rupert and Vancouver. It was clever marketing because it enabled eastern tourists to

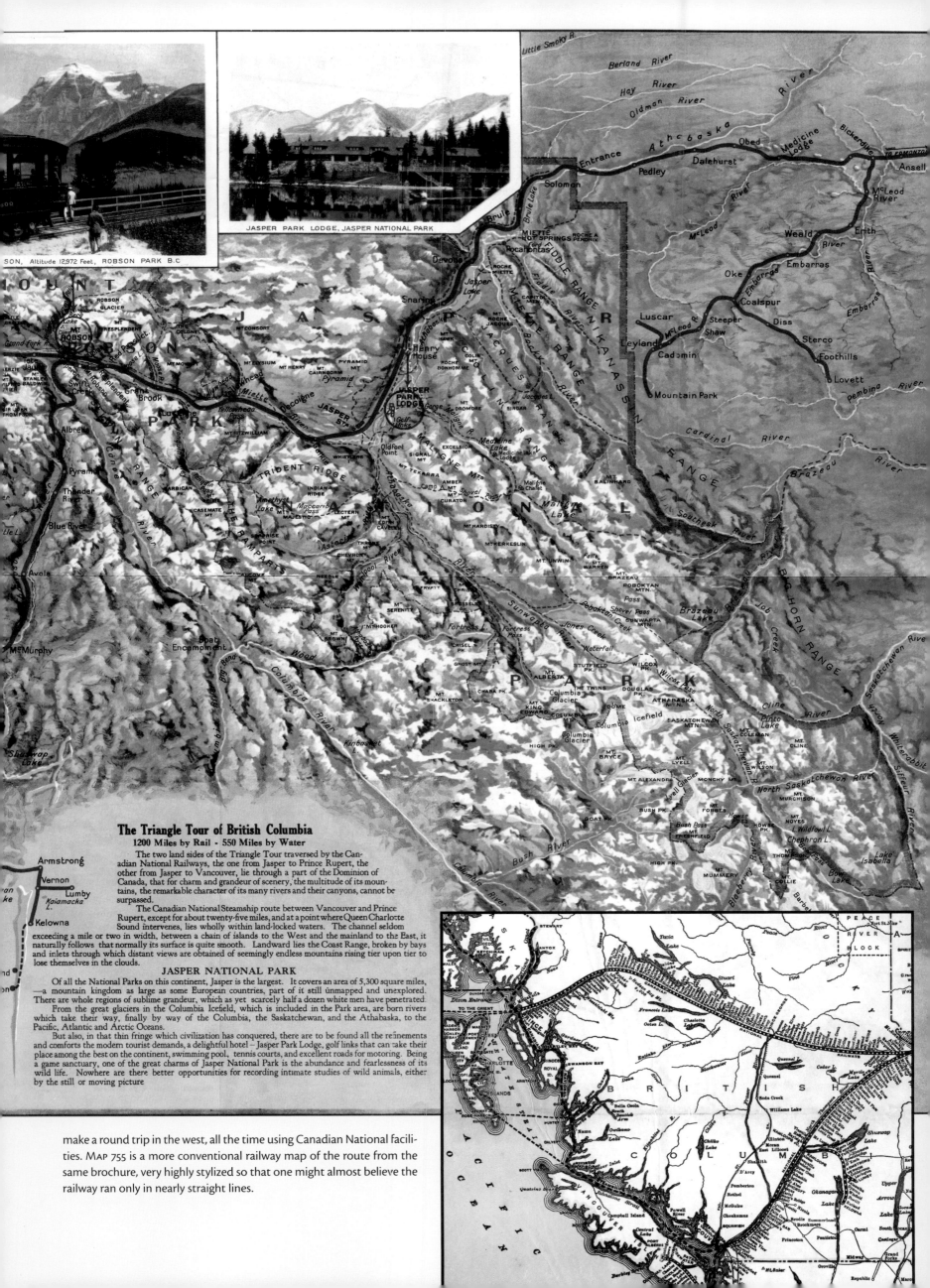

...SON, Altitude 12,972 Feet, ROBSON PARK B.C.

JASPER PARK LODGE, JASPER NATIONAL PARK

The Triangle Tour of British Columbia
1200 Miles by Rail - 550 Miles by Water

The two land sides of the Triangle Tour traversed by the Canadian National Railways, the one from Jasper to Prince Rupert, the other from Jasper to Vancouver, lie through a part of the Dominion of Canada, that for charm and grandeur of scenery, the multitude of its mountains, the remarkable character of its many rivers and their canyons, cannot be surpassed.

The Canadian National Steamship route between Vancouver and Prince Rupert, except for about twenty-five miles, and at a point where Queen Charlotte Sound intervenes, lies wholly within land-locked waters. The channel seldom exceeding a mile or two in width, between a chain of islands to the West and the mainland to the East, it naturally follows that normally its surface is quite smooth. Landward lies the Coast Range, broken by bays and inlets through which distant views are obtained of seemingly endless mountains rising tier upon tier to lose themselves in the clouds.

JASPER NATIONAL PARK

Of all the National Parks on this continent, Jasper is the largest. It covers an area of 5,300 square miles, —a mountain kingdom as large as some European countries, part of it still unmapped and unexplored. There are whole regions of sublime grandeur, which as yet scarcely half a dozen white men have penetrated.

From the great glaciers in the Columbia Icefield, which is included in the Park area, are born rivers which take their way, finally by way of the Columbia, the Saskatchewan, and the Athabaska, to the Pacific, Atlantic and Arctic Oceans.

But also, in that thin fringe which civilization has conquered, there are to be found all the refinements and comforts the modern tourist demands, a delightful hotel – Jasper Park Lodge, golf links that can take their place among the best on the continent, swimming pool, tennis courts, and excellent roads for motoring. Being a game sanctuary, one of the great charms of Jasper National Park is the abundance and fearlessness of its wild life. Nowhere are there better opportunities for recording intimate studies of wild animals, either by the still or moving picture.

make a round trip in the west, all the time using Canadian National facilities. MAP 755 is a more conventional railway map of the route from the same brochure, very highly stylized so that one might almost believe the railway ran only in nearly straight lines.

Roads Left and Right

We take paved roads for granted these days, especially in urban areas, but they are a relatively recent phenomenon. Only after World War II were roads paved as a matter of course. With the development of the automobile and truck after 1900 came a rising demand for better roads—roads that would permit the higher speeds that were becoming possible, like the sizzling 20 km/h a Surrey motorist was fined for doing in 1910.

In 1911 citizens of the new town of Hazelton, child of the Grand Trunk Pacific Railway, anxious to make their town accessible by road to the outside world, offered $1,000 for the first car to reach it under its own power. The Pacific Highway Association added a gold medal, for Hazelton was the proposed northern terminus of a Pacific Highway. A car dealer from Seattle, P.E. Sands, took up the challenge and managed to drive between most of the disconnected roads overland or by using streambeds, arriving in Hazelton to popular acclaim in late September. However, it was later revealed that for part of the journey beyond Quesnel, where no roads existed at all, the car had been broken down and taken through on a pack train. An old prospector who discovered Sands's secret blackmailed him out of half his prize money.

BRACKENDALE & CHEAKAMUS STAGE.

There is a further twist to the story: it was later further revealed that the Hazelton organizing committee knew perfectly well what had happened but was not about to penalize Sands—for that would have destroyed the myth that their town was accessible by road!

The Pacific Highway, envisaged as running up the west coast of the United States, never reached Hazelton, but it did reach Vancouver. The highway from New Westminster south to Blaine was completed in 1913 (MAP 760, *page 298*) and was even paved in 1923.

MAP 757 (*right, centre*).
The sales pitch and the reality. This is Vancouver's Kitsilano about 1907. The map was printed to sell real estate—lots were selling for $2,000 to $3,000—while in the photo two real estate salesmen are trying to extricate themselves from a rather large rut in the "road." The photo was taken on *Fourth* by the *Public Parks*, now McBride Park.

Above. An early bus. This one ran a short distance north from Squamish—then Newport—in 1910.

Left. The first car to cross the Rockies did not use roads at all. Replacing the wheels with flanged railcar ones, Charles Glidden arrived in Vancouver on 18 September 1904. He had to carry a conductor (seen sitting in the back seat), as he was designated a special train. Feats like this brought the potential of automobiles to the attention of the public.

Right. A 1912 ad touting the benefits of bitulithic paving, with before and after photos of Granville Street in Vancouver.

Expedition Under the Auspices of the Pacific Highway Association

MAP 756 (*above*).
Getting to Hazelton by "road" in 1911 required a very circuitous path. This is Sands's route from Seattle in his Studebaker 1911 "Flanders 20."

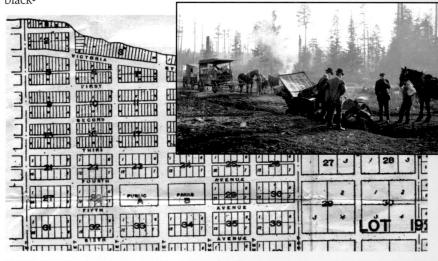

The Rule of the Road in British Co

British Columbia—it was *British* after all—drove on the left. But it was connected by land to the United States, which drove on the right, and so, on 15 July 1920, the province implemented a change to the right, but only for that part outside the southwestern corner shown in red on MAP 758, *right.* Then, on 1 January 1922, the more difficult change was implemented for the rest of the province—more difficult because this was where most of the urban roads were, and where there were roads that could actually carry two lanes of traffic moving in opposite directions. The two-stage change was possible because at that time there was no road connection between the two areas, the Cariboo

MAP 758 (*right*).
This 1920 map advertises the change of the rule of the road, but only for British Columbia outside the southwestern portion.

MAP 759 (*below*).
A road map of the Lower Mainland published in 1919 and intended for American tourists. It contained the banner (*bottom*) reminding them that in British Columbia you "Keep to the Left." Most of the roads outside the centre of Vancouver are not paved but were "improved"—meaning gravel. *Inset.* A car on the Seymour River Bridge in 1919. Outside the city it hardly mattered what side of the road one drove on; staying left was more just a rule for passing.

CHANGE OF RULE OF ROAD
IN
BRITISH COLUMBIA
1920

CAUTION
RULE OF THE ROAD
On and after 15th July, 1920.
KEEP TO THE **RIGHT**
when meeting
OUTSIDE the Red Area.

Till further notice
KEEP TO THE **LEFT**
when meeting
INSIDE the Red Area.
(Highway Act, 1920)

NOTE: There is no road connection in British Columbia
between the red and white areas.

On entering British Columbia
from the United States at Sumas
Mountain, Wash., and all points
of entry West thereof,
KEEP TO THE LEFT.

On entering at all other
points,
KEEP TO THE RIGHT.

MAP OF THE PROVINCE OF
BRITISH COLUMBIA
1920
Miles

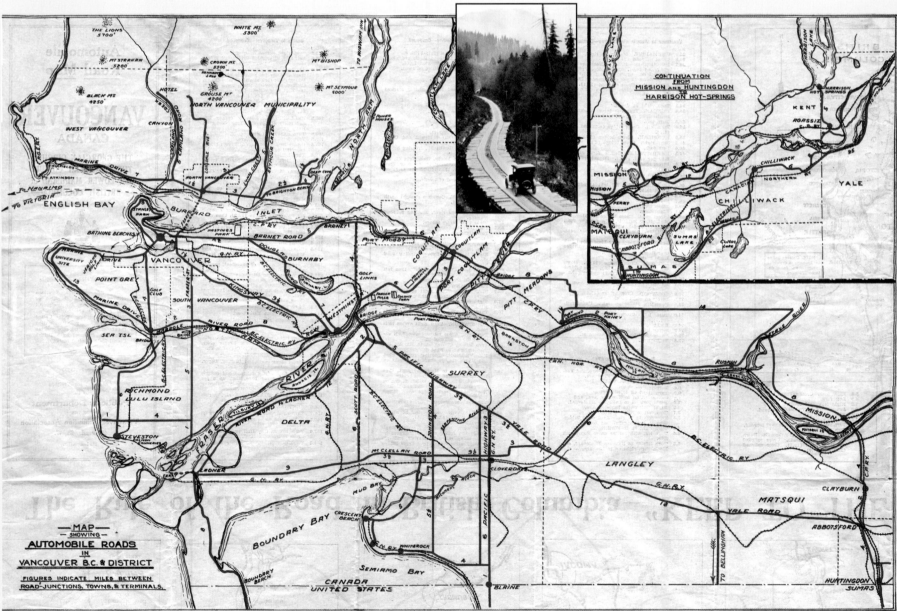

MAP
SHOWING
AUTOMOBILE ROADS
IN
VANCOUVER B.C. & DISTRICT
FIGURES INDICATE MILES BETWEEN
ROAD-JUNCTIONS, TOWNS, & TERMINALS.

CONTINUATION FROM
MISSION AND HUNTINGDON
TO
HARRISON HOT-SPRINGS

lumbia—"KEEP TO THE LEFT"

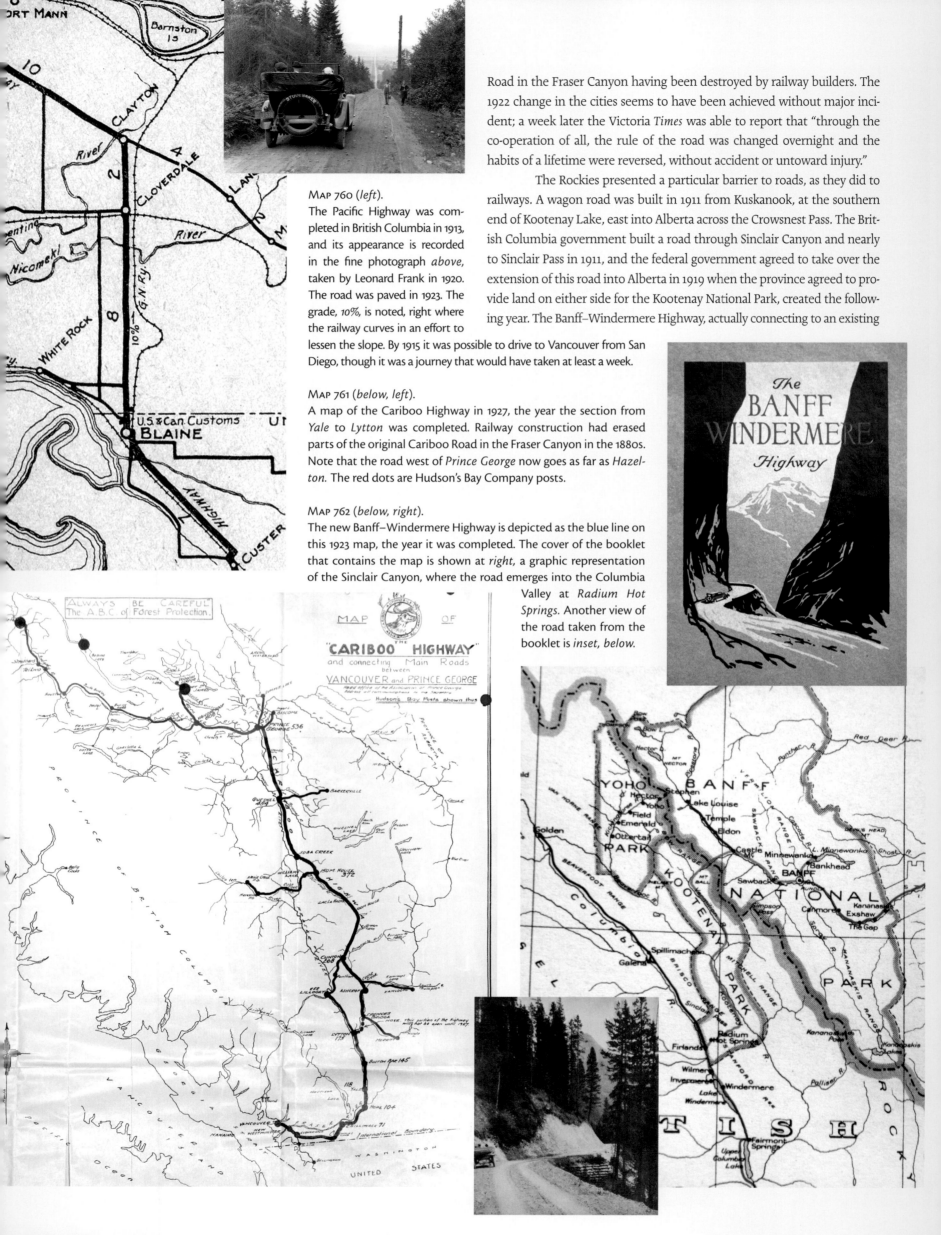

Road in the Fraser Canyon having been destroyed by railway builders. The 1922 change in the cities seems to have been achieved without major incident; a week later the Victoria *Times* was able to report that "through the co-operation of all, the rule of the road was changed overnight and the habits of a lifetime were reversed, without accident or untoward injury."

The Rockies presented a particular barrier to roads, as they did to railways. A wagon road was built in 1911 from Kuskanook, at the southern end of Kootenay Lake, east into Alberta across the Crowsnest Pass. The British Columbia government built a road through Sinclair Canyon and nearly to Sinclair Pass in 1911, and the federal government agreed to take over the extension of this road into Alberta in 1919 when the province agreed to provide land on either side for the Kootenay National Park, created the following year. The Banff–Windermere Highway, actually connecting to an existing

MAP 760 (*left*).
The Pacific Highway was completed in British Columbia in 1913, and its appearance is recorded in the fine photograph *above*, taken by Leonard Frank in 1920. The road was paved in 1923. The grade, *10%*, is noted, right where the railway curves in an effort to lessen the slope. By 1915 it was possible to drive to Vancouver from San Diego, though it was a journey that would have taken at least a week.

MAP 761 (*below, left*).
A map of the Cariboo Highway in 1927, the year the section from *Yale* to *Lytton* was completed. Railway construction had erased parts of the original Cariboo Road in the Fraser Canyon in the 1880s. Note that the road west of *Prince George* now goes as far as *Hazelton*. The red dots are Hudson's Bay Company posts.

MAP 762 (*below, right*).
The new Banff–Windermere Highway is depicted as the blue line on this 1923 map, the year it was completed. The cover of the booklet that contains the map is shown at *right*, a graphic representation of the Sinclair Canyon, where the road emerges into the Columbia Valley at *Radium Hot Springs*. Another view of the road taken from the booklet is *inset, below*.

MAP 763 (*below*).
A road was built into Yoho National Park, established in 1911, connecting Lake Louise with *Golden,* through *Field,* in 1927. It was a section of road that, much widened, straightened, and improved, would later become part of the Trans-Canada Highway. The road improved access to the Rockies for Albertans, but from British Columbia the Rogers Pass section was still missing. A connection was made using the route around the Big Bend of the Columbia in 1940. The map shows (in red) a convenient touring triangle for Albertans via the Banff–Windermere Highway and Kicking Horse Trail. The map was in an illustrated booklet published in 1927, with fine cover graphics (*right*) and spectacular photographs (*left*).

The **Kicking Horse Trail**

MAP 764 (*below*).
Published in an early tourist brochure by Kelowna, this 1918 map demonstrates that the Okanagan was more accessible from the United States than it was from the Lower Mainland. A *Proposed Trans-Prov Road* marks the gap between *Hope* and *Princeton.*

MAP 765 (*below*).
The provincial government long ago realized the tourist potential of British Columbia's scenery. This is a page from a booklet published in 1940 that married maps and photos on a route-specific basis. The tourist industry, however, was about to be severely impacted by the United States' entry into the war.

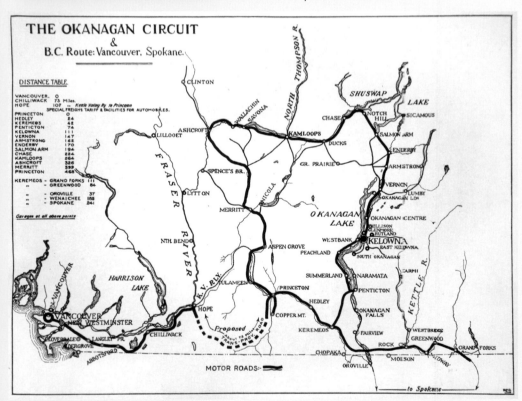

road at Castle Mountain, was completed in 1923. The route became the primary route for travellers to Calgary from Washington State. Four years later a road through Kicking Horse Pass was opened, paralleling the Canadian Pacific (MAP 763, *above, top*).

As early as 1909 the four western provinces had discussed building a transprovincial through road to connect them, and the Hope–Princeton route was considered one possible link (MAP 764, *above*). In 1924 the decision was made to build through the Fraser Canyon, in the process restoring a link in the Cariboo Road broken when the Canadian Pacific built its line there in the 1880s. A road from Yale to Lytton was completed in 1927 and a further section in the Thompson Valley the following year.

The Hope–Princeton link would remain uncompleted until 1949 (see page 330). In the meantime the completion of the Canadian Pacific's Kettle Valley Railway allowed cars to be shipped across this gap, albeit by a circuitous route up the Coquihalla.

A further break in the east–west link was Kootenay Lake, where a steamer connected Nelson with Kuskanook, at the south end of the lake. In 1931 Canadian Pacific withdrew service, having completed its rail line up the west side of Kootenay Lake to Proctor. The same year the government completed a road on the east side of the lake to Gray's Creek (also connecting with Kootenay Bay), chartering the railway's steamer *Nasookin* to connect with Fraser's Landing (Balfour), on the west side of the lake. Today a Department of Highways ferry still runs across the lake from Kootenay Bay to Balfour.

Above. The Kootenay Lake government ferry *Nasookin*, a Greyhound bus perched precariously across its bow, about 1931.

MAP 767 (*below*).
A proposal for a Trans-Canada Highway using the southern British Columbia route, shown on a map published by the Good Roads Association in 1928. The *Hope–Princeton* and *Kootenay Lake* gaps are shown, the latter also erroneously labelled *Lower Arrow Lake*. The northern route uses the Banff–Windermere Highway. Lytton to Spences Bridge is shown as incomplete on this map; the road was finished the same year.

MAP 766 (*above*).
This 1923 map with, in red, the only route across the province by road, shows the route of the Lake Steamer between Nelson and *Kuskonook (sic)* and the *Inter-Provincial* route through Crowsnest Pass. The blue line is a recommended route for Americans going to Calgary, using the Banff–Windermere Highway.

Right.
Cars at Hope about to be shipped by rail to Princeton, sometime in the 1920s. In 1927 the charge to transport a car by rail from Hope to Princeton was 77 cents per 100 pounds (45 kg), minimum 5,000 pounds (2,268 kg).

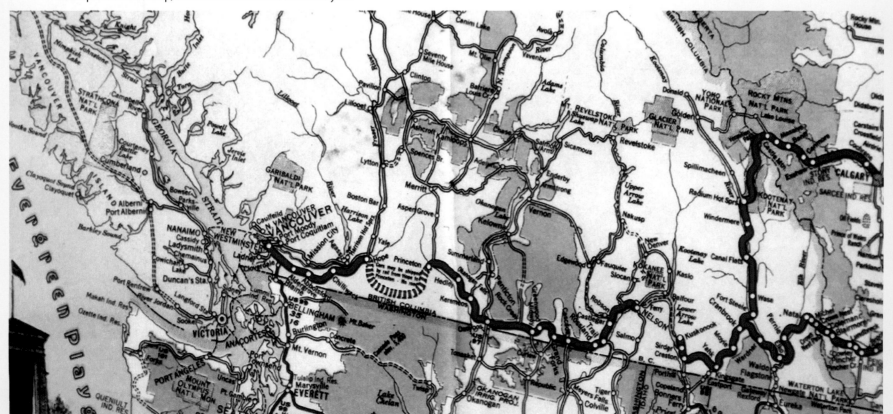

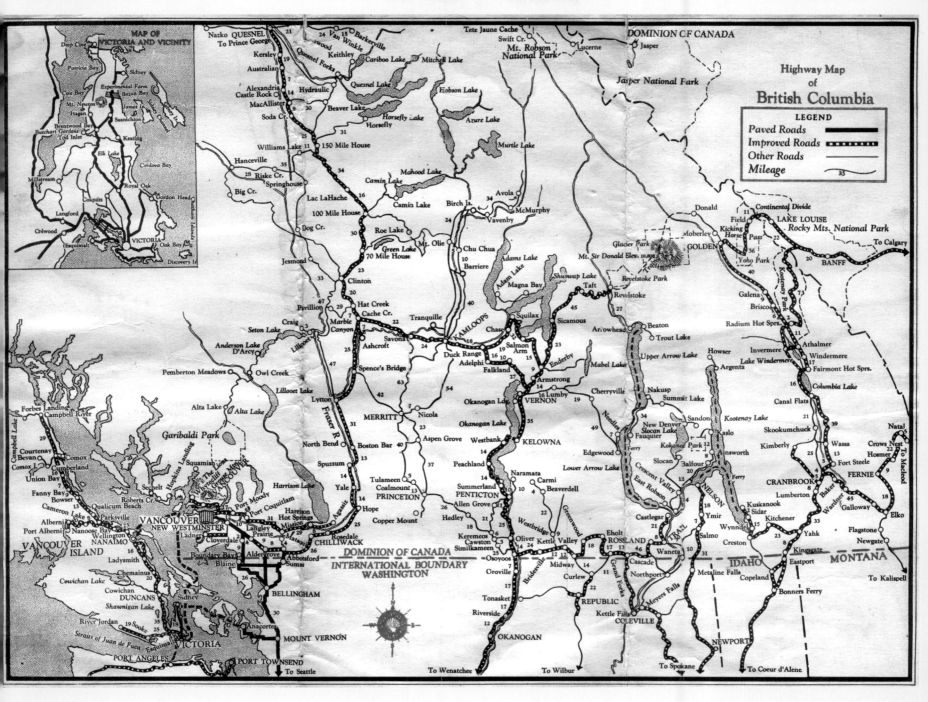

LEGEND
Paved Roads
Improved Roads
Other Roads
Mileage

MAP No 6

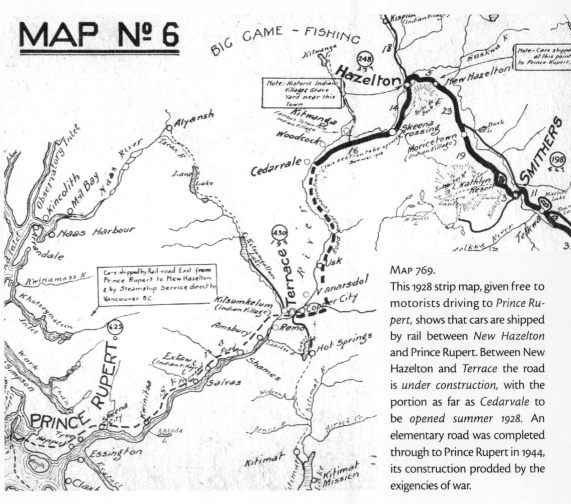

MAP 768 (above).

The highway system of southern British Columbia as it was in June 1929 on a gas company map given free to motorists. Outside of the Lower Mainland and Victoria there are no paved roads, only "improved" ones, perhaps gravel but also hard dirt roads that might not be so hard in the rainy season. But there is a road all the way from *Vancouver* to Prince George—the Cariboo Road route—and a way, albeit very circuitous, of driving from Vancouver to Alberta. The latter route is interrupted by a ferry across *Okanagan Lake* at *Kelowna*, a stretch of unimproved road just west of *Nelson*, and a steamboat connection from Nelson to *Kuskanook* on *Kootenay Lake*. But it was a major improvement over just a few years before. There are two egregious gaps: a direct route between *Hope* and *Princeton*, which would be rectified in 1949 (see page 330), and through Rogers Pass between *Revelstoke* and *Golden*. This latter gap would be filled in 1940 with the completion of another circuitous route following the Columbia around its Big Bend, and directly, with the building of the Trans-Canada Highway; the road through Rogers Pass was completed in 1952. On this map the Rogers Pass route is marked by a thin line and the word *Train*. On Vancouver Island the "improved" road reaches as far north as Campbell River.

MAP 769.

This 1928 strip map, given free to motorists driving to *Prince Rupert*, shows that cars are shipped by rail between *New Hazelton* and Prince Rupert. Between New Hazelton and *Terrace* the road is *under construction*, with the portion as far as *Cedarvale* to be *opened summer 1928*. An elementary road was completed through to Prince Rupert in 1944, its construction prodded by the exigencies of war.

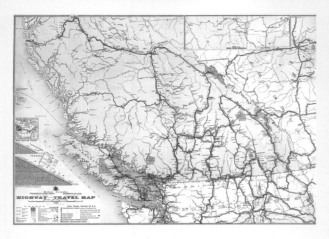

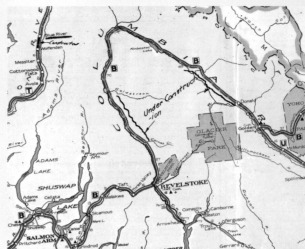

Map 770 (*above, and all details on this page*).

This detailed highways map was published by the provincial government's Department of Public Works in 1930. *Above* is the whole map, and the other maps on this page are details of the same map. A key is at *right*; note the minister, Nelson S. Lougheed, responsible for a road through Coquitlam called the Central Arterial Road, now the Lougheed Highway. He also improved the Dewdney Road in Maple Ridge; it too now forms part of the Lougheed Highway. The map was published at the beginning of the Depression, and many intended projects were put on hold because of poor economic conditions.

The road to Prince Rupert (*below*) is shown as far as *Kitwanga*, with the section as far as *Terrace* projected. The road would not be completed into *Prince Rupert* until 1944. The road north of *Campbell River* on Vancouver Island (*left, bottom*) is shown dashed to *Sayward*, with a part under construction; this road would not be completed to Sayward until the late 1940s—and on to *Port Hardy* by 1979. The Big Bend Highway (*right, top*) between *Revelstoke* and *Golden* is shown as a dashed line with parts marked *under construction*, but it would not be completed for another ten years. The *Hope* to *Princeton* highway (*right, below key*) also has sections *under construction* but would not be completed until 1949. The south-central portion of the province (*right, second below key*) shows the road on the east side of *Kootenay Lake* as *Constructed this year*; in fact it took another year to be completed. Also shown is part of a road between *Nakusp* and *Rosebery* to be opened in 1931. A road connecting *Prince George* with *Tête Jaune Cache* up the Upper Fraser Valley (*right, second from bottom*) is shown dashed; this road was not completed until 1963. The road from Kamloops to Tête Jaune Cache (north of Avola) took until 1971. The roads of the Peace River district (*right, bottom*) are disconnected from any others in British Columbia. In 1952 the Hart Highway would be completed north of Prince George to remedy this (see page 330).

Above. Brochures advertising the opening of the Big Bend Highway in 1940 and promoting Glacier National Park as the "Switzerland of America" about 1938.

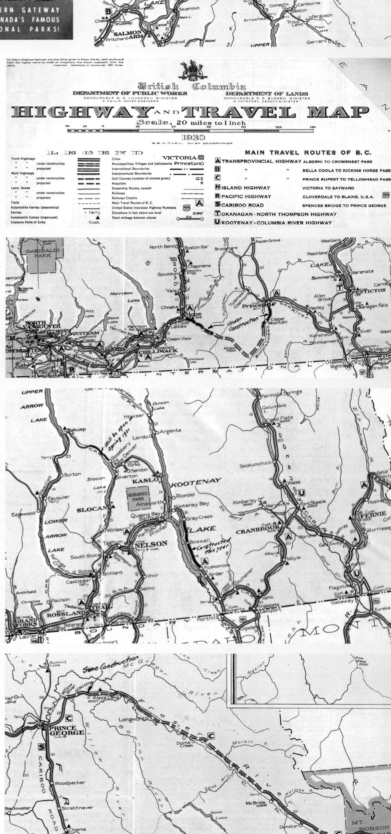

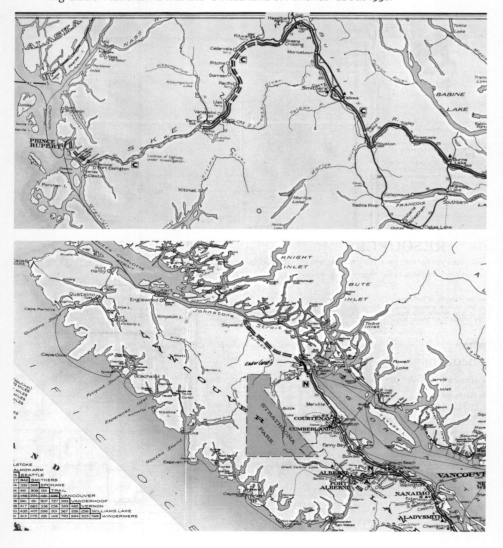

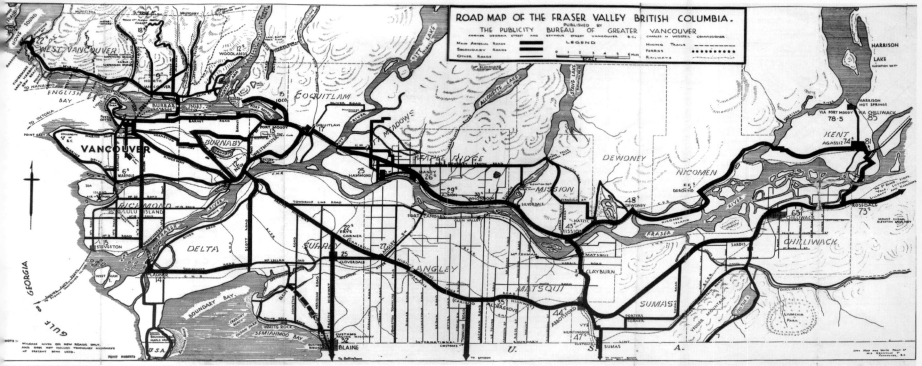

ROAD MAP OF THE FRASER VALLEY BRITISH COLUMBIA.
PUBLISHED BY
THE PUBLICITY BUREAU OF GREATER VANCOUVER

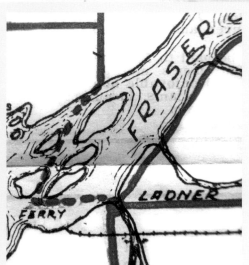

MAP 772 (*above*).
The *ferry* between Woodward's Landing and *Ladner,* shown on a 1925 map. It was replaced by the Deas Island Tunnel (now George Massey Tunnel) in 1959 (see page 332).

MAP 774 (*below*).
The new *King George VI Highway* (now King George Boulevard) was planned to provide easier access to the United States. This is the map from the opening ceremony booklet in 1940.

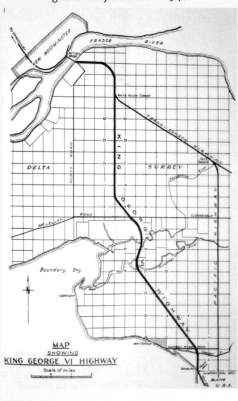

MAP 771 (*above*).
A road map of the Lower Fraser Valley published in 1936. The southern part of what became King George Highway in Surrey is shown as *Proposed Road.* Four years later this would connect with the new Pattullo Bridge at New Westminster. A *Summer Auto Ferry* connects *Steveston* with "Sydney"—Sidney, on Vancouver Island. The first Second Narrows Bridge is shown but not named; this combined rail and road bridge was completed in 1926 but from 1930 until 1934 was closed after a ship collided with it and funds were not immediately available to rebuild it. *Marine Drive* now reaches west to *Horseshoe Bay.* A road reaches a *Chalet* on *Grouse Mountain.* This road extended north from North Vancouver's *Grouse Mountain Highway* (now Mountain Highway) and is today the skiing facility's service road. Sumas Lake is no more (see overleaf).

MAP 773 (*right, centre*).
This 1935 map shows a *Bridge* across *First Narrows* but no connecting road through *Stanley Park.* The Lions Gate Bridge (shown above in an illustration from a booklet produced for the opening ceremonies) was under construction and would be completed in November 1938 (though it was officially opened again in May 1939 by King George VI during the royal visit of that year). The project overcame some initial opposition that arose especially because of the necessary road through Stanley Park, but the promise of jobs in these Depression years won the public's favour. The bridge was a project of the British Guinness brewing family, which had purchased large tracts of land on the North Shore and built the bridge to allow easy access from Vancouver, thus vastly increasing their value. Originally called Capilano Estates and laid out by the famous landscape architects Olmsted Brothers, the project's name was later changed to British Properties. MAP 789, *page 307,* shows the plan.

MAP 775 (*right*).
An initiative of Premier Thomas Dufferin Pattullo, who fought federal opposition to it during the Depression years, the Pattullo Bridge at New Westminster was opened on 15 November 1937. This 1937 oil company map shows a *Br. under const.* just downstream of the 1904 combined railway and road bridge. Note the *Auto Camp* in *Queens Park.* The photo (*inset*) shows the two sides of the bridge, built from each bank of the river, about to be joined in the centre.

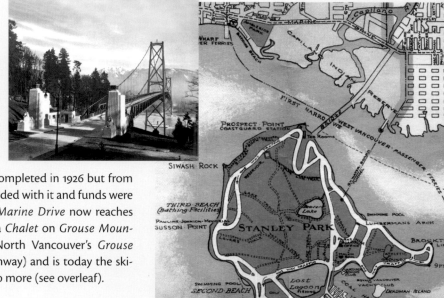

Reclaiming Sumas Lake

Sumas Lake was a large but shallow mass of water in the Lower Fraser Valley just east of Abbotsford. It provided a fine breeding ground for mosquitoes and made the eastern part of the valley very uncomfortable at times. The reports of the boundary commissioners in 1858 and 1859 are full of references to hordes of these insects.

The eastern valley was also quite restricted in its amounts of agricultural land, being increasingly hemmed in by the mountains as one progressed east. The Fraser River often flooded the area around the lake in the spring, when the Sumas River reversed its flow, with water flowing into Sumas Lake from the Fraser. It was therefore not surprising that when the best agricultural land had been taken up, the notion of reclaiming the shallow lake seemed like a good idea.

There had been attempts to drain the margins of the lake from as early as 1877. In 1905 the Sumas Development Company was created to drain the lake. The British Columbia Electric Railway purchased the company three years later, intending to lay its interurban track across the lake and sell land on either side. But these plans proved too costly, and the line was laid around the lake's southern edge (see page 270).

A new plan to drain the lake was drawn up in 1919 by the provincial government's new Land Settlement Board (see page 288), and work began the next year. The Vedder Canal was dug to divert water from the Vedder River north towards the Fraser instead of flowing into the lake. Engineering works were completed in the summer of 1923, and pumping began on 4 July; within a year the lake bed was dry.

The draining of Sumas Lake destroyed a great deal of wetland habitat along its margins but creating 12,000 ha of new, very fertile, and easily cultivated farmland. The drainage work was so well done that when the valley experienced its next major flood, in 1948, the old lake area avoided inundation.

MAP 776 (*below, centre*).
Lake Sumass on an 1862 map produced by the boundary survey, whose camp at *Camp Sumass*, on the Sumas River, was plagued by mosquitoes breeding in and around the lake.

MAP 777 (*below, right*).
This 1888 map already shows *Reclaimed Lands* on *Matsqui* Prairie on the other side of *Sumas Mountain*. The red lines are roads or trails.

MAP 778 (*below, left*).
Sumas Lake is flanked by the reversing *Sumas River* and *Sumas Mountain* to its west on this 1905 map; *Vedder Creek* (later renamed Vedder River) flows into the lake on its eastern side.

MAP 779 (*right*).
The drainage works and subdivision of the old lake bed are well shown on this 1931 map. The Vedder Canal now directs the flow of *Vedder Cr.* into the *Sumas River* and then into the Fraser. The Sumas River is contained, instead of flowing into the lake and out again, and a straight north–south canal in what was the middle of the lake also directs water into the Sumas River. New roads now cross the old lake bed.

Vancouver between the Wars

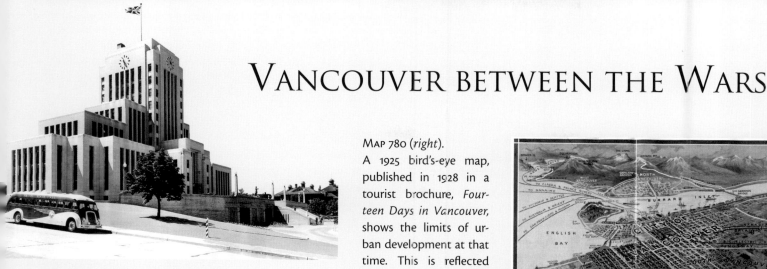

It took Vancouver a few years to emerge from the drag of bankruptcies and decline in land values after World War I, but the twenties was a period of growth. The 1921 population (within the 1931 area) of 163,000 had increased to 365,000 by 1931. Then the Depression, a worldwide phenomenon, hit, and by 1934 over 27,000 persons were on government relief (MAP 788, *overleaf*).

Vancouver grew in another way, too. In 1929 the City of Vancouver amalgamated with South Vancouver, a separate municipality created in 1891, and Point Grey, created in 1907 (see page 142), forming the city with today's boundaries. Planning for the merger included a massive and detailed report from Harland Bartholomew, then the leading urban planner in North America (see overleaf), and the building of a fine new city hall (*above*) in keeping with the now-enlarged city.

MAP 780 (*right*).
A 1925 bird's-eye map, published in 1928 in a tourist brochure, *Fourteen Days in Vancouver*, shows the limits of urban development at that time. This is reflected in the extent of the city lights in the late 1920s photo (*right*) taken from Grouse Mountain. The classic twenties brochure cover is *inset, right.*

MAP 781 (*right, centre*).
In 1920 12 km² of Crown land on Point Grey was set aside for sale, with the proceeds going to the *University of British Columbia.* This map is a plan for the subdivision of what became known as the University Endowment Lands (UEL). Some of the streets in the eastern part of the map were built, as was the *University Hill* area and two of the main boulevards into the university. Today's College High Road, in two separate sections, is all that remains of the middle, never-completed boulevard shown on the map. The onset of the Depression slowed development, and much of the land reverted to the government in the 1950s. In 1988 Pacific Spirit Regional Park was created on 763 ha of the UEL.

MAP 783 (*below*).
This innovative pictorial map was published at the height of the Depression and gives no inkling of the economic struggle of those years.

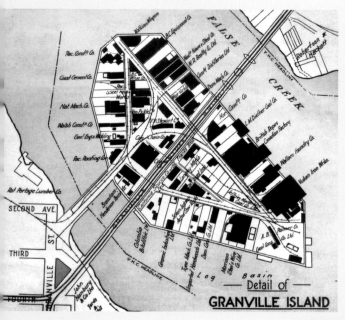

— Detail of —
GRANVILLE ISLAND

MAP 782 (*above*).
Originally named Industrial Island, by 1930, the date of this map, Granville Island was living up to its original name. Created in 1915–16 by adding dredged fill to an existing sandbar, the island was completely changed in 1979 with the opening of Granville Market, the first phase of a continuing transition from industrial to public and retail uses. Note the line of the *Granville Street Bridge.* This was the second one, opened in 1909, replacing a low trestle bridge opened in 1889. The modern one, the third, which replaced the bridge again in 1954, is built immediately above the causeway to the island shown on this map.

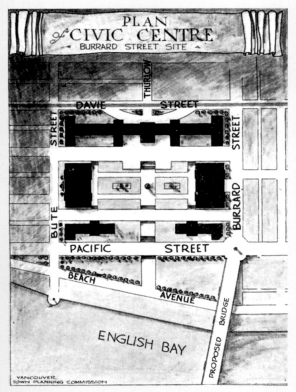

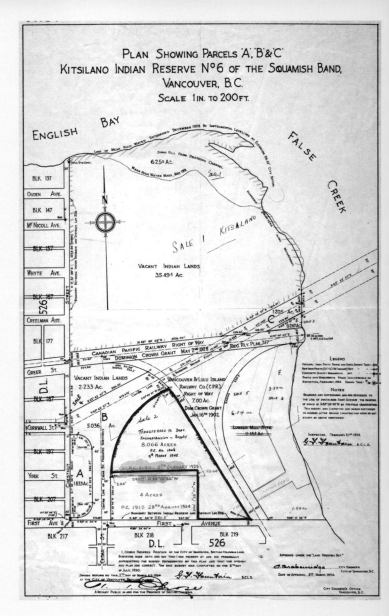

Map 784 and Map 785 (*left*), and Map 786 (*below, left centre*).

Town planning was all the rage in the 1920s. In 1925 the province allowed municipalities to establish town planning commissions, which Vancouver did the following year. Point Grey hired an American town planning expert, Harland Bartholomew, to produce an overall plan for the municipality, and Vancouver signed on later that year. South Vancouver was added in 1929 when the three amalgamated, though it had been included in the regional plan for Point Grey and Vancouver. Bartholomew's plan, published in 1929, contained many innovative ideas such as boulevards, ring roads, and cutoffs (such as, for example, the connection between West 12th and West 10th avenues at Macdonald Street and the widening of Broadway), but the city was unable to implement the entire plan because of the economic constraints introduced by the Depression. One expensive dream that was not fulfilled was that of a Civic Centre beside the Burrard Bridge, shown here. The bridge, not a Bartholomew idea, was built, however, and opened in July 1932. After much debate the city finally decided on a site at West 12th Avenue and Cambie Street for its new city hall, similar in concept to the one depicted here, and it opened in 1937. Map 786 shows a proposed system of linked boulevards and parks. Parts of this planned system that were built include West King Edward Avenue and south Cambie Street.

Map 787 (*above*).

This 1930 survey shows various removals of land from what was the Kitsilano Indian Reserve No. 6, the site of Sen'ákw (Snauq), one of several Aboriginal villages around Burrard Inlet and designated a reserve in 1877. In 1913, the village was deemed to be in the way of progress, and the provincial government paid $11,250 to the head of each family. The following day all of the inhabitants and all of their possessions, including remains from a grave site, were barged to North Vancouver. Further alienations continued both from Canadian Pacific (the original rail right-of-way here updated to 1928) and the Vancouver & Lulu Island Railway, leased to B.C. Electric for its interurban line; as well as 2.5 ha for the Burrard Street Bridge right-of-way, expropriated by the City of Vancouver. In 2002, after years of litigation and negotiation the Squamish regained control of a remnant of the reserve land, the railway rights-of-way shown. It is on this land that digital billboards were erected in 2009.

Map 788 (*left*).

Vancouver's first archivist, Major James Skitt Matthews, liked to create historical records as well as save them. He drew this map in 1934 showing the location of men on government relief (no women are shown), and it reveals that the majority of them lived outside the west side of the city, formerly Point Grey; blue collar workers were more likely to be affected by the Depression than white collar. It is notable that *Shaughnessy Heights*, at the upper end of the city's social hierarchy, has almost no relief cases. Times were so bad that many British Columbia cities went bankrupt, including North Vancouver, both city and district, and Burnaby. In April 1935 seven thousand unemployed men from relief camps descended on Vancouver demanding "work and wages." Mayor Gerry McGeer, convinced he was facing a Communist revolution, called out four hundred police, creating a confrontation that led to a riot. Later, a thousand men climbed aboard freight cars and set off to take their case to Ottawa. Stopped in Regina by a government afraid of revolution, it led to another, more serious riot. More such riots occurred in Vancouver in 1938.

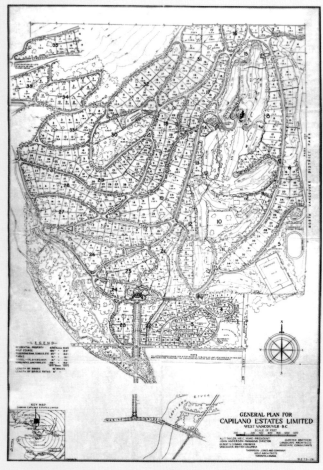

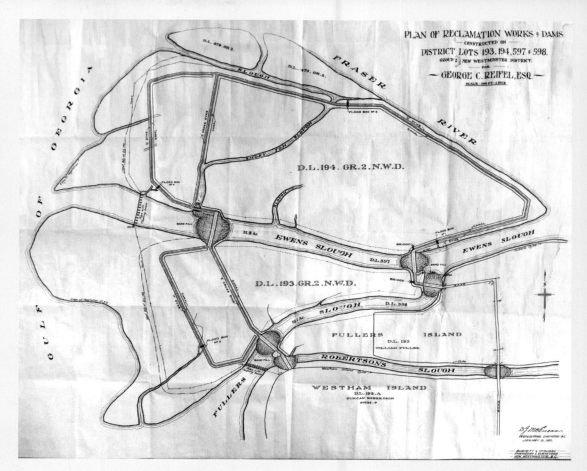

Map 789 (above).

A group of British investors formed British Pacific Properties in 1931 and purchased 1,416 to 1,618 ha of tax sales land from the municipality of West Vancouver for $75,000, promising to spend another $1 million to develop the land. The same group also built the Lions Gate Bridge to provide access to their land. The first plan for what was called Capilano Estates and later British Properties was drawn up in grand style, with large lots and boulevards by the leading landscape architects Olmsted Brothers. Though lot sizes had to subsequently be reduced, the investment certainly paid off—but only in the long term, for there is still land awaiting development.

Map 791 (below).

Vancouver celebrated fifty years of existence in 1936, and despite the Depression, parades and other events were held. The city also published a commemorative booklet, which included this map that showed the world just where Vancouver was. The map was surrounded by a panoply of the most modern transportation serving the city and one showing *Paved Highways to Vancouver*, something to be proud of!

Map 790 (above).

George Reifel was a brewer and liquor importer who made much of his fortune during the Prohibition era south of the border. He created a large farm at the western end of Westham Island, at the entrance to the *Fraser River*. This map, drawn in 1930, shows the plans for reclamation channels and dikes. In 1972 Reifel's grandson, who had farmed the land, donated 348 ha to create the George C. Reifel Migratory Bird Sanctuary. The pattern of dikes and channels survives in large part intact today.

Map 792 (below).

The Canadian Broadcasting Corporation (CBC) was created in 1936 and took over a radio station started by the Canadian National Railway Radio Department in 1925 as CNRV. The map/globe/cartoon announced the beginning of 5,000-watt broadcasting from the CBC station, then with the call sign CRCV. The same CBC station today broadcasts at 50,000 watts during the day.

SERVING THE COAST

British Columbia's mountainous and indented coastline has always lent itself to water transportation. In 1836 the Hudson's Bay Company's little *Beaver* became the first steamship to ply its coastal waters, and once the transcontinental railway was complete, steamships provided both the trans-Pacific connections and those on the local coast, and especially to Victoria and Nanaimo.

In 1887, the same year the first train from the east pulled into Vancouver, Canadian Pacific chartered three ocean liners from Cunard and began service to Yokohama and Hong Kong. Four years later the railway created Canadian Pacific Steamship Company and put into service the *Empress of China,* the *Empress of Japan,* and the *Empress of India.* Empresses would cross the Pacific from Vancouver until 1941.

Canadian Pacific at first used other steamship companies to provide a coastal service but in 1901 purchased Canadian Pacific Navigation Company (an independent, despite the name) and began its British Columbia Coast Service, creating a fleet of ships known as the Princesses to complement their larger sisters. The routes they operated are shown on MAP 794, *right.*

When the Grand Trunk Pacific Railway began building its line to Prince Rupert, it too got into the coastal shipping business, often in deliberate and direct competition with Canadian Pacific. And to emphasize the alternative choice, it named its ships Princes. *Prince Rupert* and *Prince George* were placed in service in 1909, at first sailing between Seattle and Prince Rupert. A seven-day cruise was offered at $48, the

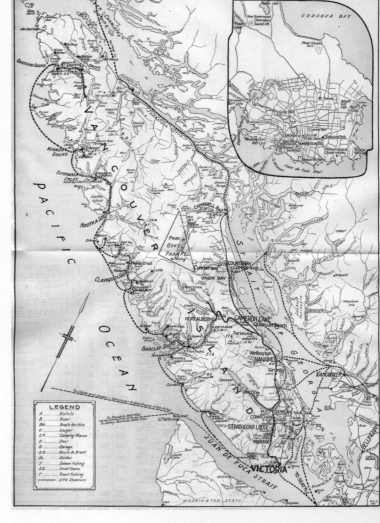

Above. Union Steamships' SS *Chilcotin,* which operated from 1947 to 1958, emerges from a coastal inlet in this illustration from the front of a route map (MAP 798, *far right, top centre*).

MAP 793 *(above, right).*
A relief-type map of Vancouver Island from about 1930 showing some of the routes taken by the Canadian Pacific's B.C. Coast Service steamers.

MAP 794 *(right).*
The extensive route network map of Canadian Pacific's B.C. Coast Service on the lower coast in 1927 also noted where wildlife could be found.

MAP 795 *(left).*
Canadian Pacific began an auto ferry service route between Vancouver, Victoria, Nanaimo, and Bellingham. The company's *Motor Princess* is illustrated here and also *below, centre,* on another page of the same 1923 booklet, showing the internal arrangement of vehicles.

MAP 796 *(below, right).*
Canadian Pacific's B.C. Coast Service steamer route between Vancouver, Victoria, and Seattle was advertised as the *Triangle Service,* complete with a stylized triangle map in 1927. The ship, *Princess Marguerite,* was built in 1924 specially for this route. The ship was sunk in wartime service in the Mediterranean in 1942; her successor, *Princess Marguerite II,* began service in 1948 and was operated by the CPR until 1974 but remained on this route until 1989.

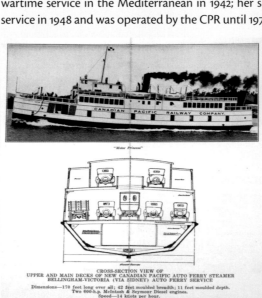

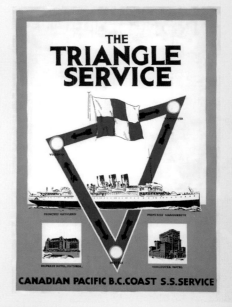

The
MAGIC DAY
SEA-TRIP
THROUGH
HOWE SOUND

*Vancouver's Glorious
Inland Sea*
BY
Union Steamships

Connecting at Squamish
for Points on the Pacific
Great Eastern Railway.

Union Steamship
Route shown in
Red ------

SERVING THE COASTAL COMMUNITIES OF
BRITISH COLUMBIA FOR OVER 50 YEARS

UNION STEAMSHIPS

The Magic Day
Sea-Trip
on Howe Sound

Connecting at Squamish for
points on
PACIFIC GREAT EASTERN RAILWAY

predecessor of modern Alaska cruises. After 1917 Canadian National continued to operate these steamship services.

Union Steamships, the other major force on the British Columbia coast for many years, was founded in 1889 and serviced communities, logging camps, canneries, and tourist resorts—many of the latter owned by the company (such as at Bowen Island, shown on the brochure *below, right*) until 1959.

The Union Steamship Company was formed in 1889 by John Darling, a retired general superintendent of an eponymous shipping company in New Zealand, who visited Vancouver in 1888 and immediately sensed an opportunity. He partnered with Captain William Webster and interested other investors. The new company took over the assets of Burrard Inlet Towing Company, which included the Moodyville Ferry and an accompanying mail contract. Webster went to England and purchased the *Cutch*, a steamship previously owned by an Indian maharajah, putting it into service in 1890 to sail between Vancouver and Steveston and Nanaimo under contract with Canadian Pacific.

The company acquired many vessels over the ensuing years and became an essential element of life on the British Columbia coast. Resorts such as Bowen Island and Selma Park near Sechelt allowed Union Steamships to offer tourist destinations as well as transportation. Increasingly uneconomic operation and competition with road and air led the company to close down in January 1959, when its assets were sold to a new company, Northland Navigation.

MAP 797 (*above, left*).
A bird's-eye map of Howe Sound showing Union Steamships' route in the 1920s. The illustration (*above, right*) is the cover of this map and shows the *Lady Alexandra*, a regular on this run and that to Bowen Island, advertised in a 1942 brochure, *right*. The *Lady Alexandra* began service in 1924.

MAP 798 (*above, centre*).
A route map for the Union Steamship Company published about 1948. The illustration, *far left, top*, is the cover of this map.

MAP 799 (*left*).
The British Columbia portion of an unusual bird's-eye map of Puget Sound published by the Puget Sound Navigation Company in 1910 to promote its routes, some of which competed with Canadian Pacific and Union Steamships.

MAP 800 (*below*).
Another unusual three-dimensional-type map showing Union Steamships' route in the 1930s along the Sunshine Coast.

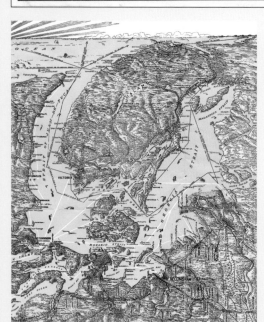

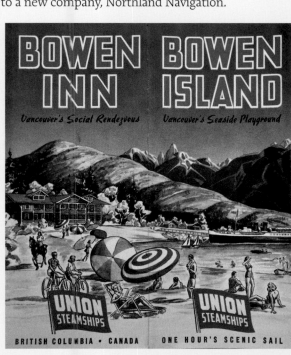

BOWEN BOWEN
INN ISLAND
Vancouver's Social Rendezvous *Vancouver's Seaside Playground*

UNION STEAMSHIPS
BRITISH COLUMBIA · CANADA

UNION STEAMSHIPS
ONE HOUR'S SCENIC SAIL

The Path of Sunshine and Sea Charm
along the
GLORIOUS GULF COAST ROUTE
of the
UNION STEAMSHIP COMPANY
A Wonderland of Mountain and Marine Scenery

The Champagne Safari

Doubtless qualifying as one of the most exotic expeditions ever to explore British Columbia was the Bedaux Expedition of 1934. Organized by Charles Bedaux, a wealthy American management consultant, it was intended to locate a path for a road through the northern part of the province west to Telegraph Creek, though this may have only been an excuse for an exploration in style.

Bedaux imported five half-track vehicles from Citroën in France and gathered a crew that included his wife and his mistress, a moviemaker—so that his glorious escapade would be recorded for present fame and history—and two surveyors enlisted from the British Columbia government, which was pleased at the idea that he would pay for the exploration and survey of a largely unknown part of the north. The two surveyors were Frank Swannell and Ernest Lamarque, authors of the two maps reproduced here as a result of the expedition.

Bedaux and his ensemble left Edmonton on 6 July 1934, first driving 800 km on a mud road to Fort St. John. When the vehicles became bogged down Bedaux showed his true colours by jettisoning 45 kg of survey equipment—but keeping his supply of French wine.

The expedition continued westward, following the route shown by Swannell and Lamarque on their maps. By August the Citroëns were abandoned, Bedaux finally admitting that they were unsuitable for the rough terrain. True to style he had his moviemaker film two of them being driven over a cliff and one floated downstream on a raft rigged with explosives—which failed to work. Two others were simply abandoned (and were later found after the Alaska Highway was built; one is in the Western Development Museum in Moose Jaw).

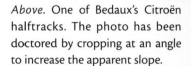

Above. One of Bedaux's Citroën halftracks. The photo has been doctored by cropping at an angle to increase the apparent slope.

Then Bedaux's horses began to develop hoof rot from the wet conditions, and Swannell advised him not to continue into the winter. In September the expedition returned to Whitewater Post, on the Finlay River, where they met power boats sent to fetch them.

The expedition was not entirely useless, for new knowledge of British Columbia's remote north was gained. Bedaux's movie remained unmade at the time, but the film was kept and finally used for a 1995 television biography of Bedaux—aptly called the *Champagne Safari*. But Bedaux was long gone by then, having committed suicide in 1944 while awaiting trial on treason charges for collaborating with the Germans in World War II.

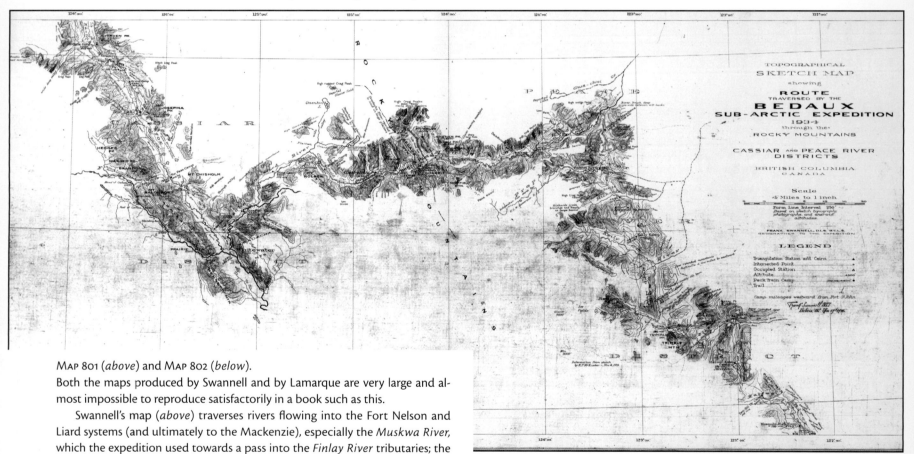

Map 801 (*above*) and Map 802 (*below*).
Both the maps produced by Swannell and by Lamarque are very large and almost impossible to reproduce satisfactorily in a book such as this.

Swannell's map (*above*) traverses rivers flowing into the Fort Nelson and Liard systems (and ultimately to the Mackenzie), especially the *Muskwa River*, which the expedition used towards a pass into the *Finlay River* tributaries; the *Kwadacha*, to *Whitewater* Post on the Finlay; and then north again up the *Fox River* to *Sifton Pass*—where Bedaux and his main party turned back—and *Citroën Peak*, named by Swannell after André Citroën, manufacturer of Bedaux's halftrack vehicles.

Lamarque's map (*below*) shows the entire intended route from *Whitewater* on the *Finlay River* (at bottom right) to *Dease Lake* and *Motor Road to Telegraph Creek* (at left). Lamarque led an advance party that did reach Telegraph Creek. His map is illustrated with numerous sketches.

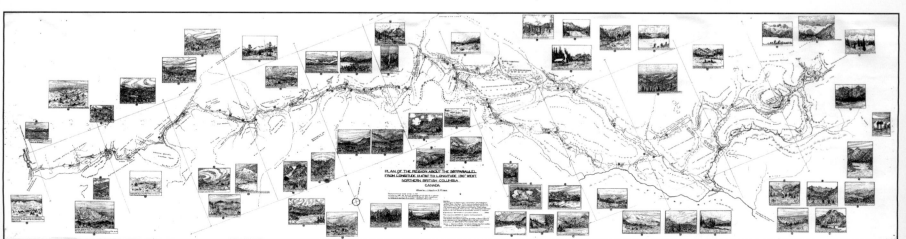

Provincial Parks

In the late 1880s William Bolton, a clergyman, explored Vancouver Island. His glowing reports caught the eye of Premier Richard McBride, and in 1910 Price Ellison, commissioner of lands, led the Strathcona Discovery Expedition to document the area's beauty, leading to creation of a government reserve that year and the creation of British Columbia's first provincial park in 1911.

This precedent began a long series of park establishments that has continued to the present day. British Columbia was the first western province to embrace the concept, taken for granted today but certainly not then.

Mount Robson Provincial Park, adjacent to the federal reserve that became Jasper National Park, was the second provincial park, created in 1913. Others followed, notably Sir Alexander Mackenzie Provincial Park in 1926, covering Mackenzie's Rock, the explorer's western terminus (illustrated on page 39; also bottom left on MAP 805, *below, right*), and Garibaldi Provincial Park the next year.

Tweedsmuir Provincial Park, the largest park at 981,000 ha, was created in 1938, and Wells Gray Provincial Park (MAP 803, *right*) the following year.

MAP 803 (*right*), with cover (*left*). This fine pictorial relief map of Wells Gray Park was published about 1940 by the British Columbia Forest Branch, Parks Section of the Department of Lands. Parks were not administered by a separate department until 1957. The park was named after Arthur Wellesley (Wells) Gray, minister of lands from 1933 to 1944.

MAP 804 (*right, centre*).
A *Prov. Gov^t. Park Res.* with *No Hunting* is shown on a 1927 map. This was Strathcona Park on Vancouver Island, British Columbia's first provincial park.

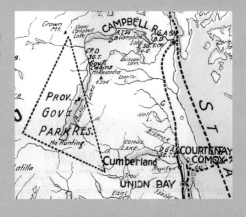

MAP 805 (*right*), with cover (inset). Tweedsmuir Park, south of *Ootsa Lake*, was British Columbia's largest provincial park, created in 1938. It is shown here on a map from about 1940. The park is now in two parts, north and south.

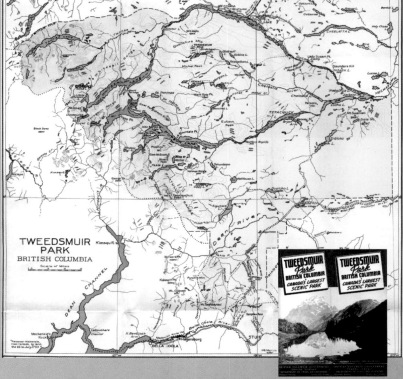

FLYING WEST OF THE MOUNTAINS

American stunt pilot Charles K. Hamilton made the first airplane flight in British Columbia in a Curtiss pusher biplane over Richmond on 25 March 1910. The following day he flew to New Westminster and back. The first homegrown aviator was William Wallace Gibson, of Victoria, who built a plane that flew in September the same year.

World War I brought the airplane into regular use once the military realized its potential, and as a result, when the war ended, there were many pilots and much larger and stronger planes available. The British government and the United States Navy even donated many surplus planes. The airplane took on a new life in British Columbia, being the ideal vehicle for communication and transport in a county largely devoid of roads. Float planes, flown by bush pilots, could land on any convenient lake.

In 1919 pilots William E. Boeing and Eddie Hubbard carried the first international airmail in North America between Vancouver and Seattle. Airmail contracts would prove to be a force behind the growth of commercial aviation.

In August 1919 Ernest Hoy flew from Vancouver to Calgary, becoming the first to conquer the Rockies, a formidable barrier for early fliers.

The federal government opened an air base at Jericho Beach on English Bay

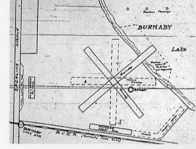
Above. A 1950 vintage Beech-craft 3NMT Expeditor at the Canadian Museum of Flight in Langley.

MAP 806 (below).
Flight paths in the early days were very broadly defined. This is the northwestern part of a 1919 map of airways and mail routes. The "All Red" Airway was, of course, entirely in Canada.

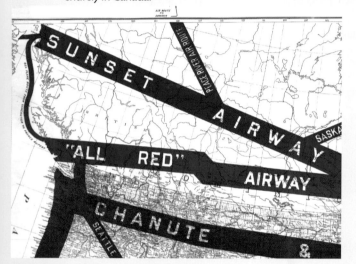

Above. An ad for a display by "bird-men" at Minoru racetrack in Richmond in April 1911. Performances were poor and the meet was not a success.

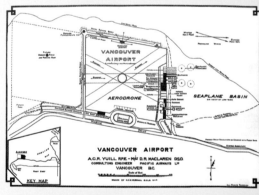

MAP 808 (right, above) and **MAP 809 (right).**
Airport sites considered for Vancouver in 1928—one on the shores of Burnaby Lake and another at Spanish Banks. Neither site was chosen.

MAP 811 (below, centre).
The 1929 plan for the first Vancouver Airport on Sea Island in Richmond.

MAP 807 (above).
This 1929 map of airmail routes shows just one in British Columbia: to Victoria from Seattle.

MAP 812 (below, right).
A map of the new Vancouver Airport was included on the cover of a booklet produced for the official opening ceremonies in July 1931.

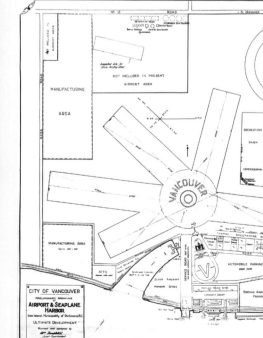

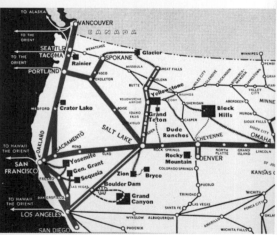

MAP 810 (above).
Pioneer commercial airline United Airlines published this map of its routes in 1936, which includes the route from Seattle to Vancouver.

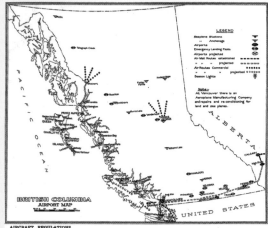

MAP 813 (*above*).
A provincial government map of airports in British Columbia published in 1930. Airmail routes are shown established to Calgary and High River in Alberta and between Vancouver, Victoria, and Seattle, but only projected across southern British Columbia from High River to Vancouver.

in 1920, and in 1928 Vancouver opened its first official airport following a refusal by pioneer transatlantic flier Charles Lindbergh to visit the city because it had "no fit field to land on." Then, following a review of a number of possible sites (two of which are shown here), the city opened an airport on Sea Island in 1931. It would expand and grow into Vancouver International Airport (YVR).

In the 1930s the federal government began building airports at regular distances right across the country to create the Trans-Canada Airway (MAP 814, *right*). United Airlines began commercial passenger service to Vancouver from Seattle in 1934. Canadian Airways also began flying this route, hoping to be awarded the mail contract. But the federal government decided to create its own Crown corporation, and Trans-Canada Airlines (TCA) took over in September 1937. In 1939 the airline began transcontinental service between Vancouver and Montréal. In 1942 Canadian Pacific established its own airline, buying ten smaller companies including Canadian Airways. Former bush pilot Grant McConachie, from Yukon Southern Airways, headed the new airline.

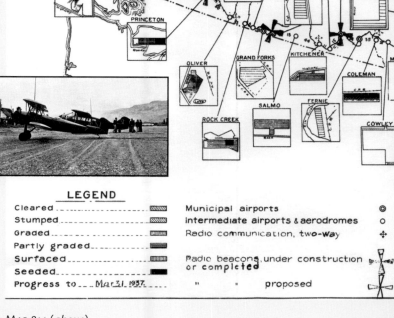

LEGEND

Cleared		Municipal airports
Stumped		intermediate airports & aerodromes
Graded		Radio communication, two-way
Partly graded		
Surfaced		Radio beacons, under construction or completed
Seeded		
Progress to Mar 31, 1937		" " proposed

MAP 814 (*above*).
The western part of a map of the Trans-Canada Airway, a series of airports similarly spaced apart to provide a path across Canada that could be used by airplanes of the day. Airports each have a little square with its own map to inform pilots what to expect, and there is a key to facilities. Some of the airports were completed as make-work schemes to provide employment during the Depression years; one is at *Oliver*, which was completed in September 1937 and is shown on opening day in the photo, *inset*.

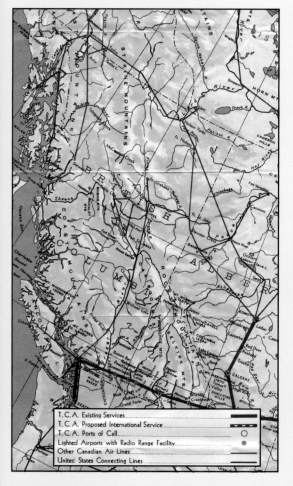

T.C.A. Existing Services	
T.C.A. Proposed International Service	
T.C.A. Ports of Call	
Lighted Airports with Radio Range Facility	
Other Canadian Air Lines	
United States Connecting Lines	

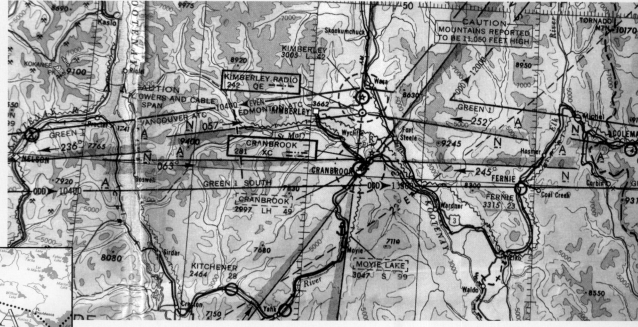

MAP 815 (*left, centre*).
Trans-Canada Airlines' first routes in British Columbia are depicted by the thicker red lines on this 1940 map: *Vancouver* to *Seattle* and *Calgary* to Vancouver via *Lethbridge*. The thinner red lines are routes of other airlines. TCA became Air Canada in 1974.

MAP 816 (*left*).
Canadian Pacific Airlines routes, shown as red dotted lines, are superimposed on a map of railway routes published in 1957 as an "all-services" map by Canadian Pacific. The airline was sold to Pacific Western Airlines in 1987, becoming Canadian Airlines International, and was taken over and merged into Air Canada in 2000.

MAP 817 (*above*).
Also from 1957 is this pilot's chart showing the Cranbrook area of southern British Columbia.

DEFENDING THE COAST

Britain and Canada declared war on Germany on 3 September 1939 following the German attack on Poland. Thousands of British Columbians went off to fight the war, and industry ramped up to supply the voracious needs of the conflict, ending the unemployment of the Depression. Shipyards, in particular, geared up to build both naval and merchant ships. At the peak, 25,000 men and women were at work. Boeing employed another 5,000 at its aircraft plant on Sea Island. Yet it was understood from the beginning that the threat to the West Coast came principally not from Germany but from the growing empire of Japan.

Canada responded to this threat by constructing gun emplacements at critical points to defend entrances to the important harbours at Vancouver, Victoria, and Prince Rupert. Seaplane bases at places like Ucluelet, on the west coast of Vancouver Island, were established to allow constant surveillance of the eastern Pacific and in particular watch for submarines.

MAP 818 (*right, top*).
Declassified only in 1997, this government summary map shows the location of defended ports and other fortifications.

MAP 819 (*above*) and MAP 820 (*right*).
One of the "other fortifications" was at Yorke Island, near Kelsey Bay, at the western tip of the larger *Hardwicke Island* at a point where *Johnstone Strait* and the entire northern approach to Vancouver could be protected. MAP 819 shows the no-fly zone around the island on a 1945 American military aeronautical chart, while MAP 820, long ago taped to mend tears, shows the elaborate gun emplacements—on the west side of the island pointing up the strait—and supporting buildings for the Yorke Island garrison.

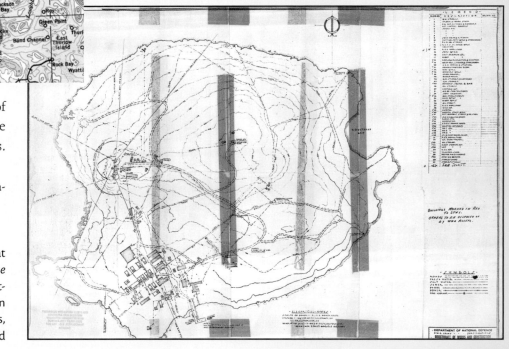

MAP 821 (*right*).
An Army plan of the guns at Point Grey, together with a 1943 aerial photo (*below*). Together with guns at Point Atkinson, on the other side of English Bay, these guns protected the outer entrance to Vancouver Harbour. Today, part of the Museum of Anthropology incorporates one of the original concrete gun emplacements.

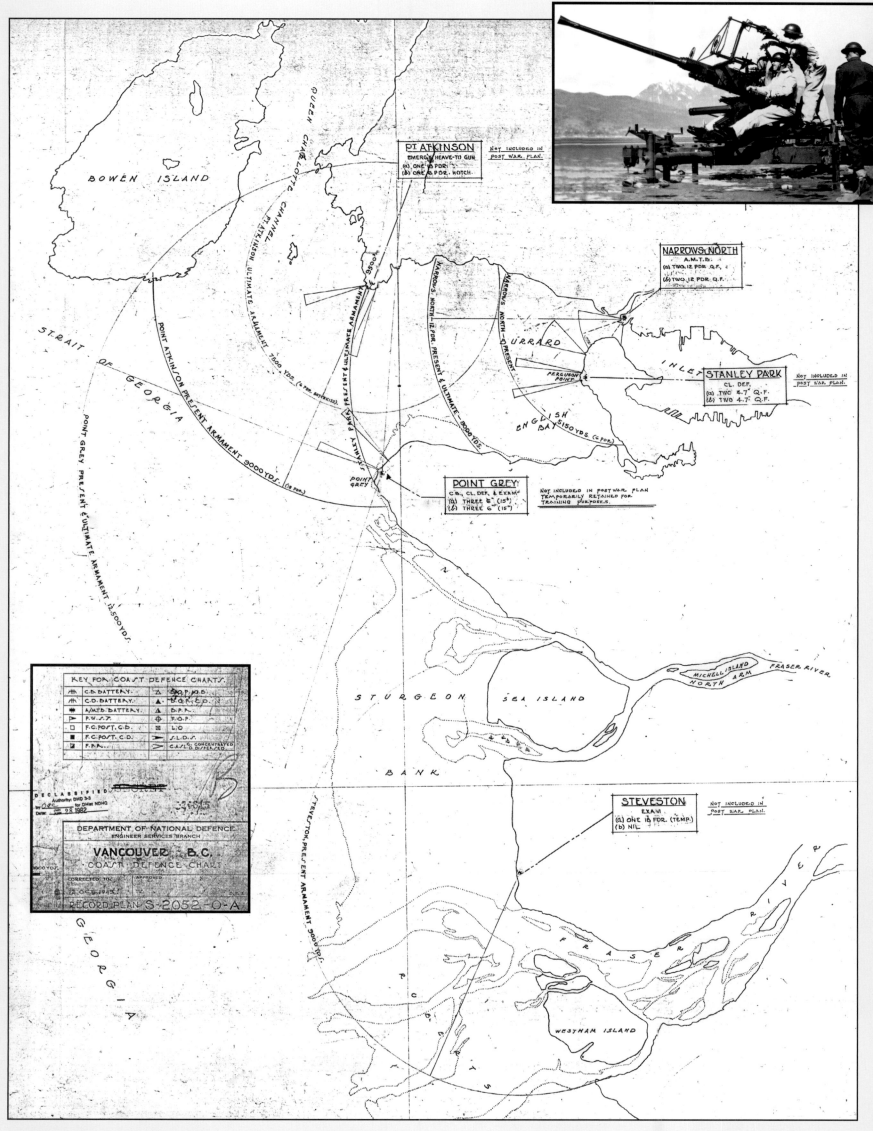

PT ATKINSON
EMERG'Y HEAVE-TO GUN
(a) ONE 18 PDR.
(b) ONE 6 PDR. HOTCH.

Not Included in
Post War. Plan.

NARROWS. NORTH
A.M.T.B.
(a) TWO. 12 PDR. Q.F.
(b) TWO 12 PDR. Q.F.

STANLEY PARK
C.L. DEF.
(a) TWO 4.7" Q.F.
(b) TWO 4.7" Q.F.

Not Included in
Post War. Plan.

POINT GREY
C.B., C.L. DEF., & EXAM'
(a) THREE 6" (15°)
(b) THREE 6" (15°)

Not Included in Postwar Plan
Temporarily Retained for
Training Purposes.

STEVESTON
EXAM
(a) ONE 18 PDR. (TEMP.)
(b) NIL

Not Included In
Post War. Plan.

KEY FOR COAST DEFENCE CHARTS.

DEPARTMENT OF NATIONAL DEFENCE
ENGINEER SERVICES BRANCH

VANCOUVER, B.C.
COAST DEFENCE CHART

RECORD PLAN S-2052-0-A

MAP 822. The defences of Vancouver are shown on this 1945 map, top secret until 1982. Gun emplacements are shown at *Pt. Atkinson, Point Grey, Stanley Park,* and *Narrows North* protecting *Burrard Inlet* and at *Steveston* protecting the entrances to the *Fraser River.* The ranges of the various gun emplacements are shown by the overlapping circles, and the types of gun are noted in the rectangles.

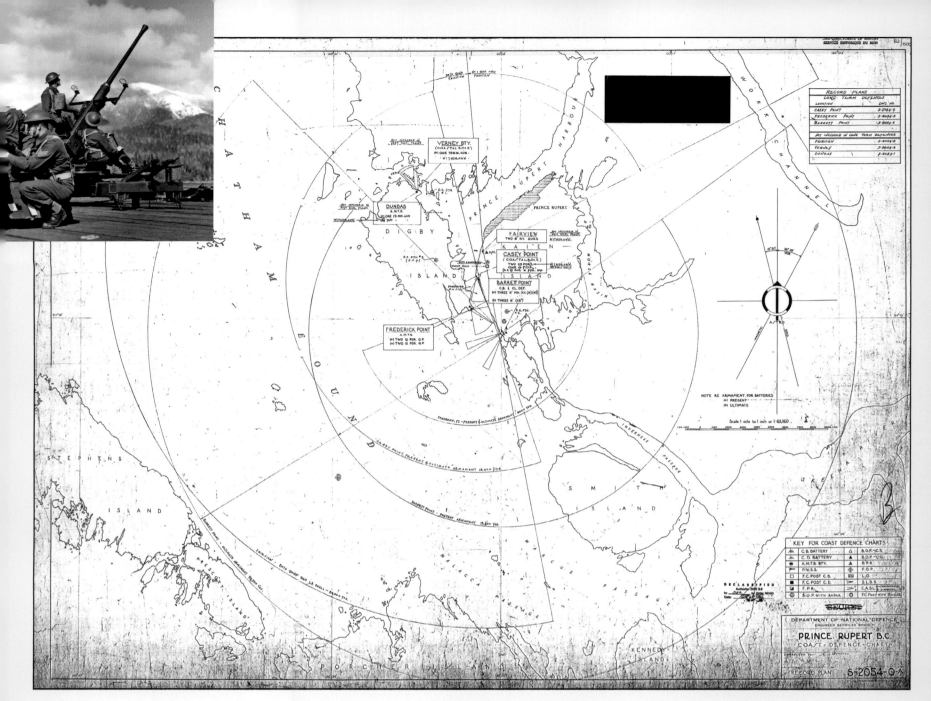

MAP 823 (*above*).
Prince Rupert, as a supply port for the American Army in Alaska, was strategically important. The Skeena Valley was thought to be a possible attack route to the interior, and so it was well defended, both with gun emplacements at the approaches to the harbour, as shown on this map, and by a unique armoured train (photo, *left*) that constantly moved up and down the Skeena Valley for two years. The train carried guns, a searchlight, and a full infantry company. The photo, *left, top*, shows an anti-aircraft gun on the Prince Rupert waterfront.

On 7 December 1941 the Japanese bombed Pearl Harbor, bringing the United States into the war. Then on Christmas Day Hong Kong, defended by Canadian troops, fell after an attack begun hours after Pearl Harbor.

There had long been anti-Asian sentiment in the province, but now it was reinforced by fear of invasion, and the public, politicians, and the press agitated alike for removal of all Japanese people from the coast. This was the same sentiment that

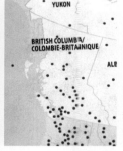

MAP 825 (*right*).
A military training camp built near Nanaimo during the war. This is a 1943 map.

MAP 824 (*above*).
Towards the end of the war, having figured out that the jet stream blew from east to west, the Japanese released hundreds of so-called fire balloons like the one shown here. After a time interval thought to place them over British Columbia or the American West, incendiaries were automatically dropped. The intention seems to have been to set the forests on fire and terrorize the population, but few worked in the normally damp environment, and information about them was censored, so few people even knew they existed. There seems to be no contemporary map of where they landed, but this modern map from the Canadian War Museum shows with red dots where they were found.

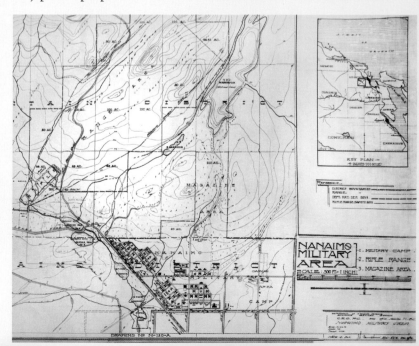

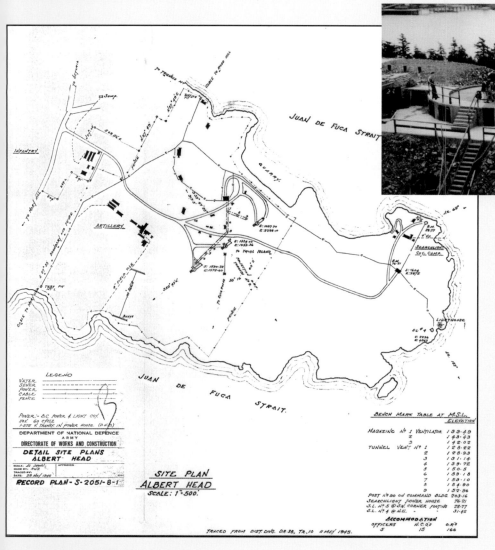

MAP 826 (*left*) and photo (*inset*).
The gun emplacements on Albert Head, guarding the approach to Esquimalt and Victoria Harbours, shown on both a Department of National Defence map and a contemporary photo. The guns are the open rectangles with a circle superimposed, while the black rectangles are buildings; a searchlight was mounted near the *Lighthouse*.

MAP 827 (*below*).
Japanese plans for the invasion of the west coast of North America, including British Columbia, were revealed in this sensational map said to have been obtained by a Korean spy; it was published in the *Los Angeles Examiner* in August 1943. After capturing islands in the South Pacific and the Aleutians and Hawaii, Japanese forces would strike the West Coast and the Panama Canal. The plan seems to have been to capture and hold the entire area west of the Rocky Mountains and then get the United States and Canada to sue for peace. It seems far-fetched now, perhaps, but did not seem so in 1943. Japan must likely have thought such a plan would work to take on countries with the resources and resolve of the United States and Canada in the first place.

was sweeping the American West, where the shelling of an oil refinery by a submarine and a perceived nighttime air attack on Los Angeles added to the hysteria. On 27 February 1942, eight days after the United States had done the same, the Canadian government ordered the removal of all persons of Japanese descent from the coastal area. One of the reasons given was the danger of anti-Japanese riots. There was also fear of fifth-column activity such as that reported in the Netherlands and Norway prior to German attack. No distinction was made between persons of Japanese citizenship and those with Canadian citizenship. The evacuation orders simply referred to "all persons of the Japanese race." Hundreds of fishing boats were impounded, property was seized by an official body known as the Custodian of Enemy Property; much was sold at fire sale prices, and little was returned. About 23,000 people were forcibly removed, first to Vancouver's Exhibition Park and then to camps hastily set up in the interior of the province, the majority of which consisted of fields full of poorly insulated shacks (MAP 828–MAP 830, *overleaf*). Some Japanese people were placed in old buildings in a number of ghost towns such as Sandon (see page 178) in the Kootenays. Others were sent to Prairie farms.

There was never any proof of Japanese espionage, but when invasion seemed a real possibility, few cared. There was only one Japanese attack on British Columbia, the shelling of Estevan Point Lighthouse on the west coast of Vancouver Island by a submarine on 20 June 1942, and even here there is some dispute because of discrepancies in witnesses' accounts. Internees' property losses were partly

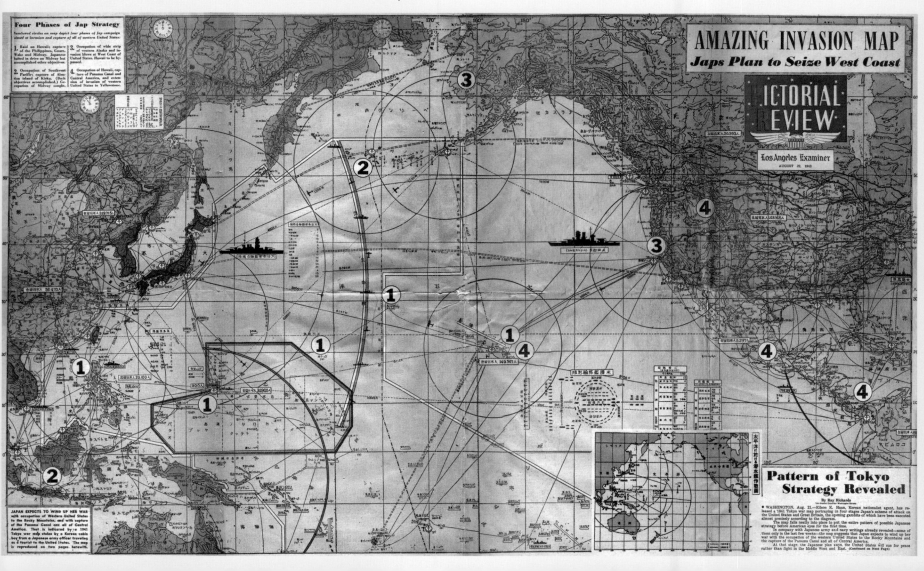

The Vancouver Sun

Japs, All Enemy Aliens, to Move From Defense Zones; Whites to Run Fish Fleet

Torrents Halt Foe In Malaya
Australians Get Ready for Big Scale Action

Above. A few of the structures that housed the Japanese internees in the Slocan Valley have been preserved and are on display at the Nikkei Internment Memorial Centre in New Denver.

Above. Japanese families unload from the back of a truck somewhere in British Columbia's interior in this 1942 photo.

compensated in 1950, but it was not until 1988 that the Canadian government recognized the injustice that had been done and offered surviving internees a formal apology and modest compensation of $21,000 each.

Fear of a Japanese attack on Alaska led to the American use of Prince Rupert as a supply port, transforming the city for the war years and finally leading to the completion, in 1944, of a road down the Skeena Valley. And the American Army built the Alaska Highway.

Above. The headline in the *Vancouver Sun* on 14 January 1942; and part of the Japanese fishing fleet tied up at Steveston.

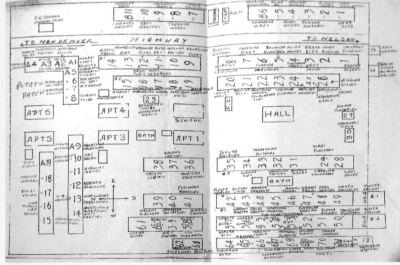

MAP 828 (*right*), MAP 829 (*left, centre*), MAP 830 (*left, bottom*) and MAP 831 (*right, bottom*).

Matsuru Kitamura, an internee, made this map in 1946 showing the location of internment camps in the Slocan Valley south of Slocan Lake: *Slocan City, Bay Farm, Popoff,* and *Lemon Creek.* The maps were part of a handmade directory he compiled. Popoff is shown, *left,* with his hand-drawn map *below.* At *bottom left* is his map of Bay Farm. Lemon Creek is *below, right,* with its map. The camps contain some two-storey apartments as well as individual houses. The latter were sometimes shared by two families. There are also halls, bath houses, temples, and schools, all to create as normal an existence as might be possible under the difficult circumstances.

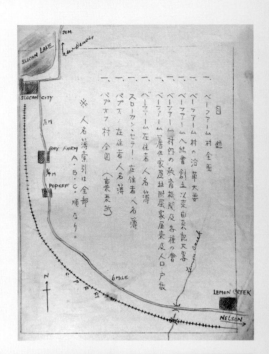

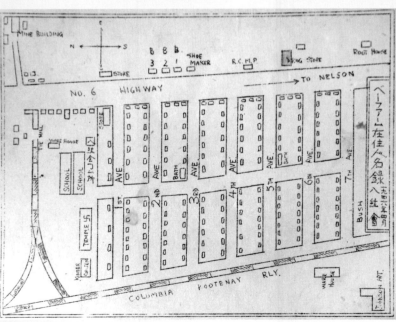

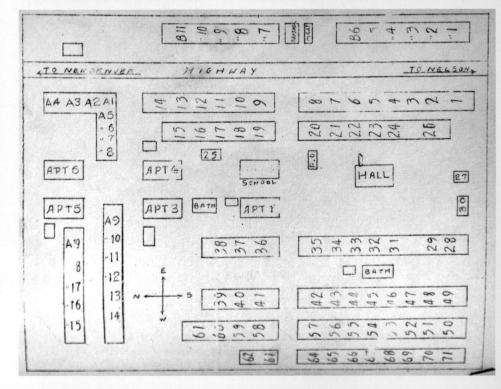

A Highway to Alaska

A road to Alaska, just like a railway, had been dreamed of for many years before the exigencies of war dictated that it should be built. And once the decision to build had been made, the road was completed in an extraordinarily short time. The 2,237 km from Dawson Creek, where it could connect with an existing road system, to Delta Junction, Alaska (160 km south of Fairbanks), where it could do likewise, was built in an astonishing 234 days, between 8 March and 28 October 1942. Road building was spurred on by the reports of Japanese landings on Kiska and Attu Islands in the Aleutian chain on 6 and 7 June. Construction took place from both ends of the road and met at what became called Contact Creek, on the British Columbia–Yukon boundary. The road when complete was far from perfect, as might be expected, but it did the job; American military supplies to Alaska started moving immediately. Upgrading of the road began right away and has continued, on and off, to the present day. Re-routing has shaved some 56 km off the Canadian portion of the highway.

The Alaska Highway's route was determined by the requirement to link a series of air strips, some of which are shown on MAP 834, *below*, that were part of another supply route—that of delivering Lend-Lease aircraft for the war effort to Russia via what was known as the Northwest Staging Route, for which rough landing places had been built in 1941 about every 150 km from Edmonton to Fairbanks. Nearly eight thousand aircraft flew this route, beginning in 1942, and were delivered to Russian flight crews in Fairbanks.

Although the road was built by the Americans and at their cost, six months after the war ended, as had been agreed in 1942, it was turned over to Canada.

MAP 834 (*below*).

A range of airports, air strips, and radio relay repeater stations, including some airports under construction, are shown on this 1943 map of the Alaska Highway; the British Columbia portion is shown here. *Dawson Creek* and the *British Columbia/Alberta* boundary line is at right, *Fort Nelson* is at centre, and the 60°N northern boundary of the province can be seen at left. Where a flight strip is complete, its configuration is shown as a mini-map within the circles, as a finding aid for pilots.

MAP 832 (*above*).

Published by the Alberta Government Travel Bureau about 1950, this map demonstrated that the Alaska Highway had to be approached from Alberta. The Hart Highway was completed north from *Prince George* in 1952, connecting British Columbia to the Peace River country—and the Alaska Highway.

MAP 833 (*below*).

Dated 6 February 1943, this air navigation map shows the *Approximate Location of Alaska Highway,* completed just months before, and the airway zone for aircraft and three *Landing Strips (Under constr).*

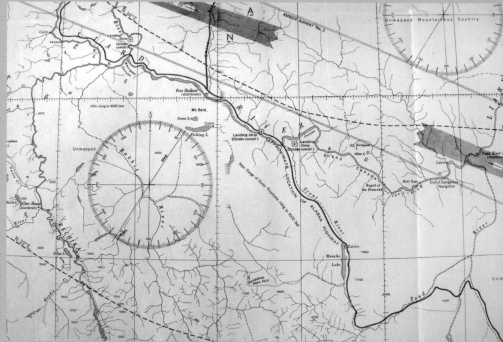

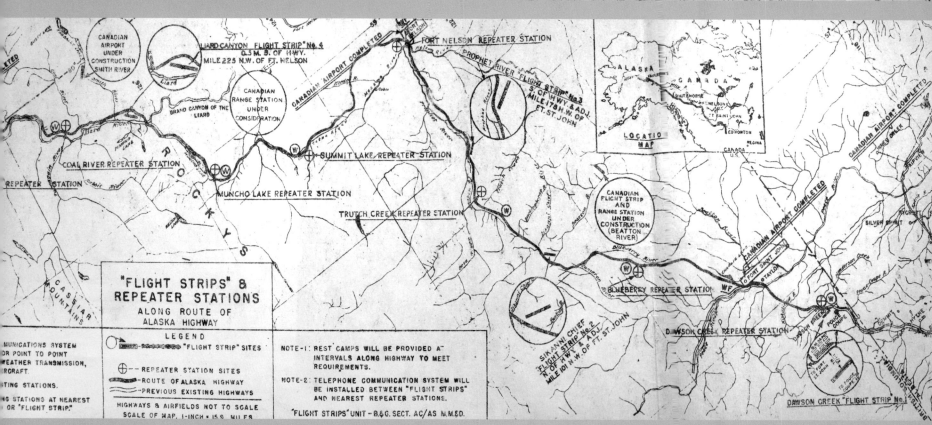

Post-War British Columbia

The Cold War

The dropping of atomic bombs on Nagasaki and Hiroshima in 1945 and the subsequent development of long-range bombers and intercontinental ballistic missiles (ICBMs) ushered in a new age of tension, this time between the Soviet Union and the United States. Civil defence organizations all over North America evaluated the location of daytime and nighttime populations and analyzed blast resistance of buildings. They also produced evacuation plans for their cities, to be implemented if a nuclear attack was considered imminent. This alarming map (Map 836, *right*) was Vancouver's effort.

Map 836 (*right*).
The evacuation plan for Vancouver and surrounding area, delivered to all households in 1957. Vancouver at the time was hampered by its peninsular nature and a paucity of river crossings. Civil defence preparedness was one of the reasons behind the building of the George Massey Tunnel, which opened two years later. All main roads were to become one way. The idea was to prevent streams of traffic from crossing each other. Residents of the pink and Lulu Island green areas were to escape using the Pattullo Bridge; the northern green area was to be evacuated via Broadway and the Barnet Highway; while the blue, brown, and white areas were to go to Horseshoe Bay and "wait for boats." Yellow-area residents likewise were to proceed to Steveston, where there might have been more boats, if they were lucky. The entire plan seems infeasible and was likely an attempt to contain panic and at least allow some to escape. Luckily it was never required.

Map 835 (*right*).
A Russian map of Victoria, 1978. The Russians produced an entire series of these detailed maps covering all of North America, and some were acquired from a retired Soviet army officer when the Cold War ended.

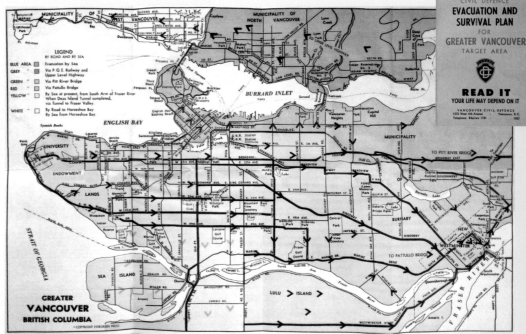

British Columbia's Railway

The province's government-owned Pacific Great Eastern Railway had run from Squamish to Quesnel since construction farther north was halted in 1922 (see page 282). After World War II, the government received several proposals to take over the railway and extend it to Alaska, but by now the province was loath to hand over its line to the Americans, and so, in 1949, in the economically revitalized later post-war period, Premier Byron "Boss" Johnson decided the government would complete the line to Prince George and then on to the Peace Country.

The line into Prince George, connecting with Canadian National, the nationalized Grand Trunk Pacific, was completed in late 1952; the first PGE train—a work train—rolled into the city on 12 September, followed by an official train on 1 November. Then, in 1954, construction began to reopen the closed section in West Vancouver—much to the horror of residents there—and extend it up the difficult side of Howe Sound to a connection at Squamish. The last spike of the southern section was driven on 10 June 1956, and the line opened with an official train (Map 837, *left, centre*) on 27 August. Two years later the line was completed to Dawson Creek and Fort St. John (Map 838, *left*).

Map 837 (*left, centre*).
Map and menu and dignitary identification ribbon from the official first train from *North Vancouver* over the new line along Howe Sound to *Prince George*, in August 1956. The map shows connections to *Fort St. John* and *Dawson Creek*, but these would not be completed for another two years.

Map 838 (*left*).
The red line on this map from about 1960 shows the PGE complete from *North Vancouver* to *Fort St. John* and *Dawson Creek*.

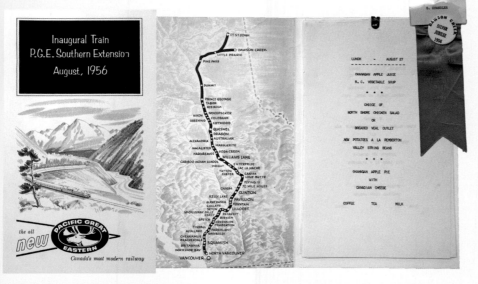

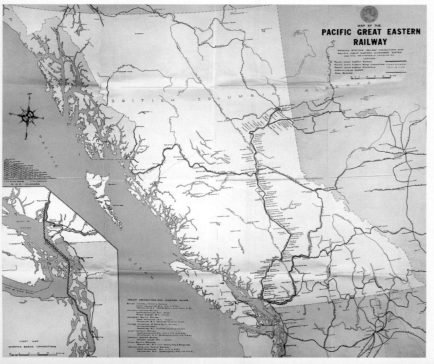

PACIFIC GREAT EASTERN RAILWAY

1095 West Pender Street, Vancouver, Canada. MUtual 1-3131

A MODERN RAILWAY SERVING THE NORTH

These extensions had been initiated by new premier W.A.C. Bennett, who had decided that the railway should reach into Yukon and Alaska. Further extensions northward were completed to Fort St. James in 1968 and under construction to Dease Lake when a decline in the price of asbestos and copper led to its demise. (The line was reopened in 1991 to a point 280 km south of Dease Lake.) Bennett did get to open a line to the north, however—the main line extension from Fort St. John to Fort Nelson in September 1971. A year later Bennett lost power; that same year the Pacific Great Eastern became the British Columbia Railway (and BC Rail after 1984).

A further extension was prompted by the potential for shipping the vast reserves of British Columbia coal to Japan. A 132-km extension to Tumbler Ridge was opened in 1983 and, unusually for a freight line, was electrified (MAP 842, *below, right*). The line was abandoned in 2003 after the mines closed. Coal is still shipped to Japan by Canadian Pacific from the Elk Valley mines via Roberts Bank, the superport opened in 1970.

A 37-km line to Roberts Bank (from an interchange with other railways) is all that remains of the British Columbia Railway; the rest was controversially sold and long-term leased to Canadian National in 2004. The Roberts Bank line had been the British Columbia Harbours Board Railway, built in 1970 to coincide with the opening of the port (see page 324).

MAP 839 (*left, top*).
The PGE's logo from 1965 to 1972 was this stylized map of British Columbia, seen here on the side of a caboose at the West Coast Railway Heritage Park in Squamish. Some unrepainted wagons with this logo survived until the 1990s.

MAP 840 (*above*).
This fine poster from about 1960 displays the PGE's freight diesel and passenger Budd cars. The railway acquired diesels in 1949 and was one of the first in North America to use them for freight. The Budd railcars were purchased in 1956.

MAP 841 (*left*).
A 1971 railway promotional booklet shows the north abundant with resources as justification for the PGE's northern lines. Here the extension to *Dease Lake* (the red line) is shown as under construction north of *Fort St. James*. The main line extends north to *Fort Nelson*.

MAP 842 (*below*).
The 132-km electrified line from Tumbler Ridge and the *Quintette Coal* mine to *Anzac*, the junction with the main line, on a 1984 blueprint map. It shows the location of the tunnels bored to cross the Continental Divide and the *Tumbler Ridge Townsite*. *Inset.* Number 6001, one of the six electric locomotives specially built for this line, preserved at the Prince George Railway and Forestry Museum.

ANZAC - TUMBLER RIDGE BRANCH LINE FIG I

Kitimat and Kemano

The power potential of the Nechako River Basin west of Prince George was revealed in 1930 after government surveyors had investigated and found a 2,600-foot difference in height between Tahtsa Lake and the Kemano River within a distance of about ten miles. The British Columbia government asked the Aluminum Company of Canada (Alcan) to investigate the power potential of the province in 1941, with a view to establishing an aluminum smelting industry on the West Coast. Power, of course, was critical, because of the immense amounts of cheap electricity required to make the smelting process economically viable.

The war sidelined this project, but it was renewed in 1947, and by 1949 the Tahtsa Lake–Kemano site was confirmed as the best location.

The flow of water through Tahtsa Lake and other lakes, part of the Nechako River system, was reversed by building the Kenney Dam, at the time the largest sloping, rock-filled, clay-core dam in the world. Some 300 feet of Nechako River water was held back at the dam, which raised the level of Tahtsa Lake 17 feet. The dam was completed in 1954.

To channel water from the lake down to the Kemano River, a 16-km, 7.6-m-diameter tunnel was bored, terminating at a generating station at Kemano. It took three years to bore the tunnel, and the project was made more difficult by the remoteness of the region, which had no railways or roads.

The Kemano site offered no room for the smelter, and so a decision was made to build it at the head of an adjacent fiord, where there was ample space for not only the smelter but also a townsite and harbour. This location could easily be connected to the existing road (Highway 16) and rail (the Prince George–to–Prince Rupert line). The separation of powerhouse and smelter necessitated construction of a 50-mile-long high-voltage transmission line between the two. The planned town of Kitimat was built to house workers at the smelter (MAP 845, *below*).

An expansion known as the Kemano Completion Project was approved in 1987 but then cancelled in 1995 because of environmental concerns. A small settlement at Kemano was removed by Alcan in 2000; most of the buildings were burned down as a fire training exercise.

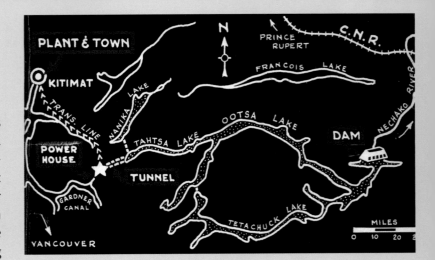

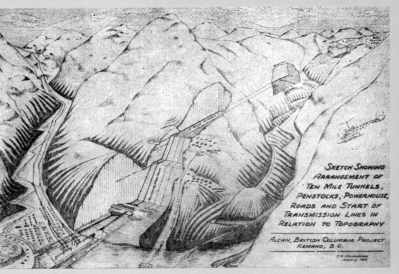

MAP 843 (*above, top*).
From a 1954 *Engineering Journal* comes this map illustrating the Kemano project in simplified form. The Nechako Canyon Dam, the Kenney Dam, raised lake levels for 250 km to the west and reversed the flow of the Nechako River.

MAP 844 (*above*).
A 1953 cutaway bird's-eye map of the Kemano tunnel and generating station. The water inlet at Tahtsa Lake is at right, and the transmission line to the Kitimat smelter runs up the Kemano River valley at left.

MAP 845 (*below*).
This was the 1952 master plan for the instant town of Kitimat, built to house the workers at Alcan's smelter, located just south of the town. The aerial photo (*right*), looking west, shows neighbourhood A-I as built.

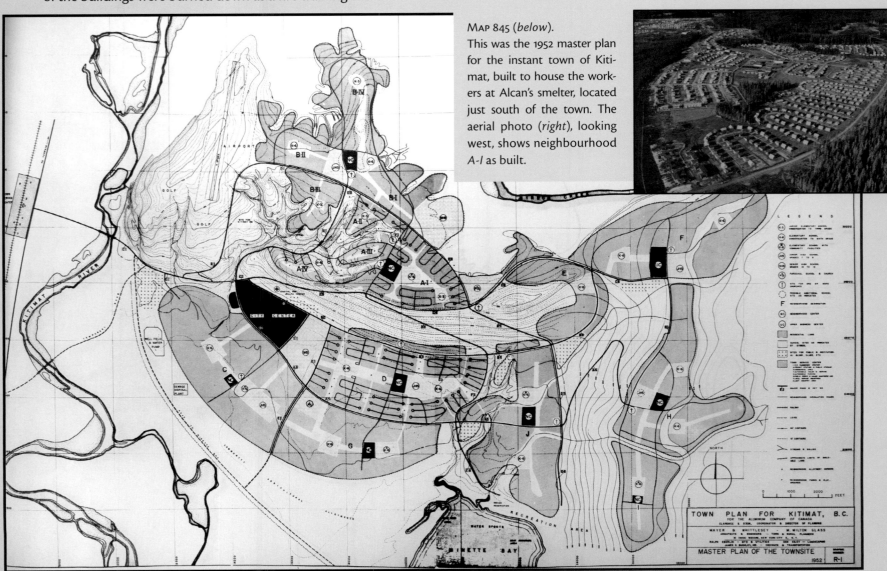

A Mountain of Copper

Mining has always been affected by metal prices, and nothing illustrates this better than Copper Mountain, south of Princeton, which began operations in 1920, with a smelter and mine closing down after a few weeks because of low copper prices. In 1925, prices having recovered, the mine reopened; this time it lasted until 1957. It was during the later part of that period that the intriguing maps of the mine shown here were produced. Copper mining resumed in 1972 and lasted until 1996, except for a period between 1993 and 1994. Now, as copper prices again climb, the mine has reopened once more.

MAP 847 (*below*).
Three dimensions reduced to two result in this huge cutaway map of Copper Mountain's underground deposits, with the surface topography resumed beyond that part cut out. Notice how the scale is given as a cube. The map, now preserved at the Princeton Museum, is about 3 m in width. *Inset* is the location key from this map. Note the *Hope–Princeton Road Uncompleted*.

MAP 848 (*below, bottom*).
This conventional map depicts the layout of the townsite and mine buildings about 1956.

MAP 846 (*below*).
Some mines, like Copper Mountain, produced three-dimensional maps like this using layers of sheet acrylic or similar transparent material. In effect it was a model of the underground combined with one of the surface. Here red has been used to represent proven ore bodies, seen descending through the layers of plastic; striped areas are inferred ore bodies. On the surface layer, moulded over a hot plate to represent the topography, are the buildings, here each a tiny model itself. The model was made in 1948–49.

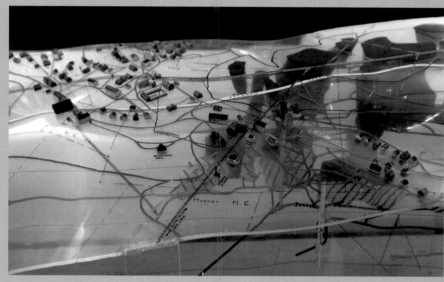

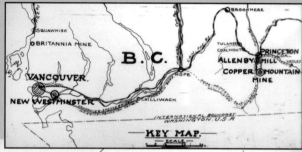

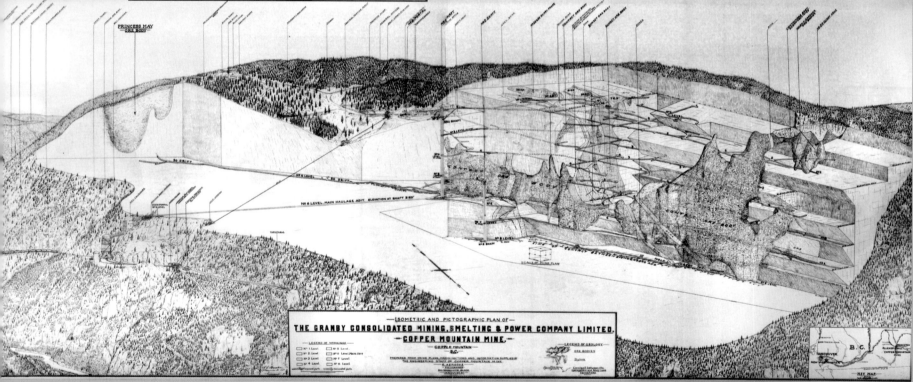

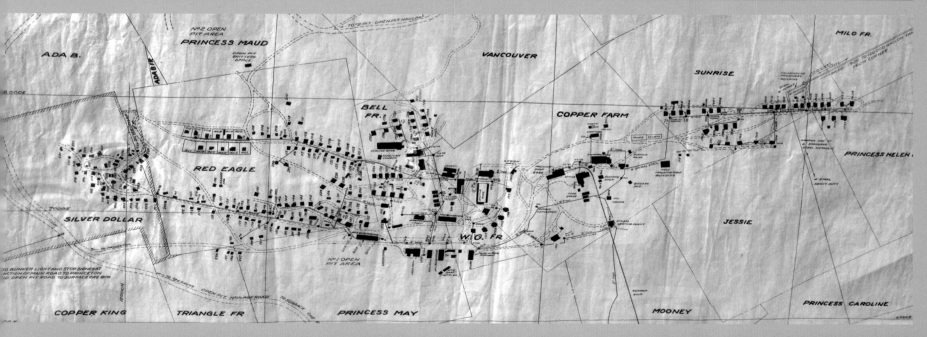

Left. The end of Ripple Rock: 9:31 am, 5 April 1958.

Map 849 (*right*).
Before (red) and after (black). The soundings (in feet) afterwards are much deeper than those before.

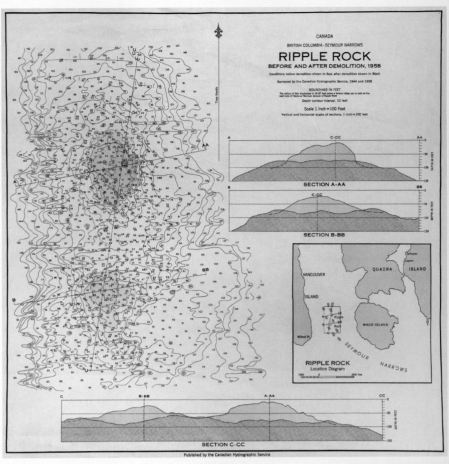

LAND LOST, LAND GAINED

Seymour Narrows, the most direct channel along the east coast of Vancouver Island, has a dangerous tidal race. Added to that, before 1958 the massive Ripple Rock lurked just beneath the waves right in the middle of the channel. Over the years it had been responsible for sinking some 119 ships.

In 1942 its removal was attempted using a series of smaller explosions to chip away at it, but this was not successful. Finally, in 1955 a concerted effort was begun to blow the rock up once and for all. It was quite an undertaking, involving drilling tunnels under the rock, which were then packed with explosives. The drilling took over two years, with seventy-five men working round the clock on three shifts. Finally, on the morning of 5 April 1958, everything was ready. The explosion was one of the largest non-nuclear blasts of all time. It had cost over $3 million, but Seymour Narrows was now a safer place.

Removal of land like this was far more expensive than adding to it. Over quite a long period the shores of Vancouver Harbour were added to, piecemeal, so the map today looks quite different from a hundred years ago, as does what remains of False Creek. The east end of False Creek was filled—much of it with garbage—by 1917, and Granville Island rose from a sandbar the year before. More recent filling has greatly reduced False Creek's size. The Fraser delta offered many opportunities for reclamation works, and some were carried out (see, for example, Map 790, *page 307*). Many others were proposed (see Map 608, *page 245*). In the 1950s yet more proposals were forthcoming (Map 853, *below, right*), including one, approved by the mayor and council of Richmond, to fill the southern part of Sturgeon bank, off Steveston, with garbage: the effect on the environment was clearly not a major consideration. Roberts Bank Superport is the one remnant of these grand plans, and even that is but a small portion of what was originally planned (Map 851, *below*).

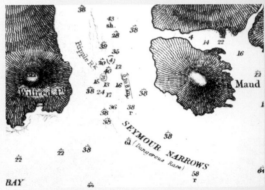

Map 850 (*above*).
Ripple Rock shown on an 1866 chart. *Seymour Narrows* is noted as a *Dangerous Race.*

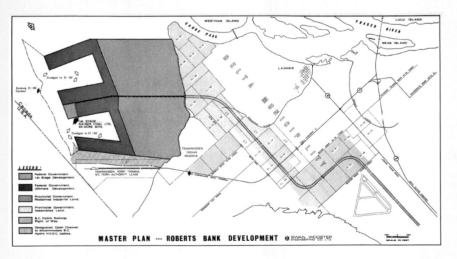

Map 851 (*left*).
The original plan for Roberts Bank, shown on a 1969 map. The project was first proposed to allow coal to be shipped from the Elk Valley to Japan. *1st Stage Kaiser Coal Ltd.* marks the part that proceeded. The project was originally intended as a federal one, but Premier W.A.C. Bennett expropriated almost 1,600 ha to prevent rail access and forced the federal government into a joint venture. It was opened on 15 June 1970—jointly by Bennett and Prime Minister Pierre Trudeau.

Map 852 (*left*).
This plan for reclaiming Burns Bog in Delta and creating a shipping basin, *Delta Port,* accessed from the *Fraser River,* was proposed in 1988.

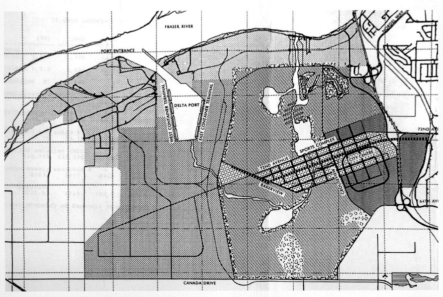

Map 853 (*below*).
A 1957 proposal to reclaim much of Sturgeon Bank by filling outward from *Sea Island* and *Lulu Island.* A similar but more restrained reclamation was considered and rejected in the 1970s to provide land for a third runway for Vancouver International Airport.

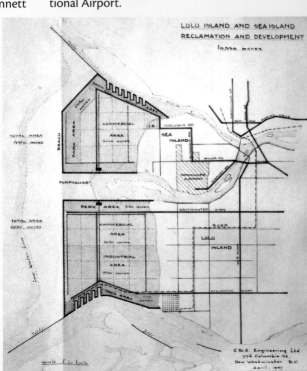

BRITISH COLUMBIA'S FLEET

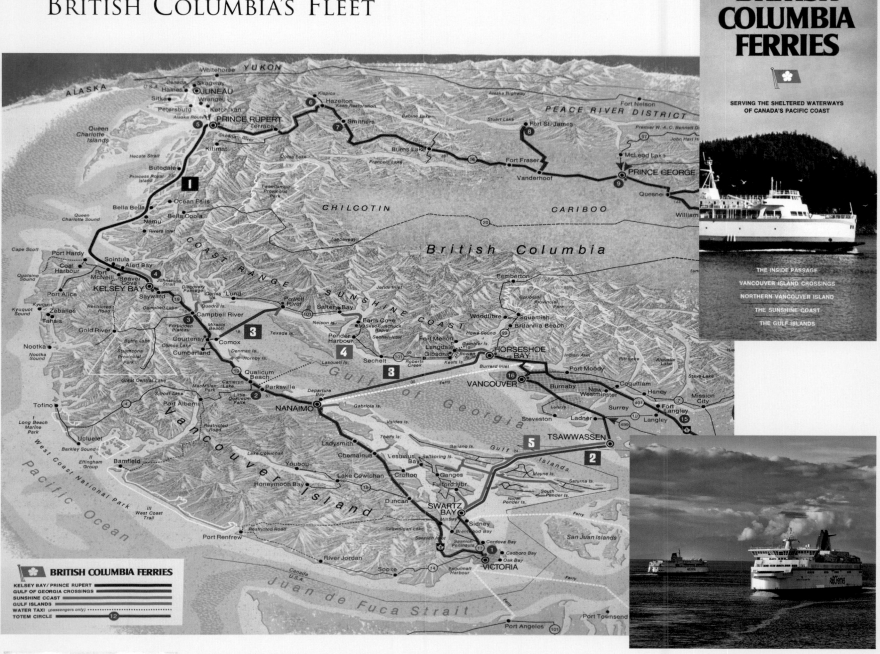

Map 855 (*above*).
There were many private ferry services in British Columbia before B.C. Ferries began taking them over. One was the Mill Bay Ferry, which ran across Saanich Inlet from *Brentwood* to *Mill Bay*, allowing vehicles to avoid the drive over the Malahat. This is a 1936 map and timetable card to hang in one's car.

British Columbia Ferries was the brainchild of Premier W.A.C. Bennett. After a 1958 strike at Black Ball Ferries and Canadian Pacific greatly inconvenienced Vancouver Island residents, Bennett, acting on advice from his friend, Captain Harry Terry of Northland Navigation, announced that the province was taking over the ferry business.

The project was given to his energetic highways minister Phil Gaglardi. New terminals and connecting roads were constructed—one at Tsawwassen, at the end of a long causeway jutting out into the strait, and another at Swartz Bay. Two new ships were commissioned, *Sidney* and *Tsawwassen*, and service began on 15 June 1960.

As B.C. Ferries grew, it took over other routes from private operators and added ships, both specially built and purchased from other jurisdictions. So it was with the *Stena Danica*, purchased from Sweden in 1973. She was refitted as the *Queen of Surrey*, and in 1980, after another refit, became the *Queen of the North* and was put into service on the Inland Passage route to Prince Rupert. Early in the morning of 22 March 2006 she ran aground on Gil Island, 135 km south of Prince Rupert, and sank, with the loss of two passengers. It was B.C. Ferries' worst accident.

Map 854 (*above*) with cover (*above, top*).
B.C. Ferries' extensive network depicted on an excellent hybrid map-bird's-eye published in 1971. The modern photo, *above*, shows the *Queen of Nanaimo*, having just left Galiano Island, slipping by the *Spirit of Vancouver Island*, which is about to enter Active Pass.

Map 856 (*below*).
An informational brochure with floor plan, cross-section, and photo for passengers on the ill-fated *Queen of the North* published in 1981 as the ship went into Inland Passage service.

MV Queen of the North

POWER FOR THE PEOPLE

Electricity, of course, caused a revolution in the way people lived, but generating it on a large enough scale to supply whole cities was never an easy task. Early power came from coal-fuelled, steam-powered generators. At suitable locations, hydro power was generated. Smaller companies merged; in 1897 West Kootenay Power & Light and B.C. Electric were created, and it was the latter that in 1961 became the British Columbia Hydro and Power Authority (B.C. Hydro).

B.C. Hydro, as a provincial crown corporation, was charged with fulfilling Premier W.A.C. Bennett's dream of cheap electric power to fuel industry and give the province a competitive edge, creating growth.

Bennett had been toying with the idea of harnessing the power of the Peace River but came to accept federally mandated power generation on the Columbia, the result of the Columbia River Treaty, first signed in 1961 but changed and not ratified until 1964. This was Bennett's "Two Rivers Policy"—develop both the Peace and the Columbia but export most Columbia power to the United States. Under the 1961 terms Canada would have been entitled to 50 per cent of the downstream power generated in the United States; instead Bennett negotiated a controversial large up-front cash payment—which he could use to build his Peace River dams.

The United States had tried to control flooding and provide power and water for irrigating interior Washington State since the construction of two large projects on the Columbia—the Bonneville Dam, completed in 1937, and the Grand Coulee Dam, completed in 1941. In 1944 the International Joint Commission (IJC) established the Columbia River Engineering Study Group to determine the optimum use of the entire Columbia Basin on both sides of the border. Its Columbia Basin Study took fifteen years to complete. The amount of water storage required to reduce peak river flows to manageable levels were calculated. On the American side of the border existing riparian rights prevented construction of more storage, but there was no such constraint in Canada. A plan for dams and reservoirs was drawn up by General A.G.L. McNaughton, head of the Canadian IJC. The plan included the Murphy Creek or Low Arrow Dam, near Trail, which would have raised water levels in the Arrow Lakes only to the normal high water level. But a reservoir would have been created in the Upper Columbia and Kootenay Valleys.

Bennett vetoed the flooding of that valley, insisting that the Arrow Lakes valley be flooded instead. It seems that he had hoped this would kill the Columbia River Treaty and allow development of his Peace River project. But the federal government, anxious not to create further delays, accepted Bennett's veto and overrode McNaughton's plan. As a result the High Arrow Dam—renamed the Hugh Keenleyside Dam after the chairman of B.C. Hydro—was built just west of Castlegar. It raised the lake level 12 m, displaced 2,000 people, and flooded 8,000 ha of arable land, and it remains a source of contention to this day.

MAP 857 (*below*).
The power-generating facilities on *Bonnington Falls,* on the Kootenay River just west of *Nelson,* are typical of the smaller-scale plants constructed early in the twentieth century where conditions allowed. The City of Nelson completed its power plant in 1907 after battling with West Kootenay Power & Light, which also completed a power plant here that year. One of the company's directors filed a mineral claim on the Nelson site to prevent the city from building, but Nelson fought this in court. West Kootenay Power & Light's plant is shown in the photo, while the map shows the *Transmission Line* from the *City of Nelson Power House* to Nelson.

MAP 858 (*below*).
The electricity transmission network and locations of hydro, diesel, and steam generating plants in the southern interior of British Columbia in 1953.

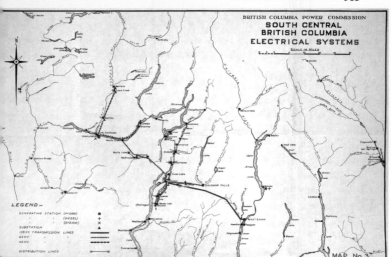

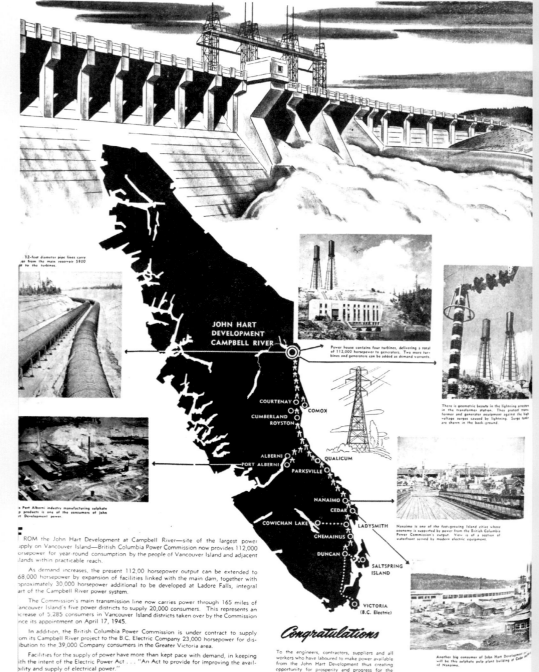

Other dams constructed to satisfy the terms of the Columbia River Treaty were the Duncan Dam, completed in 1967, which almost doubled the length of Duncan Lake (north of Kaslo), and the Mica Dam, completed in 1973, 135 km north of Revelstoke. Other dams have been added to the Columbia system for power generation (and power stations have been added to dams, such as the Mica, initially only intended for storage) but these are not part of the treaty. They include the Revelstoke Dam, completed in 1983.

The Peace River project that W.A.C. Bennett was so keen on grew from a 1956 proposal by a private company controlled by Swedish Electrolux millionaire Axel Wenner-Gren to develop a huge tract of land in the northeastern part of British Columbia, almost 10 per cent of the total land area of the province. It was an almost ridiculous idea that included a railway to the Yukon and a 250-km/hr monorail—all of which predictably came to nothing—but did include a huge hydroelectric project. This Bennett liked, and he realized it gave him a good negotiating position with the federal government in relation to the Columbia scheme. It would prevent, he said, in typical style, the development of the province from being "held back while the U.S. and Ottawa hold pink teas."

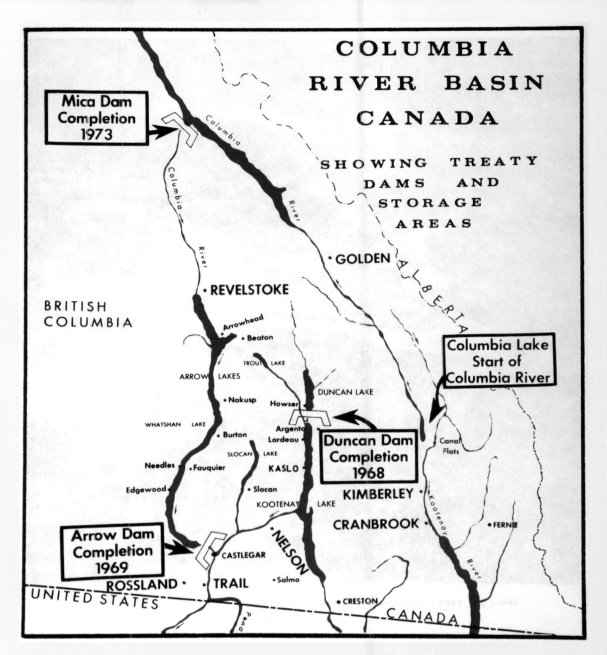

Map 859 (left).
By 1949 hydro projects had become larger in scale. This graphic ad was placed that year by the B.C. Power Commission (which, grouped with B.C. Electric, became B.C. Hydro in 1961). It shows the location and transmission lines from the John Hart Generating Station, which came online in 1947 but was not fully completed until 1953. It was built on the Campbell River. Along with other plants built in 1945, 1955, and 1958, the Campbell River hydroelectric system today supplies about 11 per cent of Vancouver Island's electrical needs.

Map 860 (right, top).
A 1965 summary map of Columbia River Treaty dams and storage areas published by B.C. Hydro in a progress report.

Map 861 (right).
An engineer's plan for the (High) Arrow (Hugh Keenleyside) Earth Dam is here superimposed on an aerial photo, an effective way of relating the future to what exists on the ground beforehand. The Columbia River flows from left to right, and the enlarged Lower Arrow Lake, shown in a modern photo, inset, will form at left. The plan extending in front of the dam is for a clay blanket, required to be placed on top of the gravel base. The railway is to be relocated, and the existing railway line can be seen on the aerial photo. The map-photo was published in a report on the dam's progress in 1965.

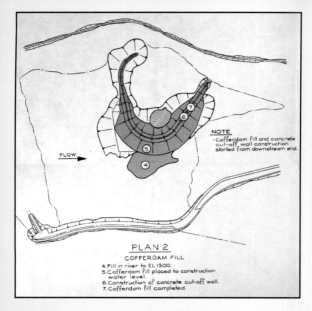

PLAN 2
COFFERDAM FILL
4. Fill in river to EL.1300.
5. Cofferdam fill placed to construction
   water level.
6. Construction of concrete cut-off wall.
7. Cofferdam fill completed.

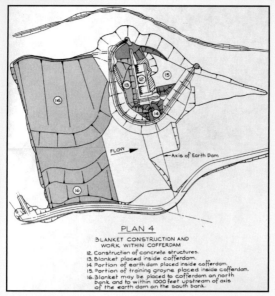

PLAN 4
BLANKET CONSTRUCTION AND
WORK WITHIN COFFERDAM
12. Construction of concrete structures.
13. Blanket placed inside cofferdam.
14. Portion of earth dam placed inside cofferdam.
15. Portion of training groyne placed inside cofferdam.
16. Blanket may be placed to cofferdam on north
    bank and to within 1000 feet upstream of axis
    of the earth dam on the south bank.

MAP 862 (*left* and *below*, five sequential maps).
The location chosen for the High Arrow Dam presented difficulties because the valley floor was composed
of only gravel rather than bedrock. This problem was
solved by placing a clay blanket around the site. These
maps, five of nine in total, show the sequence of construction of the Arrow (Keenleyside) Dam. The main
flow control structure was built within a coffer dam on
the north side of the river (Plan 2); clay blankets were
laid at all stages to create a suitable site for the dam
on the gravel valley surface. The river was then diverted to the north side of the river, through the control
structure, while the dam—composed of earth fill—was
completed.

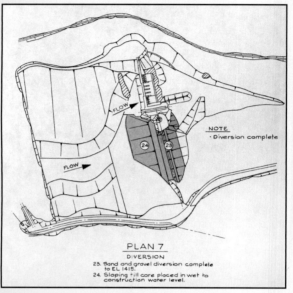

PLAN 7
DIVERSION
23. Sand and gravel diversion complete
    to EL.1415.
24. Sloping fill core placed in wet to
    construction water level.

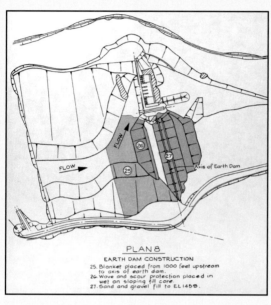

PLAN 8
EARTH DAM CONSTRUCTION
25. Blanket placed from 1000 feet upstream
    to axis of earth dam.
26. Wave and scour protection placed in
    wet on sloping fill core.
27. Sand and gravel fill to EL.1458.

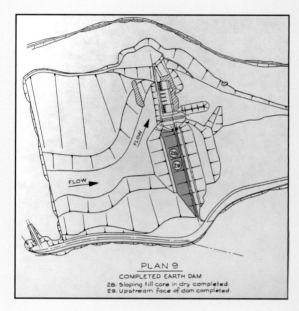

PLAN 9
COMPLETED EARTH DAM
28. Sloping fill core in dry completed.
29. Upstream face of dam completed.

In 1961 Bennett nationalized B.C. Electric, which was a private company, because it had
refused to agree to purchase Peace River power,
and created B.C. Hydro, though the takeover was
challenged in the courts, and he had to pay more
than he had hoped. B.C. Hydro was instructed
to proceed with building an enormous dam near
Hudson Hope, which would create a lake so large
Bennett bragged that it would likely change the
climate of northern British Columbia. The Portage Mountain Dam—which was, at the instiga-
tion of the lieutenant governor, renamed the
W.A.C. Bennett Dam just before Bennett himself
released the last 80 tons of earth to complete it
in September 1967—was indeed a major project.
At the time it was the largest earth fill dam in
the world, and it created the largest artificial lake
in the world—named Lake Williston after Ray
Williston, Bennett's trusty minister of lands and
forests. Electricity from the dam came onstream
the following year, with Bennett "pushing the
button" in front of three thousand spectators
crammed into the underground power station.
Another dam, the Peace Canyon Dam, this time
made of concrete, was completed in 1980 23 km
downstream.

Today power from Bennett's "two riv-
ers" provides 64 per cent of all the electricity
generated in British Columbia.

MAP 863 (*below*).
Flood control was also necessary on the Fraser River. In 1948 floods had inundated considerable areas of
the Lower Fraser Valley and cut off railways and roads. The Fraser River Board was created to find ways
of controlling flooding. This map shows possible diversion routes that could be created to control Fraser
River runoff in times of flood. It appeared in a 1958 report from the board. It is interesting because the route
embodies some that were proposed previously: *Route 1* appears on an 1859 map by Royal Engineer Colonel Richard Moody (MAP 194, *page 70*) as a possible channel to access the new capital, *New Westminster*,
without negotiating the difficult delta. *Route 3* is the route taken by Hudson's Bay Company men up the
Nicomekl River and down the *Salmon R.* to reach *Fort Langley* from the south and was the route taken in
1824 by the Hudson's Bay Company's James McMillan during the initial reconnaissance of the location (see
page 49). *Route 4* was the route of a proposed canal linking *Port Moody* with *Pitt River* in 1910 (MAP 605,
page 245), again to enable large ships to access the Fraser without passing through the delta.

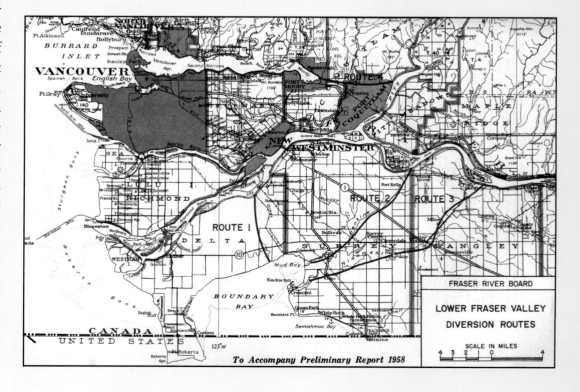

FRASER RIVER BOARD

LOWER FRASER VALLEY
DIVERSION ROUTES

SCALE IN MILES

To Accompany Preliminary Report 1958

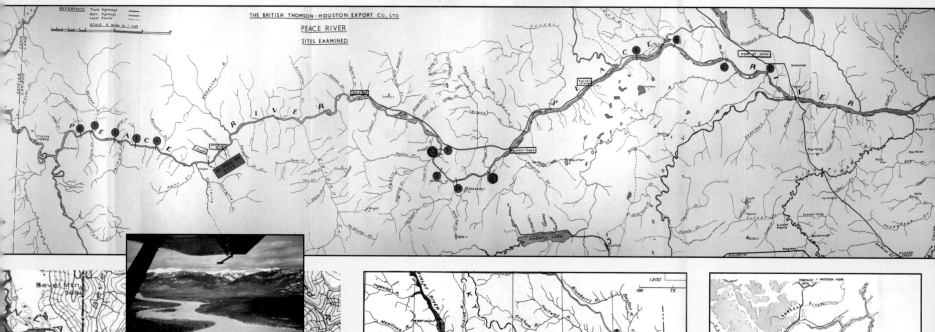

PEACE RIVER

SITES EXAMINED

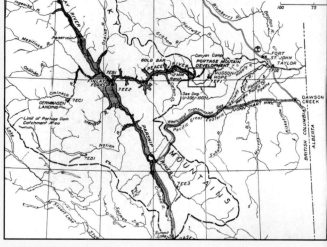

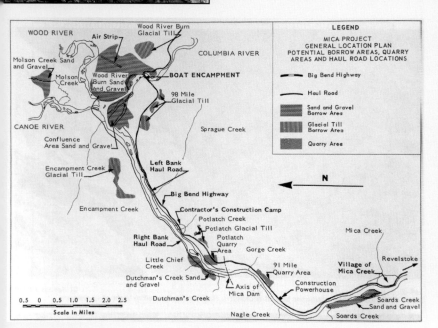

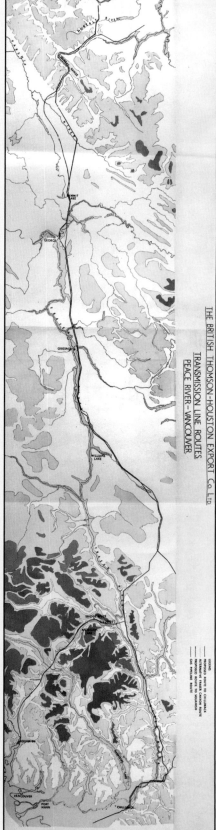

MAP 864 (*above, top*).

From a Wenner-Gren report in 1960 comes this map showing the location of sites considered for dams on the Peace River in British Columbia. *3A* is the one ultimately chosen and is today the location of the massive W.A.C. Bennett Dam. *Finlay Forks* is at left, *Fort St. John* at right. Of interest is one of the other sites considered, *C*, just west of Fort St. John. Site C, as it is named, moved slightly east to include the *Moberly R.* in its catchment, is the proposed location of a future third dam on the Peace River.

MAP 865 (*above, left*).

Finlay Forks, at the confluence of the Peace (right), *Finlay* (left), and Parsnip (bottom) as it was before the creation of Lake Williston behind the W.A.C. Bennett Dam. The photo, *inset,* is an aerial view, looking northeast down the Peace, from B.C. Hydro files.

MAP 866 (*above, centre*).

This 1961 map shows the expected reservoir size and shape after project completion. *Portage Mountain Development* is the dam site.

MAP 867 (*above, left*).

This engineer's map, published in 1961, shows the proposed location of the materials that would be required to construct the earth fill W.A.C. Bennett Dam. In a project of superlatives, fill was even moved from a moraine to the dam by what was said to be the world's longest unbroken conveyor belt. *Inset* is a photo of the dam from the west side.

MAP 868 (*right*).

The relative remoteness of the Peace River project required high-voltage transmission lines, and the route for these had to be determined. This is a map of their route from *Portage Mountain*, at top, as far as *Chilliwack*. The map is from the 1960 Wenner-Gren report.

MAP 869 (*left*).

The Mica Dam, another of the Columbia River Treaty dams, is at the Big Bend of the Columbia. It was completed in 1965. This engineering map shows the location of the fill area that would be used to provide materials for the dam, which is one of the largest earth fill dams in the world. Behind it is Kinbasket Lake.

MAP 870 (*above*).
The cover of a government map of public carriers—buses and trains, ships, and planes—in British Columbia, published in 1950.

ROADS TO FREEWAYS

With a few exceptions, like the war-essential extension of the road to Prince Rupert—and, of course, the Alaska Highway—road building in British Columbia was on hold for the duration of World War II. When the war ended, there was also a brief recessionary period, and it was anticipating this that Premier John Hart announced in 1945 a spending program that included completion of the Hope–Princeton road and the building of a link to the Peace River to finally connect the province's orphan region directly to the coast. The latter was complete by the end of 1949, but the Hart Highway, named after the premier, was not finished until 1952.

The federal government resolved to fund a road across the width of Canada in 1948, and construction began two years later on the Trans-Canada Highway. One of the first sections was a considerable upgrade to the road through the Fraser Canyon, but equally important for British Columbia was the decision to build the direct road from Revelstoke to Golden, through Rogers Pass, opened in 1962. The freeway section of Highway 1 in the Lower Fraser Valley, connecting into Vancouver over the new Port Mann Bridge, opened in 1964.

In the Vancouver area, where population and automobile growth was greatest, many other new roads and upgrades were built. The Deas Island Tunnel, the brainchild of Ladner resident George Massey, for whom the tunnel would later be renamed, was completed, with its approach from a new Oak Street Bridge, in 1959 (MAP 884, *page 332*), and officially opened by Queen Elizabeth II. This new route permitted the location of the new B.C. Ferries' terminal at Tsawwassen with the addition of a short connector, Highway 17. The road to the border, where it connected with the under-construction Interstate 5, was completed in 1962.

A new Second Narrows Bridge was completed in 1960 and a North Shore connection to it in 1962; farther west, in West Vancouver, an existing two-lane Upper Levels Highway, built in the 1950s, was upgraded to freeway standard by 1975 all the way to Horseshoe Bay and its ferry connection.

The East–West Connector road, Highway 91, was completed in 1989 in Richmond to access the Alex Fraser Bridge, completed three years before, and a connection to Highway 99 south of the bridge was finished at the same time.

Continued on page 334.

MAP 871 (*above*).
The road north of *Horseshoe Bay* is shown as *under construction (open late 1958)* on this 1958 road map, but the *PGE Ry* here was completed two years before. *Inset.* A Pacific Great Eastern ad published in a centennial year booklet in Squamish.

MAP 872 (*left, centre*).
This 1945 government map shows the proposed route of the Hart Highway north of Prince George as a dashed red line. The Peace River country was finally connected to the rest of British Columbia—without having to drive through Alberta—in 1952.

MAP 873 (*below, left*), MAP 874 (*below, centre*), and MAP 875 (*below, right*).
The intended route of the Hope–Princeton highway shown as a *Proposed Rd* on a 1940 road map (MAP 873) and just a dashed double red line on a 1931 map (MAP 874). Parts of the road at each end had been roughly constructed in the 1930s, but the interruptions of Depression and war foiled its completion. The road was officially opened on 2 November 1949, and the government published this ad (MAP 875), complete with a map showing the new through route to the Okanagan.

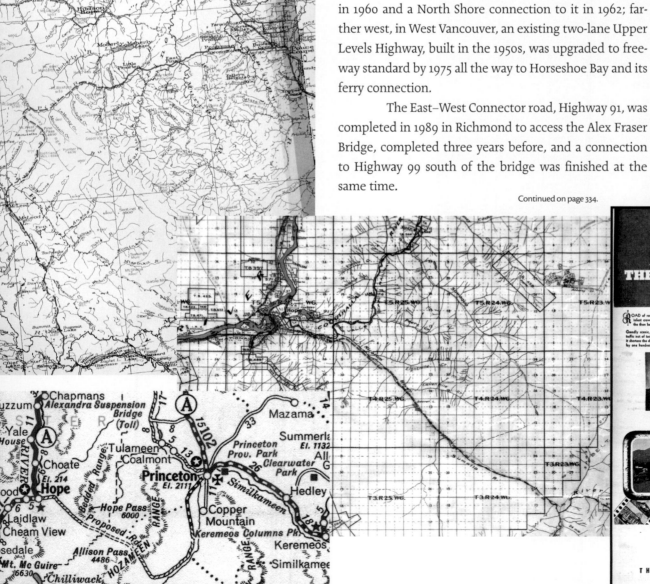

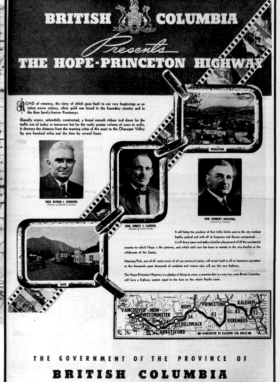

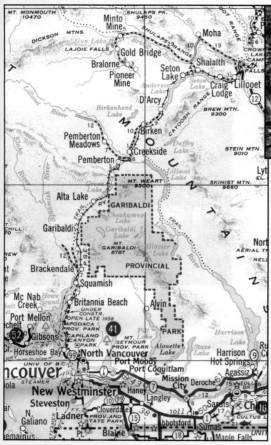

Map 876 (*left*).

The road between West Vancouver and *Lillooet* was piecemeal right up to the opening of the Duffey Lake Road north of *Pemberton* in 1990. On this 1958 government road map the *Howe Sound* section of the road is *under constr.*; other existing sections were gravel. The section between *Britannia Beach* and *Squamish* remained gravel until the early 1970s. The gap between *Shalalth* and *Lillooet* is particularly interesting because, like the Hope–Princeton before 1949 and Revelstoke–Golden before 1952, vehicles could be shipped by rail. The photo (*above*) shows cars being towed by a Pacific Great Eastern railcar along the shore of *Seton Lake* about 1948.

Left, centre. The cover of this government road map, published in British Columbia's centennial year, showed Captain George Vancouver contemplating his eponymous city.

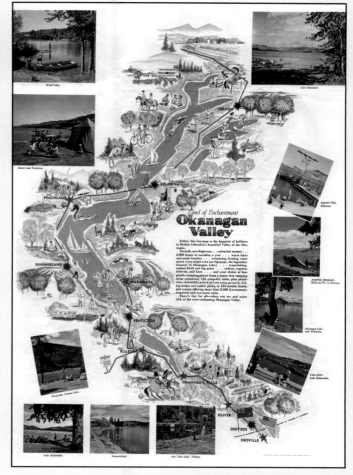

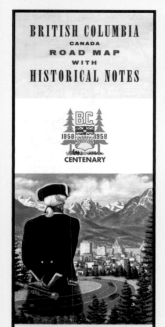

Map 877 (*right, top*).

A fine illustrated tourist road map of the Okanagan Valley, published in 1957. The road crossed the lake at *Kelowna* by ferry until 1958, when it was replaced by a unique-in-Canada floating bridge modelled on one crossing Lake Washington in Seattle. It was replaced by the new, conventional William R. Bennett Bridge in 2008.

Map 878 (*right*).

The 1958 centennial map also showed the road between *Revelstoke* and *Golden* over Rogers Pass as under construction—the dashed red line. Here also was a rail link operated during the winter months when the Big Bend route was closed.

Map 879 (*below*).

The federal government published this road map of the new Trans-Canada Highway now complete across Rogers Pass in 1962. Although announced as complete across Canada and officially opened—in Rogers Pass—in 1962, a number of sections outside British Columbia were not in place until 1971. Construction had begun in 1950 following passage of the Trans-Canada Highway Act in 1949. The photo shows construction of the Trans-Canada in the Fraser Canyon in 1952. The existing road is being widened and improved.

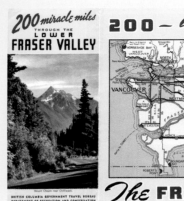

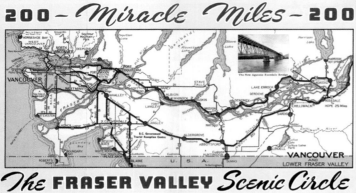

200 – *Miracle Miles* – 200

The **FRASER VALLEY** *Scenic Circle*

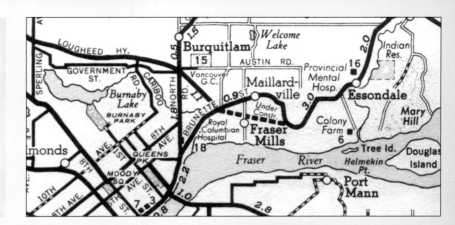

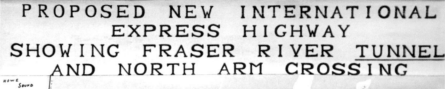

PROPOSED NEW INTERNATIONAL
EXPRESS HIGHWAY
SHOWING FRASER RIVER TUNNEL
AND NORTH ARM CROSSING

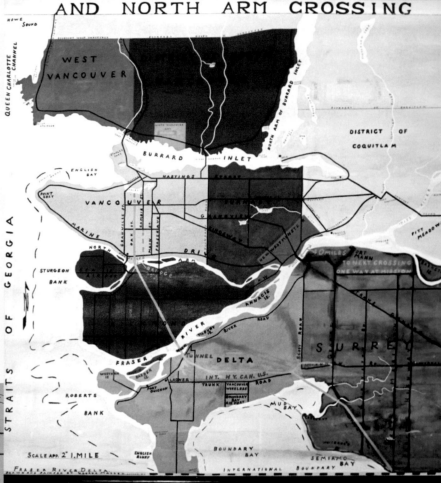

Map 880 (*above, left*).

The pre-freeway, pre–Deas Island Tunnel and Port Mann Bridge road system of the Lower Fraser Valley is shown on this 1957 tourist map.

Map 881 (*above, right*).

A 1954 road map shows the extension of the *Lougheed Hy.* shown as *Under Constr.* through Coquitlam; today that section of the highway parallels Highway 1.

Map 882 (*below*).

George Massey advocated the building of a tunnel to replace the Ladner ferry at Woodward's Landing for many years after learning how the Maastunnel in Rotterdam, completed in 1942, had been built using a sunken tube method. His initial map, dated 3 March 1947, locates a *Proposed Tunnel* at *Woodwards* connecting No. 5 Road in *Richmond* to the *Ladner Trunk Road* in *Delta* and then continuing south to the U.S. border via one of two possible connectors near and across *Mud Bay*. Another *Proposed Bridge or Tunnel* is located where the Oak Street Bridge was completed in 1957.

Map 883 (*right*).

As Massey developed his ideas, he painted this large wall map that he used to promote his tunnel at public meetings. The route of the *Int. Hy. Can. U.S.* (in yellow) follows the route taken by Highway 99 except for the shortcut across *Mud Bay*. The photo (*right*) shows George Massey with this map at a public meeting about 1952.

Map 884 (*right*).

On 21 May 1959, two days before the *Deas Island Tunnel* opened, this map appeared in local newspapers. The tunnel opened on 23 May but was officially opened again on 16 July by Queen Elizabeth II. At that time the new freeway was incomplete; several intersections remained north of the tunnel and the freeway connected to the *Ladner Trunk Road* about 7 km south of the tunnel. The *Oak St. Bridge* was open, having been completed in 1957. The rest of the freeway to the U.S. border was opened in May 1962, though even then the crossing of the Great Northern (now BNSF) railway line remained at grade. Massey died the year after his tunnel opened, but in 1969 it was officially renamed the George Massey Tunnel. Note the *Rejected Route,* almost identical to the Highway 91 route completed in 1989 across the Alex Fraser Bridge, which opened in 1986.

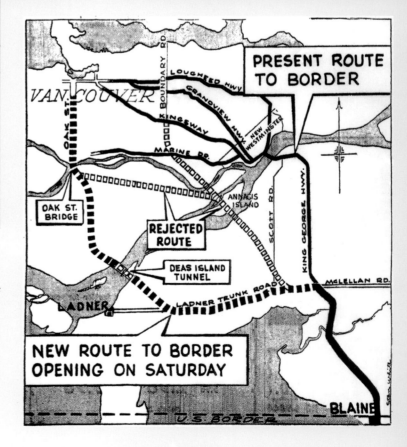

MAP 886 (*below*).

This 1963 road map shows the path of the Highway 1 freeway, complete with interchanges, through Vancouver, Burnaby, and Coquitlam as a double dashed red line *Under Constr. Open Late 1963*.

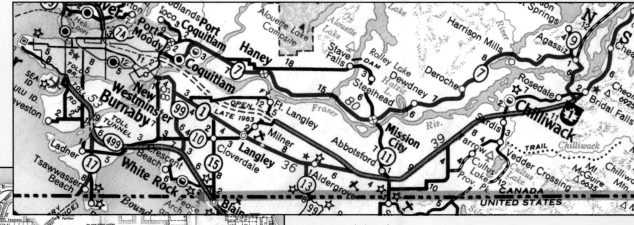

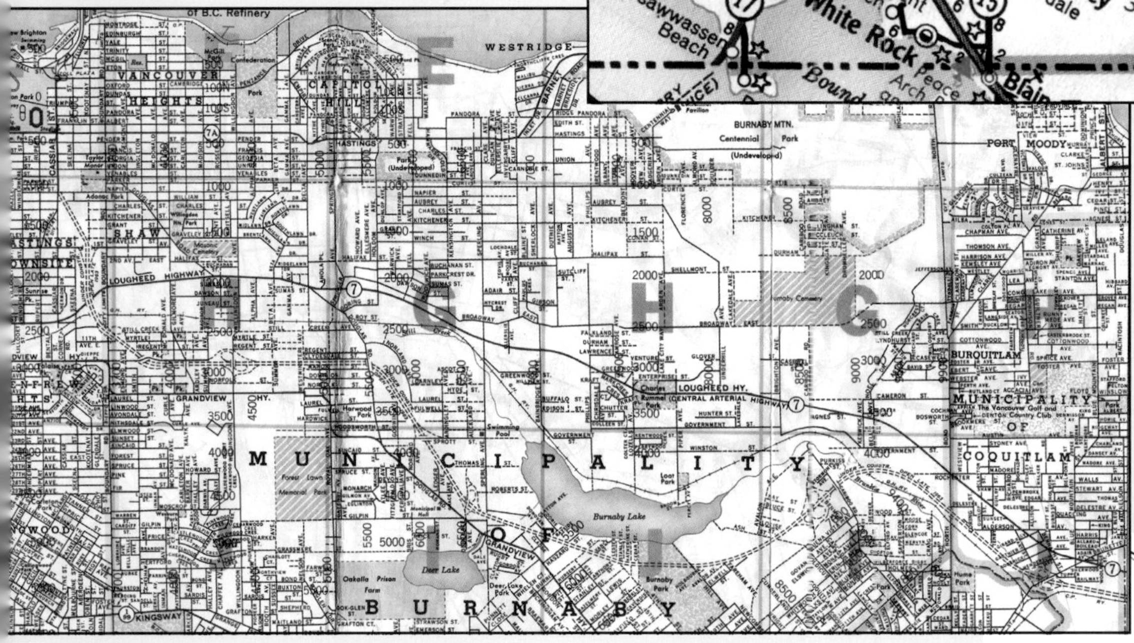

MAP 885 (*above*).

The new Lower Fraser Valley freeway, Highway 1, is complete to near Fort Langley on this 1963 road map. Its westward continuance across the Port Mann Bridge and into Vancouver would be completed the following year. Construction had begun in 1959 and was slated for completion in 1962, but unexpected problems with the bridge footings, plus some labour issues, delayed completion. The bridge and the freeway were officially opened on 12 June 1964 by Premier W.A.C. Bennett and his highways minister Phil Gaglardi, who then proceeded to drive at high speed all the way to Chilliwack.

MAP 887 (*right*).

This was the proposed freeway system for Vancouver published in a 1964 report. In its essentials, except for the portions in the City of Vancouver, it is not much different from what actually happened. Highway 1 crossed the Second Narrows to a North Shore freeway, the Upper Levels Highway; Highway 99 fed into the city at the Oak Street Bridge; and the east–west and southern connectors to Highway 99 on the yellow-coloured highways on this map were simply moved south to Richmond, where the East–West Connector, Highway 91, was completed to the Alex Fraser Bridge in 1989 (though one traffic light remained), three years after the bridge itself opened.

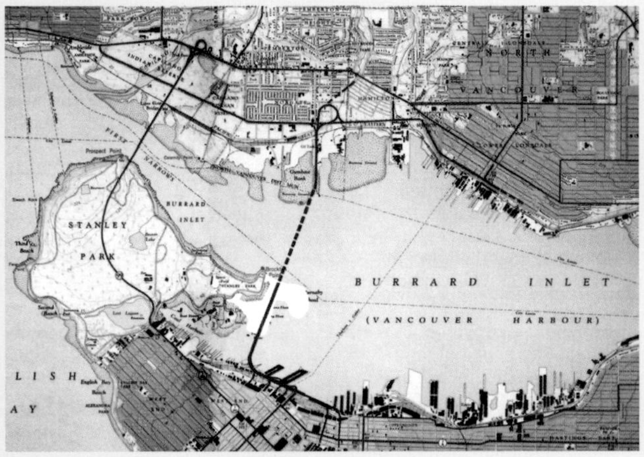

MAP 888 (*left*).

The proposed Third Crossing of *Burrard Inlet*, shown on a map created in 1963. A bridge carried the freeway to *Brockton Point* and then into a tunnel under the harbour. Brockton Point was to be augmented with (saleable) reclaimed land, shown in white. One proposal at the time even included an air strip here.

MAP 889 (*below*).

The controversial proposed freeway system for downtown Vancouver, published in the Vancouver Transportation Study of October 1968. Ultimately killed by citizen opposition, the system included a waterfront freeway approach to the Third Crossing and the so-called Chinatown freeway (top right), which would have irrevocably changed the character of Vancouver's Chinatown. One part of this system was built, however: the two sections of the Georgia Viaduct, completed in 1972 and also shown on this map, connecting with *Georgia* and *Dunsmuir* Streets.

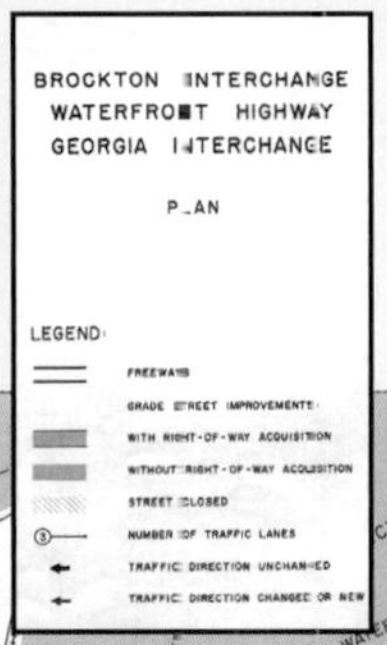

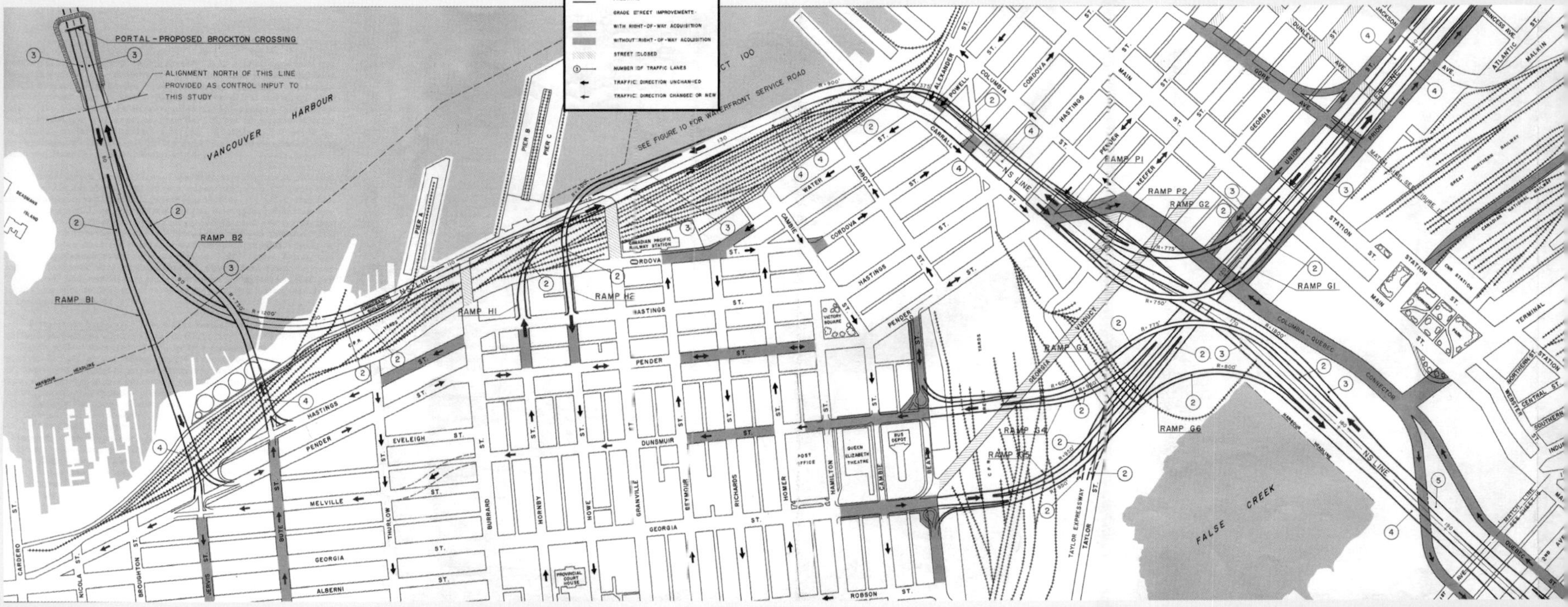

A long-dreamed-of but extremely difficult road was completed through the Coquihalla Valley in 1986, just in time for Expo 86; it ended temporarily at Merritt but the following year was completed through to Kamloops, allowing drivers to bypass the Fraser Canyon if they wished. The Okanagan Connector, Highway 97C, from Highway 97 to the Okanagan Valley near Kelowna, was completed in 1990.

On Vancouver Island a completely new freeway-standard road was constructed between Parksville and Campbell River between 1996 and 2001. This Inland Island Highway now forms part of an upgrade of Highways 1 and 19 connecting Victoria with Port Hardy.

In Vancouver, the Albion Ferry across the Fraser at Fort Langley was replaced by the Golden Ears Bridge and its connecting roads in 2010, and work continues apace on a new Port Mann Bridge and extensive upgrades to the freeway into Vancouver to cope with a considerable increase in commuter traffic; completion is scheduled for 2013.

Map 890 (right).
This government poster announced the opening of the *Coquihalla Highway* in 1986. Originally toll booths were placed near the Coquihalla Summit, but all tolls were removed after a surprise announcement in 2009. The map on the poster shows the location of the *Toll Booth*, roughly equidistant from *Hope* and *Merritt*. The map also shows *Phase 2 Open 1987* in white between *Merritt* and *Kamloops*, and *Phase 3* to the Okanagan Valley as a dotted black line; this was the Okanagan Connector, completed in 1990.

Below, left. The new Port Mann Bridge, on 2 September 2011, halfway to completion, dwarfs the 1964 bridge behind it.
Below, right. Taken on 14 June 2010, this photo shows the Albion Ferry *Kulleet* in its last weeks of operation, having been replaced by the Golden Ears Bridge. The ferry operated from 1957 to 2010. The sign was part of an abortive campaign to save the ferry.

THE COQUIHALLA
... a whole new point of view in British Columbia

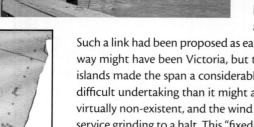

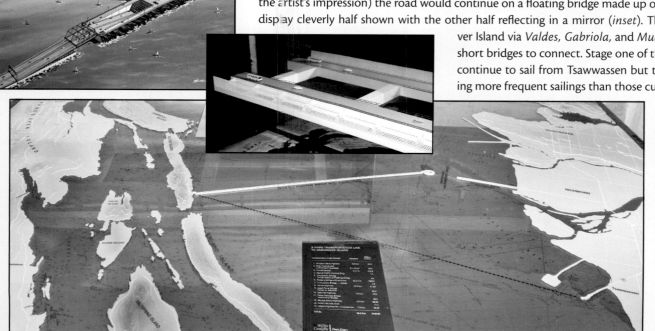

Map 891 (below, bottom), with detail (left).
Perhaps the most interesting transportation proposal in British Columbia in many years was this idea for a "fixed link"—a bridge-tunnel across the Strait of Georgia, linking Vancouver Island to the mainland. Such a link had been proposed as early as the 1870s, when the western terminus of the transcontinental railway might have been Victoria, but the crossing was much farther up the coast, opposite Bute Inlet, where islands made the span a considerably shorter distance. In the southern strait location the link is a far more difficult undertaking than it might at first appear, for the water there is deep, the current fast, the bedrock virtually non-existent, and the wind and waves at times a challenge, enough even to periodically bring ferry service grinding to a halt. This "fixed-link" was first proposed by Pat McGeer, a provincial cabinet minister in the Bill Bennett government of 1975–86 and a respected neuroscientist. A call for proposals from engineering firms in 1980 led to the construction of this model of the most feasible option, which was exhibited in the British Columbia pavilion at Expo 86. The cost of the project was estimated in 1985 to be $2.144 billion. A causeway similar to the one at the Tsawwassen ferry terminal was to be constructed from the end of Steveston Highway in *Richmond*. This would lead to a tunnel that created a channel for shipping to pass. Emerging at an artificial island (shown in the artist's impression) the road would continue on a floating bridge made up of dual pontoon-like sections, in the model display cleverly half shown with the other half reflecting in a mirror (*inset*). The road would then continue to Vancouver Island via *Valdes*, *Gabriola*, and *Mudge* Islands, which would require relatively short bridges to connect. Stage one of the construction would have seen the ferry continue to sail from Tsawwassen but to a terminal on Valdes Island, thus allowing more frequent sailings than those currently to Swartz Bay. The idea of the fixed link has never progressed beyond the concept stage because of the formidable and perhaps currently insurmountable engineering difficulties, not to mention an extraordinary cost that would seem hard to justify given the still relatively low population densities on both sides of the strait. For the moment it is an idea before its time, but Pat McGeer still believes a fixed link will be built one day.

NOW YOU KNOW US—EH?

As the twentieth century drew to a close, Vancouver was emerging as a world-class city. The world's fair, Expo 86, was held on the previously industrial land and railway yards lining False Creek (MAP 892, *right*), and in the decades that followed the area emerged as a jewel of city planning, with well-designed high-rises and low-rises, green spaces, and pathways, all abutting waterfront, the envy of the world. Vancouver, indeed, has a place always at or near the top in those surveys of the most livable cities in the world, and it is big news if she slips very far.

The growth had a downside, of course, as real estate prices mushroomed, fed in large part by immigration, now predominantly from Asia and additionally fuelled in 1997 by the reversion of Hong Kong to China. Sky-high housing prices remain a favourite talking point.

The city's rapid transit system, SkyTrain, began operating in 1986 in time for Expo and has been expanded several times (MAP 894 and MAP 895, *overleaf*), including the Canada Line linking Richmond and the airport to downtown in time for the hosting of the 2010 Winter

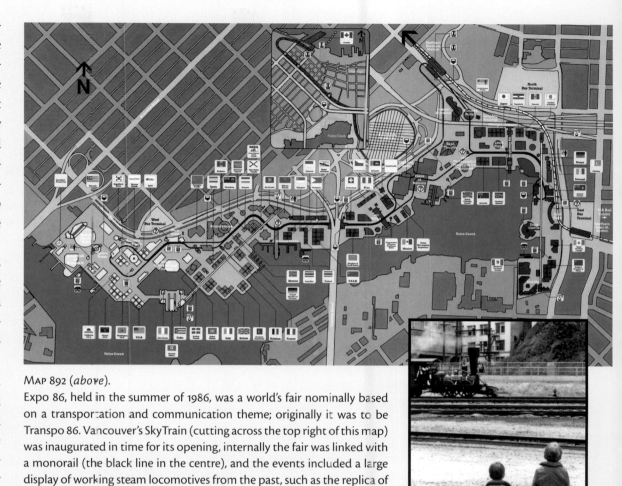

MAP 892 (*above*).
Expo 86, held in the summer of 1986, was a world's fair nominally based on a transportation and communication theme; originally it was to be Transpo 86. Vancouver's SkyTrain (cutting across the top right of this map) was inaugurated in time for its opening, internally the fair was linked with a monorail (the black line in the centre), and the events included a large display of working steam locomotives from the past, such as the replica of 1833 American locomotive *John Bull* (inset).

New Treaties

The 1763 Royal Proclamation bound Canadian governments to negotiate the sale of land from the Aboriginal peoples before it could be settled by EuroCanadians, and this principle has been carried forward to the present day. Although treaties were negotiated right across the Prairies in advance of settlement, only the northeastern corner of British Columbia was covered; this was Treaty 8 in 1899 (see page 233). Other than a few treaties covering small areas negotiated by James Douglas in the 1850s, no other treaties were signed in British Columbia.

The Nisga'a of the Nass Valley first created a Land Committee in 1890, but their claims were ignored. Between 1927 and 1951 it was even illegal for them to raise money to advance land claims. After this law was repealed, the Land Committee reconstituted itself as the Nisga'a Tribal Council and in the late 1960s began legal action to advance their land claim. Because of legal delays, the Nisga'a and the federal government did not begin negotiations until 1976. But the process was very drawn out. They were joined by the provincial government in 1990, and preliminary agreement was reached in 1996. The Nisga'a Final Agreement, usually referred to as the Nisga'a Treaty, was signed by the federal and provincial governments and the Nisga'a First Nation in May 1999 and was ratified by Parliament in April 2000.

The groundbreaking agreement gave the Nisga'a control of 2,019 square kilometres of land; $196 million; an allocation of salmon, their traditional food supply, and limited rights to other wild animals; and funding for health care, social services, and education. Although they would have self-government, they would still be subject to Canadian laws and the Canadian Constitution.

Another groundbreaking treaty was signed with the Tsawwassen First Nation in 2009. A 290-ha reserve was expanded to 724 ha by the controversial removal of land from the Agricultural Land Reserve, and the Tsawwassen were given control equivalent to that of an autonomous municipality; with this they can begin to develop the land and create an economic base.

Another treaty, with the Maa-nulth group of five First Nations on Vancouver Island, was finalized in 2011. Sixty other British Columbia treaties are still being negotiated.

MAP 893 (*right*).
The extent of the lands awarded to the Nisga'a is shown in this map, a summary and index map to more detailed sheets that accompanied the Final Agreement. The *Nass R.* flows down the centre of the land. The purple-coloured area is the Nisga'a Memorial Lava Bed Park, created as part of the treaty.

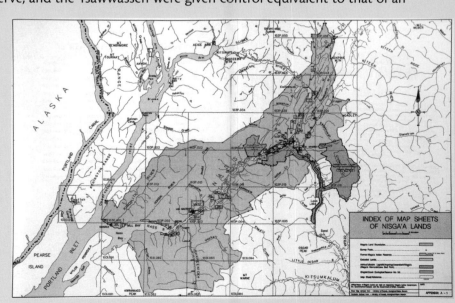

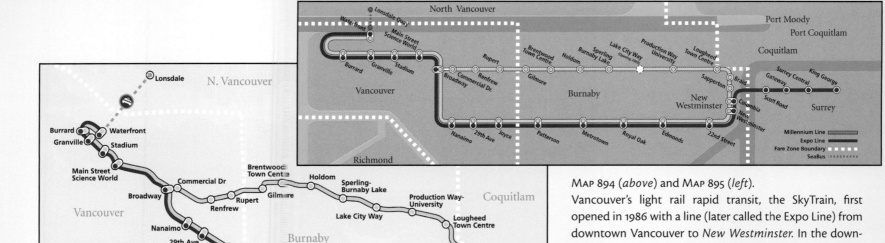

Map 894 (*above*) and Map 895 (*left*).

Vancouver's light rail rapid transit, the SkyTrain, first opened in 1986 with a line (later called the Expo Line) from downtown Vancouver to *New Westminster*. In the downtown it utilized the old tunnel that allowed Canadian Pacific to access its yards on the north side of False Creek. The line (shown here in blue) was extended across the Fraser in 1990 and to *King George* station in Surrey in 1994. The Millennium Line (in yellow) was completed in 2006. Another line, the Canada Line, opened from downtown to Richmond and the airport in 2009 (Map 906, *page 338*). These two maps are both schematic, but Map 894 illustrates the extreme simplification (into a *diagram*) pioneered in 1933 for the London Underground by Harry Beck.

Map 896 (*left*). A selection of upcoming transportation improvements planned by Metro Vancouver's TransLink in 2004. The Canada Line to the airport opened in 2009, as did the *Golden Ears Bridge* over the Fraser between Langley and Maple Ridge. The South Fraser Perimeter Road (not on map) would soon be added.

Map 897 (*left*).

This very large illustrated map, photographed in 2010, covered the entire side of a semi-trailer. It is a map of the B.C. Electric interurban line to Chilliwack (see page 270), advocating the route's resurrection as a modern rail transit line. Some have considered it contradictory that an ad for rail transit would be placed on the side of a semi-trailer, but, of course, that is the only way around anti-billboard bylaws. The rails still exist, for the most part, used by mainline railways. The main issue would seem to be the lack of grade separation, resulting in limited speeds.

Map 900 (*below*).

Housing and house prices have been a topic of conversation in Vancouver, and to a lesser extent in much of southern British Columbia, for many years. The popularity of Vancouver, with its moderate climate and pivotal position on the Pacific Rim, has made it a world-class city. But the limited amount of land available, because of the sea, the rivers, and the mountains, plus the removal of much land into the Agricultural Land Reserve (ALR), the latter introduced in 1973, has ensured that housing costs go ever skyward, with just the occasional downturn such as in 1913–18 and 1981. This is a typical billboard in a new subdivision, with a map displaying the layout of the lots for sale, with houses built by four companies. This one was at Morgan Heights in South Surrey in 2008. A period of low interest rates added to the momentum of rising house prices.

Map 898 (*below*) and Map 899 (*right*).

This excellent pair of maps show the non-stop routes flown out of Vancouver International Airport (YVR) in 2009. Modern technology has allowed Vancouver a long reach, though this means that some long-haul flights that used to pause in Vancouver now can fly through without stopping. It is interesting to compare this map with the partial fantasy of the 1936 Vancouver Jubilee map (Map 791, page 307).

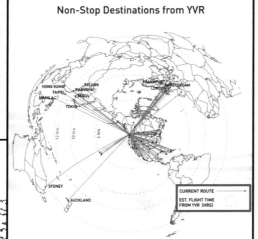

Non-Stop Destinations from YVR

Map 901 (*right*).

The devastation caused by the mountain pine beetle in the previous decade is evident from this 2007 computer-generated map from the provincial Ministry of Forests and Range. The large grey area covering a vast swath of land between *Prince George* and *Williams Lake* and west to the Coast Range represents forest lands that are completely overrun. The various colours on the margins of this area (see key) represent lower degrees of infestation. The pine beetles have infested over 300,000 ha of principally lodgepole pine forest since 1997, though the outbreak can be traced back to 1993. They are considered second only to fire as the enemy of the province's forests. Larvae feed on the tree under the bark and hence are difficult to get at. Their attack is accompanied by a fungus that has a distinctive blue stain, which gives the wood a quite attractive appearance that is marketable for some uses. Recent evidence indicates that the beetle population, unable to reproduce within increasingly younger trees, is crashing, having literally eaten itself out of a home, and allowable cuts have been reduced in some areas as a result. However, the pine beetle is moving east into Mount Robson Provincial Park and Jasper National Park, leaving dead trees that are increasingly becoming a fire hazard.

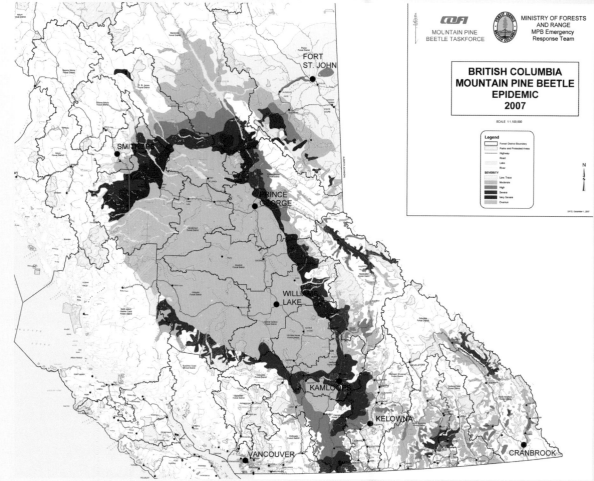

Map 902 (*right*).

The summer of 2003 was the worst season of forest fires the province had ever seen, with over 2,500 separate fires. But every year there are many fires, caused principally by lightning strikes but also careless humans. Environment Canada measures lightning strikes with special detectors. The magnitude of the problem is well illustrated by this digital map showing lightning strikes over southern British Columbia on 26 July 2009, when over a hundred new forest fires started. And as it was happening, an estimated 300,000 people in Vancouver watched fireworks that evening in English Bay. Clearly they could have seen a "Celebration of Light" elsewhere! The lightning strikes are differentiated in time by symbol shape and colour as follows: white triangles, flash after 2, 3, and 4 pm; circles, purple, 5 pm; green, 6 pm; light blue, 7 pm; dark blue, 8 pm; white, 9 pm; yellow, 10 pm; red, 11 pm; black, 12 pm. The movement of lightning-carrying storms can be seen from the time progression.

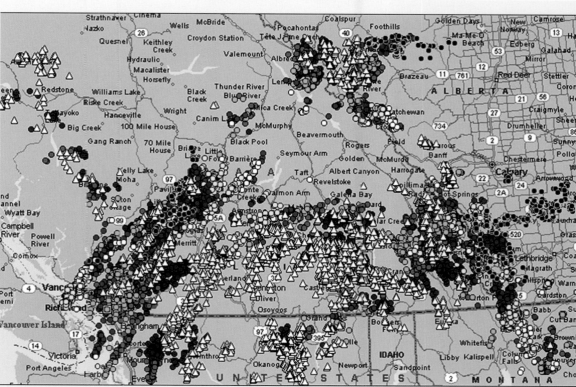

Map 903 (*left*).
Skiing has a long history in British Columbia, as might be expected in a land with many mountains and lots of snow. This is an ad for Red Mountain, at Rossland, where local volunteers in 1948 created a ski run with a 425-m elevation, complete with a ski lift, and invited community participation in its financing.

Olympics (**Map 905**, *overleaf*). The Olympics proved to be the province's seminal event of many decades. The games put Vancouver and Whistler—and British Columbia—on the world map as no other event can. As John Furlong, CEO of the Olympic Vancouver Organizing Committee (VANOC), said at the closing ceremonies—with a humorous nod to a supposed Canadian form of speech—"Now you know us—eh?"

The gateway to British Columbia, YVR, now managed by the independent and non-profit Vancouver International Airport Authority, consistently comes up near the top of world airport rankings and links the province to the far corners of the world (**Map 898** and **Map 899**, *left*). On 11 September 2001 the airport's abilities were put to the test as planes were grounded and diverted in the wake of the New York terrorist attacks. Thirty-four passenger jets, carrying 8,000 passengers—and 8,000 bags—were diverted to Vancouver, and there were 110 planes on the ground at YVR that day.

Tourism to the province has become a major industry, as the attractive potential of this scenic land was recognized, helped by an innovative campaign to market "Beautiful British Columbia" and "Super, Natural British Columbia." There have been setbacks, such as the massive out-of-control forest fires that swept much of the southern part of the

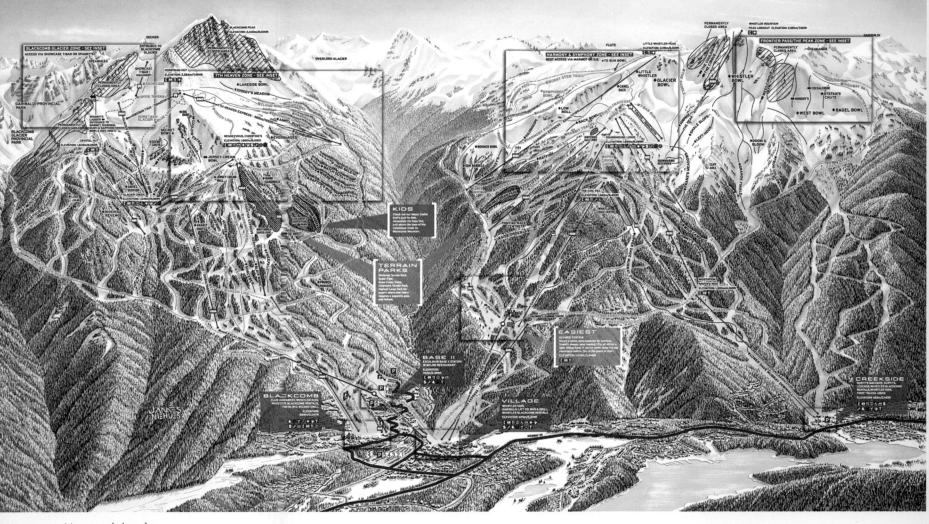

MAP 904 (*above*).
This superb bird's-eye map of the ski runs at Blackcomb (at left) and Whistler (at right) was published in 2004. On a markedly three-dimensional surface such as these mountains, the bird's-eye map is clearly superior to any regular map. Whistler Village is at centre, bottom, and Alta Lake is at bottom right. Whistler was incorporated as British Columbia's first resort municipality in 1975. It had at first been a stop on the Pacific Great Eastern Railway (see MAP 720, *page 281*).

MAP 905 (*below*).
The 2010 Winter Olympics were preceded by the Olympic Torch Relay right across Canada. Lit in Olympia, Greece, on 22 October 2009, the torch criss-crossed Canada for 106 days and the province for 27 days before landing at B.C. Place in downtown Vancouver on 12 February for the opening ceremonies. Every night BCTV carried maps like this one, superimposed on a satellite photo, to document the day's travels.

This was the torch's journey from Kamloops to Williams Lake on 28 January. The photo, *left*, shows the Olympic flame on Marine Drive in White Rock on 9 February, as dawn breaks under the looming silhouette of Mount Baker just across the border. Later that morning a ceremony, with Torch Relay (*right*), was held under the Peace Arch, at Peace Arch Park, in which the premier of British Columbia and the governor of Washington State participated, as did local Aboriginal people. Also present as a spectacular backdrop was the province's own steam locomotive, the *Royal Hudson,* sending plumes of steam into the chill morning air just for show. The Olympics and the Paralympics that immediately followed dramatically raised the profile of Vancouver and the Province of British Columbia in the eyes of the world.

THE TORCH IN BC
JANUARY 28
Williams Lake

100 Mile House

Kamloops

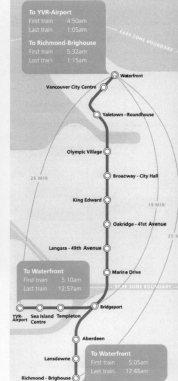

MAP 906 (*right*).
The Canada Line extension to Vancouver's SkyTrain transit system began operating on 17 August 2009, well in time for the Olympics the following February. The system proved itself during the games, carrying capacity loads. All the bus routes from places south of the Fraser that used to terminate downtown were rerouted to terminate at the Canada Line's *Bridgeport* Station.

Base: © 2010 Google
©2010 Terra Metrics

MAP 907 (*right*).
The vast beauty of British Columbia is well illustrated by this beautiful computer-generated map, based on a satellite photo, of Haida Gwaii, until recently known as the Queen Charlotte Islands, as named by George Dixon in 1786 (MAP 69, *page 28*).

MAP 908 (*below*).
The extreme remoteness of a great deal of the province is humorously illustrated by this detail from a 2000 hiking map of the Stein Valley published by a Vancouver mapmaking company. The red star indicates that *you are two days from any help in any direction. This is the middle of nowhere.*

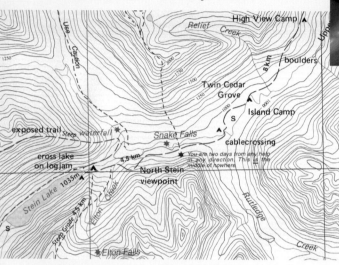

province in the summer of 2003, even encroaching on the south part of Kelowna. British Columbia, as a densely forested province, regularly suffers from hundreds of forest fires every summer, but 2003 was especially bad. From late spring to early fall more than 2,500 forest fires caused untold damage and heartbreak. Some entire communities were destroyed, as were more than 250,000 ha of forest.

An epidemic of pine beetles decimated the lumber industry of the central interior from about 1997, and, although the worst is over, it will take many years for recovery (MAP 901, *page 337*).

The maps selected for these last few pages illustrate just some of the advances and setbacks of British Columbia up to 2010.

And finally we end—not where we started, but close—with a computer-generated, digitally rendered map of the point of contact of Europeans with the Aboriginal peoples (MAP 910, *below*). The world has come a long way in the intervening years. One wonders what Juan Pérez would have thought of the modern global positioning system—GPS—map that would have been available to him had he been able to wait another two hundred years or so!

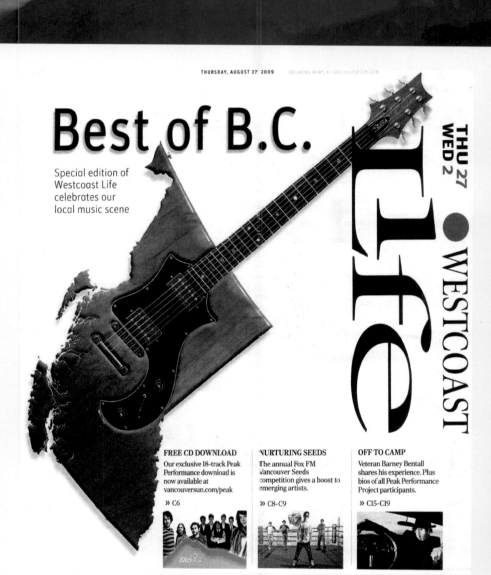

MAP 909 (*above*).
British Columbia as a guitar. The front page of the *Westcoast Life* section in the *Vancouver Sun* on 27 August 2009. This ingenious portrayal of the map of the province is a Photoshop creation. Computers now make maps—of any sort.

MAP 910 (*left*).
This is a map of *St. Margaret Point*—Juan Pérez's Punta de Santa Margarita—on the northern shore of Langara Island, at the northwest tip of Haida Gwaii. It is displayed on the screen of the Coastguard Auxiliary Zodiac rescue boat stationed at Masset; maps such as this are now familiar to most boaters and hikers. In theory it is almost impossible to get lost—you will always know where you are—but that doesn't prevent trouble; it just means you'll know where you were when you sank. But for a rescue boat such as this, GPS is invaluable. If the coordinates of a vessel in distress are known, then it can be located very much more swiftly than before. Compare this map with the photo on page 24.

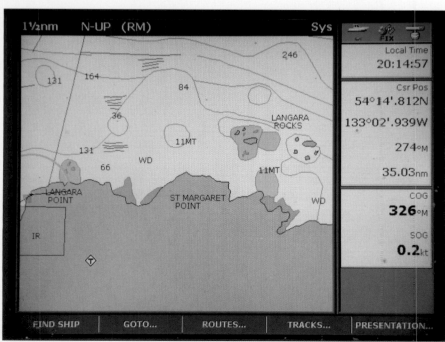

Map Catalogue & Sources

Maps that do not have sources quoted are from the collection of the author or certain private collections.

MAP 1 *(half-title page)*.
Over 1,000 Miles of Coast Line
From: *Prince Rupert and a Dawning Empire*,
Prince Rupert Publicity Club and Board of Trade, 1910
British Columbia Archives Library NWp 971.1Pr P957pu

MAP 2 *(title page)*.
Map of North America from 20 to 80 Degrees North Latitude Exhibiting the Recent Discoveries, Geographical and Nautical
James Wyld, 1823
David Rumsey Collection

MAP 3 *(page 5)*.
Map of British Columbia with neoclassical figure
H.S. Study, 1912
Cover of *Annual British Columbia Development Edition, [B.C.] Saturday Sunset*, 1912
Library and Archives Canada e10800078

MAP 4 *(contents page)*.
Map of the Province of British Columbia
Edward Mohun, 1884
Library and Archives Canada NMC 17944

MAP 5 *(page 8)*.
British Columbia (New Caledonia)
Edward Weller, 1858
University of British Columbia Rare Books and Special Collections G3510 1859 W4 B7

MAP 6 *(page 9)*.
British Columbia, Canada
British Columbia Travel Bureau, c. 1954

MAP 7 *(page 9)*.
Prop'd. Removal of Tracks #4 to #10 inclusive at Pender Street Yard, Vancouver, B.C.
Office of District Engineer, Great Northern Railway, 1925
Henry Ewert Collection

MAP 8 *(page 10)*.
Aboriginal Map of North America denoting the Boundaries and the Locations of various Indian Tribes.
John Arrowsmith, 1857
Glenbow Archives G3301 E1 1857 A779

MAP 9 *(page 11)*.
Skidegate Inlet
Daniel Pender, 1866; published as Admiralty Chart, 1872; revised 1899, 1912
Land Title and Survey Authority 24T1 Old Charts

MAP 10 *(page 11)*.
Map of British Columbia and Vancouvers Id
Francis Young, 1858
From: William Carew Hazlitt, *British Columbia and Vancouver Island*, 1858

MAP 11 *(page 11)*.
Vancouver Island and the Gulf of Georgia
Henry Kellett, 1849
Admiralty Chart 1917, 1849

MAP 12 *(page 11)*.
Head of North Bentinck Arm
James Turnbull, 1862
Land Title and Survey Authority 5T2 Miscellaneous

MAP 13 *(page 12)*.
Sprachenkarte von Britisch-Columbien
Franz Boas, 1896
From: *Petermanns Geographische Mitteilungen*, 1896
National Oceanographic and Atmospheric Administration Central Library

MAP 14 *(page 12)*.
Ethnological Map of the Province of British Columbia
Provincial Museum, 1900
Land Title and Survey Authority 4T5 Old Maps

MAP 15 *(page 12)*.
Alert Bay
Surveyed by HMS *Egeria*, 1901; Admiralty Chart, 1907
Land Title and Survey Authority 78T3 Old Charts

MAP 16 *(page 12)*.
Fort Simpson
Surveyed by HMS *Egeria*, 1901; Admiralty Chart, 1906
Land Title and Survey Authority 84T3 Old Charts

MAP 17 *(page 13)*.
Map of British Columbia Being a Geographical Division of the Indians of the Province According to their Nationality or Dialect
Office of the Superintendent of Indian Affairs, Victoria, 1872
Library and Archives Canada NMC 119561

MAP 18 *(page 14)*.
Map of the Dominion of Canada Showing Indian Reserves
From: *Annual Report of the Department of Indian Affairs*, 1891
Library and Archives Canada G3401.E1 1891 c212

MAP 19 *(page 14)*.
Rough Sketch for a Waggon Road
John Jane, 1878
Land Title and Survey Authority 1T2 Roads and Trails

MAP 20 *(page 14)*.
Index to Townsite No. 1, New Westminster District, shewing the page of the Field Book
William Ralph, 1873
Land Title and Survey Authority PH 1, Group 2, 7/73

MAP 21 *(page 15)*.
Untitled map of the Lower Fraser Valley and adjacent area
Thiusoloc, c. 1859
From: Keith Thor Carlson (ed.), *A Stó:lo-Coast Salish Historical Atlas*, 2001

MAP 22 *(page 16)*.
A New Map of North America, with the British, French, Spanish, Dutch & Danish Dominions on that Great Continent; and the West India Islands, done from the latest geographers, with great improvements from the Sieurs d'Anville & Robert
Robert Sayer, 1750
Library of Congress G3300 1750 .S3 Vault

MAP 23 *(page 16)*.
Universalis Cosmographia Secundum Ptholomaei Traditionem et Americi Vespucci Aliou[m]que Lustrationes
Part of main map and inset map
Martin Waldseemüller and Matthias Ringmann, 1507
Library of Congress G3200 1507 .W3 Vault

MAP 24 *(page 17)*.
Universale Descittioni di tutta da terra conosciuta fin qui
Paolo Forlani, c. 1565
Library of Congress G3200 1565.F6 Vault

MAP 25 *(page 17)*.
Americae Sive Novi Orbis Nova Descriptio
From: Abraham Ortelius, *Theatrum Orbis Terrarum*, 1570
Library of Congress G1006 .T5 1570 Vault

MAP 26 *(page 17)*.
Tartariae Sive Magni Chami Regni
From: Abraham Ortelius, *Theatrum Orbis Terrarum*, 1570
Library of Congress G1006 .T5 1570 Vault

MAP 27 *(page 17)*.
Universi Orbis Sev Terreni Globi In Plano Effigies
From: Gerard de Jode, *Speculum Orbis Terrarum*, 1578

MAP 28 *(page 17)*.
Nova Totius Terrarum Orbis iuxta Neotericorum Traditiones Descriptio
Abraham Ortelius, 1564
British Library Maps.C.2.a.6

MAP 29 *(page 18)*.
Chinese world map, c. 1650 (reproduction)

MAP 30 *(page 18)*.
Map known as "Map with ship"
Anon., attributed to or from Marco Polo, c. 1300 (date highly questionable)
Library of Congress G7800 coll .M3 copy 1, Marcian F. Rossi Collection

MAP 31 *(page 18)*.
Nouveau Systeme Géographique par lequel on concilie les anciennes connaissances sur le Pays Nord-Ouest de l'Amérique avec les nouvelles découvertes des Russes au Nord de la Mer du Sud. Par Mr. de Vaugondy 1774.
Bill Warren Collection

MAP 32 *(page 18)*.
An Exact Map of North America from the Best Authorities
John Lodge, 1778

MAP 33 *(page 18)*.
A Map of the Discoveries made by the Russians on the North West Coast of America
Thomas Jefferys, 1768
From: *A General Topography of North America and the West Indies*, 1768
Library of Congress G1105 .J4 1768

MAP 34 *(page 18)*.
A Chart of the Interior Part of North America Demonstrating the very great probability of an Inland Navigation from Hudsons Bay to the West Coast
John Meares, 1790
From: John Meares, *Voyages Made in the Years 1788 and 1789 from China to the North West Coast of America*, 1790

MAP 35 *(page 18)*.
Sketch of the Entrance of the Strait of Juan de Fuca
Charles Duncan, 1788
Published by Alexander Dalrymple, 1790
University of British Columbia Rare Books and Special Collections, W. Kaye Lamb collection

MAP 36 *(page 19)*.
La Herdike (sic; Heroike) Enterprinse Faict par de Signeur Draeck d'avoir Cirquit Toute la Terre
Nicola van Sype, 1583
British Library Maps Cs.a7 (1)

MAP 37 *(page 19)*.
Illustri Viro, Dimino Phillipo Sidnaes Michael Lok Civis Londinensis Hanc Chartum
Michael Lok, 1582
From: Richard Hakluyt, *Divers Voyages touching the Discoverie of America*, 1582

MAP 38 *(page 19)*.
Carta particolare della stretto do Iezo Pra 'America e l'Isola Iezo D'America Carta XXXIII
Robert Dudley, 1647
From: *Dell' Arcano del Mare*, 1647

MAP 39 *(page 19)*.
Vera Totius Expeditionis Nauticae descriptio D. Franc. Draci
Jodocus Hondius, 1589
Maritiem Museum Prins Hendrik, Rotterdam

MAP 40 *(page 19)*.
Detail of terrestrial globe
Emery Molyneux, 1603
Middle Temple Library, London

MAP 41 *(page 20)*.
Limes Occidentis Quivira et Anian. 1597
From: Cornelius Wytfliet, *Descriptionis Ptolemaicae Augmentum sive Occidentis Notitia Breui commentario illustrata studio et opera*, 1597.

MAP 42 *(page 20)*.
Quiveræ Regnu cum alijs versus Borea
From: Cornelis de Jode, *Speculum Orbis Terrae*, 1593
Newberry Library

MAP 43 *(page 21)*.
Map of the North Pacific Ocean
Girolamo de Angelis, 1621
Archivum Romanum Societatis Iesu (Jesuit Archives), Rome

MAP 44 *(page 21)*.
Bankoku Sozu
Japanese world map, 1645
University of British Columbia Rare Books and Special Collections G3200 1645.z.S5

MAP 45 *(page 21)*.
Imaginary map of the Northwest
From: Jonathan Swift, *Gulliver's Travels*, 1726

MAP 46 *(page 21)*.
Carte de L'Amerique Septentrional
Jacques-Nicolas Bellin, 1743
Library and Archives Canada NMC 98179

MAP 47 *(page 21)*.
A New Map of North America from the Latest Discoveries
From: Jonathan Carver, *Travels through the Interior Parts of North America in the Years 1766, 1767 and 1768*, 1778

MAP 48 *(page 21)*.
Carte Representant la Situation et la Distance de la Tartarie Orientale jusqaux Terres les plus voisines de L'Amerique
Joseph-Nicolas De L'Isle, 1731
A.V. Efimov, 1964, Map 78

MAP 49 *(page 21)*.
Map of Aleksei Chirikov's voyage in 1741
Ivan Elagin, c. 1742
A.V. Efimov, 1964, Map 98

MAP 50 *(page 21)*.
Carte Reduite des Parties Septentrionales du Globe Situées Entre L'Asie et L'Amerique
Jacques-Nicolas Bellin, 1758
Beach Maps, Toronto

MAP 51 *(page 22)*.
A Map of North America with Hudson's Bay and Straights Anno 1748
Richard Seale, 1748
Hudson's Bay Company Archives G4/20b

MAP 52 *(page 22)*.
Amerique Septentrionale
Gilles and Didier Robert de Vaugondy, 1762

MAP 53 *(page 22)*.
Amerique Septentrionale
Jean Janvier, 1762
Library of Congress G3300 1762 J31 vault

MAP 54 *(page 23)*.
Carte Generale des Découvertes de l'Amiral de Fonte
Gilles and Didier Robert de Vaugondy, 1755
From: Denis Diderot, *Encyclopédie*, 1755

MAP 55 *(page 23)*.
Amerique Septentrionale
From: Jean B. Nolin, 1783, *Atlas Général a L'Usage des Colleges et Museums d'Education*
Library of Congress G1015 .N68 1783

MAP 56 *(page 23)*.
Mappe Monde ou Globe Terrestre en deux Plans Hemispheres
Jean Covens and Corneille Mortier, c. 1780

MAP 57 *(page 24, also with detail enlarged)*.
Carta Reducida del Oceano Asiatico ô Mar del Sur aue contiene la Costa de la California comprehendida desde el Puerto de Monterrey. hta la Punta de Sta Maria Magdelena hecha segun las observaciones y Demarcasiones del Aljerez de Fragata de la Rl. Armada y Primer Piloto de este Departamento Dn. Juan Perez por Dn. Josef de Cañizarez.
José de Cañizares, 1775
U.S. National Archives RG 77, "Spanish maps of unknown origin," No. 67

Map 58 (page 25).
A Chart Containing the Coasts of California, New Albion, and Russian Discoveries to the North
From: An American Atlas
Thomas Jefferys/Robert Sayer, 1775
Library and Archives Canada NMC 27718 (section 3/6)

Map 59 (page 25).
Carta Reducida de las Costas y Mares septentrionales de California
Anon., Spanish, 1775
Bancroft Library, University of California

Map 60 (page 25).
Untitled English map of Bodega y Quadra and Hezeta voyages
From: Francisco Antonio Mourelle de la Rúa, Journal of a Voyage in 1775, to explore the coast of America, northward of California, 1780

Map 61 (page 26).
Track from first making the Continent, March 7th, to Anchoring in King George's Sound
From: Journal of James Burney, 1778
U.K. National Archives ADM 51/4528

Map 62 (page 26).
Carte de L'Ocean Pacifique au Nord de l'Equateur/ Charte des Stillen Weltmeers im Nördlichen Aequator
Tobias (Tobie) Conrad Lotter, 1781
Library and Archives Canada NMC 8607

Map 63 (page 26).
Sketch of King George's Sound
From: Journal of James Burney, 1778
U.K. National Archives ADM 51/4528

Map 64 (page 27).
Chart of part of the N W Coast of America Explored by Capt. J. Cook in 1778
James Cook, 1778
U.K. National Archives MPI 83

Map 65 (page 27).
General Chart exhibiting the discoveries made by Capt. James Cook in this and his preceding two voyages, with the tracks of the ships under his command.
Henry Roberts; William Faden, 1784
Library and Archives Canada NMC 27702

Map 66 (page 27).
[Carte de la Côte Ouest de l'Amérique du Nord, de Mt. St. Elias à Monterey, avec la trajectoire l'expédition de La Pérouse et la table des données de longitude compilées par Bernizet and Dagelet]
Joseph Dagelet and Gérault-Sébastien Bernizet, 1786
Archives nationales de France 6 JJ1: 34B

Map 67 (page 28).
Chart of Part of the N.W. Coast of America by Capt. James Hanna in Snow Sea Otter 1786
James Hanna, 1786
Published by Alexander Dalrymple, 1789

Map 68 (page 28).
A Chart Exhibiting the Route of the Experiment Snow
S. Wedgborough, 1786
From: James Strange's Journal and Narrative of the Commercial Expedition from Bombay to the North-West Coast of America, 1928
Vancouver Public Library 970P S89j1

Map 69 (page 28).
Chart of the North West Coast of America with the Tracks of the King George and Queen Charlotte in 1786 & 1787
George Dixon, 1788
From: George Dixon, A Voyage Round the World; But More Particularly to the North-West Coast of America, Performed in 1785, 1786, 1787 and 1788, 1789

Map 70 (page 28).
Chart of the N.W. Coast of America and N.E. Coast of Asia, explored in the Years 1778 & 1779 by Capt. Cook & Further Explored in 1788 & 1789
From: John Meares, Voyages Made in the Years 1788 and 1789, from China to the North West Coast of America, 1790

Map 71 (page 29).
A Sketch of Port Cox in the District of Wicananish
From: John Meares, Voyages Made in the Years 1788 and 1789, from China to the North West Coast of America, 1790

Map 72 (page 29).
Russian map of the Northwest Coast incorporating the 1790 map of John Meares, c. 1792
From: A.V. Efimov, 1964

Map 73 (page 29).
Chart of the World on Mercator's Projection Exhibiting All the New Discoveries to the present Time
Aaron Arrowsmith, 1790

Map 74 (page 29).
A Map of Hudson's Bay and of the Rivers and Lakes between the Atlantick and Pacifick Oceans
Alexander Dalrymple, 1790
U.K. National Archives CO 700 Canada 42

Map 75 (page 30).
N.W. America Drawn by J.C. from his own Information & what could be collected from the Sloop Pr Royal & Boats in the Years 1787 1788
James Colnett, 1787–88
U.K. Hydrographic Office

Map 76 (page 30).
Untitled map of Oregon and the Northwest
Hall Jackson Kelley, 1839
U.S. National Archives

Map 77 (page 30).
Untitled map usually known as "Territory of Oregon and High California"
Hall Jackson Kelley, 1839
U.S. National Archives

Map 78 (page 31).
A Sketch of the passage between Nootka and Ahasset Sound's
Robert Gray, 1789
Massachusetts Historical Society

Map 79 (page 31).
A Sketch of Hancocks Rivr on the North Side of Washingtons Isla.
Robert Haswell, 1792
Massachusetts Historical Society

Map 80 (page 31).
Quadra's Isles
Joseph Ingraham, 1792
From: Journal of the Hope, 1792
Library of Congress

Map 81 (page 31).
Chart on Mercators projection exhibiting the Tracks of Maldonado and De Fonte in 1598 and 1640; Compared with the Modern Discoveries
William Goldson, 1793
From: William Goldson, Observations on the Passage between the Atlantic and Pacific Oceans, 1793
University of British Columbia Rare Books and Special Collections 610.3at G636 1793a

Map 82 (page 31).
Washingtons Isles
Joseph Ingraham, 1792
From: Journal of the Hope, 1792
Library of Congress

Map 83 (page 32).
Plano del Estrecho de Fuca Reconocido y le bantado en el año de 1790
Gonzalo López de Haro (attrib.), 1790

Map 84 (page 32).
Carta que comprehende
José María Nárvaez, 1791
Library of Congress G3351.P5 1799.C vault, Map 12

Map 85 (page 33).
Numero 3. Continuacion de los reconocimientos hechos En La Costa No. De America Por Los Buques de S.M. en varias Campañas desde 1774 a 1792
From: Relación del Viaje hecho por las Goletas Sutil y Mexicana en el año 1792, 1802
Vancouver Public Library Special Collections

Map 86 (page 33).
Plano Del Archipelago de Clayocuat
José María Nárvaez, 1791
Library of Congress G3512.C5 1791 .P5 Vault

Map 87 (page 33).
Plano de la Cala de Los Amigos Situada en la Parte Ocidental de la entrada de Nutka Año 1791
From: Relación del Viaje hecho por las Goletas Sutil y Mexicana en el año 1792, 1802
Vancouver Public Library Special Collections

Map 88 (page 33).
Carta Esferica de la parte de la Costa No. de America Comprehendida entre la Entrada de Juan de Fuca y la Salidas de las Goletas con algunos Canales interiores
Dionisio Alcalá Galiano, 1792
U.K. National Archives FO 925 1650 (13)

Map 89 (page 34).
Map of the southern Strait of Georgia
Dionisio Alcála Galiano, 1792
Museo Naval Borradores No. 7 MS 2456 (John Crosse papers)

Map 90 (page 34).
Map of part of the Strait of Georgia including Porlier Pass
Dionisio Alcála Galiano, 1792
Museo Naval Borradores No. 8 MS 2456 (John Crosse papers)

Map 91 (page 34).
Carta general de quanto asta hoy se ha descubierto y examinado por los Espanoles en la Costa Septentrional de California, formada . . . por D. Juan Francisco de la Bodega y Quadra Año de 1791
Juan Francisco de la Bodega y Quadra, 1791
Museo Naval

Map 92 (page 34).
Carta Reducida de la Costa Septentrional de California
Juan Francisco de la Bodega y Quadra, end of 1791 or beginning of 1792
Library of Congress G3351.P5 1799.C vault, Map 1

Map 93 (page 34).
Carta de los Descubrimientos hechos en la Costa N.O. America Septentrional
Juan Francisco de la Bodega y Quadra, 1792
Library of Congress G3350 1792.B6 TIL vault

Map 94 (page 35).
Plano del estrecho de Juan de Fuca descuvierto el ano 1592, reconocido en 1789 por Dn Jose Narvaez en el de 90 . . . Dn Manuel Quimper, en 91 . . . Dn Franco Eliza y concluido en . . . el Comandame Vancouver y Dionisio Galiano en el qual sedenetan con el color negro los descubrim los hechos . . . con el ecarnado los Vancouver y con elazul los de Galiano
Juan Francisco de la Bodega y Quadra, 1792
Oregon Historical Society

Map 95 (page 35).
Numo. 2 Carta Esferica de los Reconocimientes hechos en la Costa N.O. de America en 1791 y 92 por las Goletas Sutil y Mexicana y otros Buques de S.M.
From: Relación del Viaje hecho por las Goletas Sutil y Mexicana en el año 1792, 1802
Vancouver Public Library Special Collections

Map 96 (page 35).
La America Septentrional desde su extremo Norte hasta 10° de Latitud segun las ultimas observaciones y descubrimientos para el Curso de Geografia de D. Isidoro de Antillon
From: Isidoro de Antillon, Carta de la America Septentrional, 1802

Map 97 (page 35).
Carta esferica en la Costa N.O. de America para la entrada de Juan de Fuca. Reducida por Pablo Barnet y Roca
Pablo Barnet y Roca, 1808
Auction sale, 2008 (now in University of Washington Library Special Collections)

Map 98 (page 36).
Preliminary chart of part of the northwest coast of North America
Joseph Baker and George Vancouver, 1792
U.K. National Archives MPG 1 557 (4)

Map 99 (page 36).
Preliminary chart of Johnstone Strait and vicinity
Joseph Baker and George Vancouver, 1792
U.K. National Archives MPG 1 557 (x3)

Map 100 (page 36).
A Chart shewing part of the Western Coast of N. America In which the Continental shore from the Latde of 42°.30′ N. and Longde 230°.30′ E. to the Latde of 52°.15′ N. and Longde 238°.03′ E. has been finally traced and determined by His Majesty's Sloop Discovery and Armed Tender Chatham under the Command of George Vancouver Esqr in the Summer of 1792
Joseph Baker and George Vancouver, 1792
U.K. Hydrographic Office 228 on 82

Map 101 (page 37).
A Chart shewing part of the Coast of N.W. America with the tracks of His Majesty's Sloop Discovery and Armed Tender Chatham; Commanded by George Vancouver Esqr. and prepared under his immediate inspection by Lieut. Joseph Baker (manuscript compilation chart for atlas plate 8 for engraver)
George Vancouver, 1798
U.K. National Archives CO 700 British Columbia 1

Map 102 (page 37).
A Chart shewing part of the Coast of N.W. America with the tracks of His Majesty's Sloop Discovery and Armed Tender Chatham; Commanded by George Vancouver Esqr. and prepared under his immediate inspection by Lieut. Joseph Baker (Plate 7, northwestern British Columbia)
From: A Voyage of Discovery to the North Pacific Ocean and round the World in Which the Coast of North-West America Has Been Carefully Examined and Accurately Surveyed, atlas volume, 1798
George Vancouver, 1798
David Rumsey Collection

Map 103 (page 37).
A Chart shewing part of the Coast of N.W. America with the tracks of His Majesty's Sloop Discovery and Armed Tender Chatham; Commanded by George Vancouver Esqr. and prepared under his immediate inspection by Lieut. Joseph Baker (Plate 8, southwestern British Columbia)
From: A Voyage of Discovery to the North Pacific Ocean and round the World in Which the Coast of North-West America Has Been Carefully Examined and Accurately Surveyed, atlas volume, 1798
George Vancouver, 1798
David Rumsey Collection

Map 104 (page 37).
A Chart shewing part of the Coast of N.W. America with the tracks of His Majesty's Sloop Discovery and Armed Tender Chatham; Commanded by George Vancouver Esqr. and prepared under his immediate inspection by Lieut. Joseph Baker (Plate 14, west coast of North America)
From: A Voyage of Discovery to the North Pacific Ocean and round the World in Which the Coast of North-West America Has Been Carefully Examined and Accurately Surveyed, atlas volume, 1798
George Vancouver, 1798
David Rumsey Collection

Map 105 (page 37).
A New Chart of the World on Wright's or Mercator's Projection in Which Are Exhibited All the Parts Hitherto Explored or Discovered
Laurie & Whittle, 1800
David Rumsey Collection

Map 106 (page 38).
A Map shewing the communication of the Lakes and the Rivers between Lake Superior and Slave Lake in North America
Peter Pond, 1790
From: Gentleman's Magazine, March 1790, page 197

Map 107 (page 38).
[Copy of a map of western Canada and the North Pacific Ocean thought to have been prepared by Peter Pond for presentation by Alexander Mackenzie to the Empress of Russia] Copied from the original signed P. Pond Araubaska 6th December 1787 (marginal notation)
Peter Pond, 1787
U.K. National Archives CO 700 America North and South 49

MAP 542 (*page 225*).
Map of Nanaimo Townsite
Anon., 1863
Hudson's Bay Company Archives G1/154

MAP 543 (*page 225*).
Map of Nanaimo
Anon., c. 1880(?)
Hudson's Bay Company Archives G1/258 (v)

MAP 544 (*page 226*).
*Plan of one hundred acres of land Preempted for
the Hon*[le]*. Hudson's Bay C*[o]*. at Fort Rupert, V.I.*
Pym Nevins Compton, 1863
Hudson's Bay Company Archives G1/231

MAP 545 (*page 226*).
*Nanaimo and Area to South Including Most of
Cranberry, Cedar, Bright & Oyster Districts, V.I.*
J.B. Davenport, c. 1926, revised 1952
British Columbia Archives Buckham 88055−0558
C22050

MAP 546 (*page 227*).
Ladysmith, Oyster Harbour, Vancouver Island
Thomas Kitchin, 1901
Ladysmith Historical Society

MAP 547 (*page 227*).
*Map of Chemainus Mining District, Mt. Sicker,
with Adjacent Camps, Vancouver Island, B.C.*
H. Fry, 1905
City of Victoria Archives B0037

MAP 548 (*page 228*).
Nanaimo
Admiralty Chart, 1899, revised 1912
Land Title and Survey Authority 51T3 Old Charts

MAP 549 (*page 228*).
Map of Northfield Mine, Nanaimo
Anon., 1941
British Columbia Archives Buckham 88055−0558

MAP 550 (*page 228*).
Wakesiah and Jingle Pot Mines, Nanaimo
Anon., c. 1940
British Columbia Archives Buckham 88055−0558

MAP 551 (*page 229*).
*Comox District Plan Showing Topographical
Features in the Vicinity of Comox Mines*
Canadian Collieries (Dunsmuir) Ltd., c. 1915
Cumberland Museum

MAP 552 (*page 229*).
*Comox District Plan Showing Topographical
Features in the Vicinity of Comox Mines*
Canadian Collieries (Dunsmuir) Ltd., c. 1915
Cumberland Museum

MAP 553 (*page 229*).
Development Plan, Cumberland Coalfield
Canadian Collieries (Dunsmuir) Ltd., 1922
Cumberland Museum

MAP 554 (*page 230*).
*Comox District Plan Showing Topographical
Features in the Vicinity of Comox Mines*
Canadian Collieries (Dunsmuir) Ltd., c. 1915
Cumberland Museum

MAP 555 (*page 230*).
Plan Comox Mine No. 4, Lower Seam Coal Workings
Canadian Collieries (Dunsmuir) Ltd., c. 1922
Cumberland Museum

MAP 556 (*page 230*).
Plan Comox Mine No. 4, Lower Seam Coal Workings
Canadian Collieries (Dunsmuir) Ltd., c. 1922
Cumberland Museum

MAP 557 (*page 230*).
Plan Comox Mine No. 4, Lower Seam Coal Workings
Canadian Collieries (Dunsmuir) Ltd., c. 1922
Cumberland Museum

MAP 558 (*page 230*).
*Structural Contours and Coal Seams of the
Cumberland and Tsable River Coal Areas*
T.B. Williams, Department of Mines, 1924
British Columbia Archives Buckham 88055−0558

MAP 559 (*page 231*).
Courtenay Sheet, Vancouver Island
Canada Department of Mines, Geological Survey, 1924
Vancouver Public Library Special Collections
SPEMAPC 912.7112 G34c 1924

MAP 560 (*page 231*).
*Structural Contours and Coal Seams of the
Cumberland and Tsable River Coal Areas*
T.B. Williams, Department of Mines, 1924
British Columbia Archives Buckham 88055−0558

MAP 561 (*page 232*).
*Plan of Subdivision of Lot 18, Gr. 2. New
Westminster District (St. Mungo Cannery).*
Albert James Hill, c. 1908.
University of British Columbia Rare Books and
Special Collections, R.C. Harris Collection

MAP 562 (*page 232*).
*Fraser River Canneries, British Columbia,
Including Steveston*
Fire insurance map, Charles Goad, 1897
Library and Archives Canada NMC 151563

MAP 563 (*page 232*).
Wadhams Cannery, Rivers Inlet, B.C.
British Columbia Fire Underwriters, 1923
Campbell River Museum and Archives

MAP 564 (*page 233*).
Map of the Dominion of Canada (with manuscript
additions)
Department of the Interior, 1887, with manuscript
additions, Department of the Interior or Depart-
ment of Indian Affairs, and Clerk of the Privy
Council, 1898
Library and Archives Canada RG2 M890327 Item 1

MAP 565 (*page 233*).
*Map showing the Territory ceded under Treaty
No. 8 and the Indian tribes therein*
Department of Indian Affairs, attributed to James
Macrae, 1900
Library and Archives Canada FG10M 78903/45

MAP 566 (*page 233*).
Indian Treaties, 1850−1912
James White, Department of the Interior, 1912
Library and Archives Canada NMC 7139

MAP 567 (*page 234*).
Part of Russian map reproduced as Map 194 in
A.V. Efimov, *Atlas geograficheskikh otkrytii v Sibiri
i v severozapadnoi Amerike XVII−XVIII vv.* (Atlas
of geographical discoveries in Siberia and North-
Western America, XVII and XVIII centuries)
University of British Columbia Library G1036 .E5

MAP 568 (*page 234*).
*Northwestern America Showing the Territory
Ceded by Russia to the United States
Reduced from the Map by the U.S.C.S. Dep*[t]*.*
Samuel Augustus Mitchell, 1879

MAP 569 (*page 234*).
*Map of the Canadian Yukon and Northern Terri-
tory of British Columbia*
Province Publishing Company, 1897
Library and Archives Canada NMC 8408

MAP 570 (*page 234*).
Alaska Boundary Tribunal southern water
boundary composite of 6 sheets: *Sheet No. 1*
to *Sheet No. 6*
(Sheet 4 signed by tribunal)
U.S. Coast and Geodetic Survey, 1903
National Oceanic and Atmospheric Administration
Central Library

MAP 571 (*page 235*).
*Map to Accompany Report on the Alaska
Boundary 1903* (MS details on base map *North
West Coast of America. Dixon Entrance to Cape Elias.*)
U.S. Coast and Geodetic Survey, 1904
National Oceanic and Atmospheric Administration
Central Library

MAP 572 (*page 235*).
Map of Southeastern Alaska (with MS additions)
U.S. Coast and Geodetic Survey, 1903
National Oceanic and Atmospheric Administration
Central Library

MAP 573 (*page 236*).
*Vancouver is the Keystone to the Greatest
Commercial Arch in the World*
B.C. Saturday Sunset, 24 April 1909
Legislative Library of British Columbia

MAP 574 (*page 236*).
To the Land of Gold
Poster promoting Vancouver as the supply base for
the Klondike Gold Fields, 1897

MAP 575 (*page 236*).
Vancouver's Strategic Position on World Trade Routes
Anon., c. 1910
British Columbia Museum

MAP 576 (*page 237*).
*Bird's Eye View of the City of Vancouver, Viewed
from the North* (caption)/*City of Vancouver, B.C.,
Canada, 1908* (title on map)
Vancouver Tourist and Information Bureau, 1908
B.C. Saturday Sunset, 5 December 1908, page 1
Legislative Library of British Columbia

MAP 577 (*page 237*).
For Birdseye Views, Maps, Designs and Etchings
Dominion Illustrating Co., Ltd., 1911
Vancouver World, 9 September 1911
Vancouver Public Library

MAP 578 (*page 237*).
Bird's-eye map of Vancouver from Burnaby
showing Vancouver Heights subdivision
Charles H. Rawson for Morden and Thornton, 1907
University of British Columbia, Rare Books and
Special Collections 616.8ap R220 1907

MAP 579 (*page 237*).
*If Vancouver's Population Doubles in Five Years
What Will These Suburban Lots Be Worth*
Ad in *B.C. Saturday Sunset*, 27 March 1909, page 16
Legislative Library of British Columbia

MAP 580 (*page 238*).
*The Long Distance Telephone Places the Whole
Coast within Easy Communication*
British Columbia Telephone Company, 1909
B.C. Saturday Sunset, 18 December 1909
Legislative Library of British Columbia

MAP 581 (*page 238*).
*Plan for a grand boulevard, Georgia Street,
Vancouver*
Thomas Mawson, 1913
Town Planning Review, 6, 1915

MAP 582 (*page 238*).
*Accepted Scheme for Coal Harbour, Stanley Park,
Vancouver*
Thomas Mawson, 1913
Town Planning Review, 6, 1915

MAP 583 (*page 239*).
Eburne, B.C.
Central Real Estate Company,
Ad in *B.C. Saturday Sunset*, 3 August 1907
Legislative Library of British Columbia

MAP 584 (*page 239*).
*Watch Eburne Grow/Map of Eburne, B.C.,
and Vicinity*
M.R. Wells, 1911
Point Grey Gazette, 21 October 1911
University of British Columbia, Rare Books and
Special Collections HR AN 5 P63 G7

MAP 585 (*page 240*).
Langara, Point Grey
Trites & Leslie, 1909

MAP 586 (*page 240*).
Broadway's Glorious Future
A.E. Higinbotham & Co., 1909
Vancouver Province, 4 September 1909

MAP 587 (*page 240*).
Marine View, Point Grey
Ad in *B.C. Saturday Sunset*, 16 December 1911
Legislative Library of British Columbia

MAP 588 (*page 240*).
Vancouver, B.C. Canada: The Commercial Centre
Vancouver Tourist Association, 1908
Ad in *B.C. Saturday Sunset*, 28 March 1908
Legislative Library of British Columbia

MAP 589 (*page 241*).
C.P.R. Land Department Shaughnessy Heights
Canadian Pacific Railway Land Department, 1917
University of British Columbia, Rare Books and
Special Collections G3514.V3 355 1917 C3

MAP 590 (*page 241*).
*Bird's-eye view of Point Grey showing Properties
(in red) For Sale by Alvo von Alvensleben, Limited*
Alvo von Alvensleben, Ltd., 1910
Man to Man Magazine, back cover, November 1910

MAP 591 (*page 241*).
Dundarave and Vicinity
Irwin and Billings Co. Ltd., July 1911
West Vancouver Museum and Archives 15 WVA Map

MAP 592 (*page 242*).
*Route map of the Canadian Pacific Railway into
North Vancouver*
B.C. Saturday Sunset, 9 March 1912
Legislative Library of British Columbia

MAP 593 (*page 242*).
*Panorama of the City of North Vancouver and
District*
Irwin & Billings, in the *Express*, 1907
North Vancouver Museum and Archives

MAP 594 (*page 242*).
A Plan of the Townsite of North Vancouver
Anon., c. 1907
City of Vancouver Archives CVA Map 621

MAP 595 (*page 242*).
City of North Vancouver, B.C.
G.H. Dawson, 1908
From: *North Vancouver, British Columbia:
The Beginnings of a Great Port*, 1908
North Vancouver Museum and Archives MP 240

MAP 596 (*page 242*).
*The Grand Boulevard, City of North Vancouver,
British Columbia*
Mahon, McFarland & Mahon, 1908
North Vancouver Museum and Archives Pamphlet
1908-2

MAP 597 (*page 243*).
1905 Map of New Westminster District
D.R. Harris, 1905
Vancouver Public Library Special Collections
912.71135 H311 1905 copy 2

MAP 598 (*page 243*).
Stanley Heights, the Beauty Spot of North Vancouver
Ad in *B.C. Saturday Sunset*, 23 October 1909
Legislative Library of British Columbia

MAP 599 (*page 243*).
Sunset Heights No. 2, North Vancouver
Ad in *B.C. Saturday Sunset*, 25 September 1909
Legislative Library of British Columbia

MAP 600 (*page 243*).
"In on the Ground Floor"/Sunset Heights
Ad in *B.C. Saturday Sunset*, 31 July 1909
Legislative Library of British Columbia

MAP 601 (*page 244*).
*Marlbank, North Vancouver's Choice Residential
Section*
Ad in *Express*, 24 May 1912
North Vancouver Museum and Archives

MAP 602 (*page 244*).
Rosslyn
Ad in *Express*, 8 July 1910
North Vancouver Museum and Archives

MAP 603 (*page 244*).
"Skyland," Vancouver's Scenic Paradise
Ad in *B.C. Saturday Sunset*, 3 June 1911
Legislative Library of British Columbia

MAP 604 (*page 245*).
*Plan of Proposed Dam, Locks & Wharves, Second
Narrows, Burrard Inlet*
Municipality of Coquitlam, 1910
City of Vancouver Archives CVA Map 12

MAP 605 (*page 245*).
*Sketch of Proposed Interurban Canal and Harbour
Development for Greater Vancouver, B.C.*
Engineer's Office, Municipality of Coquitlam, 1910
City of Vancouver Archives Map 11

MAP 606 (page 245).
No. 82. Plan of Foreshore at Point Grey Applied
for by Vancouver Terminals Co./ Proposed
Improvements of Spanish Bank at Point Grey
by the Vancouver Terminals Co.
Vancouver Terminals Company, 1912
Copy of original plan, 1928
City of Vancouver Archives PD2895

MAP 607 (page 245).
Greater Vancouver, Its Suburbs, and the Great
Docks to be Built for the Vancouver Harbour and
Dock Extension Company Ltd.
Vancouver Harbour and Dock Extension Company,
1912
Annual Development Issue, B.C. Saturday Sunset.
1912

MAP 608 (page 245).
Proposed West Richmond Docks Scheme
Vancouver Harbour and Dock Extension Company,
1912
British Columbian, 27 November 1912

MAP 609 (page 246).
Bird's-Eye View Vancouver Harbour, B.C.
Proposed Kitsilano Terminal and Free Port
(Undecipherable, consulting engineer), for Chi-
cago, Milwaukee, St. Paul & Pacific Railroad, 1917
City of Vancouver Archives CVA 380

MAP 610 (page 246).
Proposed new docks, New Westminster
British Columbian, 27 November 1912

MAP 611 (page 246).
Bird's-eye map of New Westminster and Vancouver
over New Westminster Bridge, 1912
City of Surrey Archives

MAP 612 (page 247).
Westminster Heights
Lovewell, 1911
Ad in B.C. Saturday Sunset, 6 May 1911
Legislative Library of British Columbia

MAP 613 (page 247).
Twin City Industrial Subdivision
Ad in B.C. Saturday Sunset, 14 May 1910
Legislative Library of British Columbia

MAP 614 (page 248).
Bird's-eye map of New Westminster, looking
towards Vancouver, c. 1912
Burnaby Village Museum

MAP 615 (page 248).
Plan of Whiterock Townsite, 1912
White Rock Museum and Archives

MAP 616 (page 249).
Burnaby on the Lake
Ad in B.C. Saturday Sunset, 27 November 1909
Legislative Library of British Columbia

MAP 617 (page 249).
Second Division Coquitlam Townsite
Ad in B.C. Saturday Sunset, 30 March 1912
Legislative Library of British Columbia

MAP 618 (page 249).
You can't afford to pick blindfolded at Coquitlam
Ad in Vancouver World, 5 March 1912
Vancouver Public Library

MAP 619 (page 249).
Buy in the Manufacturers' Subdivision: The Cream
of the Townsite
Ad in B.C. Saturday Sunset, 24 February 1912
Legislative Library of British Columbia

MAP 620 (page 249).
Cutting the Melon
Coquitlam Star, 24 April 1912

MAP 621 (page 249).
If you are one of those who have not yet profited
in Coquitlam
Ad in Vancouver World, 16 March 1912
Vancouver Public Library

MAP 622 (page 249).
Coquitlam Mercantile Center
Ad in Vancouver World, 15 January 1913
Vancouver Public Library

MAP 623 (page 250).
The Grand Trunk Pacific Railway. The Only All
Canadian Line
From: Christmas Globe, 1906, page 20

MAP 624 (page 250).
Map of the Grand Trunk Pacific Railway and
Branches
Stovel Co., 1906

MAP 625 (page 250).
Grand Trunk Pacific Railway
From a Liberal Party banquet card, 1906
Toronto Public Library

MAP 626 (page 251).
Map Showing Grand Trunk Pacific Railway and
the Lines of the Grand Trunk Railway System in
Canada
Grand Trunk Pacific, 1903
Library and Archives Canada NMC 48975

MAP 627 (page 251).
From Edmonton to Prince Rupert
From: Victoria Colonist, 30 May 1909

MAP 628 (page 252).
Prince Rupert and a Dawning Empire
(brochure cover)
Prince Rupert Publicity Club and Board of Trade, 1910
British Columbia Archives Library NWp 971.1Pr P957pu

MAP 629 (pages 252–53).
Over 1,000 Miles of Coast Line, Through the
Myriads of Beautiful Islands Forming the "Inland
Passage"/A Vast New Empire, of Which Prince
Rupert Is the Threshold
From: Prince Rupert and a Dawning Empire, Prince
Rupert Publicity Club and Board of Trade, 1910
British Columbia Archives Library NWp 971.1Pr P957pu

MAP 630 (page 253).
Map of the Grand Trunk Pacific Railway in British
Columbia Showing Terminus at Prince Rupert/
Map of Prince Rupert and Vicinity/Latitude Map
Poole Brothers, 1910
David Rumsey Collection

MAP 631 (page 254).
Map of Prince Rupert, British Columbia
Grand Trunk Pacific Town and Development
Company, 1909 (Brett & Hall 1908 plan)
Library and Archives Canada e008439686

MAP 632 (page 254).
Scarborough's New Map of British Columbia
Scarborough Co., 1914
Windermere Valley Museum

MAP 633 (page 254).
General Plan for the Development of Prince
Rupert, B.C.
Brett & Hall, 1908
From: Prince Rupert, British Columbia: The
Pacific Coast Terminus of the Grand Trunk Pacific
Railway, 1911

MAP 634 (page 254).
Sketch showing Rattenbury plan for Prince Rupert
harbour
Francis Mawson Rattenbury, 1913
British Columbia Archives Rattenbury papers,
RP(8), #5

MAP 635 (page 255).
The Grand Trunk Pacific Railway Has a Grade of
4/10 of 1%, Practically a Level Grade
(Map showing position of Prince Rupert in North
America)
From: Prince Rupert Journal, 4 July 1911

MAP 636 (page 255).
Map of Grand Trunk Pacific
Prince Rupert Daily News, 22 June 1912

MAP 637 (page 255).
Map of the Northern Interior of British Columbia
Shewing Undeveloped Areas
Provincial Bureau of Information Bulletin No. 22, 1908
Library and Archives Canada e10689801

MAP 638 (page 256).
B.C. Government Will Back Railroad to Fort
George
Natural Resources Security Company, 1912
From: B.C. Saturday Sunset, 2 March 1912
Legislative Library of British Columbia

MAP 639 (page 256).
Map showing central location of Prince George
Prince George Board of Trade, 1919

MAP 640 (page 256).
Study the Map—It Tells the Story
Natural Resources Security Company, 1913
From: B.C. Saturday Sunset, 7 May 1913
Legislative Library of British Columbia

MAP 641 (page 257).
The Grand Trunk Pacific Railway Co.'s Townsite of
Prince George/General Plan for the Development
of Prince George, B.C.
Brett & Hall, 1913
Fort George Herald, 12 April 1913
Prince George Public Library

MAP 642 (page 257).
Key Plan of Fort George Townsite and Additions
Inset map in Townsite of Fort George
Natural Resources Security Company, 1912

MAP 643 (page 258).
Plan Showing Provincial Government Property in
the Subdivisions of Lots 343, 936, 937, 938 & 1429,
Group 1 Cariboo District, Fort George and Prince
George, British Columbia
J.T. Armstrong, 1914
Land Title and Survey Authority 15T10 Old Maps

MAP 644 (page 258).
Fort George Townsite
Natural Resources Security Company, c. 1913
Fraser Fort George Museum/Exploration Place
A980-16-1

MAP 645 (page 258).
Plan Showing Provincial Government Property in
the Subdivisions of Lots 933, & 934, Group 1 Cari-
boo District, South Fort George, British Columbia
J.T. Armstrong, 1914
Fraser Fort George Museum/Exploration Place

MAP 646 (page 259).
Map showing location of Willow River
Ad in The Globe, 12 October 1912
University of British Columbia Rare Books and
Special Collections

MAP 647 (page 259).
Scarborough's New Map of British Columbia
Scarborough Company, 1914
Windermere Valley Museum

MAP 648 (page 259).
Close-in Acreage
Ad in Fort George Herald, 12 April 1913
Prince George Public Library

MAP 649 (page 259).
Map of Birmingham Townsite Destined to be the
Second Greatest City in British Columbia
Birmingham Townsite Company, 1910
Ad in Victoria Daily Colonist, 1 April 1910 (!)

MAP 650 (page 259).
Central British Columbia is attracting the
attention of the whole of America
Birmingham Townsite Company, 1910
From: British Columbia Centennial Committee, It
Happened in British Columbia, 1971, p.73
(no original source given)

MAP 651 (page 260).
Fort Fraser
Dominion Stock & Bond Corporation, 1913
Ad in B.C. Saturday Sunset, 4 October 1913
Legislative Library of British Columbia

MAP 652 (page 260).
Fort Fraser (The Hub of B.C. on the G.T.P.)
Fort Fraser Townsite, 1911
Ad in B.C. Saturday Sunset, 2 December 1911
Legislative Library of British Columbia

MAP 653 (page 260).
Fort Fraser, The Heart of British Columbia
Dominion Stock & Bond Corporation, c. 1912
Copy, source unknown

MAP 654 (page 260).
Fort Fraser, B.C., Land of Big Crops/A Sure Road
to Wealth
Dominion Stock & Bond Corporation, 1913
Ad in B.C. Saturday Sunset, 31 May 1913
Legislative Library of British Columbia

MAP 655 (page 260).
When the Last Spike Has Been Driven/Fort Fraser
(The Hub of B.C. on the G.T.P.)
Dominion Stock & Bond Corporation, 1912
Ad in B.C. Saturday Sunset, 13 January 1912
Legislative Library of British Columbia

MAP 656 (page 260).
Fort Fraser, A City Made Necessary by Nature
Dominion Stock & Bond Corporation, 1912
Ad in B.C. Saturday Sunset, 22 June 1912
Legislative Library of British Columbia

MAP 657 (page 261).
Vanderhoof Part of Sec. 4 & 9, T.P. 11, Range 5,
Coast District, B.C.
Blueprint, no author, no date (c. 1914); right inset
map
Vanderhoof Community Museum

MAP 658 (page 261).
Vanderhoof, British Columbia, Resurvey and
New Residence Sections
Walter Burley Griffin and Francis Barry Byrne,
c. 1914 or 1915
Vanderhoof Community Museum

MAP 659 (page 262).
Fraser Lake Townsite
Macmillam & Vollans, c. 1912
British Columbia Archives CM A1826

MAP 660 (page 262).
Map of Hubert
From a real estate brochure or ad, c. 1911
Bulkley Valley Museum

MAP 661 (page 262).
Amundsen, The G.T.P. Divisional Point at Auction,
May 11, 1912
Ad in B.C. Saturday Sunset, 27 April 1912
Legislative Library of British Columbia

MAP 662 (page 262).
Terrace Littleton Townsite
Townsite plan for George Little, 1911
Land Title and Survey Authority 12T4 Townsites

MAP 663 (page 262).
Grand Trunk Pacific Railway Official Townsite of
Smithers Range 5, Coast District, B.C.
Aldous & Murray, 1913
University of British Columbia Rare Books and
Special Collections Spam 9824

MAP 664 (page 262).
Smithers Official G.T.P. Town in the Heart of the
Bulkley Valley
Ad in B.C. Saturday Sunset, 13 September 1913
Legislative Library of British Columbia

MAP 665 (page 263).
A Birdseye View of Northern and Central British
Columbia Showing Country Tranversed by the
Grand Trunk Pacific Railway from Prince Rupert
to Fort George/Grand Trunk Pacific Railway
Official Townsite of Smithers
Aldous & Murray, 1913
University of British Columbia Rare Books and
Special Collections Spam 9824

MAP 666 (page 263).
Grand Trunk Pacific Railway Townsite at Hazelton
Ad in Prince Rupert Daily News, 23 November 1911

MAP 667 (page 263).
British Columbia 1913 Mining Divisions
British Columbia Department of Lands, 1913
Penticton Museum and Archives

MAP 668 (page 264).
Grand Trunk Pacific Townsite at Hazelton
Natural Resources Security Company, 1911
Ad in B.C. Saturday Sunset, 28 October 1911
Legislative Library of British Columbia

MAP 669 (page 264).
Only One Hazelton
Natural Resources Security Company, 1911
Ad in B.C. Saturday Sunset, 30 December 1911
Legislative Library of British Columbia

MAP 670 (page 264).
Put Ten Dollars into the New City of Hazelton
Standard Securities, 1911
Ad in B.C. Saturday Sunset, 30 December 1911
Legislative Library of British Columbia

MAP 671 (page 264).
Untitled bird's-eye map of Hazelton area with proposed Grand Trunk Pacific townsite of South Hazelton
From: *GTP South Hazelton: Northern Interior Metropolis*, Grand Trunk Pacific Railway, 1913
Frank Leonard; copy of Vancouver Public Library item now lost

MAP 672 (page 265).
Map of Jasper Park, Alberta, Mount Robson Park, British Columbia. Canadian Rocky Mountains. Grand Trunk Pacific Railway
Grand Trunk Pacific Railway, 1916
From: Grand Trunk Pacific Railway, *The North American Alps, Canadian Rockies, Mount Robson Route*, 1916
Library and Archives Canada FC3807 G73

MAP 673 (pages 264–65).
Map of the Central Section of British Columbia Shewing the Country Served by the Grand Trunk Pacific Railway
Grand Trunk Pacific Railway/Poole Brothers, 1911
Library and Archives Canada NMC 135838

MAP 674 (page 266).
Map of the Northern Interior of British Columbia
Adrien-Gabriel Morice, published by Department of Lands and Works, 1907
Library and Archives Canada H2/602/1907

MAP 675 (page 266).
Carte des Sources et du Bassin Supérieur de la Nétchakhoh
Adrien-Gabriel Morice, *Bulletin de la Société Neuchâteloise de Géographie*, Vol. 15, 1904
Land Title and Survey Authority 25T3 Old Maps

MAP 676 (page 267).
Christian Community of Universal Brotherhood/ A Brilliant Orchard
Jam label, c. 1920
Nanaimo Museum

MAP 677 (page 267).
Columbia River Fruit Lands West Kootenay Dist. British Columbia
J.D. Anderson, 1916
British Columbia Archives 90/55/46-1487 Nelson

MAP 678 (pages 268–69).
Thompson River at Pennys, B.C.
Robert H. Lee, 1909
Kamloops Museum and Archives

MAP 679 (page 269).
British Columbia, Kamloops Sheet West of Sixth Meridian
Surveyor General, Ottawa, 1916

MAP 680 (page 270).
Sweet Briar Subdivision, Chilliwack, B.C.
Cawley and Carmichael, ad in *B.C. Saturday Sunset*, 18 March 1911
Legislative Library of British Columbia

MAP 681 (page 270).
Choice Apple Land on the Chilliwack Tram
John D. Kearns, ad in *B.C. Saturday Sunset*, 12 November 1910
Legislative Library of British Columbia

MAP 682 (page 270).
Ad for land on Matsqui and Sumas prairies
Lindsay Russell, *B.C. Saturday Sunset*, 10 September 1910
Legislative Library of British Columbia

MAP 683 (page 271).
Bird's Eye View of Country Covered by B.C. Electric Railway System
From: *Tips for Tourists. Interurban Trips over B.C. Electric Railway System in Vicinity of Vancouver, B.C.*, B.C. Electric Railway Company, 1913
University of British Columbia Rare Books and Special Collections SPAM287

MAP 684 (page 271).
B.C. Electric Ry. Co. Ltd. Mainland System
B.C. Electric Railway Company, 1923
City of Vancouver Archives Map 812

MAP 685 (page 271).
B.C. Electric Railway Company Limited. Vancouver Island System
B.C. Electric Railway Company, 1923
City of Vancouver Archives Map 812

MAP 686 (page 271).
Map of Vancouver and vicinity
Canada Department of Mines, 1923 (1919–20 survey)
Delta Archives

MAP 687 (page 272).
British Columbia
G.G. Aitken, Department of Lands, 1912
Vancouver Public Library Special Collections 912.711 B8630b 1912

MAP 688 (page 272).
Official Map of Port Mann, Pacific Terminus of the Canadian Northern Railway
From: *Vancouver Province*, 20 February 1912
City of Surrey Archives

MAP 689 (page 272).
Map of the Port Mann Townsite
Anon., c. 1909

MAP 690 (page 272).
Survey map of the Port Mann Townsite
Anon., 1818
Land Title and Survey Authority 150484 and 150485

MAP 691 (page 273).
Key Map of District Showing Location of Port Mann, the Pacific Terminal of the Canadian Northern Railway/Plan of Centre Portion of Port Mann Townsite
Hubert Nevile Smith, 1912
Library and Archives Canada NMC 5974

MAP 692 (page 273).
Plan of Part of New Westminster District
D.R. Harris, updated to 1912 by Elliott & Hewett, 1913
White Rock Museum and Archives

MAP 693 (page 273).
Port Mann Official Townsite/A Dog-On-Good-Proposition/Gentlemen, it is a Good Proposition: Port Mann
Ad in *B.C. Saturday Sunset*, 30 March 1912
Legislative Library of British Columbia

MAP 694 (page 274).
Scarborough's New Map of British Columbia
Scarborough Co., 1914
Library and Archives Canada F/600/1914

MAP 695 (page 274).
Canadian Northern Railway System
From: *Hints to Intending Emigrants*, Canadian Northern Railway, 1910

MAP 696 (page 274).
Indexed Guide Map of the City of Vancouver and Suburbs
Vancouver Map and Blueprint Co., 1914
University of British Columbia Library 624.9a Vancouver 1914

MAP 697 (page 274).
Proposed Terminus for Canadian Northern Pacific Railway
Anon., 1912
City of Vancouver Archives Add MSS 44 502 C7 File 2

MAP 698 (page 275).
Plan of Former Songhees Reserve, Victoria, Showing Railway Terminals, Wharves, and Street Connections
D.O. Lewis, Canadian Northern Railway, c. 1916
University of British Columbia Rare Books and Special Collections 616.9(88) B8 1916

MAP 699 (page 275).
Canadian National Railways
Canadian National Railways, 1929

MAP 700 (page 275).
British Columbia
G.G. Aitken, Department of Lands, 1912
Vancouver Public Library Special Collections 912.711 B8630b 1912

MAP 701 (page 276).
Field Hill Spiral Tunnels (cartoon with John E. Schwitzer)
W.W.S., 1911
Canadian Pacific Railway Archives A.15510

MAP 702 (page 277).
Plan of possible track configurations in Kicking Horse Canyon
Blueprint, John E. Schwitzer, 1906 or 1907
University of British Columbia Rare Books and Special Collections, Chung Collection

MAP 703 (page 276).
Hector-Field "Grade Reduction"
From: *B.C. Saturday Sunset*, 8 March 1908
Legislative Library of British Columbia

MAP 704 (page 276).
Bird's-eye map of the Spiral Tunnels
Anon., c. 1950s
West Coast Railway Heritage Park

MAP 705 (page 277).
Bird's-eye map of the Spiral Tunnels
Canadian Pacific Railway, 1930s
University of British Columbia Rare Books and Special Collections, Chung Collection

MAP 706 (page 277).
Reconnaissance Map of the Northern Selkirk Mountains and the Big Bend of the Columbia River, British Columbia
Howard Palmer and Robert H. Chapman, 1915
Revelstoke Museum

MAP 707 (page 277).
Connaught Tunnel—Canadian Rockies
Postcard, c. 1916

MAP 708 (page 278).
Sketch Map Showing Location of Right of Way of H.S. & N. Ry.
From: *Howe Sound & Northern Ry. Memorandum*, 1911
British Columbia Archives NWp 385 H858m

MAP 709 (page 278).
Plan of Subdivision of Part of D.L. 486 N.W.D. (Newport Townsite)
Ernest Albert Cleveland, Cleveland & Cameron, 1912
Land Title and Survey Authority, New Westminster Land Title Office, Plan 3960, DL 486

MAP 710 (page 279).
British Columbia
G.G. Aitken, Department of Lands, 1912
Vancouver Public Library Special Collections 912.711 B8630b 1912

MAP 711 (page 279).
Pacific Great Eastern Railway Plan of Location . . Mile 3 to Mile 11.2 (Ambleside portion)
John Callaghan, Chief Engineer, Pacific Great Eastern Railway, August 1912
Land Title and Survey Authority 5Tube131 Rlys.

MAP 712 (page 279).
Southern British Columbia. Map Showing Disposition of Lands
F.C.C. Lynch, Canada Department of the Interior, 1914
University of British Columbia Rare Books and Special Collections G3511.G4 1914 .C3

MAP 713 (page 279).
Pacific Great Eastern Railway from Vancouver to Fort George, British Columbia
Pacific Great Eastern Railway, c. 1914
Library and Archives Canada Amicus 9831690

MAP 714 (page 279).
Map of the City and Municipality of North Vancouver. B.C.
Dawson & Elliott, 1910
Land Title and Survey Authority 10T1 Vancouver Town

MAP 715 (page 280).
Map Showing (Approximate) Holdings of the Grand Trunk Lands Co.
From: *Waiting: British Columbia's Inland Empire*, Grand Trunk Lands Company, c. 1910
Library and Archives Canada FC3808.2

MAP 716 (page 280).
Map of Pacific Great Eastern (built and projected line)
From: *Pacific Great Eastern: The Short Route to the Goldfields*
British Columbia Provincial Bureau of Information, c. 1913
University of British Columbia Rare Books and Special Collections spam9107

MAP 717 (page 280).
Map of Pacific Great Eastern Railway Vancouver to Fort George
John Callaghan, Chief Engineer, Pacific Great Eastern Railway, June 1915 (with updates to 1920)
University of British Columbia Rare Books and Special Collections G3511.P3 1915 .P3

MAP 718–MAP 722 (page 281).
Map of Pacific Great Eastern Railway Vancouver to Fort George
John Callaghan, Chief Engineer, Pacific Great Eastern Railway, June 1915 (with updates to 1920)
University of British Columbia Rare Books and Special Collections G3511.P3 1915 .P3

MAP 723 (page 282).
Sketch Route Map P.G. & E. Rly
From: J.G. Sullivan, *Report on the Engineering and Economic Features of the Pacific Great Eastern Railway*, 1922
University of British Columbia Rare Books and Special Collections HE2810.P2 S8 1922

MAP 724 (page 282).
The Route of the Pacific Great Eastern
From: *British Columbia Farm Lands along the line of the Pacific Great Eastern Railway*
British Columbia Legislative Assembly, 1925
Library and Archives Canada C.O.P. cop. BC. 7226

MAP 725 (page 283).
Map of British Columbia shewing Existing & Projected Railways
From: *A Townsite Most Interesting: Fort Salmon*
Ad in *B.C. Saturday Sunset*, 14 December 1912
Legislative Library of British Columbia

MAP 726 (page 283).
Map showing Fort George's commercial control of Central British Columbia
Ad in *B.C. Saturday Sunset*, 3 May 1913
Legislative Library of British Columbia

MAP 727 (page 283).
Colossus of Roads
Cartoon in *This Week*, 9 March 1912

MAP 728 (page 283).
Brownlee's New Indexed Map of British Columbia South of 54° North Latitude
Brownlee Company, 1898
Land Title and Survey Authority 107544

MAP 729 (page 284).
Triangulation Survey of Portion of East Kootenay. Plan Shewing Position of Mineral Monuments
Lands and Works, 1898
Land Title and Survey Authority 7T9 Old Maps

MAP 730 (page 284).
Triangulation Survey of Portion of East Kootenay. Plan Shewing Position of Mineral Monuments
Ernest Albert Cleveland, 1898
Fort Steele Archives

MAP 731 (page 284).
Outline Map of Part of the Railway Belt British Columbia
From: *Report of the Triangulation of the Railway Belt of British Columbia Between Kootenay and Salmon Arm Bases Surveyed by P.A. Carson and M.P. Bridgeland*, 1915

MAP 732 (page 285).
Southwestern Districts of British Columbia (with MS triangulation survey added)
Lands and Works, 1912
Land Title and Survey Authority 1T8 Old Maps

MAP 733 (page 285).
Reduction of Surveyor's Sketch Plan accompanying Exploratory Survey Report
Frank Swannell, Surveyor General Department, 1913

MAP 734 (page 285).
Interprovincial Boundary Commission. Boundary Between Alberta and British Columbia, Index Map
Topographical Survey of Canada, 1925
Land Title and Survey Authority: Part III, B.C.–Alberta Boundary (Atlas), Sheets 29–54

Map 735 (page 285).
Interprovincial Boundary Commission. Boundary Between Alberta and British Columbia, Sheet No. 29
Topographical Survey of Canada, 1925
Land Title and Survey Authority: Part III, B.C.–Alberta Boundary (Atlas), Sheets 29–54

Map 736 (page 286).
Sketch Plan Salmon Valley Lands Sayward District Open for Pre-emption May 18th 1914
Anon., 1914
Campbell River Museum

Map 737 (page 286).
Map Shewing Portion of the West Coast of Vancouver Island
Lands and Works, 1905
Vancouver Public Library Special Collections

Map 738 (page 286).
Guide Map to the Peace River for Land-Seekers and Settlers
W.R. Stevenson, 1911
Land Title and Survey Authority

Map 739 (page 287).
Pre-emptor's Map, Fort George Sheet
Department of Lands, 1914
Library and Archives Canada NMC 96013

Map 740 (page 287).
Southern British Columbia. Map Showing Disposition of Lands
Department of the Interior, 1914
University of British Columbia Rare Books and Special Collections G3511.G4 1914 .C3

Map 741 (page 288).
Southern Okanagan Lands. Area for Proposed Townsite
Anon. (Department of Lands, Water Rights Branch?), 1916
British Columbia Archives Water Rights Plans Box 1: 90-55-46 1778-Fairview

Map 742 (page 288).
Plan of the Townsite of Oliver
Department of Lands, c. 1921
Oliver Archives (from Penticton Museum 1994)/Oliver and District Heritage Society, Dr. D.B. Robinson Memorial Archives

Map 743 (page 288).
Southern Okanagan Irrigation Project Subdivision of Portion of D.L. 2450S. Similkameen Division, Yale District
Department of Lands, c. 1921
Oliver Archives Map 119a, 2002.017/Oliver and District Heritage Society, Dr. D.B. Robinson Memorial Archives

Map 744 (page 289).
Mineral Reference Map (Showing Surveyed Claims) Covering Portions of Greenwood & Osoyoos Mining Divisions, Similkameen District
Department of Mines, 1934
Oliver Archives Map 35, M992-134/Oliver and District Heritage Society, Dr. D.B. Robinson Memorial Archives

Map 745 (page 289).
Water Resource Investigation to Accompany Report on Southern Okanagan Lands Project
British Columbia Department of Lands and Forests, Water Rights Branch, 1955
University of British Columbia Library G3512.O36N44 1963 .B7

Map 746 (pages 290–91).
Canadian Rockies Showing Main Line of Canadian Pacific Ry. with Branch Lines & Steamship Connections
From: Resorts in Canadian Pacific Rockies, Canadian Pacific Railway, 1917

Map 747 (page 291).
Part of the Selkirk Range, British Columbia
W.S. Green, 1888 (survey), 1905 (printed)
Library and Archives Canada G3512.S4 1905 .W4

Map 748 (page 291).
The Canadian Pacific Rockies
Arthur O. Wheeler and Canadian Pacific Railway, 1927
Hudson's Bay Company Archives G3/85

Map 749 (page 292).
Bungalow Camps in the Canadian Pacific Rockies
Canadian Pacific Railway, 1923
Glenbow Museum and Archives

Map 750 (page 292).
Canadian Rockies Showing Main Line of Canadian Pacific Ry. with Branch Lines & Steamship Connections
From: Resorts in Canadian Pacific Rockies, Canadian Pacific Railway, 1917

Map 751 (page 292).
The Canadian Pacific Rockies
Arthur O. Wheeler and Canadian Pacific Railway, 1927
Hudson's Bay Company Archives G3/85

Map 752 (pages 292–93).
Panorama of the Canadian Rockies
Canadian Pacific Railway, 1936

Map 753 (page 293).
Bird's-eye map of the Upper Columbia and Upper Kootenay valleys showing the location of Radium Hot Springs
From: Radium Hot Springs British Columbia: A Fascinating Proposition (Investor prospectus), 1913
Glenbow Museum and Archives

Map 754 (pages 294–95).
The Triangle Tour of British Columbia
From: Canadian National Railways, Map of Canadian Rockies and the Triangle Tour of British Columbia, 1929

Map 755 (page 295).
Untitled map of Canadian National Railways routes in British Columbia
Inset in: Routes of the Canadian National Steamships Pacific Coast Service
From: Canadian National Railways, Map of Canadian Rockies and the Triangle Tour of British Columbia, 1929

Map 756 (page 296).
Expedition Under Auspices of the Pacific Highway Association
Pacific Highway Association, 1911

Map 757 (page 296).
Sinnott & Hodgson's Vest Pocket Map of Vancouver
Sinnott & Hodgson, c. 1907

Map 758 (page 297).
Change of Rule of Road in British Columbia 1920/Map of the Province of British Columbia 1920
Department of Highways/Geographic Branch, 1920
Rossland Museum

Map 759 (page 297).
Map Showing Automobile Roads in Vancouver, B.C., & District
From: Vancouver, Canada. The Georgian Circuit, Publicity Department of the Vancouver Exhibition Association, 1919

Map 760 (page 298).
Automobile Road from Bellingham to Vancouver, B.C.
Automobile Club of Western Washington, 1917

Map 761 (page 298).
Map of the Cariboo Highway and connecting Main Roads between Vancouver and Prince George
Cariboo Automobile Association, 1927
Hudson's Bay Company Archives A92.18.37 folio 89

Map 762 (page 298).
Canadian Rockies Circle Tour
From: The Banff-Windermere Highway, Canadian National Parks Branch, 1923

Map 763 (page 299).
Untitled map showing location of Kicking Horse Trail
From: The Kicking Horse Trail: Scenic Highway from Lake Louise, Alberta, to Golden, British Columbia
Department of the Interior, Canadian National Parks Branch, 1927
University of British Columbia Rare Books and Special Collections spam4466

Map 764 (page 299).
The Okanagan Circuit & B.C. Route: Vancouver, Spokane
From: Kelowna, the Orchard City by the Lake, 1918
Kelowna Public Archives

Map 765 (page 299).
Map 3
From: British Columbia's Picturesque Highways
British Columbia Government Travel Bureau, 1940

Map 766 (page 300).
Motor Roads to the Canadian Rockies
Calgary Good Roads Association, 1923
Fort Steele Archives 2009.001.099

Map 767 (page 300).
Good Roads Everywhere/Touring Canada/Showing Trans-Canada Highway
Good Roads Association/National Highways Association, 1928
Library and Archives Canada G3301.P2 1928.N37

Map 768 (page 301).
Highway Map of British Columbia
Shell Oil, 1929

Map 769 (page 301).
Map No. 6/B.C. Strip Map No. 6
British Columbia Advertisers, 1928

Map 770 (page 302).
British Columbia Highway and Travel Map
Department of Public Works and Department of Lands, 1930
Library and Archives Canada NMC44051

Map 771 (page 303).
Road Map of the Fraser Valley, British Columbia
Publicity Bureau of Greater Vancouver, 1936
Vancouver Public Library Special Collections 912.71137 C5812r 1936

Map 772 (page 303).
Map of Ladner ferry route, c. 1925

Map 773 (page 303).
Guide Map Vancouver—New Westminster Burnaby and North Shore Municipalities
City Map and White Print Company, 1935

Map 774 (page 303).
Map Showing King George VI Highway
From: opening day booklet, 1940
City of Surrey Archives 89.67

Map 775 (page 303).
Vancouver and Vicinity
Imperial Oil road map, 1937

Map 776 (page 304).
Untitled map of the area south of the Fraser River
British Boundary Commission, 1862
Land Title and Survey Authority 26T1 Roads and Trails

Map 777 (page 304).
Map of Part of Westminster District, B.C., and Part of Washington Territory
Canadian Pacific Railway Engineering Office, 1888
University of British Columbia, Rare Books and Special Collections G3512.L68 1888.C36

Map 778 (page 304).
1905 Map of New Westminster District
D.R. Harris, 1905
Vancouver Public Library Special Collections 912.71135 H31n 1905 copy 2

Map 779 (page 304).
Lower Fraser Valley, Parts of New Westminster and Yale Districts/Preliminary Map
G.G. Aitken, Department of Lands, 1931
Vancouver Public Library Special Collections 912.71137 A31L 1931

Map 780 (page 305).
Bird's-eye map of Vancouver
From: Fourteen Days in Vancouver, Greater Vancouver Publicity Bureau, 1928

Map 781 (page 305).
Government Lands, Point Grey, B.C.
[Illegible signature] (Consulting Engineer), 1925
Land Title and Survey Authority (no reference number)

Map 782 (page 305).
Commercial Map of Greater Vancouver and District
Vancouver Map and Blueprint, 1930
City of Vancouver Archives Map 429

Map 783 (page 305).
Pictorial Map of British Columbia
Vancouver Map & Blue Print Company, 1934
Vancouver Public Library Special Collections 338.209711 B8623v 1934

Map 784 (page 306).
The Civic Centre, City of Vancouver
From: A Plan for the City of Vancouver British Columbia Including Point Grey and South Vancouver and a General Plan of the Region, Harland Bartholomew, 1929

Map 785 (page 306).
Plan of Civic Centre Burrard Street Site
From: A Plan for the City of Vancouver British Columbia Including Point Grey and South Vancouver and a General Plan of the Region, Harland Bartholomew, 1929

Map 786 (page 306).
Greater Vancouver, British Columbia, Large Parks & Proposed Pleasure Drive System
From: A Plan for the City of Vancouver British Columbia Including Point Grey and South Vancouver and a General Plan of the Region, Harland Bartholomew, 1929

Map 787 (page 306).
Plan Showing Parcels 'A,' 'B,' & 'C,' Kitsilano Indian Reserve No. 6 of the Squamish Band, Vancouver, B.C.
George F. Fountain, 1930/countersigned 1934 with some additions to 1948
© Government of Canada. Reproduced with the permission of the Minister of Public Works and Government Services Canada (2011)
Library and Archives Canada / Department of Indian Affairs and Northern Development fonds/RG10M 76703/9, Item 501/NMC 13444

Map 788 (page 306).
Map of Vancouver. Unemployment and Relief. May, 1934
James Skitt Matthews and Frank C. Turner, City of Vancouver Archives, 1934
City of Vancouver Archives Map 36

Map 789 (page 307).
General Plan for Capilano Estates Limited West Vancouver, B.C.
Olmsted Brothers, 1937
Vancouver Public Library Special Collections 912.71133 G32b 1937

Map 790 (page 307).
Plan of Reclamation Works and Dams Constructed on District Lots 193, 194, 597 & 598, Group 2, New Westminster District for George C. Reifel, Esq.
Burnett & McGuigan, 1930
Delta Municipal Archives

Map 791 (page 307).
Untitled map showing the position of Vancouver in North America.
From a brochure issued for the Golden Jubilee of Vancouver, 1936

Map 792 (page 307).
This is CRCV Vancouver, the Voice of British Columbia
Cartoon, Jack Boothe, Province newspaper, February 1937

Map 793 (page 308).
Circuit Tour, Vancouver Island
Canadian Pacific Railway B.C. Coast Service, c. 1930

Map 794 (page 308).
Map of Canadian Pacific Railway coastal steamer routes
Canadian Pacific Railway B.C. Coast Service, 1927

Map 795 (page 308).
Auto Ferry Services to Vancouver Island
Canadian Pacific Railway, 1923
Vancouver Public Library Special Collections 912.7112 C2124a 1923

Map 796 (page 308).
The Triangle Service
Canadian Pacific Railway B.C. Coast Service, 1927

MAP 797 (page 309).
The Magic Day Sea-Trip through Howe Sound,
Vancouver's Glorious Inland Sea by Union
Steamships
Union Steamship Company, c. 1920s

MAP 798 (page 309).
Union Steamships route map
Union Steamship Company, c. 1948

MAP 799 (page 309).
Birdseye View of Puget Sound Country and
Vicinity
Puget Sound Navigation Company, 1910
Chris Warner Collection

MAP 800 (page 309).
The Path of Sunshine and Sea Charm along the
Glorious Gulf Coast Route of the Union Steamship
Company
Union Steamship Company, c. 1930s
Sechelt Community Archives

MAP 801 (page 310).
Topographical Sketch Map Shewing the Route
Traversed by the Bedaux Sub-Arctic Expedition
1934 through the Rocky Mountains, Cassiar and
Peace River Districts, British Columbia, Canada
Frank Swannell, 1935
Glenbow Archives G3512.N876 1934 .S972

MAP 802 (page 310).
Plan of the Region about the 58th Parallel
from Longitude 124° 30´ to Longitude 130° West,
Northern British Columbia, Canada
Ernest C.W. Lamarque, 1936–37
Land Title and Survey Authority

MAP 803 (page 311).
Wells Gray Park
C.P. Lyons/British Columbia Government Travel
Bureau, c. 1940

MAP 804 (page 311).
Strathcona Park reserve
Canadian Pacific Railway, 1927

MAP 805 (page 311).
Tweedsmuir Park, British Columbia
British Columbia Government Travel Bureau, c. 1940

MAP 806 (page 312).
Official Map of American and Canadian Airways
and Aerial Mail Routes
Aeronautic Maps Association, 1919
National Oceanic and Atmospheric Administration
Central Library

MAP 807 (page 312).
Map of Air Mail Routes
Unknown, 1929

MAP 808 (page 312).
Map of proposed airport at Burnaby Lake
Anon., c. 1928

MAP 809 (page 312).
Vancouver Airport (at Spanish Banks)
A.C.R. Yuill and D.R. MacLaren, 8 February 1928
City of Vancouver Archives Map 377

MAP 810 (page 312).
United Airlines Route Map
United Airlines, 1936

MAP 811 (page 312).
City of Vancouver. Preliminary Drawing of Airport
& Seaplane Harbor, Sea Island, Municipality of
Richmond, B.C.
William Templeton, with approval signatures,
c. 1929
City of Richmond Archives 1985 94 1

MAP 812 (page 312).
Vancouver Civic Airport and Seaplane Harbour
Official Opening July 22nd–25th, 1931
City of Richmond Archives Reference files

MAP 813 (page 313).
British Columbia Airport Map
Lands and Works, 1930
Library and Archives Canada NMC 44051

MAP 814 (page 313).
Trans-Canada Airways
Department of Transport, 1937

MAP 815 (page 313).
Trans-Canada Airlines (route map)
Trans-Canada Airlines, 1940

MAP 816 (page 313).
Canadian Pacific Railway All Service Map
Canadian Pacific Railway, 1957
Canadian Pacific Railway Archives

MAP 817 (page 313).
Aeronautical Route Chart 1:1,000,000/Vancouver-
Medicine Hat Arc 1
Department of Transport, 1957
Penticton Museum and Archives

MAP 818 (page 314).
Pacific Coast Defences, 1939–1940
Historical Section, Army General Staff, 1945(?)
15th Field Artillery Regiment Museum and Archives

MAP 819 (page 314).
AAF Aeronautical Chart
United States Air Force, 1945

MAP 820 (page 314).
Plan of the fortifications on Yorke Island
Department of National Defence, c. 1945
15th Field Artillery Regiment Museum and Archives

MAP 821 (page 314).
Fortifications/Point Grey Battery, Vancouver, B.C.
Department of National Defence Record Plan
S-2052-2, 1945
15th Field Artillery Regiment Museum and Archives

MAP 822 (page 315).
Vancouver, B.C. Coast Defence Chart
Department of National Defence, Record Plan
S-2052-0-A, 1945
15th Field Artillery Regiment Museum and Archives

MAP 823 (page 316).
Prince Rupert, B.C. Coast Defence Chart
Department of National Defence, Record Plan
S-2054-0-A, 1945, corrected to 1946
15th Field Artillery Regiment Museum and Archives

MAP 824 (page 316).
Map of landing places of Japanese fire balloons
Canadian War Museum, c. 2008

MAP 825 (page 316).
Nanaimo Military Area
Department of National Defence, 1943
British Columbia Archives, Buckham Collection

MAP 826 (page 317).
Detail Site Plans, Albert Head
Department of National Defence, Record Plan
S-2051-8-1, 1945
15th Field Artillery Regiment Museum and Archives

MAP 827 (page 317).
Amazing Invasion Map: Japs Plan to Seize West Coast
Reproduced in the Los Angeles Examiner,
22 August 1943

MAP 828 (page 318).
Map of Japanese internment camp locations in the
Slocan Valley
Matsuru Kitamura, 1946
Japanese Canadian National Museum

MAP 829 (page 318).
Map of Popoff Japanese internment camp
Matsuru Kitamura, 1946
Japanese Canadian National Museum

MAP 830 (page 318).
Map of Bay Farm Japanese internment camp
Matsuru Kitamura, 1946
Japanese Canadian National Museum

MAP 831 (page 318).
Map of Lemon Creek Japanese internment camp
Matsuru Kitamura, 1946
Japanese Canadian National Museum

MAP 832 (page 319).
Alaska Highway and Connections
Alberta Government Travel Bureau, c. 1950

MAP 833 (page 319).
National Topographic Series, Air Navigation
Edition, Liard River, British Columbia
Department of Mines and Resources, 1943
Library and Archives Canada G3401.P6 s501C36
94 NW1943

MAP 834 (page 319).
"Flight Strips" and Repeater Stations Along Route
of Alaska Highway
Alaska Road Commission (attrib.), 1943
Glenbow Archives G3507 A323 1943 F621

MAP 835 (page 320).
Russian topographic map of Victoria, 1978

MAP 836 (page 320).
Greater Vancouver British Columbia. Civil
Defence Evacuation and Survival Plan for
Greater Vancouver Target Area
Vancouver Civil Defence, 1957
City of Vancouver Archives PAM 1957-28

MAP 837 (page 320).
Inaugural Train, P.G.E. Southern Extension,
August, 1956
Pacific Great Eastern Railway, 1956
University of British Columbia Rare Books and
Special Collections spam 16644

MAP 838 (page 320).
Map of the Pacific Great Eastern Railway Showing
Stations, Railway Connections and Pacific Great
Eastern Microwave System
Pacific Great Eastern Railway, c. 1960
Vancouver Public Library Special Collections

MAP 839 (page 321).
Pacific Great Eastern map logo on caboose
West Coast Railway Heritage Park

MAP 840 (page 321).
Pacific Great Eastern Railway/A Modern Railway
Serving the North
Pacific Great Eastern Railway poster, c. 1960
West Coast Railway Heritage Park

MAP 841 (page 321).
Pacific Great Eastern Railway
From: Patterns of Growth, Pacific Great Eastern
Railway, 1971
University of British Columbia Library

MAP 842 (page 321).
Anzac-Tumbler Ridge Branch Line
From: Construction of the Tumbler Ridge Branch
Line in Northeastern British Columbia, Prepared
for the American Railway Engineering Association
by British Columbia Railway Engineering Depart-
ment, March 1984
University of British Columbia Library

MAP 843 (page 322).
General map of the Kemano Project
Reprinted from The Engineering Journal, April 1953
Vancouver Public Library

MAP 844 (page 322).
Sketch Showing Arrangement of Ten Mile Tunnels,
Penstocks, Powerhouse, Roads and Start of
Transmission Lines in Relation to Topography
C.W. Abrahamson, 1953
Reprinted from The Engineering Journal, April 1953.
Vancouver Public Library

MAP 845 (page 322).
Town Plan for Kitimat, B.C.
Master Plan of the Townsite
Mayer B. Whittlesey and M. Milton Glass, 1952
Aluminum Company of Canada
Vancouver Public Library

MAP 846 (page 323).
Acrylic model/map of Copper Mountain Mine
Frank Burgess, Harry W. Day, and Keith Fahrni,
1948–49
Princeton Museum

MAP 847 (page 323).
Isometric and Pictographic Plan of the Granby
Consolidated Mining, Smelting & Power Company
Limited. Copper Mountain Mine
R.A. Brooke, c. 1947
Princeton Museum

MAP 848 (page 323).
Copper Mountain. Plan of Townsite Showing
Waterlines
Granby Consolidated Mining, Smelting & Power
Company, 1956
Princeton Museum

MAP 849 (page 324).
Ripple Rock Before and After Demolition, 1958
Canadian Hydrographic Service, 1958
University of British Columbia Map Library
G3512.R5 C31 1958 C3

MAP 850 (page 324).
Vancouver Island/Discovery Passage/Seymour
Narrows
George Henry Richards, 1860 and Daniel Pender
British Admiralty Chart, 1866
University of British Columbia Rare Book and Special
Collections G3511 .P5 svar G7 Chart 538 1867a

MAP 851 (page 324).
Master Plan—Roberts Bank Development
Swan Wooster Engineering for the National Harbours
Board, 1969

MAP 852 (page 324).
Western Delta Lands Inc., Delta Centre, B.C., Deep
Sea Port & New Community
Leman Group Inc., 1988
Delta Archives

MAP 853 (page 324).
Lulu Island and Sea Island Reclamation
and Development
CBA Engineering, 1957
City of Richmond Archives Reference files

MAP 854 (page 325).
British Columbia Ferries (route map)
British Columbia Ferry Corporation, 1971

MAP 855 (page 325).
The Scenic Circle/Mill Bay Ferry
Mill Bay Ferry advertising card and timetable, 1936

MAP 856 (page 325).
MV Queen of the North (floor plan)
British Columbia Ferry Corporation, 1981

MAP 857 (page 326).
Sketch Plan Showing Power-House and
Transmission Line, Hydro-Electric Power Plant,
City of Nelson
A.L. McCulloch, city engineer, 1905
British Columbia Archives 91 55 2 1507 Nelson

MAP 858 (page 326).
South Central British Columbia Electrical Systems
British Columbia Power Commission, 1953
University of British Columbia Map Library G3512.
S678N4 1953 .B7

MAP 859 (page 326).
Power for Progress
Ad by British Columbia Power Commission in
Journal of Commerce, 29 October 1949
Vancouver Public Library

MAP 860 (page 327).
Columbia River Basin Canada Showing Treaty
Dams and Storage Areas
B.C. Hydro Progress Report 1, 1965
British Columbia Archives GR 0880, Box 21

MAP 861 (page 327).
Arrow Project
Plan on aerial photo, April 1965
B.C. Hydro Progress Report 2, 1965
British Columbia Archives GR 0880, Box 21

MAP 862 (page 328).
Construction plan for Arrow Dam
Arrow Project Progress Report for May, June, July, 1965
B.C. Hydro, 1965
British Columbia Archives GR 0880, Box 21

MAP 863 (page 328).
Lower Fraser Valley Diversion Routes
From: Fraser River Board, Preliminary Report on
Flood Control and Hydro-Electric Power in the
Fraser River Basin, 1958
University of British Columbia Library

Map 864 (page 329).
Peace River Sites Examined
From: Wenner-Gren British Columbia Develop-
ment Co., Submitted by British Thomson-Houston
Export Co., *Report on the Feasibility of Building
Dams on the Peace River*, c. 1960
British Columbia Archives GR 0880, Box 57

Map 865 (page 329).
Map of Peace River, British Columbia
British Columbia Department of Lands, c. 1939

Map 866 (page 329).
*Portage Mountain Development Map of
Catchment Area and River Profiles*
B.C. Hydro, 1961
British Columbia Archives GR 0880, Box 57

Map 867 (page 329).
*Portage Mountain Development Peace River
Canyon Area Plan Showing Location of Materials
for Construction*
B.C. Hydro, 1961
British Columbia Archives GR 0880, Box 57

Map 868 (page 329).
Transmission Line Routes Peace River–Vancouver
From: Wenner-Gren British Columbia Develop-
ment Co., Submitted by British Thomson-Houston
Export Co., *Report on the Feasibility of Building
Dams on the Peace River*, c. 1960
British Columbia Archives GR 0880, Box 57

Map 869 (page 329).
*Mica Project General Location Plan Potential
Borrow Areas, Quarry Areas, and Haul Road
Locations*
From: B.C. Hydro, Progress Report 2, 1965
British Columbia Archives GR 0880, Box 21

Map 870 (page 330).
British Columbia Canada Public Carrier Map
(cover)
British Columbia Government Travel Bureau, 1950

Map 871 (page 330).
British Columbia (Shell road map)
H.M. Gousha Company, 1958

Map 872 (page 330).
British Columbia
Department of Lands and Forests, 1945
Hudson's Bay Company Archives RG1/87/102/10/i-iv

Map 873 (page 330).
Shell Official Road Map of British Columbia
H.M. Gousha Company, 1940

Map 874 (page 330).
*Lower Fraser Valley, Parts of New Westminster
and Yale Districts/Preliminary Map*
Department of Lands, 1931
Vancouver Public Library Special Collections
912.71137 A31L 1931

Map 875 (page 330).
*British Columbia Presents the Hope–Princeton
Highway*
Ad in *Journal of Commerce*, 5 November 1949

Map 876 (page 331).
British Columbia, Canada
British Columbia Government Travel Bureau/
H.M. Gousha Company, 1958

Map 877 (page 331).
Land of Enchantment Okanagan Valley
British Columbia Government Travel Bureau, 1957

Map 878 (page 331).
British Columbia, Canada
British Columbia Government Travel Bureau/
H.M. Gousha Company, 1958

Map 879 (page 331).
The Trans-Canada Highway
From: *Adventure along the Trans-Canada Highway*,
Canadian Government Travel Bureau, 1962

Map 880 (page 332).
200-Miracle Miles-200/The Fraser Valley Scenic Circle
British Columbia Government Travel Bureau, 1957

Map 881 (page 332).
Vancouver and Vicinity
Home Oil/H.M. Gousha Company, 1954

Map 882 (page 332).
Map of Greater Vancouver Showing Proposed Tunnel
George Massey, 3 March 1947
Delta Archives 1987-36, folder 10-20

Map 883 (page 332).
*Proposed New international Express Highway
Showing Fraser River Tunnel and North Arm
Crossing*
George Massey, c. 1949
Delta Archives 1987-36 (roll)

Map 884 (page 332).
*Present Route to Border/New Route to Border
Opening on Saturday*
Vancouver Sun, 21 May 1959

Map 885 (page 333).
Street Map of Vancouver
H.M. Gousha, Chevron road map, 1963

Map 886 (page 333).
Street Map of Vancouver
H.M. Gousha, Chevron road map, 1963

Map 887 (page 333).
Freeway System Levels
From: Stanford Research Institute and Wilbur
Smith and Associates, *Review of Transportation
Plans, Metropolitan Vancouver, B.C.*, 1964
City of Vancouver Archives PD 273

Map 888 (page 333).
Proposed Burrard Inlet Tunnel Crossing
From: Christiani & Nielsen, *Burrard Inlet Tunnel
Crossing. A Comprehensive Proposal for Handling
Traffic across Burrard Inlet*, 1963
City of Vancouver Archives PD 682

Map 889 (page 333).
*Brockton Interchange, Waterfront Highway,
Georgia Interchange*
From: PBQ&D Inc. Engineers, Vancouver
Transportation Study, 18 October 1968
City of Vancouver Archives PD 381

Map 890 (page 334).
*The Coquihalla . . . a whole new point of view in
British Columbia*
British Columbia government poster, 1986
Nicola Valley Museum and Archives

Map 891 (page 334).
A Fixed Transportation Link to Vancouver Island
Model, 1986
With thanks to Pat McGeer

Map 892 (page 335).
On Site Attractions (Map of Expo 86)
From: *Expo 86 Official Souvenir Guide*
Expo 86 Corporation, 1986

Map 893 (page 335).
Index of Map Sheets of Nisga'a Lands
Nisga'a Treaty map, 1993
Land Title and Survey Authority Treaties Drawer

Map 894 (page 336).
Schematic diagram of Expo and Millennium
SkyTrain Lines
TransLink, 2004

Map 895 (page 336).
Schematic map of Expo and Millennium SkyTrain Lines
TransLink, 2004

Map 896 (page 336).
Proposed transportation improvement, 2004–08
TransLink, 2004

Map 897 (page 336).
Bring Community Rail Back to the Fraser Valley Now!
Ad on side of semi-trailer, 2010
Valley Transportation Advisory Committee

Map 898 (page 336).
British Columbia and North American destinations
from Vancouver International Airport
Vancouver International Airport, 2009
© Vancouver International Airport Authority

Map 899 (page 336).
Non-Stop Destinations from YVR
International destinations from Vancouver
International Airport
Vancouver International Airport, 2009
© Vancouver International Airport Authority

Map 900 (page 337).
Map of Morgan Heights subdivision for sale in
South Surrey
Advertising board, 18 October 2008

Map 901 (page 337).
British Columbia Mountain Pine Beetle Epidemic,
2007
Online computer-generated map, Ministry of
Forests and Range, 2007

Map 902 (page 337).
*British Columbia cumulative lightning strikes
map, 26–27 July 2009*
Online computer-generated map, Environment
Canada, Kelowna office, 2009

Map 903 (page 337).
Conception of Proposed Red Mountain Ski Centre
Red Mountain Ski Club, 1948
Rossland Museum

Map 904 (page 338).
Bird's-eye map of ski runs, trails, and ski lifts at
Whistler and Blackcomb
James Niehues, Intrawest, 2004

Map 905 (page 338).
The Torch in B.C.
Television screenshot, CTV British Columbia,
28 January 2010
Route map: BCTV
Base: © 2010 Google/© 2010 Terra Metrics

Map 906 (page 338).
Map of the Canada Line
From: TransLink, Canada Line opening brochure,
September 2009

Map 907 (page 339).
Gwaii Haanas Orbital
J. Broadhead and D. Leveresee, Gowgaia Institute,
2009
From: *The Story of Gwaii Haanas Marine: A
Proposed National Marine Conservation Area
Reserve*, Gwaii Haanas National Park Reserve and
Haida Heritage Site, 2008

Map 908 (page 339).
Stein Valley
International Travel Maps, 2000
ITMB Publishing Ltd.

Map 909 (page 339).
Best of B.C. (British Columbia as a guitar)
Front page of *Westcoast Life* section, *Vancouver
Sun*, 27 August 2009
Courtesy of the *Vancouver Sun*

Map 910 (page 339).
Computer-generated map of St. Margaret Point,
Langara Island
Photo of screen on Coast Guard Auxiliary rescue
boat, Massett, August 2009

Map 911 (page 354).
Untitled single western sheet of a multi-sheet map
showing the route of the Canadian Pacific Railway
across Canada
Burland Lithographic, 1883

Map 912 (page 355).
Western North America
Juan Pedro Walker, c. 1817
Huntington Library

Map 913 (pages 358–59).
Map of British Columbia to the 56th Parallel
James Launders, Department of Lands and Works,
1871
University of British Columbia Rare Books and
Special Collections G3510 1871.B7 copy 1

Map 914 (page 368).
Carte de Comparaison
Jean-Nicolas Buache, 1775
Bill Warren Collection

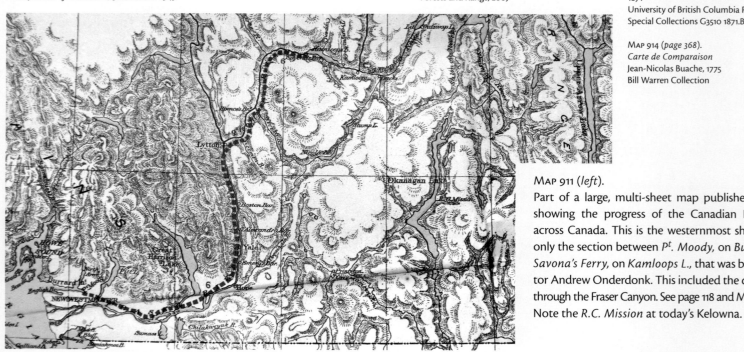

MAP 911 (*left*).
Part of a large, multi-sheet map published in May 1883
showing the progress of the Canadian Pacific Railway
across Canada. This is the westernmost sheet and shows
only the section between *P^t. Moody*, on *Burrard Int.*, and
Savona's Ferry, on *Kamloops L.*, that was built by contractor Andrew Onderdonk. This included the difficult section
through the Fraser Canyon. See page 118 and MAP 311 (*page 121*).
Note the *R.C. Mission* at today's Kelowna.

Other Illustrations

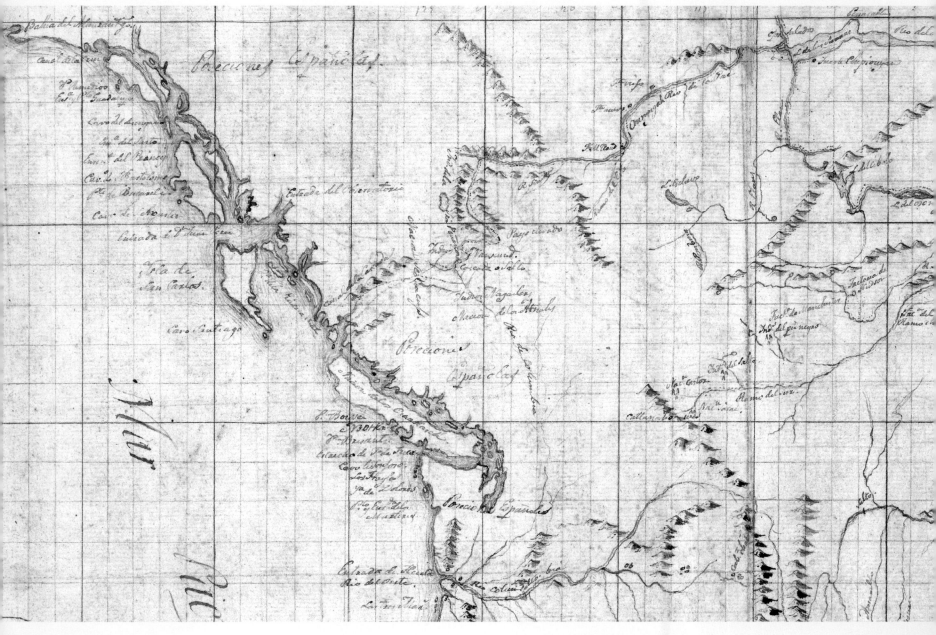

MAP 912 (*above*).

This is the British Columbia part of a large map of western North America drawn about 1817 by Juan Pedro Walker. As Spain lost its claims to the Northwest two years later (see page 35), this may be considered the last Spanish map of the region. Despite a detailed knowledge of the coast (though drawn only in sketch form), the interior is still largely a blank, and what information there is comes from published British sources, notably Alexander Mackenzie's 1802 book and map (see page 38). The confusion between the Fraser and the Columbia is shown here as the northernmost *Rio de la Columbia*.

Bibliography

Akrigg, G.P.V., and Helen B. Akrigg. *British Columbia Place Names*. Victoria: Sono Nis, 1988.

Alcan Aluminum Ltd. *Kitimat-Kemano: Five Years of Operation, 1954–1959*. Kitimat: Aluminum Company of Canada, 1959.

Anderson, Charles, and Lori Culbert. *Wildfire: British Columbia Burns*. Vancouver: Greystone, 2003.

Anderson, Nancy Marguerite. *The Pathfinder: A.C. Anderson's Journeys in the West*. Victoria: Heritage House, 2011.

Bannerman, Gary, and Patricia Bannerman. *The Ships of British Columbia: An Illustrated History of the British Columbia Ferry Corporation*. Surrey: Hancock House, 1985.

Barman, Jean. *The West beyond the West: A History of British Columbia*. Toronto: University of Toronto Press, 1991.

Bartroli, Tomas. *Brief Presence: Spain's Activity on America's Northwest Coast (1774–1796)*. Vancouver: self-published, 1991.

Basque, Garnet. *British Columbia Ghost Town Atlas*. Langley: Sunfire Publications, 1982.

———. *West Kootenay: The Pioneer Years*. Langley: Sunfire Publications, 1990.

———. *Ghost Towns & Mining Camps of the Boundary Country*. Langley: Sunfire Publications, 1992.

Basque, Garnet (ed.). *Frontier Days in British Columbia*. Langley: Sunfire Publications, 1993.

B.C. Hydro Power Pioneers. *Gaslights to Gigawatts: A Human History of B.C. Hydro and its Predecessors*. Vancouver: Hurricane Press, 1998.

Beals, Herbert K. (translation and annotation). *Juan Pérez on the Northwest Coast: Six Documents of His Expedition in 1774*. Portland: Oregon Historical Society Press, 1989.

Belshaw, John Douglas. *Becoming British Columbia: A Population History*. Vancouver: UBC Press, 2009.

Berton, Pierre. *The National Dream: The Great Railway, 1871–1881*. Toronto: McClelland and Stewart, 1970.

———. *The Last Spike: The Great Railway, 1881–1885*. Toronto: McClelland and Stewart, 1971.

Brown, Justine. *All Possible Worlds: Utopian Experiments in British Columbia*. Vancouver: New Star/Transmontanus, 1995.

Cail, Robert E. *Land, Man, and the Law: The Disposal of Crown Lands in British Columbia, 1871–1913*. Vancouver: University of British Columbia Press, 1974.

Carlson, Keith Thor (ed.). *A Stó:lo-Coast Salish Historical Atlas*. Vancouver: Douglas & McIntyre, 2001.

Cook, Warren L. *Flood Tide of Empire: Spain and the Pacific Northwest, 1543–1819*. New Haven, Connecticut: Yale University Press, 1973.

Cotton, Hugh Barrington, Robert William Allen, and Gordon McKay Thomson. *The L.S. Group: British Columbia's First Land Surveyors*. Sidney: Association of British Columbia Land Surveyors, 2007.

Coull, Cheryl. *A Traveller's Guide to Aboriginal B.C.* Vancouver: *Beautiful British Columbia Magazine*/Whitecap, 1996.

Cross, Francis E., and Charles M. Parkin, Jr. *Sea Venture: Captain Gray's Voyages of Discovery, 1787–1793*. No location given, Valkyrie Publishing House, 1981.

Davies, David Ll., and Lorne Nicklason. *The CPR's English Bay Branch: The Intended Terminus of the Canadian Pacific Railway?* Vancouver: Pacific Coast Division, Inc., Canadian Railroad Historical Association, 1993.

Dendy, David. "The Development of the Orchard Industry in the Okanagan Valley, 1890–1914." Okanagan Historical Society, *Annual Report* 38, pp. 68–73, 1974.

Doeksen, Corwin, and Gerry Doeksen. *Railways of the West Kootenay. Part 1, Railways of Western Canada*. Montrose: self-published, 1991.

Dorman, Robert (compiler). *A Statutory History of the Steam and Electric Railways of Canada, 1836–1937*. Ottawa: Canada Department of Transport, 1938.

Dorman, Robert (compiler), and D.E. Stoltz (reviser and updater). *A Statutory History of Railways in Canada, 1836–1986*. Kingston: Canadian Institute of Guided Ground Transport, Queen's University, 1887.

Downs, Art. *Wagon Road North: The Story of the Cariboo Gold Rush in Historical Photos*. Surrey: Foremost Publishing, 1969.

———. *British Columbia-Yukon Sternwheel Days*. Surrey: Heritage House, 1992.

Downs, Art (ed.). *Pioneer Days in British Columbia*. Surrey: Heritage House, 4 vols. 1973, 1975, 1977, and 1979.

Duff, Wilson. *The Indian History of British Columbia: The Impact of the White Man*. Victoria: Royal British Columbia Museum, 1997.

Duffy, Dennis. *Imagine Please: Early Radio Broadcasting in British Columbia*. Sound Heritage Series. Victoria: Provincial Archives of British Columbia, 1983.

Duffy, Dennis, and Carol Crane (eds.). *The Magnificent Distances: Early Aviation in British Columbia, 1910–1940*. Sound Heritage Series. Victoria: Provincial Archives of British Columbia, 1980.

Efimov, A.V. *Atlas geograficheskikh otkrytii v Sibiri i v severozapadnoi Amerike XVII–XVIII* (Atlas of geographical discoveries in Siberia and North-Western America XVII and XVIII centuries). Moscow: Nauka, 1964.

Engineering Institute of Canada. *Alcan Nechako-Kemano-Kitimat Development*. Reprint from the *Engineering Journal*, 1954.

Ewert, Henry. *The Story of the B.C. Electric Railway Company*. North Vancouver: Whitecap, 1986.

Ficken, Robert E. *Unsettled Boundaries: Fraser Gold and the British-American Northwest*. Pullman, Washington: WSU Press, 2003.

Fitzgerald, Kathleen. "Collins Overland Telegraph." In: *The History of the Canadian West*, No. 1, pp. 23–63. Langley: Sunfire Publications, 1982.

Ford, Helen, Dorrit Macleod, and Gene Joyce (eds.). *Place Names of the Alberni Valley*. Port Alberni: Alberni District Museum and Historical Society, 1978.

Forester, Joseph E., and Anne D. Forester. *Fishing: British Columbia's Commercial Fishing History*. Saanichton: Hancock House, 1975.

Forsythe, Mark, and Greg Dickson. *The Trail of 1858: British Columbia's Gold Rush Past*. Madeira Park, B.C.: Harbour Publishing, 2007.

Fraser River Board. *Preliminary Report on Flood Control and Hydro-Electric Power in the Fraser River Basin*. Victoria: Fraser River Board, 1958.

Gibson, James R. *The Lifeline of the Oregon Country: The Fraser-Columbia Brigade System, 1811–47*. Vancouver: UBC Press, 1997.

Goodacre, Richard. *Dunsmuir's Dream: Ladysmith, the First Fifty Years*. Victoria: Porcépic Books, 1991.

Gordon, Katherine. *Made to Measure: A History of Land Surveying in British Columbia*. Winlaw: Sono Nis, 2006.

Griffin, Walter Burley. *The Writings of Walter Burley Griffin*. Cambridge, U.K.: Cambridge University Press, 2008.

Hacking, Norman R., and W. Kaye Lamb. *The Princess Story: A Century and a Half of West Coast Shipping*. Vancouver: Mitchell Press, 1974.

Hall, Ralph. *Pioneer Goldseekers of the Omineca*. Victoria: Morriss Publishing, 1994.

Harris Rees, Charlotte. *Secret Maps of the Ancient World*. Bloomington, Indiana: AuthorHouse, 2008.

Harvey, R.G. *The Coast Connection: A History of the Building of Trails and Roads between British Columbia's Interior and Its Lower Mainland from the Cariboo Road to the Coquihalla Highway*. Lantzville: Oolichan Books, 1994.

———. *Carving the Western Path: By River, Rail and Road through B.C.'s Southern Mountains*. Surrey: Heritage House, 1998.

———. *Carving the Western Path: By River, Rail, and Road through Central and Northern B.C.* Surrey: Heritage House, 1999.

———. *Carving the Western Path: Routes to Remember*. Surrey: Heritage House, 2006.

Hayes, Derek. *Historical Atlas of British Columbia and the Pacific Northwest*. Delta: Cavendish Books, 1999.

———. *Historical Atlas of the North Pacific Ocean*. Vancouver: Douglas & McIntyre, 2001.

———. *Historical Atlas of Canada: Canada's History Illustrated with Original Maps*. Vancouver: Douglas & McIntyre, 2002.

———. *Historical Atlas of Vancouver and the Lower Fraser Valley*. Vancouver: Douglas & McIntyre, 2005.

———. *Historical Atlas of Washington & Oregon*. Berkeley: University of California Press, 2011.

Hayman, John (ed.). *Robert Brown and the Vancouver Island Exploring Expedition*. Vancouver: UBC Press, 1989.

Hind, Patrick O. *The Pacific Great Eastern Railway Company: A Short History of the North Shore Subdivision, 1914–1928*. North Vancouver: North Vancouver Museum & Archives Commission, 1999.

Hinde, John Roderick. *When Coal Was King: Ladysmith and the Coal-Mining Industry on Vancouver Island*. Vancouver: UBC Press, 2003.

Howay, F.W., E.O.S. Scholefield, and William G.R. Hind. *Cariboo Gold Rush*. (Reprinted excepts of writings.) Surrey: Heritage House, 1999.

Inglis, Robin. *Historical Dictionary of the Discovery and Exploration of the Northwest Coast of America*. Lanham, Maryland: Scarecrow Press, 2008.

Jordan, Mabel E. "The Kootenay Reclamation and Colonization Scheme and William Adolph Baillie-Grohman." *British Columbia Historical Quarterly*, Vol. XX, pp. 187–220.

Joyce, Art. *Hanging Fire and Heavy Horses: A History of Public Transit in Nelson*. Nelson: City of Nelson, 2000.

Karamanski, Theodore J. *Fur Trade and Exploration: Opening the Far Northwest, 1821–1852*. Norman, Oklahoma: University of Oklahoma Press, 1988.

Kendrick, John. *The Men with Wooden Feet: The Spanish Exploration of the Pacific Northwest*. Toronto: NC Press, 1985.

Kendrick, John (translation and introduction). *The Voyage of Sutil and Mexicana, 1792: The Last Spanish Exploration of the Northwest Coast of America.* Spokane, Washington: Arthur H. Clark Co., 1991.

Koroscil, Paul M. *British Columbia: Settlement History.* Burnaby: Simon Fraser University Department of Geography, 2000.

———. *The British Garden of Eden: Settlement History of the Okanagan Valley, British Columbia.* Burnaby: Simon Fraser University Department of Geography, 2000.

Lamb, W. Kaye (ed.). *The Letters and Journals of Simon Fraser, 1806–1808.* Toronto: Macmillan, 1960.

Lavallee, Omer. *Van Horne's Road: An Illustrated Account of the Construction and First Years of Operation of the Canadian Pacific Transcontinental Railway.* Montréal: Railfare, 1974.

Leonard, Frank. *A Thousand Blunders: The Grand Trunk Pacific Railway and Northern British Columbia.* Vancouver: UBC Press, 1996.

Lillard, Charles. *Seven Shillings a Year: The History of Vancouver Island.* Ganges: Horsdal & Schubart, 1986.

———. *The Ghostland People: A Documentary History of the Queen Charlotte Islands, 1859–1906.* Victoria: Sono Nis, 1989.

Little, C.H. *18th Century Maritime Influences on the History and Place Names of British Columbia.* Madrid: Editorial Naval/Museo Naval, 1991.

McDonald, J.D. *Storm Over High Arrow: The Columbia River Treaty (A History).* Rossland: Rotary Club of Rossland, n.d. (1993).

Mackay, Donald. *The Asian Dream: The Pacific Rim and Canada's National Railway.* Vancouver: Douglas &McIntyre, 1986.

Mackie, Richard Somerset. *Trading beyond the Mountains: The British Fur Trade on the Pacific, 1793–1843.* Vancouver: UBC Press, 1997.

MacLachlan, Donald F. *The Esquimalt & Nanaimo Railway: The Dunsmuir Years.* Victoria: British Columbia Railway Historical Association, 1986.

Meldrum, Pixie. *Kitimat: The First Five Years.* Kitimat: The Corporation of the District of Kitimat, 1958.

Merk, Frederick. *The Oregon Question: Essays in Anglo-American Diplomacy & Politics.* Cambridge, Massachusetts: Belknap Press of Harvard University Press, 1967.

Mitchell, David. J. *W.A.C. Bennett and the Rise of British Columbia.* Vancouver: Douglas & McIntyre, 1983.

Mitchell, David, and Dennis Duffy. *Bright Sunshine and a Brand New Country: Recollections of the Okanagan Valley, 1890–1914.* Aural History Program, Province of British Columbia: Sound Heritage series, Vol. VIII, No. 3, 1979.

Molyneux, Geoffrey. *British Columbia: An Illustrated History.* Vancouver: Polestar, 1992.

Moogk, Peter N., assisted by Major R.V. Stevenson. *Vancouver Defended: A History of the Men and Guns of the Lower Mainland Defences, 1859–1949.* Vancouver: Antonson Publishing, 1978.

Morwood, William. *Traveller in a Vanished Landscape: The Life and Times of David Douglas.* London: Gentry Books, 1973.

Muckle, Robert J. *The First Nations of British Columbia.* Vancouver: UBC Press, 1998.

Muirhead, Richard H. "Vancouver, Victoria and Eastern Railway." Okanagan Historical Society, *Annual Report* 38, 1974, pp. 25–28.

Murton, James. *Creating a Modern Countryside: Liberalism and Land Resettlement in British Columbia.* Vancouver: UBC Press, 2007.

Nechako Valley Historical Society. *Vanderhoof: The Town That Wouldn't Wait.* Vanderhoof: Nechako Valley Historical Society, 1979.

Neering, Rosemary. *A Traveller's Guide to Historic British Columbia.* North Vancouver: Whitecap, 2009.

Nicola Valley Museum Archives Association. *Merritt & the Nicola Valley: An Illustrated History.* Merritt: Sonotek Publishing, 1998.

Nokes, J. Richard. *Almost a Hero: The Voyages of John Meares, R.N., to China, Hawaii and the Northwest Coast.* Pullman, Washington: WSU Press, 1998.

Ormsby, Margaret A. *British Columbia: A History.* Toronto: Macmillan, 1958.

Paterson, T.W. *Okanagan-Similkameen.* British Columbia Ghost Town Series. Langley: Sunfire Publications, 1983.

———. *Lower Mainland.* British Columbia Ghost Town Series. Langley: Sunfire Publications, 1984.

Paterson, T.W., and Garnet Basque. *Ghost Towns & Mining Camps of Vancouver Island.* Surrey: Heritage House, 2006.

Prince Rupert City and Regional Archives Society. *Prince Rupert: An Illustrated History.* Prince Rupert: Prince Rupert City and Regional Archives Society, 2010.

Ramsey, Bruce. *Ghost Towns of British Columbia.* Vancouver: Mitchell Press, 1975.

Rich, E.E. *The History of the Hudson's Bay Company, 1670–1870.* London: The Hudson's Bay Record Society, 2 vols., 1959.

Riegger, Hal. *The Kettle Valley and Its Railways: A Pictorial History of Rail Development in Southern British Columbia and the Building of the Kettle Valley Railway.* Edmonds, Washington: Pacific Fast Mail, 1981.

Ringuette, Janis. *Beacon Hill Park History 1849–2009.* Online: *http://www.beaconhillparkhistory.org/.* Accessed 2011.

Robin, Martin. *The Rush for Spoils: The Company Province, 1871–1933.* Toronto: McClelland and Stewart, 1972.

———. *Pillars of Profit: The Company Province, 1934–1972.* Toronto: McClelland and Stewart, 1973.

Roy, Patricia E., and John Herd Thompson. *British Columbia: Land of Promises.* Oxford Illustrated History of Canada. Don Mills, Ontario: Oxford University Press, 2005.

Runnalls, F.E. *A History of Prince George.* Prince George: self-published, 1946.

Rushton, Gerald A. *Whistle Up the Inlet.* North Vancouver: Douglas & McIntyre, 1978.

Ruzesky, Jay, and Tom Carter. *Paying for Rain: A History of the South East Kelowna Irrigation District.* Kelowna: South East Kelowna Irrigation District, 1990.

Sanford, Barrie. *McCulloch's Wonder: The Story of the Kettle Valley Railway.* North Vancouver: Whitecap, 1977.

———. *Steel Rails & Iron Men: A Pictorial History of the Kettle Valley Railway.* North Vancouver: Whitecap, 2003.

———. *Royal Metal: The People, Times and Trains of New Westminster Bridge.* Vancouver: National Railway Historical Society, 2004.

———. *Railway by the Bay: 100 Years of Trains at White Rock, Crescent Beach and Ocean Park, 1909–2009.* Vancouver: National Railway Historical Society, 2009.

Sterne, Netta. *Fraser Gold 1858! The Founding of British Columbia.* Pullman, Washington: WSU Press, 1993.

Taylor, G.W. *The Automobile Saga of British Columbia, 1864–1914.* Victoria: Morriss Publishing, 1984.

Tovell, Freeman. *At the Far Reaches of Empire: The Life of Juan Francisco de la Bodega y Quadra.* Vancouver: UBC Press, 2008.

Turnbull, Elsie. *Trail: A Smelter City.* Langley: Sunfire Publications, 1985.

Turner, Robert D. *West of the Great Divide: An Illustrated History of the Canadian Pacific Railway in British Columbia, 1880–1986.* Victoria: Sono Nis, 1987.

———. *Those Beautiful Coastal Liners: The Canadian Pacific's Princesses.* Victoria: Sono Nis, 2001.

———. *Vancouver Island Railroads.* Victoria: Sono Nis, 2005.

Turner, Robert D., and J.S. David Wilkie. *Steam along the Boundary: Canadian Pacific, Great Northern and the Great Boundary Copper Boom.* Winlaw: Sono Nis, 2007.

Twigg, Alan. *First Invaders: The Literary Origins of British Columbia.* Vancouver: Ronsdale, 2004.

Vancouver, George./W. Kaye Lamb (ed.) *A Voyage of Discovery to the North Pacific Ocean and Round the World.* London: Hakluyt Society, 4 vols., 1984 (1798).

Ward, W. Peter, and Robert A.J. McDonald (eds.). *British Columbia: Historical Readings.* Vancouver: Douglas & McIntyre, 1981.

Webber, Jean. *A Rich and Fruitful Land: The History of the Valleys of the Okanagan, Similkameen and Shuswap.* Madeira Park, B.C.: Harbour Publishing, 1999.

West, Willis J. *Stagecoach and Sternwheel Days in the Cariboo and Central B.C.* Surrey: Heritage House, 1985.

White, Bob. *Bannock and Beans: A Cowboy's Account of the Bedaux Expedition.* Victoria: Royal BC Museum, 2009.

Wilson, Charles. *Mapping the Frontier: Charles Wilson's Diary of the Survey of the 49th Parallel, 1858–1862, While Secretary of the British Boundary Commission.* Toronto: Macmillan, 1970.

Woodward, Frances. "The Influence of the Royal Engineers on the Development of British Columbia." *B.C. Studies,* 24, 1974–75, pp. 3–51.

Wuest, Donna Yoshitake. *Coldstream: The Ranch Where It All Began.* Madeira Park, B.C.: Harbour Publishing, 2005.

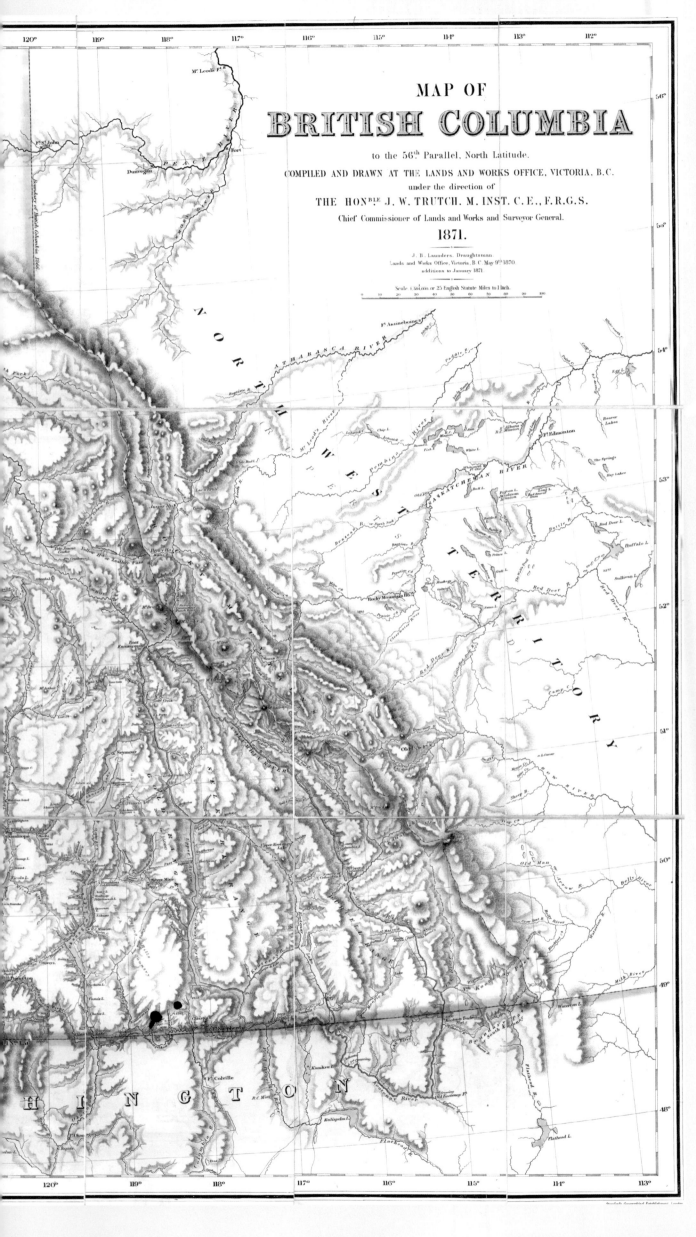

MAP OF

BRITISH COLUMBIA

to the 56th Parallel, North Latitude.

COMPILED AND DRAWN AT THE LANDS AND WORKS OFFICE, VICTORIA, B.C.

under the direction of

THE HON.BLE J. W. TRUTCH. M. INST. C. E., F. R. G. S.

Chief Commissioner of Lands and Works and Surveyor General.

1871.

J. B. Launders, Draughtsman.
Lands and Works Office, Victoria, B.C. May 9th 1870.
additions to January 1871.

Scale 1:580,000 or 25 English Statute Miles to 1 Inch.

MAP 913.

Now considered one of the landmark maps of British Columbia is this January 1871 compilation of all available knowledge at the time. Drawn by James Launders, an ex–Royal Engineer, the map was created under the supervision of Joseph Trutch, chief commissioner of Lands and Works, and is for this reason popularly known as the "Trutch map." Trutch became lieutenant governor later in 1871 when British Columbia joined Confederation. Trutch had been a private surveyor working under Colonel Richard Moody (see MAP 195, *page 70*). He had been appointed chief commissioner of Lands and Works and surveyor general in 1864. He is reviled by some for his adamant denial of Aboriginal title; he worked to reduce the size of a number of Indian reserves established by James Douglas. This map is full of detail, some of which is reproduced elsewhere in this book (see, for example, MAP 285, *page 105*, and MAP 368, *page 144*).

Index

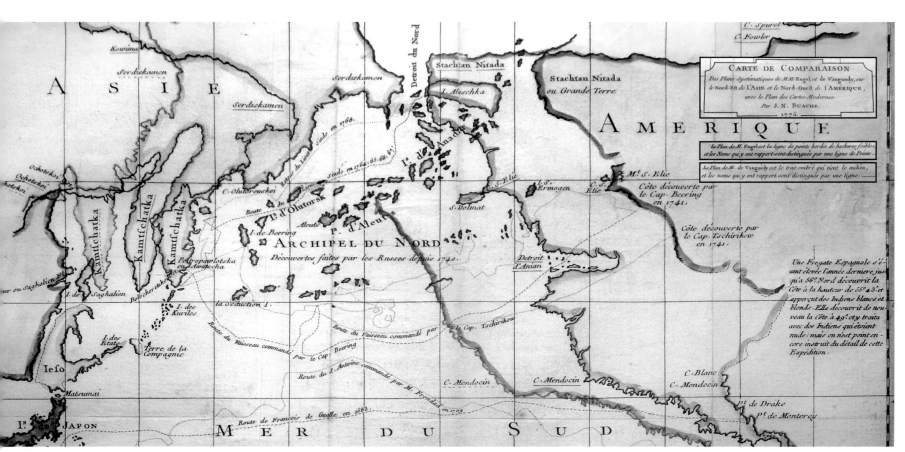

Map 914.
The fact that no one really knew where the West Coast was before the arrival of the famous British explorer James Cook in 1778 is well illustrated by this map published three years before his visit by French mapmaker Jean-Nicolas Buache, who assembled competing theories onto a single sheet. Here the possible position of British Co-lumbia varies longitudinally by thousands of kilometres. Cook and his chronometers finally ended this longitudinal confusion (see page 26). The tracks of several navigators are shown, including the voyages of Vitus Bering (*Cap. Beering*) and Aleksei Chirikov (*Cap. Tschirkow*), who reached the Northwest Coast from Russia in 1741.